Palaeomagnetism

Palaeomagnetism

Principles and Applications in
Geology, Geophysics and Archaeology

D. H. TARLING

LONDON NEW YORK
CHAPMAN AND HALL

First published 1983 by
Chapman and Hall Ltd
11 New Fetter Lane, London EC4P 4EE
Published in the USA by
Chapman and Hall
733 Third Avenue, New York NY 10017

© 1983 D. H. Tarling
Softcover reprint of the hardcover 1st edition 2007

ISBN-13: 978-94-009-5957-6   e-ISBN-13: 978-94-009-5955-2
DOI: 10.1007/ 978-94-009-5955-2

British Library Cataloguing in Publication Data

Tarling, D. H.
Palaeomagnetism.
1. Paleomagnetism
I. Title
523.01'887   QE501.P35

Library of Congress Cataloging in Publication Data

Tarling, D. H. (Donald Harvey)
Palaeomagnetism: principles and applications in
geology, geophysics, and archaeology.

Bibliography: p.
Includes index.
1. Palaeomagnetism.   I. Title.
QE501.4.P35T37   1983   538'.7   83-5176

Contents

Contents

Contents

Preface

Palaeomagnetism and archaeomagnetism are fascinating specialized studies because they are applicable to such a wide range of problems in geology, archaeology and geophysics. They can also be undertaken cheaply, when compared with most other geophysical techniques, and, at first sight, simply. In fact, real comprehension of the magnetic processes that have occurred in rocks and other types of material over several thousands or many millions of years is still extremely difficult to assess and measure. On this basis, this book cannot explain all such features, nor can it attempt to cover all the actual and potential applications of the method. All that can be attempted is to give an impression of the ways in which such techniques can be used in a wide variety of fields, and how these techniques are usually applied. The magnetization of rocks is, in fact, one of the earliest of the true sciences, but we are still not in a position to answer many of the problems posed. Consequently some of the examples given of applications are, essentially, state-of-the-art comments, rather than being a review as such. The changing position of the geomagnetic poles with time is still not adequately defined, for example, and some of the more recent conventional views are given, although the emphasis is placed on more subjective, probably more controversial, evaluations. In these, I have possibly been too pessimistic, although, as with any pessimist, I think that the assessments are realistic! In any case, the data base is expanding rapidly in most fields and the evidence evaluated here must be under constant review.

It is difficult to know where to start with the acknowledgements. Obviously I have probably learnt more from past and present students than they have learnt from me, and many colleagues have been involved in discussions and reading some of the sections. Specifically to mention A. Stephenson, W. O'Reilly, D. W. Collinson, R. Thompson, E. A. Hailwood and F. J. Lowes runs the risk of omission of many others of my mentors such as J. A. Clegg, P. M. S. Blackett, K. M. Creer, S. K. Runcorn and, above all, E. Irving. On the technical side, Lynn Whiteford, Dorothy Cooper and Marie Summersby were always co-operative with the typescript, drawings and computing – even when I was not!

Newcastle upon Tyne
May 1982

Chapter One

Introduction

1.1 SCOPE OF THE BOOK

Magnetism is one of the oldest of the true sciences (Section 1.2) although its remarkable properties have only recently achieved recognition and many of the ways in which it can be applied to a range of geological, geophysical and archaeological problems have still to be assessed and developed. Its unique feature is that it is the only geophysical property of the Earth that can be satisfactorily measured and evaluated throughout time. Seismicity, gravity and electrical properties are transient features that leave no clear trace of their previous values. The strength and direction of the geomagnetic field can, however, be studied over archaeological, geological and even cosmic time-scales. Studies within an archaeological context, *archaeomagnetism*, do not differ in principle from longer-term studies, *palaeomagnetism*, and so these terms mainly reflect the purpose of the research rather than any intrinsic difference. The general principles of magnetization (Chapter 2) and the minerals that retain ancient magnetizations (Chapter 3) are therefore considered before the specific materials that may be of interest to geologists, geophysicists, and archaeologists (Chapter 4). Methods of sampling and measurement (Chapter 5) and of statistical analyses (Chapter 6) are again of common interest, but the interests of archaeologists, geophysicists and geologists tend to differ in the ways in which they can apply such palaeomagnetic observations. Archaeological interests are specifically concerned with the last few thousand years (Chapter 7) and are considered before geomagnetic observations (Chapter 8). Most of the knowledge of geomagnetic field behaviour over periods longer than 3–400 years is based on archaeomagnetic observations. However, the reliability of some archaeological and most geological studies depends on the validity of models for the long-term behaviour of the geomagnetic field and so the geological and cosmic time-scales are discussed last (Chapter 9).

1.2 HISTORICAL PERSPECTIVE

Magnetized stones showing attraction and repulsion may have been known to Early Man, but such properties would be difficult to see unless the rocks were almost entirely formed of magnetite and had previously been struck by

1

lightning. Even then it is likely that such properties could not be studied until the discovery of iron. The ability of such magnetic stones, lodestones, to induce a magnetization by rubbing and stroking iron needles would certainly be regarded as magical and such attractive and repulsive forces were probably discovered very early in the Iron Age. The phenomenon of magnetic repulsion was certainly known in Ancient Egypt and both repulsive and attractive properties were described by Thales of Miletus, c. 624–565 BC (Hesse, 1961). Greek legends also included, obviously exaggerated, stories of the ability of some rocks to pull the nails out of passing vessels. However, the ability of magnetized needles to point in specific directions was not known in Western Europe, although such properties were well known to the Chinese in the 1st century AD (Needham, 1962) and had almost certainly been known for at least 300 years previously. The Chinese were also aware, by at least 720 AD, that compasses did not point due south (the Chinese prime meridian was south), but at an angle to it.

In Europe, the earliest record of compasses is by Neckham (1187) and it seems probable that such vital navigational instruments reached Europe via Arabic traders during the 12th century. During the European 'Dark Ages', it was thought that the compass pointed towards the Pole star, i.e. along the axis of the Universe, but the 'Dark Ages' also saw the beginnings of experimental science, even though the main development was delayed until the 17th century. In particular, the work of Roger Bacon (1267) established the principles of experimental science and Petrus Peregrinus (Pierre de Maricourt), a French military engineer, described his experiments with magnets in a widely circulated letter (1269). These experiments included the construction of spheres of lodestone on which he defined lines of equal force and also found that such lines of force converged to two diametrically opposite points, which he termed poles. He also went on to show how like poles repelled and unlike attracted (Chapman, 1967; Smith, 1970). Peregrinus's *Epistola* has, in fact, been described as 'the first original scientific work of western Christendom' (Bernal, 1965).

The European 'discovery' of declination, i.e. that the compass does not point to true north but at an angle to it, is not clear. The Flemish cartographer, Gerhard Mercator, and the Portuguese explorer, João de Castro, reported differences between the magnetic direction and true north in the mid-16th century (Hellmann, 1896), but some sundials of around 1450 AD are thought to have markings indicating compass settings that differ from true north, as do some pre-1500 road maps (Chapman, 1967). Similarly, the 'discovery' of inclination, the angle which a suspended magnetic needle will make to horizontal, is unclear. It was recorded by Hartmann (1544) and independently discovered by Norman in 1576 (Norman, 1581). Nonetheless, the philosophy of the 'Scientific Revolution' was laid by Francis Bacon (1605), and William Gilbert (1600) is generally recognized as the founder of experimental science, although much of his work, *De Magnete*, probably relied on the earlier observations by Peter Peregrinus, but was extended to include both inclination and declination values over the sphere. Gilbert termed such spheres 'terellas' and likened the magnetic field of the Earth to that of one of these uniformly magnetized spheres, i.e. *magnus magnes ipse est globus terrestris* (the earth globe itself is a great magnet). Shortly after, Gellibrand in 1839 (Hellmann, 1896) demonstrated that the long-term changes in the direction of magnetic north were real – at the same time demonstrating that Gilbert's concept was an inadequate model of the actual geomagnetic field. (Strangely enough, it was not until 1799–1803 that the latitudinal variation in strength of the geomagnetic field was actually measured in South America by von Humboldt (1858).)

During the 17th century, many observations of declination, and some of inclination, were made during most of the voyages of discovery, culminating in the compilation of declination charts for the North and South Atlantic by Halley in 1701 (Chapman and Bartels, 1940), followed by charts of inclination (Wilcke, 1777). These, and subsequent data, provided the basic information that was used by Gauss (1833, 1839) in his utilization of spherical harmonic analysis to define the geomagnetic field, thus demonstrating mathematically its dipolar nature and internal origin. Subsequent geomagnetic studies have been increasingly influenced by palaeomagnetic and archaeomagnetic observations (Chapter 8), although 20th-century studies have, so far, largely been mathematical attempts to model the geomagnetic field using magnetohydrodynamic equations (Larmor, 1919; Cowling, 1934; Bullard, 1949a,b, 1955; Alfvén, 1950; Elsasser, 1955, 1956a,b; Rikitake, 1958, 1966) with such equations being modelled by laboratory experimentation confirming the physical reality of self-coupled dynamos as an explanation of the ability of the Earth's magnetic field to reverse polarity (Lowes and Wilkinson, 1963, 1968).

Early compasses were obviously less sensitive than their modern equivalents, but were clearly attracted by locally magnetized rocks. Alexander von Humboldt, the pioneer of many explorations, related this attraction to lightning effects while studying the high ground in the Palatine states (von Humboldt, 1797), but the first detailed studies of the magnetic properties of volcanic rocks were by Delesse (1849) and Melloni (1853), with the conclusion that they acquire their magnetization on cooling. These early studies were vastly extended by Folgerhaiter (1894, 1895, 1899a,b). He concluded that not only did the volcanic rocks acquire their magnetization on cooling, but also the direction of the acquired magnetization paralleled the geomagnetic field for younger volcanics, and some older volcanics were magnetized in the opposite direction (1894, 1895). Folgerhaiter also studied fired pottery, as did Mercanton (1918), showing that the remanence of the bases of vases buried for 2000 years was randomized, i.e. each vase had preserved its initial direction of magnetization for that period of time. Boyle (1691) not only showed that lodestones have identical properties to sedimentary ironstones but also that bricks, on cooling in the Earth's field, become magnetized along their long axes. Bricks were also studied by Gheradi (1862) and David (1904) and Brunhes (1906) studied naturally fired materials, such as the clays heated by lava extrusions. They found that these clays had the same direction of remanence as the lavas and that reversely magnetized lavas also had reversely magnetized baked clays associated with them. This led them to propose that the Earth's magnetic field reversed polarity. The development of improved magnetic theories (Weiss, 1907) was preceded by a suggestion for a remanence that may be acquired chemically (Maurain, 1901), although this aspect was largely neglected until the 1950s (Doell, 1957; Nagata and Kobayashi, 1958).

In Japan, Nakamura and Kikuchi (1912) had also shown that volcanic rocks acquired a magnetization parallel to the ambient field as they cooled and detailed studies of Tertiary and Quaternary volcanic rocks from Japan, Korea and Manchuria were made by Matuyama (1929), with the proposal that all young rocks were magnetized parallel to the present geomagnetic field during the Late Quaternary, but rocks formed during the Early Quaternary were reversely magnetized and somewhat older rocks were again of the same polarity as today. Chevallier (1925) published detailed results from studies on lavas from Mt Etna, with the construction of a record of secular variation changes during the last 2000 years. Similar secular variation studies were later carried out in Japan (Kato and Nagata, 1949), and much Japanese work has subsequently been aimed at monitoring these variations in even greater detail. In Europe, Mercanton (1926, 1931, 1932) studied a range of rocks from areas in the North Atlantic region (Greenland, Spitsbergen, Iceland, Faeroes, Scotland, Jan Mayen) and also Australia, with evidence for both normal and reversed polarities occurring equally frequently and that the average direction of the geomagnetic field was coincident with an axial geocentric dipole. On this basis, Mercanton proposed that such observations could be used to test theories of polar wandering and continental drift. The rock magnetic studies of Koenigsberger, culminating in a review (1938), were highly influential in post-war palaeomagnetic rock and magnetic studies, although it was the studies of Thellier (1936, 1937a,b, 1938a,b), mainly using fired archaeological materials, that were fundamental in establishing archaeomagnetic studies as well as the basis for a physical understanding of the acquisition of thermal remanence. The fundamental physical basis (Chapter 2) was established by Néel (1948, 1952a,b) and widely disseminated by Nagata (1953, 1961) and Stacey (1963). The establishment of a sound physical basis was followed shortly by the establishment of a statistical model (Fisher, 1953) in order to allow analysis of the increasing amount of data that were becoming available. (This model had, in fact, been developed by Fisher some 25 years previously, but had remained in a drawer until he was approached about the need for such an analysis in palaeomagnetic work.)

The early history of palaeomagnetism and archaeomagnetism was dominated by studies of naturally or artificially heated materials as the available instrumentation was too insensitive to detect the magnetization of more weakly magnetized materials, such as sediments. Nonetheless, improvements in spinner magnetometers (Section 5.3) allowed the intensity and direction of remanence in more weakly magnetized rocks to be measured, with much of the early work concentrating on Recent glacial sediments, varves in North America (McNish and Johnson, 1938; Johnson *et al.*, 1948) and in Europe (Rusnak, 1957; Granar, 1958). The observations showed that records of secular variation of the geomagnetic field appeared to be preserved, although it was also noted that the magnetization was sometimes shallower than expected. Studies of the acquisition of remanence in sedimentary rocks have been carried out on various types of unconsolidated rock, Nagata (1943) using crushed basalt in some depositional experiments, but the vast majority of the work on inclination and bedding errors (Section 4.3) as well as the magnetic fabric (Section 4.3), were initiated using varved clays (Griffiths, 1953, 1954, 1955; Griffiths and King, 1954; King, 1955; Griffiths *et al.*, 1957, 1958; 1962; Rees, 1961; King and Rees, 1962, 1966; Hamilton, 1963; Hamilton and King, 1964) as these have fairly uniform grain sizes and the conditions of deposition are better known and more reliably duplicated than for some other forms of sediment. Post-depositional effects, although postulated at an earlier stage (Irving, 1957a), were not studied until somewhat later (Irving and Major, 1964). Investigation of the magnetic properties of consolidated sedimentary rocks was delayed, partly because of uncertainty about their reliability as indicators of the geomagnetic field and partly because they were generally more weakly magnetized. The development of the astatic magnetometer, which is still one of the cheapest, most sensitive magnetometers available (Section 5.3.1), was therefore of major importance in the development of palaeomagnetism. Blackett (1952) originally designed an improved version of this instrument to test the possibility that magnetization was a fundamental property of rotating matter. The experiment showed that this was not the case but Blackett went on to recommend that the instrument could then be used for other purposes, such as the measurement of the magnetization of rocks (Collinson *et al.*, 1957).

In Britain, two groups emerged. One small group led by P. M. S. Blackett, and including J. Clegg and P. M. Stubbs, had a geological bias (Clegg *et al.*, 1954a,b) and extended studies into Spain (Clegg *et al.*, 1957) and India (Clegg *et al.*, 1956; 1958). Another group, led by S. K. Runcorn, comprised D. W. Collinson, K. M. Creer, E. Irving and F. J. Lowes, and was initially more concerned with geomagnetic investigations. Earlier studies of Permo–Triassic red sandstone were extended to Pre-Cambrian sandstones in northwestern Britain (Irving, 1957a) and to rocks of a wide variety of ages throughout Britain (Creer, Irving and Runcorn, 1954; Nairn, 1956, 1957, 1960a; Creer, 1957a,b) and America (Runcorn, 1955, 1956a; Creer *et al.*, 1957, 1958). (The study of the Pre-Cambrian Torridonian Sandstones was particularly fortuitous as these rocks have proved to be amongst the most stable rocks yet identified and gave hope that rocks of all ages could thus be used for such studies.)

The comparison of the palaeomagnetic directions for Permo–Triassic rocks in Britain and North America (Runcorn, 1956b,c; Collinson and Runcorn, 1960; Nairn, 1960a,b) provided the first clear geophysical evidence for the former unity of these continents as the observations could be best interpreted by closing the Atlantic at least 200 million years ago, later shown, on palaeo-magnetic and radiometric grounds, to be 180 million years (Hailwood and Mitchell, 1971). Following these findings, palaeomagnetic research expanded into most other continents, such as South America (Creer, 1958; Creer and Valencio, 1969; Creer et al., 1969) – subsequently largely developed by Valencio – into Africa, extending the earlier work by Graham (Nairn, 1960b; McElhinny and Gough, 1963; McElhinny and Opdyke, 1964; McElhinny et al., 1968), and Australia (Irving, 1956, 1957b; Irving and Green, 1958) and subsequently by McElhinny (McElhinny and Luck, 1970) and Embleton (1972a,b).

In the United States, the early work on glacial sediments had ceased, but Graham (1949, 1954a,b, 1955) and Graham and Torreson (1951) developed spinner magnetometers and initiated work on consolidated sediments as well as igneous rocks, while Doell and Cox (Doell, 1955, 1956; Cox, 1957) tended to concentrate on igneous rock studies, culminating in their major review (Cox and Doell, 1960). Graham had meanwhile established palaeomagnetic studies in southern Africa (Graham and Hales, 1957, 1961). In Russia, early interest in the susceptibility of rocks and magnetic surveying was not followed by an interest in palaeomagnetism until much later (Al'thausen, 1949; Grabovsky, 1952, 1953; Grabovsky and Pushkov, 1954; Khramov, 1958; Khramov and Sholpo, 1967). In France, archaeomagnetic work continued under the direction of Thellier and was extended to igneous rocks (Roche, 1950a,b, 1951, 1953, 1954), with archaeomagnetic studies commencing in Britain (Aitken, 1958). Meanwhile Dutch workers, particularly J. Hospers, had begun to extend the earlier studies of Icelandic lavas as well as starting studies into the Mediterranean region – the Icelandic work leading towards the construction of a polarity time-scale (discussed below).

Most of these earlier studies were based only on the natural remanence as methods of partial demagnetization had not yet been developed. However, Thellier (1937a) had introduced thermal demagnetization in order to deter-mine the different components of thermal remanence, as well as to measure the past intensity of the geomagnetic field. Routine procedures for studying the directional properties of geological samples did not develop until the later 1950s (Irving et al., 1961b). Alternating magnetic field demagnetization was undertaken by Thellier and Rimbert (1954) who also reported on the import-ance of anhysteretic magnetizations (Section 4.6) in such procedures. Improved demagnetizers (Creer, 1959) were then developed, some using tumbling systems to reduce the effects of anhysteretic magnetizations (Cox, 1961), with the routine procedures being outlined by Irving *et al.* (1961a). In Russia, d.c. demagnetization was developed by Khramov and Andreyeva (1964).

Most of the initial interest in such palaeomagnetic studies was essentially geomagnetic as the early experiments in the 1950s were undertaken at a time when it was not thought that the continents could move through a rigid mantle. As a part of such geomagnetic studies, Hospers (1951, 1953a,b, 1954a,b) began to determine the magnetostratigraphy of the Icelandic lavas. This work, continued by Einarsson and Sigurgeirsson (1955), Einarsson (1957a,b) and Sigurgeirsson (1957), indicated that the geomagnetic field remained of constant polarity for periods of some 250 000 to 500 000 years, and that transitions of polarity took some 2000–20 000 years to take place. Although such estimates were based on dating of lava flows that are now drastically different, the time-scales were remarkably close to those postulated at present (Section 8.4). The polarity scale was complicated by the discovery of natural self-reversal properties in the Haruna dacite in Japan (Nàgata *et al.*, 1951) and the discovery of an apparent correlation between petrology and polarity (Balsley and Buddington, 1958; Ade-Hall and Wilson, 1963) – thus suggesting that not all observed polarity inversions were due to geomagnetic changes (Section 8.4). However, the main difficulty was that the Plio–Pleistocene lavas being examined were not datable by existing techniques until the development of the all-glass spectrometer (Reynolds, 1956), following which improvements in potassium–argon dating eventually allowed the measurement of the age of potassium-rich rocks of young ages. This culminated in the almost simultaneous publication of radiometric dated time-scales by Cox *et al.* (1963a,b) and McDougall and Tarling (1963), and the discovery of very brief polarity periods, such as the Olduvai 'event' at 1.9 million years (Grommé and Hay, 1963). The available data were collated and extensively supplemented to produce revised time-scales (Cox *et al.*, 1964a,b; Doell *et al.*, 1966) and were confirmed shortly afterwards by the polarity sequences found in deep-sea sediments (Opdyke *et al.*, 1966). These then formed the basis for dating the magnetic anomalies of the ocean floor.

Oceanic magnetic anomalies had been known for many decades prior to their use for dating the ocean floors (Section 9.2), but it was not realized that such anomalies were linear until precise navigation methods were available (Mason, 1958; Menard and Vacquier, 1958; Menard, 1959; Mason and Raff, 1961). The offsets between such anomalies off the western United States were soon interpreted in terms of major horizontal displacements (Vacquier, 1959), but the origin of the anomalies remained enigmatic, being attributed to linear block faulting giving rise to linear rifts filled with low susceptibility material. Vine and Matthews (1963) studied the northwestern Indian Ocean and found that the magnetic anomalies were more readily interpreted as due to normally and reversely magnetized strips of the igneous ocean floor, rather

7

than to susceptibility differences. They also proposed that the observations were consistent with Hess's concept (1960, 1962) that new oceanic crust was continually being generated at the crest of the oceanic ridges and that this new crust then spread away from the ridges (L. W. Morley proposed the same idea, but his articles were rejected by both *Nature* and the *Journal of Geophysical Research* – Cox, 1973). This concept was not immediately welcomed, but a previous arch-opponent of continental drift, J. T. Wilson, realized that this could also account for the restricted location of seismic activity on major fault lines cutting the oceans, which he termed 'transform' faults. The 'sudden' acceptance of sea-floor spreading, however, came about partly because of the world-wide distribution of earthquake activity (Sykes, 1967) but mainly from relating the magnetic anomaly patterns on the mid-Atlantic Ridge, just south of Iceland, with the newly appearing polarity time-scale (Vine and Wilson, 1965; Vine, 1966). Since then, the concept of plate tectonics and associated sea-floor spreading have become widely accepted, although many of the details of this concept still remain ambiguous.

Since the mid 1960s, there has been a tremendous expansion in the number of centres studying archaeomagnetism and palaeomagnetism. This reflects, to some extent, the relatively low costs of these techniques and their importance throughout the whole range of the Earth Sciences. Standard references are now available on different aspects – Irving (1964), Collinson *et al.* (1967), Strangway (1970), Tarling (1971a), McElhinny (1973a) and Collinson (1983) – but the range of applications for which such studies can have future application is still enormous. As will be indicated in this book, many assumptions in the subject have yet to be properly tested and the technological advances mean that many aspects must be re-examined. Space technology, for example, has now brought back to Earth samples of the Moon, and studies of the magnetic fields of other planets have revolutionized ideas on the origin and development of the Solar System; at the same time, palaeomagnetic techniques are being applied on an increasingly small scale to understand the processes of rock formation and deformation.

1.3 DATA PRESENTATION, MAGNETIC UNITS AND GEOLOGICAL AGES

1.3.1 Data presentation

Directions of magnetization are normally expressed in terms of the polar co-ordinates, declination and inclination. The *declination* (Dec) is the direction of magnetization, in the horizontal plane, measured in degrees clockwise from north, while the *inclination* (Inc) is the angle of dip of the magnetization measured in degrees from the horizontal plane (Fig. 1.1(a)). However, magnetization has a strength as well as a direction, i.e. it is a vector, and it is often convenient to express this in terms of the magnitude of its

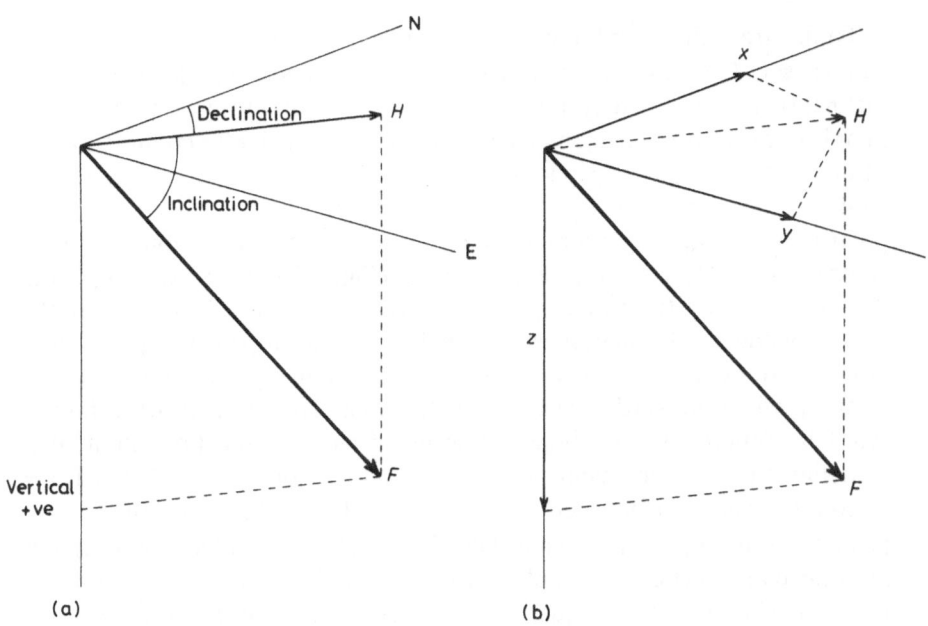

Figure 1.1 Magnetic vectors. A magnetic field has a certain strength and a certain direction. The strength is expressed in terms of the forces between two magnetic poles, while the directions are given in various co-ordinate systems. Generally the direction is given in terms of (a) polar co-ordinates, declination and inclination (equivalent to longitude and latitudes on a sphere). The *declination* is the angle, from north, of the horizontal component, *H*, of the total field direction and strength *F*. The vertical component is then given as the *inclination* of the vector from horizontal, positive downwards. Where the direction of north and horizontal are not known, or are irrelevant, modified cartesian co-ordinates can be used (b) in which the magnitudes of the components along the three mutually perpendicular axes, *x*, *y*, and *z*, are given. Generally *x* corresponds to north, but unlike standard co-ordinates, *z* is positive vertically downwards.

components in cartesian co-ordinates, *x*, *y* and *z*. In magnetic work, these co-ordinates normally correspond to *x* towards the north and *y* eastwards, as in conventional cartesian co-ordinates, and *z* is the vertical component, regarded as positive in a downward direction (Fig. 1.1(b)). The conversion from polar to cartesian co-ordinates can now be undertaken with the simplest pocket calculators, the equations relating them, assuming the length of the vector to be 1, being as follows:

$$x = \cos \mathrm{Dec} \cos \mathrm{Inc}; \quad y = \sin \mathrm{Dec} \cos \mathrm{Inc}; \quad z = \sin \mathrm{Inc}$$

$$\mathrm{Dec} = \tan^{-1} \frac{y}{x}; \qquad \mathrm{Inc} = \sin^{-1} z$$

(The magnitude of the magnetization along each axis can then be determined by multiplying *x*, *y* and *z* by the intensity of the total vector.)

Palaeomagnetism

To illustrate the distribution of palaeomagnetic directions, either system can be used. Conventionally, the stereographic projection of declination and inclination has been widely used (Phillips, 1971). (This projection was chosen mainly because most palaeomagnetic statistics are circular functions (Chapter 6) and circles project as circles on this projection (C and D in Fig. 1.2(a)) while they project as ovals on other projections.) The usual projection angle is polar in which the circumference of the circle represents horizontal and the centre of the projection represents vertical (Fig. 1.2(a)). The distance from the circumference (primitive) therefore corresponds to the inclination of the vector, while its declination is the angle from the top of the projection. Downward, i.e. positive, inclinations are conventionally marked with a solid symbol, while upward, negative, inclinations are marked with hollow symbols. Where the directions are grouped close to the horizontal, it is sometimes more convenient to use an equatorial projection in which the 'east–west' diameter now represents horizontal, and the centre of the projection is some appropriate value of declination. With the advent of computer plotting, ovals can be as easily plotted as circles and there has been a tendency for increasing use of the equal area projection for such illustrations. (The distribution of pole positions (Section 6.2) can similarly be displayed in which latitudes are plotted in the same way as inclinations, and longitudes are plotted as declinations. In such polar plots, it should normally be specified whether the projection is of the northern or southern hemisphere.)

The magnetic vector can also be illustrated in terms of its cartesian co-

---

Figure 1.2 Stereographic and cartesian projections. (a) Individual directions, but not intensities, can be plotted on this projection. Declinations are measured clockwise from north, and inclinations are measured from the equator, which represents horizontal, i.e. 0° inclination, with the centre of the projection as ±90°. Conventionally, the projection is of the lower hemisphere, i.e. downward inclinations are plotted as solid symbols, and upward (negative) inclinations are plotted as hollow symbols. Sample A is therefore plotted as a solid dot, while B is plotted as a circle. The projection is an equal-angle projection so that circles project as circles, but their sizes differ according to their location, e.g. the error circle on the mean directions C and D have the same magnitude. (On an equal area, such error limits would have the same area, but would not be circular.) Changing directions plot as a succession of points, sometimes passing from one hemisphere to another, e.g. the changes in direction during demagnetization, E. (b) A cartesian (As–Zijderveld) projection. This projection allows components of the remanence to be plotted, thereby including both intensities and directions of a vector as it is progressively demagnetized. Usually the components chosen are the horizontal component, H, and the vertical, V, component (usually x vs. z). The horizontal and vertical scales should be identical and are usually chosen as appropriate to the range of intensity of remanences involved (Section 6.4). This projection is not generally suitable for plotting groups of directions.

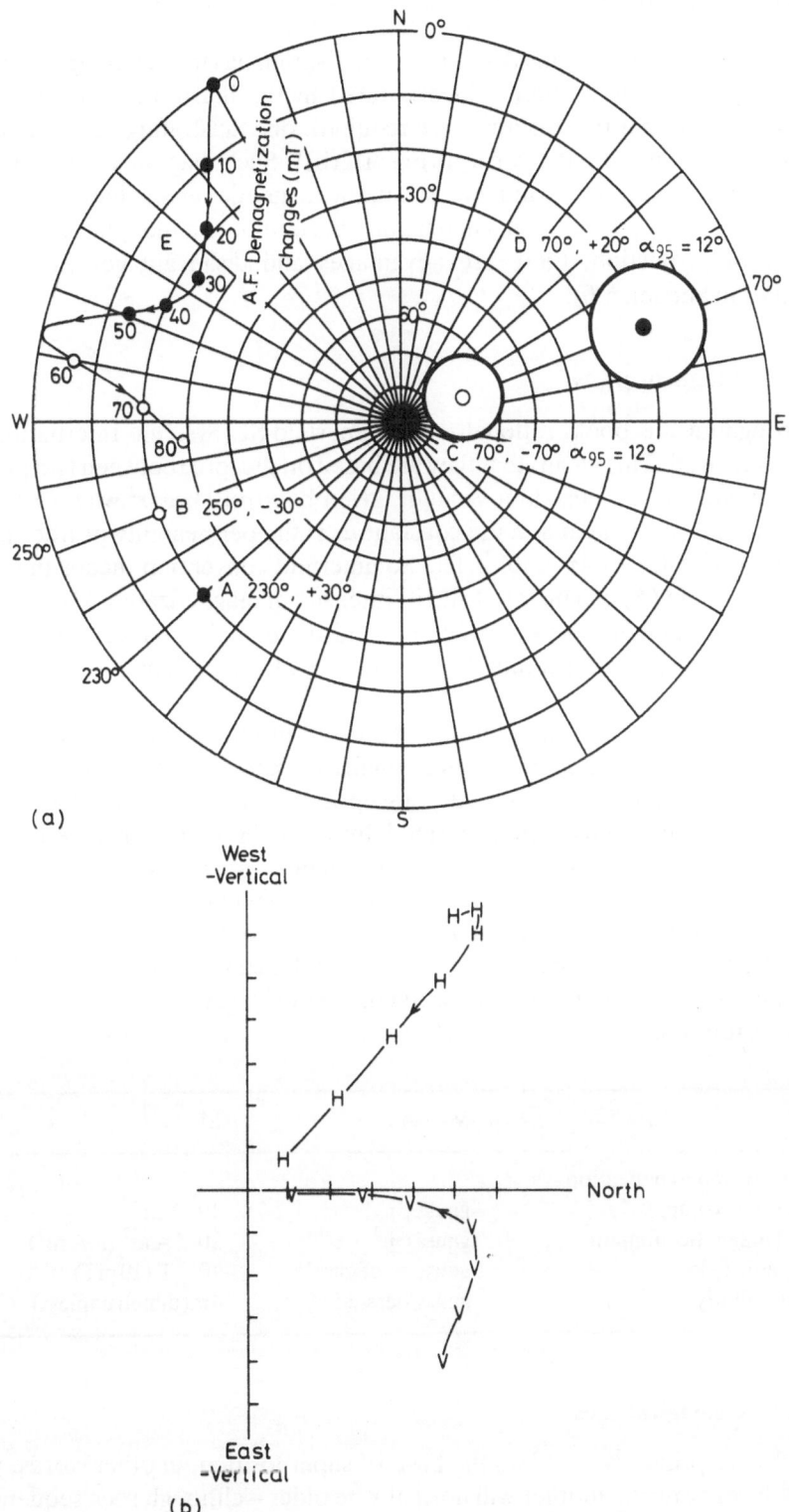

(a)

(b)

ordinates (As, 1960). This As–Zijderveld diagram has the advantage that the intensity of magnetization is incorporated in the illustration, but has the disadvantage that two diagrams are required, one illustrating the horizontal component, and one the vertical (Fig. 1.2(b)). Such diagrams are of use in seeing how the direction and intensity of the magnetic vector changes when it is subjected to partial demagnetization (Section 5.4) although, as with all types of projection, there are advantages and disadvantages (discussed further in Section 6.6).

1.3.2 Magnetic units

Throughout this book, rationalized Sommerfeld SI (Système International) units are used. This means that they are based on the force between two poles, of strengths m_1 and m_2, in a vacuum, given by $m_1m_2/4\pi\mu_0r^2$ where r is the distance between them and the constant μ_0 is the permeability of free space with an SI value of $4\pi \times 10^{-7}$. The Sommerfeld convention means that the magnetic field (F) of a material and its intensity of magnetization (J) are both measured in ampères per metre ($A\,m^{-1}$) and are related to magnetic induction (B), in a magnetized medium, by $B = \mu_0(F+J)$, which is measured in tesla (T).

Magnetic fields will therefore be expressed in terms of tesla (T) and the intensity of magnetization per unit volume is expressed in terms of $A\,m^{-1}$. This convention means that the susceptibility constant (k) relating the intensity of magnetization generated by a material in a magnetic field becomes dimensionless, but is 4π greater than it would have been in cgs or MKSA unrationalized units. Similarly, useful quotients, such as the ratio of the intensity of magnetization to its susceptibility would have unit value in cgs or MKSA units if both had the same value, but this converts to 4π in SI units.

The conversion of non-rationalized units (emu = gauss or oersted) to SI is thus as follows:

	MKSA	SI
Intensity of magnetization per unit volume	gauss	$10^3\,A\,m^{-1}$
Total magnetic moment	gauss cm^3	$10^{-3}\,A\,m^2$ ($mA\,m^2$)
Magnetic field	gauss = oersted	$10^{-4}\,T$ ($10\,mT$)
Susceptibility	gauss/oersted	4π (dimensionless)

1.3.3 Geological ages

Rocks are primarily dated on the basis of superposition, in other words, the rock layer beneath another will normally be older – although rock sequences can, in tectonically disturbed areas, be inverted. They can also be dated from the fossil content or radiometrically. Fossil dating comprises a knowledge of the sequence of evolution of different organisms, plants and animals, which can be identified, usually down to at least a species level, from their fossilized remains. Such fossils are abundant in sediments younger than some 580 million years and their relative ages are often known with a very high precision. This enables very precise relative dating, i.e. that a rock containing a certain fossil, or group of fossils, is either younger, older or of the same age as another fossiliferous rock. Rock sequences can thus be placed into stratigraphic groups which represent a specific range of time on the basis of their

Table 1.1 Geological and radiometric time-scales

Geological period	Radiometric age of base	Geological period	Radiometric age of base
Cenozoic		*Upper Palaeozoic*	
QUATERNARY		PERMIAN	
Pleistocene	2.0	Upper	258
TERTIARY		Lower	286
Pliocene	5.1	CARBONIFEROUS	
Miocene	24.6	Upper	320
Oligocene	38	Lower	380
Eocene	55		
Palaeocene	65.0	DEVONIAN	
		Upper	374
Mesozoic		Middle	387
CRETACEOUS		Lower	408
Upper	97.5		
Lower	144	*Lower Palaeozoic*	
JURASSIC		SILURIAN	438
Upper	163	ORDOVICIAN	505
Middle	188	CAMBRIAN	590
Lower	213		
TRIASSIC			
Upper	231		
Middle	243		
Lower	248		

The Phanerozoic comprises the last 590 million years, i.e. the Cambrian and younger. Pre-Cambrian ages are geologically poorly defined, reflecting the paucity of the fossil record. The major divisions are:

> Proterozoic = Upper (1650 million years base) and
>
> Middle Pre-Cambrian (2500 million years base)
>
> Archaean = older than 2500 million years

changing fossil content, but the actual time represented by such a sequence is not known from the fossil evidence alone.

Radiometric dating is based on the decay rates of a range of radioactive isotopes and their decay products – notably potassium–argon, rubidium–strontium and the uranium–lead series, with even rarer isotopes, such as niodymium, becoming of increasing importance. Most radiometric techniques can only be applied, reliably, to igneous rocks. It is thus often difficult to correlate radiometrically dated igneous rocks with fossil-dated sedimentary rocks. A radiometric scale, giving the age of the oldest rocks, i.e. the basal rocks of each major stratigraphic unit, is given in Table 1.1, based on recent recommended revision on the decay constants for potassium used in radiometric dating methods (Steiger and Jager, 1977) and based on Harland *et al.* (1982). It should be emphasized that all these ages are subject to dispute and may be in error 2–3%, possibly up to 5–6%.

Chapter Two

The physical basis

2.1 MAGNETIZATION ON AN ATOMIC SCALE

A magnetic field is produced by the movement of an electrical charge. So, at the lowest level, the movement of an electron results in the creation of a magnetic field. Electrons will normally spin about their axes and also orbit their nucleus and therefore have two types of motion that can produce magnetic fields. All substances can thus be regarded as being magnetic at an atomic level and can be classified into two types. In *diamagnetic* substances the electron shells are full and the precession of electron orbits when placed in a magnetic field results in the creation of a magnetic field in the opposite direction to the applied field. The magnetization acquired per unit field applied, the magnetic *susceptibility*, is small, less than 10^{-5} SI. In substances in which the electron shells are incomplete, *paramagnetic* materials, each atom has a magnetic moment due to the uncompensated electron spins. When placed in a magnetic field, the electron orbits precess but the magnetic moment is aligned in the same direction as the applied field and is generally stronger than that of diamagnetic substances, having susceptibilities of the order of 10^{-3} to 10^{-5} SI.

In the iron group of elements (the first transitional series), the electron orbital motions are quenched by crystal lattice fields and the external magnetic field arises almost entirely from the spins of unpaired electrons in the 3d orbital shell. Such magnetization can be coupled between adjacent atoms (direct exchange) or they can interact via an intermediate anion, usually oxygen, by means of an indirect (superexchange) reaction. These exchange reactions mean that the electron spins are coupled either parallel or anti-parallel to each other, depending on the degree of overlap of the electron orbits. In *ferromagnetic* materials, the couplings are parallel to each other and they attain a very strong spontaneous magnetization, even in the absence of an external magnetic field (Fig. 2.1(a)). In materials in which the coupling is antiparallel, the magnetization can be visualized as two separate magnetic lattices, each of which is magnetized in the opposite direction to the other. In *antiferromagnetic* substances, the magnetization of the two lattices is exactly balanced, so there is no external magnetization and hence no external spontaneous magnetization (Fig. 2.1(b)), although each lattice has its internal magnetic coupling. In *ferrimagnetic* materials, the two magnetic lattices are

15

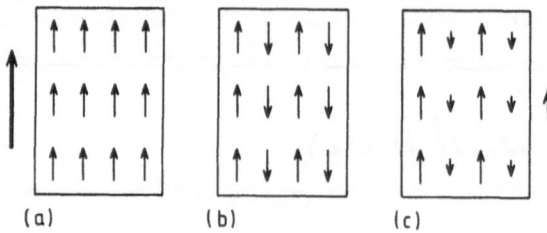

Figure 2.1 Ferromagnetism, antiferromagnetism and ferrimagnetism. (a) In ferromagnetic materials, all the individual spin magnetizations are coupled parallel to each other by internal exchange forces, resulting in a very strong external field being generated even in the absence of an applied field. (b) In antiferromagnetic substances, the internal superexchange forces cause the spin magnetizations to be coupled anti-parallel to each other so that they have no external magnetic field unless crystal lattice defects prevent perfect antiparallel alignment. (c) In ferrimagnetic substances, the spin magnetizations are antiparallel to each other, but have different magnitudes so that the materials have a weaker magnetization than a ferromagnetic material, but similarly acquire this magnetization even in the absence of an applied magnetic field.

not exactly equal so they have a spontaneous magnetization in the same way as ferromagnetic materials, but this is much weaker as it depends on the difference between the two lattices (Fig. 2.1(c)).

Paramagnetic and diamagnetic magnetizations are only temperature dependent in that they depend on the radius of the electron orbits, but ferro-, antiferro- and ferrimagnetism depend strongly on the interatomic distances. On heating, there is thus a specific temperature at which this distance is exceeded and the exchange and super-exchange coupling breaks down. In ferro- and ferrimagnetic substances, this temperature is the *Curie temperature*, while in antiferromagnetic substances, the coupling between the lattices breaks down at the same temperature as the interatomic coupling, the *Néel temperature*. Above the Curie or Néel temperature, all such substances behave paramagnetically.

On a larger scale, substances that are commonly termed 'magnetic' are normally either ferromagnetic or ferrimagnetic materials and it is their properties that are of importance in palaeomagnetic and archaeomagnetic studies. On this basis, the term magnetic will be subsequently taken to refer to this type of behaviour and not to diamagnetism or paramagnetism. (The physical basis is also covered in more detail by Nagata (1961), Stacey and Banerjee (1974) and O'Reilly (1976).)

2.2 MAGNETIC DOMAINS AND ANISOTROPY

As a crystal is an ordered network of atoms, their exchange forces also have a structure relative to the crystal lattice so there are directions, within any

16

specific crystal lattice, along which it is easier for a substance to become magnetized than along another, although it is equally easy to magnetize the substance in either direction along such magnetic 'easy' axes. Energy is, however, required for a change from one direction to another and this depends on the composition and also on the size of the crystalline grain. In small grains, the spontaneous magnetization can 'flip' from one easy direction to another, depending on the balance between the temperature and the internal forces. Even in the presence of an external field, the magnetization of each crystalline grain will be along the 'easy' direction that has a component of the applied field along it, with thermal agitations only occasionally being sufficient to 'flip' the direction into an antiparallel alignment – unless the applied field is very strong.

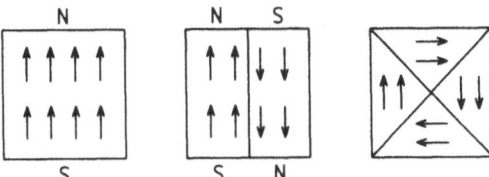

Figure 2.2 Domain patterns. In large grains, the exchange and superexchange forces cause internal alignment of the spin magnetizations, resulting in the formation of strong magnetostatic poles on the outside of the grain. To balance these two kinds of force, the domains may form antiparallel pairs or closure domains – the latter showing virtually no magnetic field as the magnetostatic forces are effectively cancelled within the grain.

In larger grains, the internal magnetocrystalline forces will still operate to cause magnetic alignments parallel or antiparallel to the 'easy' axes, but the magnetostatic forces are increased as the two poles on the surface of the grain, north and south, become larger and attract each other. Under such conditions, the magnetization within each large grain breaks down to form individual volume elements, *domains*, within which the individual electron spins are coupled parallel to each other, but the different domains become oriented in such a way that the magnetostatic forces on adjacent domains are reduced by mutual interaction (Fig. 2.2). The boundary between two domains, the *Bloch wall*, is a narrow zone within which the directions of the magnetization of the electron spins cant over from that of one domain to the next (Fig. 2.3). The dimensions of domain and Bloch walls in substances of palaeomagnetic and archaeomagnetic interest are difficult to determine experimentally, but theory and observation indicate that the domain walls are some 0.1 µm thick, while the domains themselves are only of similar size, 0.1–0.05 µm, in magnetite (Stacey and Banerjee, 1974), but possibly up to 1.5 µm in haematite (Banerjee, 1971). Individual domains, in natural substances, are likely to be very variable, reflecting the presence of impurities, defects, etc. within the crystal lattices, but it is probable that such effects are

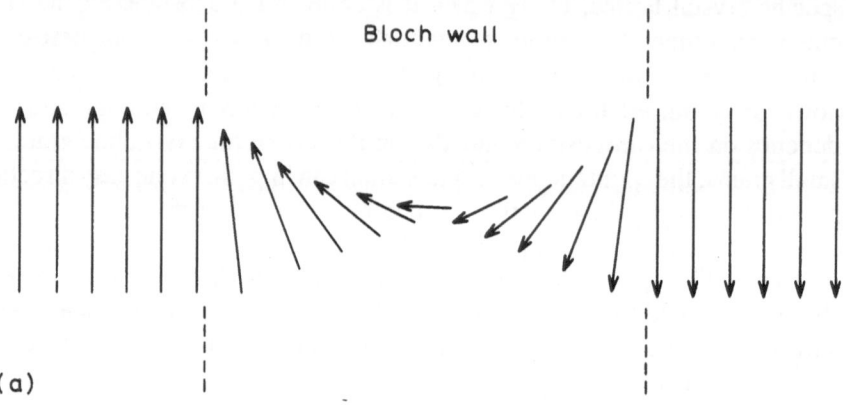

Bloch wall

(a)

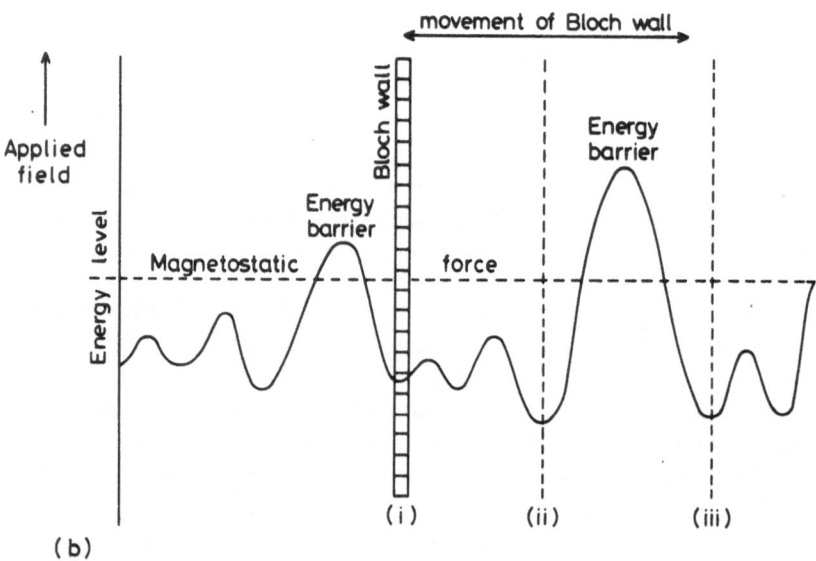

movement of Bloch wall

Applied
field

Energy level

Bloch wall

Energy
barrier

Energy
barrier

Magnetostatic

force

(i)   (ii)   (iii)

(b)

Figure 2.3 Domain wall and wall movements. (a) In a domain wall, the individual directions of the spin magnetizations gradually cant over from the direction in one domain to that of the bordering domain. When a field is applied to the grain, the domain wall unrolls so that the domain with a magnetization parallel to the applied field increases in volume. The unrolling of the wall, (b), takes it through different energy barriers within the crystal. If these are small, then the domain wall can roll back to its previous position after the externally applied field is removed. However, if the applied field is strong, the wall may roll through a large energy barrier and is then unable spontaneously to roll back to its previous position after removal of the applied field. However, if left for long enough, thermal agitations gradually allow the wall to unroll, although it may take a very long period of time to return to its previous state.

dominated by the actual shape of the individual grains as an elongate particle could well behave as a single domain even if its volume were some 100 times greater than that of a spherical single domain sized grain. The presence of lattice imperfections, without which the crystal could not have grown, also means that there are often locations within the lattice, particularly at the surfaces of the grain, at which the domain walls become locked. In such circumstances, large grains may well contain zones that behave, magnetically, as if they were single domains. The nature and role of such pseudo-single domain particles is likely to be important, but is still poorly understood (Shive, 1969; Merrill, 1981), as both single and pseudo-single domain particles behave differently from multidomain particles.

In a multidomain particle, the application of an external magnetic field causes the domain walls to unroll, causing a growth in the volume of domains with a magnetization in the direction of the applied field at the expense of domains that are at angles to the applied field (Figs. 2.2 and 2.3). If the applied field is weak, the domain walls roll back to their previous position when the field is removed. However, as stronger fields are applied, the domain walls eventually jump past the impurities, vacancies, etc. (*Barkhausen jumps*) that form energy barriers within the crystal lattice. On removal of the applied field, not all domain walls are capable of rolling back through the energy barriers to their previous location. This means that domains aligned with the external magnetic field remain in that direction and the grains therefore have a remanent magnetization, an *isothermal remanence*, acquired by the application of magnetic fields at a constant temperature. As even higher fields are applied, more domain walls cross even larger energy barriers, and the intensity of isothermal remanence increases. In very strong fields, the magnetization of domains originally antiparallel to the applied field eventually 'flips' into alignment and the isothermal magnetization saturates at a peak value, M_{sat}, when measured after removal of the external field. In extremely strong magnetic fields, the magnetization within each domain may be forced into greater alignment with the field, but this relaxes back to the saturation values after removal of the field. The application of increasing magnetic fields, first in one direction and then in the reverse direction, therefore causes the magnetization to behave in a systematic manner, forming a *hysteresis loop* (Fig. 2.4(b)). Such a loop also defines some important rock magnetic parameters, such as the *saturation moment*, M_{sat}, and the field required to attain it, H_{sat}, as well as the back field required to reduce the isothermal remanence to zero, H_{cr}, the *back field coercivity*.

If a magnetic field is applied to a collection of randomly oriented single domain particles, the direction of magnetization in domains that have a magnetization exactly antiparallel to the applied field will not change their direction of magnetization until the applied field exceeds the internal energy, when the direction will suddenly flip into alignment. Such domains will each have a square hysteresis loop (Fig. 2.4(a)). Where the domains have a

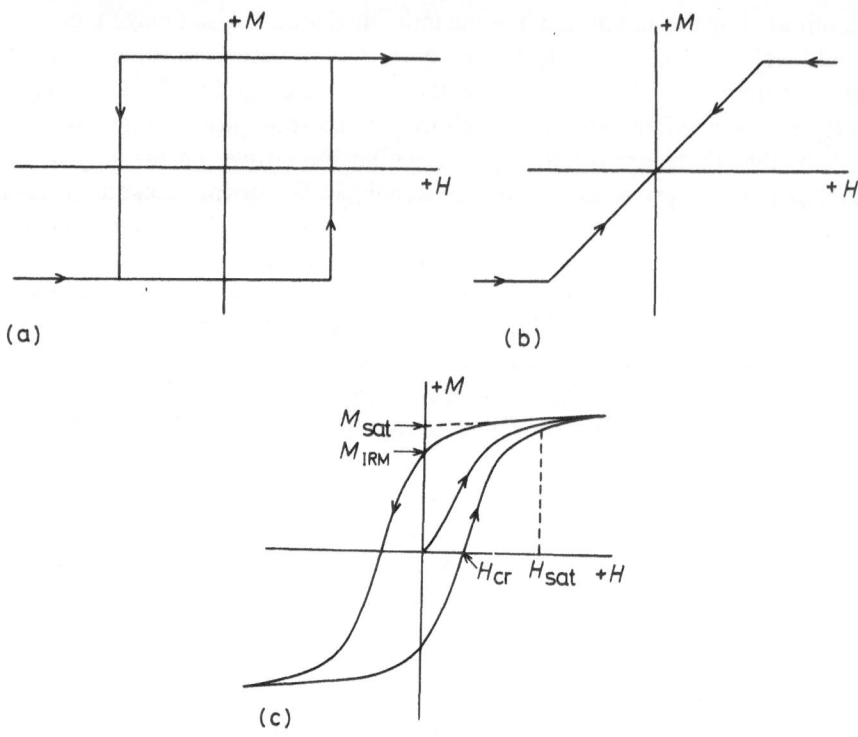

Figure 2.4 Hysteresis loops. An external field, H, applied along the 'easy' axis of a single domain particle has no effect on the intensity and direction of magnetization, M, until the internal forces are exceeded, when it suddenly changes direction, resulting in (a) square hysteresis loops when M is plotted against H. A field applied at an angle to the 'easy' axis (b) induces a magnetic field but this is immediately lost on removal of the field, i.e. it has no hysteresis. The intensity of magnetization, M, therefore increases linearly until all spin moments are aligned, after which there can be no further increase in its magnetization. In an assembly of single domain particles in which the 'easy' axes are randomly oriented, the summation hysteresis loop is similar but not quite identical to that of the hysteresis loop for multidomain particles (c) in which there is an initially linear increase as a weak field is applied to a previously demagnetized sample. As the domain walls unroll, magnetizations of the domains eventually all become aligned and the material becomes magnetically saturated, M_{sat}, in a specific applied field, H_{sat}. If the applied field is then removed, some domain walls unroll, but others remain locked, so that the sample has a remanence, M_{IRM}, in zero field. This isothermal remanence is then destroyed as the reversed applied field (back field) exceeds a specific value, H_{cr}, which is the coercive field.

direction at an angle to the applied field, then 'flips' will occur when the component of the field along the 'easy' direction exceeds the internal energy. The hysteresis loop for a collection of randomly oriented single domain particles will thus comprise a summation of all such behaviour and this will be similar to, but not identical to, the hysteresis loop of multidomain particles.

20

Very small grains, such as haematite less than 0.03 μm in diameter and magnetite grains less than 0.05 μm, show a strong susceptibility when placed in a magnetic field, but thermal vibrations rapidly destroy such alignment as soon as the field is removed. The moment acquired may, however, take several minutes before becoming randomized. This property is termed *superparamagnetism* as it is comparable to a paramagnetic material with a large moment when placed in a field, but also has the additional ability to retain the magnetization for a short time after the field has been removed (Néel, 1949). Small grains may also show unusual effects as statistical fluctuations in, for example, the distribution of cations, which may mean that the magnetic lattices in an antiferromagnetic material may not be exactly balanced and therefore show weak ferrimagnetic properties, termed *superantiferrimagnetism* (Néel, 1962a). The effect of grain size is therefore fundamental in all magnetic studies, but the shape and crystalline properties are also critical.

The existence of 'easy' directions of magnetization within individual ferromagnetic or ferrimagnetic materials means that any alignment of the crystalline axes will result in a magnetic *crystalline anisotropy*, depending on the strength of the anisotropy for individual particles and their degree of alignment with each other. The internal anisotropy forces in magnetite are weak, so that the crystalline anisotropy is low, but the crystalline anisotropy in haematite is very high (Section 3.2) as it is some 100 times more readily magnetized within the basal plane of its crystals than in any other direction (at room temperature) – Uyeda *et al.* (1963). (In high applied fields the degree of crystalline anisotropy may be smaller in haematite, but the behaviour of all ferro- and ferrimagnetic materials tends to be more complex – Porath and Chamalaun (1966).) However, the anisotropic properties of these two minerals is reversed when their *shape anisotropy* is considered (Fig. 2.5). Although the magnetic field within a magnetic grain is uniform, the magnetostatic forces due to the magnetic poles at its surface depend on their separation (Stoner, 1945). In spherical grains, the separation of the surface poles is constant in all directions, so that a sphere has zero shape anisotropy (Fig. 2.5). All other shapes can be essentially represented as ellipsoids (Stacey, 1960; Uyeda *et al.*, 1963) and will be more readily magnetized along their long axes than in any other direction. The difference in susceptibility, $\Delta\chi$, is given by:

$$\Delta\chi = \frac{\chi_i^2(N_a - N_b)}{(1 + N_a\chi_i)(1 + N_b\chi_i)}$$

where χ_i is the intrinsic susceptibility, and N_a and N_b are the maximum and minimum demagnetizing factors along the different axes (Stacey, 1963). The dependence on χ_i^2 is particularly important as it means that shape anisotropy is most important in mineral grains of high susceptibility, such as iron and magnetite (Sections 3.2.1 and 3.2.5), and weak in low susceptibility materials such as haematite (Section 3.2.2).

21

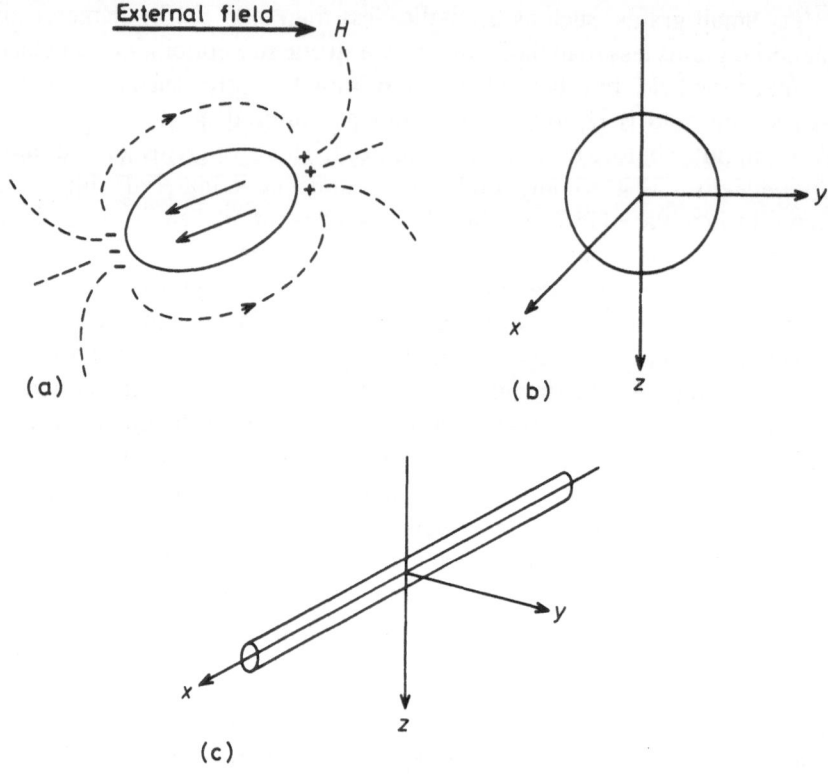

Figure 2.5 Shape anisotropies. (a) An elongated particle will always be more readily magnetized along its long axis as the internal magnetic field due to the magnetostatic poles at its surface is weakest in this direction. (b) For a sphere, there is no shape anisotropy as the demagnetizing fields are identical in all directions, while (c) a rod has a very high anisotropy. (N is the demagnetizing factor due to the separation of the surface magnetostatic poles.)

Interactions between neighbouring single domain particles have only occasionally been considered (Dunlop and West, 1979) within single domain theories (Néel, 1949, 1955; Stacey, 1958, 1960, 1963). In general such interactions are only likely to be important in two geological or archaeological contexts:

1. Where a string of single domain particles occurs, each of which is just in contact with its neighbour, then a strong shape anisotropy will arise, even though the individual grains may be spherical.
2. When one magnetic grain forms in close proximity to an already existing, magnetized grain, then two possibilities arise. If the grain is in atomic contact, such as may occur when haematite forms within the crystalline structure of maghemite (Sections 3.2.1 and 3.2.2), then superexchange interactions may occur, causing the magnetization of the newly forming

mineral to be directly coupled to that of the pre-existing mineral. If the new mineral is not quite so close that the electrons are shared, then the new mineral may form in a location where the magnetic field of the pre-existing grain may be greater than that of the ambient magnetic field, such as that of the Earth.

The effects of pressure on the magnetization of magnetic grains are not clearly understood. *Hydrostatic pressure*, such as occurs when rocks become deeply buried (Fuller, 1970), seems to have little effect on the Curie temperature, but appears to be associated with a decrease in high-field susceptibility. Kim and Carmichael (1977) found a 30% drop in the saturation magnetization at pressures of 0.1 GPa (equivalent to c. 12 km depth in the Earth). Pearce and Karson (1981) found that under hydrostatic pressure magnetite and haematite particles of diameter 10–20 μm became saturated in weaker fields than under normal pressures and that the saturation magnetization drop was greatest in grains of lowest coercivity ($H_{cr} < 30$ mT) so that magnetite was affected more than haematite. Nonetheless, most experiments indicate that all such effects are reversible as the pressure is removed and are, in any case, small for magnetic field strengths of the order of that of the Earth. The effects of brief, very high pressures, such as during meteoritic impact, are complicated because such natural phenomena are also accompanied by very high local heating and also regional heating (Section 9.8), in which the thermal effects often appear dominant. *Uniaxial stresses* may cause a permanent deformation of lattice fabrics with associated distortion of their magnetization. If such deformations are not reversible on removal of the stress, then the material may be left with a *piezoremanent magnetization*. In most natural materials, however, the main effect of stress is the change in shape or crystalline properties, with consequent effects on the anisotropic properties, rather than a direct magnetic effect.

2.3 TEMPERATURE, GRAIN SIZE AND TIME

When a magnetic field is applied to a single domain particle, there will usually be a component of the field which will lie along one of the easy directions of the domain. The direction of magnetization within the domain will therefore try to 'flip' into the direction of this easy axis, but can only do so if the internal energy barriers are exceeded, i.e. the total energy available exceeds Kv, where K is the anisotropy constant for that composition and lattice, and v is the volume of the domain. The energy is available directly from the strength of the field applied, reduced by the angle which it makes to the easy direction ($\sin^2 \theta$, where θ is the angle to the easy axis). In strong fields, such flips therefore take place almost immediately on application of the external field. In weak fields, however, the magnetic energy may be insufficient to cause a flip until thermal vibrations, kT (k is Boltzmann's constant and T is the

23

absolute temperature), allow the internal barrier to be exceeded. As temperature is a statistical measure of thermal vibrations, a few electrons, at any one time, will have sufficient energy to exceed the energy barrier so that, if the domain is kept at the same temperature, all electrons will eventually exceed the internal energy barriers and the domain takes up an alignment along the easy direction that has a component in the direction of the applied field. Each domain therefore has a specific *relaxation time* that it will take for it to acquire a magnetization with a component in the direction of the external field. This concept, introduced by Néel (1949, 1955), is fundamental to any understanding of laboratory and even longer-term behaviour of magnetic materials. In the case of a sample containing single domain particles, the domains will acquire a magnetization in the direction of an applied field. On removal of the field, the thermal vibrations will cause the magnetization in very small domains to flip backwards and forwards along the easy directions, but larger domains will take longer for such flips to occur – eventually, however, all domains will have a magnetization along easy axes that have no relationship with the original field direction. If the easy directions have a random orientation, then their individual magnetizations will cancel each other so that the sample has no externally observed magnetization. Similar considerations also apply to multidomain systems as the ability of domain walls to unroll past energy barriers within the crystal lattice is similarly affected by temperature considerations and hence by the effect of time.

For any given material, such as magnetite or haematite, the anisotropy 'constant' (K) is only slightly affected by temperature, especially near the Curie temperature, but will be affected by impurities, etc. Nonetheless, this constant does not change very much, so that the relaxation time for such materials is almost entirely dependent on the volume and temperature of the magnetic grains. More precisely, the relaxation time τ is given by:

$$\tau = \frac{1}{C}\exp\left(\frac{\nu K}{kT}\right) = \frac{1}{C}\exp\left(\frac{\nu H_a M_s}{2kT}\right)$$

where C is the frequency factor which is of the order of $10^{10}\,\mathrm{s}^{-1}$, ν is the volume of the grain, H_a is the anisotropy field and M_s is the saturation magnetization. For any given grain, therefore:

$$\log \tau \propto \frac{\nu}{T}$$

A single domain particle of constant diameter will, on cooling, therefore show a logarithmic increase in its relaxation time as it cools from a temperature at or below its Curie temperature. Similarly a minute particle, as it grows at constant temperature, will show a logarithmic increase in its relaxation time.

In order to have a measurable remanence, the relaxation time must be of the order of a few minutes so such remanence can be acquired in two separate ways:

24

1. By cooling in a magnetic field over temperature ranges below the Curie point, in which case the remanence is termed a *thermoremanent magnetization*, and
2. By crystal growth within a magnetic field, when the remanence is termed either a *chemical remanent magnetization* or a *crystalline remanent magnetization*

In both cases, if the direction of the field changes, then particles with low relaxation time will begin to relax into the new field direction even when there is no change in temperature or grain size. This new remanence is termed a *viscous remanent magnetization* and, for any given field strength, depends on the duration of the applied field.

In natural materials, the relationship between thermal, chemical and viscous remanences can be extremely complex (Section 4.6) as many geological and archaeological processes involve simultaneous thermal and chemical changes that may also be extended over extremely long periods of time. However, at this stage it is convenient to describe each process separately, mainly in terms of single and pseudo-single domain theories (Néel, 1949, 1955; Stacey, 1963) although these theories do not completely account for all observations (Merrill, 1981; Sugiura, 1981) and the behaviour of pseudo-single domains, in particular, is poorly understood. Nonetheless, most departures from Néel's theories occur either very close to the Curie temperature or in very small grain volumes – in both situations the assumption of an anisotropy constant is probably invalid. Such departures are, in most cases, small and only become relevant under abnormal conditions.

2.3.1 Thermoremanent magnetization (TRM)

In a material containing a range of grain sizes, the magnetic grains will all behave paramagnetically at temperatures above the Curie temperature (Section 2.2). Just below the Curie temperature, the larger single domain particles will have relaxation times of the order of a few minutes, enabling their remanence to be measured, but smaller particles will have shorter relaxation times and hence become rapidly randomized by thermal fluctuations. On cooling to room temperature, the larger single domain particles will have relaxation times of the order of many thousands or millions of years (Fig. 2.6) and smaller particles will show a range of relaxation times, some of which may still be of the order of a few seconds. Such fine grained particles will therefore behave superparamagnetically in that they have a high susceptibility in an applied field, but this decays away very rapidly after removal of the field. The definition of the relaxation times of such superparamagnetic grains is not established, but is of the order of some 10^2 s, enabling the remanence to decay during actual measurement following removal of the field. However, there will also be grains with somewhat longer relaxation times, c. 500 s, that will retain their remanence until all normal measurements, following removal of the field, are completed. For such grains, the thermoremanence is therefore blocked within them for the duration of the laboratory experiment. This relaxation time will occur at different temperatures, with larger single domain

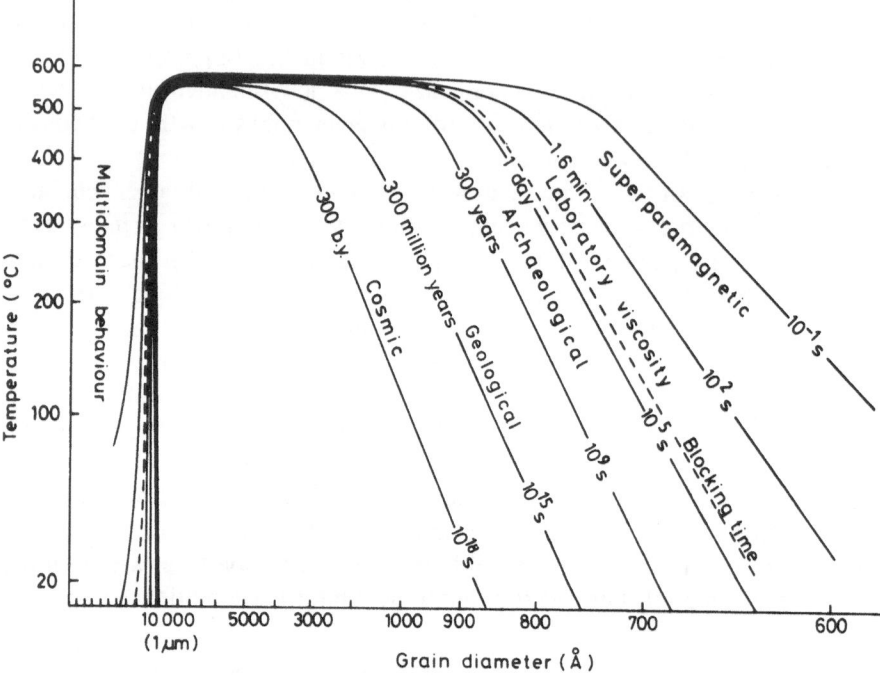

Figure 2.6 The relationship between temperature, volume and relaxation time. This graph is based on a titanomagnetite, although the general shape and relationships are valid for all magnetic minerals, but with different values for different minerals. The behaviour as the grain size changes from single domain to multidomain is very strongly controlled by the presence of imperfections in the crystal lattice and makes actual physical determinations difficult to measure. The estimated values are thus shown dashed. The onset of such multidomain behaviour may also take place at smaller grain sizes than indicated here.

particles reaching this relaxation time at higher temperatures. The temperature at which such a relaxation time is reached is therefore termed the *blocking temperature* (T_B) and is thus the temperature at which the remanence acquired during cooling in a magnetic field becomes blocked within it for the duration of the experiment. The blocking temperature of any individual domain obviously depends on its size and composition, but is also affected by the strength and duration of the ambient field – a higher field

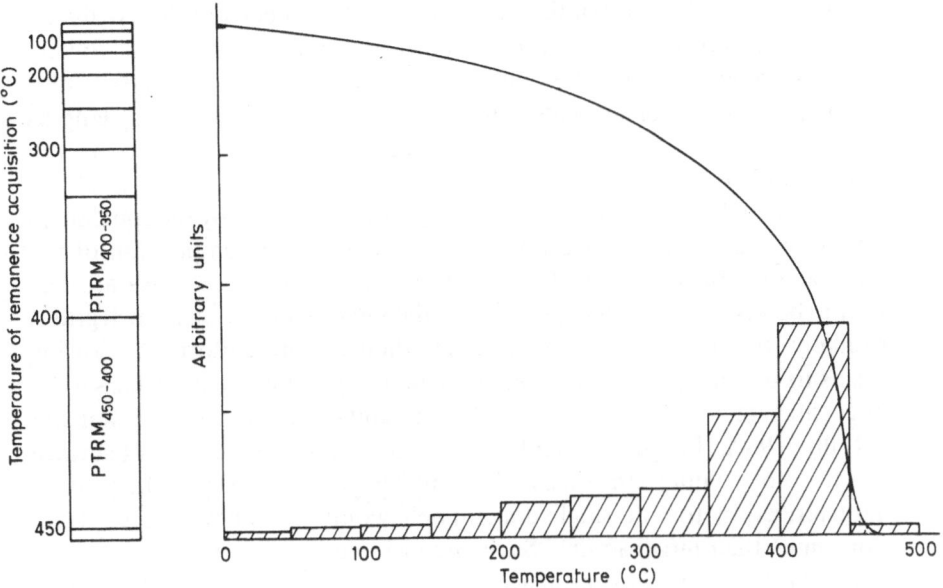

Figure 2.7 Thermal remanence and partial thermal remanence. As a rock or other heated material cools, it acquires a magnetization over different temperature ranges below its Curie temperature. The final magnetization is the summation of all components acquired at higher temperatures up to the Curie temperature.

strength or a longer duration corresponding to an effectively higher blocking temperature (Sugiura, 1980).

Most natural materials have a wide range of grain sizes and hence a wide range of blocking temperatures. On cooling over a specific temperature range below the Curie temperature (T_c), only specific grain sizes will acquire a magnetization in an applied field. Grains with blocking temperatures in excess of the peak temperature will retain the random directions they acquired at higher temperatures when there was no ambient field. All particles with blocking temperatures corresponding to the temperature range ($T_1–T_2$) over which the field is applied will become magnetized in the direction of the field, while those with blocking temperatures below that of the lowest temperature at which the field was applied will be unable to retain any memory of the applied field and hence will take up random directions as further cooling takes place. This means that the total thermal remanence at room temperature will always be the summation of the magnetization acquired over all higher temperature ranges (Thellier, 1937a; Thellier and Thellier, 1959), i.e. the *partial thermoremanence* (PTRM) acquired over specific temperature ranges can be summed to give the total observed remanence (Fig. 2.7):

$$\text{TRM}_{T_c}^{T_n} = \text{PTRM}_{T_c}^{T_1} + \text{PTRM}_{T_1}^{T_2} + \text{PTRM}_{T_2}^{T_3} \ldots + \text{PTRM}_{T_{n-1}}^{T_n}$$

27

Palaeomagnetism

Although such theories of thermoremanence have been developed almost exclusively in terms of single domain particles, similar concepts, involving relaxation times, can also be applied to pseudo-single domain and multi-domain particles as domain walls can obviously be blocked for varying lengths of time, depending on the strength of the energy barriers pinning down the domain wall movements. However, such a situation is less amenable to theoretical analysis as the effects are dominantly controlled by random lattice defects and these are not closely related to the size or composition of the magnetic materials concerned. Nonetheless, long relaxation times are less likely to be associated with large grains of the same composition as such grains are likely to have grown more slowly and hence more uniformly, allowing defects to become partially annealed. Similarly, pinning often appears to be closely associated with grain surfaces (Stacey and Banerjee, 1974; Tucker and O'Reilly, 1980a; Halgedahl and Fuller, 1980) and hence the ratio of surface area to total volume also means that smaller grains are more likely to be characterized by pseudo-single domain behaviour and to have longer relaxation times than larger grains. Such suppositions are also indicated by the ability of different grain sizes to acquire stronger or weaker magnetization in the same weak magnetic field (less than 1 mT). Magnetite grains about 1 μm in diameter acquire an intensity of thermoremanence of the order of 10^4 A m^{-1} on cooling in a field of some 0.1 mT, but this intensity decreases in direct relation to grain diameter, being some 10^2 A m^{-1} for grains of diameter 50 μm cooling under identical conditions (O'Reilly, 1976; Tucker and O'Reilly, 1980a,b; Sugiura, 1980). For larger grains, the magnetization acquired remains approximately constant, irrespective of grain size, indicating that normal multidomain behaviour is dominant. Haematite grains have not been studied in detail, but grains of some 0.1 μm, cooling in the same field, 0.1 mT, acquire a much lower remanence, c. 50 A m^{-1}.

The dependence of the remanence on the strength of the applied field and on the duration for which it is applied have received very little attention. Day (1977) found that the strength of the remanence acquired was essentially linear with applied strengths between 0.6 and 0.8 mT and that saturation occurred when cooling in fields of 1.5 mT or greater. Laboratory cooling rates are naturally much more rapid than those that may occur in archaeological and geological contexts. At any one temperature, the intensity of magnetization acquired in a constant field will gradually increase (Dodson and McClelland-Brown, 1980) because more domains will 'relax' into the direction of the ambient field – a *high-temperature viscous remanence*. Determination of the original strength of the field in which a thermoremanence was originally acquired therefore also requires a knowledge or reliable estimate of the probable rate of cooling. Even in the archaeological context, where pottery may take one or two days to cool to ambient temperatures, differences between such cooling rates and those used in the laboratory can introduce errors of up to 20% (mostly <10%) in determinations of past geomagnetic field strengths (Sections 5.6 and 7.2.1) and these may be even greater under geological conditions (Weaver, 1970; Tanguy, 1975; Fox and Aitken, 1980; Walton, 1980; Dodson and McClelland-Brown, 1980; Halgedahl et al., 1980).

2.3.2 Chemical remanent magnetization (CRM)

The acquisition of a remanence by means of crystal growth at constant temperature is exactly analogous to the acquisition of a thermoremanence. As a crystallite grows after its nucleation, it behaves initially in a superparamagnetic manner. Further growth sees a logarithmic increase in its relaxation time until it reaches its *blocking volume* (V_B) at which size its relaxation time is of the order of some 500 s. Further growth results in a continued increase in its relaxation time until it reaches dimensions where multidomain behaviour commences and the relaxation time becomes less predictable. In haematite, the blocking volume, at room temperature, is of the order of 0.003 μm (McNab et al., 1968; Banerjee, 1971; Dunlop et al., 1974; O'Reilly, 1976), but is less well defined in titanomagnetites and titanomaghaemites, probably being of the order of 0.015 to 0.04 μm (Dunlop, 1973a; O'Donovan, 1975; O'Reilly, 1976). The transition to multidomain behaviour is even less well defined, but in both cases, probably occurs well below 1 μm (O'Reilly, 1976; Westcott-Lewis and Parry, 1971). Obviously the nature of chemical remanence will be affected by the temperature at which the chemical growth takes place and the duration for which the field is applied over any given volume range. The rate of growth is also likely to affect the nature of the pseudo-single domain and multidomain grains and hence their ability to retain a chemical remanent magnetization.

2.3.3 Viscous remanent magnetization (VRM) and partial thermal demagnetization

For any sample containing magnetic grains, the acquisition of a viscous remanence, being time-dependent, is determined by its spectrum of relaxation times and the duration of the magnetic field (Dunlop, 1973b; Walton, 1980). The spectrum of relaxation time, for any given composition, depends on the temperature and grain sizes involved. Under normal circumstances, the magnitude of viscous magnetization, M_{vrm}, acquired at a constant temperature in a weak magnetic field is thus dependent on the log of time

$$M_{vrm} = S_a \log t$$

and its acquisition is thus linear when plotted on a log scale (Fig. 2.8). The constant, S_a, is dependent on the size of the grain, its composition, etc., and can be altered by several factors, including the record of any previous treatment, such as previous demagnetization (Tivey and Johnson, 1981).

29

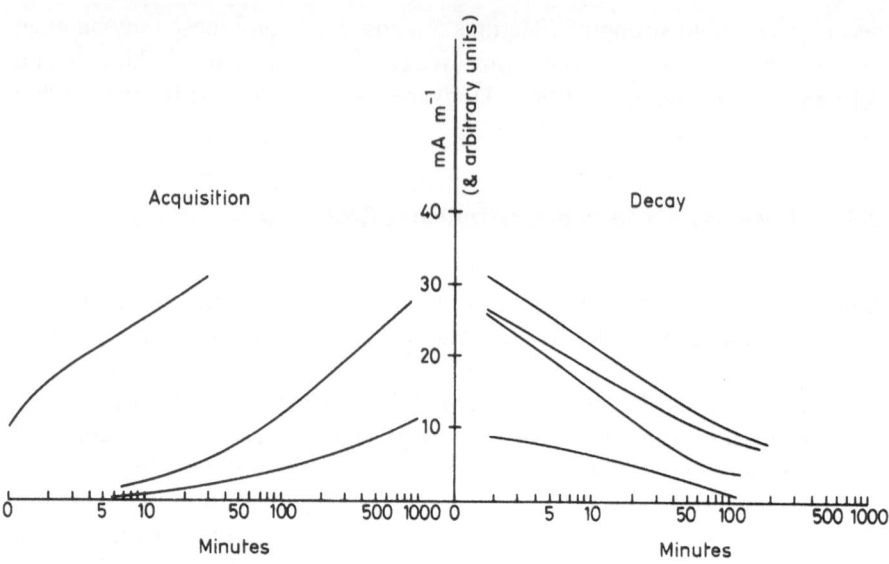

Figure 2.8 The short-term acquisition of viscous remanence. Although the acquisition of viscous magnetization is theoretically linear with log time, there are frequently significant changes at first before this acquisition rate is attained. (Longer term acquisition is illustrated in Fig. 5.9.)

Theoretically the rate of decay of a pre-existing remanence, when placed in zero magnetic field, should be similar to the rate of acquisition, but these often differ when examined over laboratory time-scales. Most of these effects seem to be confined to such short-term studies (Urrutia-Fucugauchi, 1981b) but it is not yet practicable to design extremely long-term experiments that would need to extend over a decade or so. In any case, such long-term behaviour, which is clearly relevant to archaeological and geological studies, may well be confused by other long-term changes, such as the gradual ex-solution of magnetic grains or increasing surface oxidation or reduction. Many of the features of short-term viscous remanence cannot be regarded as fully understood (Dunlop, 1973a,b; Daly, 1981) and most studies have been confined to either oceanic basalts or lunar samples (Gose *et al.*, 1972; Lowrie and Kent, 1978) – both types of material showing indications of chemical instability under normal laboratory conditions. Single domain sized magnetite particles appear to be least sensitive to their previous handling (Tivey and Johnson, 1981) with multidomain and very fine grained single domain particles showing the most complex behaviour.

Although the theoretical and experimental basis of viscous remanence acquisition and decay is not adequately established, the ability of a magnetization, however acquired, to relax in the absence of an external field is fundamental to most palaeomagnetic and archaeomagnetic investigations. If a sample is heated, for example, all the relaxation times of its constituent grains are rapidly reduced. If kept in a zero field at the peak temperature, grains with blocking temperatures less than that temperature will relax into random directions, assuming their easy axes are randomly oriented. However, grains with very large relaxation times at room temperature will still have relaxation times greater than 500 s, and will retain their original remanence. On cooling in zero magnetic field, the relaxation times of all grains will increase logarithmically, leaving those magnetizations that became randomized still in random positions. Grains with long relaxation time, i.e. those whose blocking temperature was not exceeded, will retain their original direction unaffected by the heating and cooling. Thus the magnetization of low relaxation time, such as that associated with viscous remanences, can be preferentially removed by partial demagnetization (Section 5.4). It is important, in such experiments, that both the duration of the peak temperature and the rate of cooling are identical when comparing the behaviour of different samples as high-temperature viscous effects may obscure or enhance differences in the apparent magnetic behaviour.

2.3.4   Coercivity, alternating magnetic fields and laboratory remanences

When a magnetic field is applied to a sample containing magnetic grains, if the field exceeds the internal forces, the magnetization of each domain will flip so that its easy direction lies with a component in the direction of the applied field (Section 2.2). If the field is alternating, then the individual domain magnetization will flip from one easy direction to the other. However, if the field does not exceed the coercivity of each easy axis, the magnetization will remain unchanged. The coercivity of each grain depends on the internal magnetocrystalline forces and on the magnetostatic forces (Section 2.2). These, in turn, are strongly influenced by the size of the particle and the temperature, as well as by its particular composition, with lower coercivities being associated with the smaller single domain sizes and higher temperatures (Fig. 2.9). The situation for pseudo-single domain and multidomain particles is less clear, although a similar temperature relationship can be expected. For any one composition, therefore, the coercivity spectrum is related to the relaxation time spectrum as both are strongly dependent on grain size. When an alternating magnetic field is applied, some grain sizes will have coercivities that are exceeded by the field while others will have higher coercivities and be unaffected by the applied field. In general, the low relaxation grains will have the lowest coercivities and are thus more readily affected by such alternating fields than are grains of long relaxation time.

   If the alternating magnetic field is rotated around the sample or the sample is tumbled within the alternating field, then grains with coercivity equal to or lower than the peak strength of the alternating magnetic field will have their magnetization changed as it follows the changing direction of the applied

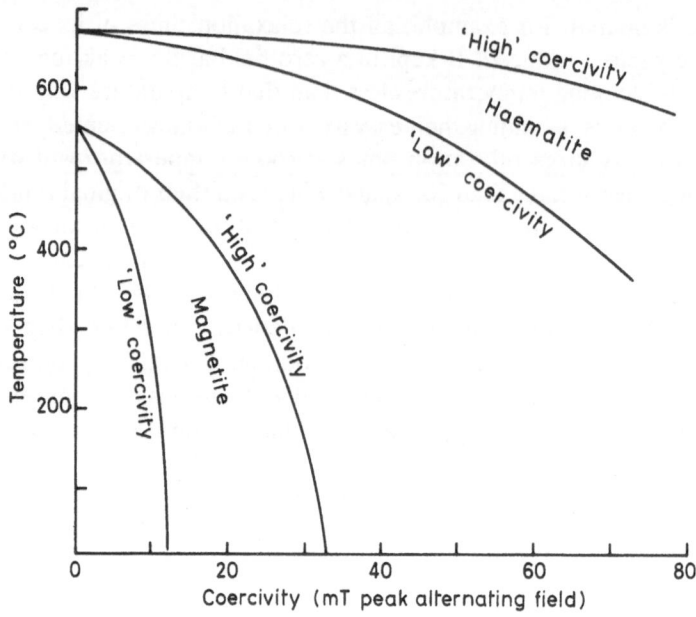

Figure 2.9 Blocking temperature and coercivity. This diagram is illustrative of the relationship between blocking temperature and coercivity for magnetite and haematite. Magnetite generally has a lower coercivity and blocking temperature range than haematite, but these properties are strongly influenced by grain sizes and the specific composition, defects, etc., in the titanomagnetites and ilmenohaematites, so that the ranges of both properties can overlap each other.

field, with the direction within each single domain flipping in opposite directions along its easy axes as the angle between the applied field and the easy axis changes. If the peak field intensity is then gradually reduced, the field will no longer affect all the grains, and those with higher coercivities will be left in the directions in which they were last lying before the field decreased below their coercivity. If the easy directions are randomly oriented within the sample, then such magnetizations will essentially cancel each other, so that by the time the applied field has been reduced to zero, all grains with a coercivity less than the peak applied field will have been effectively demagnetized in that they contribute nothing to the external measured magnetic field of the sample.

As with thermal partial demagnetization, the duration of the applied field is important as viscous components could occur. However, this can be controlled quite easily and most demagnetizers (Section 5.4) operate at 50 or 60 Hz and higher. The problem is that, as with thermal methods, the ambient field must be cancelled. If there is a magnetic field over the sample at the same time as it is being demagnetized by alternating magnetic fields, then the flips of the individual particle magnetizations will take place in a preferred direction, i.e. in the direction of the ambient field, and the sample will thus acquire an *anhysteretic remanent magnetization* (ARM), the magnitude of which depends on the strength of the direct magnetic field, the amplitude of the alternating magnetic field and the susceptibility of the sample, i.e. its composition, grain size, etc. In fact, the properties of anhysteretic and thermal remanences are extremely similar (Dunlop and West, 1969) and allow the properties of thermal remanence to be evaluated from study of the anhysteretic magnetization (Sections 7.2.1 and 8.3) without the complication of chemical changes induced by the heating necessary to study thermal remanence. (The properties of both remanences are not, in fact, exactly comparable and are affected, for example, by the concentration of magnetite – Sugiura (1981).) In general, however, such anhysteretic magnetizations are introduced inadvertently, particularly at high alternating magnetic field strengths, partially reflecting problems in completely cancelling the ambient geomagnetic field in the presence of strong alternating magnetic fields (eddy currents, for example, being induced in the shielding devices), but mainly because of the need for the applied field to be extremely pure – odd harmonics in the field resulting in a d.c. component over the sample, and the samples themselves generally not containing absolutely randomly oriented grains within them.

The behaviour of magnetic grains within an alternating magnetic field is not very well understood. It is, for example, practically impossible completely to demagnetize a natural rock sample by this method. Even when the coercivity of all its constituent grains must have been exceeded, the residual magnetization is greater than would be expected from a summation of the randomly oriented moments of its individual grains. It seems probable that the actual alternating magnetic field demagnetization process itself induces magnetizations, even when anhysteretic components have been eliminated. A major source for such a magnetization may be related to a *rotational remanence* (Wilson and Lomax, 1972; Edwards, 1980a,b). This appears to arise only in some apparatus in which the samples are tumbled within an alternating magnetic field and is acquired along the axis of rotation of the samples. (This effect is not present in all samples nor in all tumbling systems.) This unusual effect has been, at least partially, explained in terms of a *gyromagnetic remanence* (Stephenson, 1980a,b,c, 1981) associated with the flip of magnetic vectors within individual grains as they are forced into new easy directions of magnetization. This effect is not dependent on the samples being rotated (Stephenson, 1980a) and can occur even when a sample is simply placed in an alternating magnetic field. Gyroremanence is not fully understood, but is clearly present in some samples in which it can be of the order of 5% of an anhysteretic magnetization that would be acquired if the geomagnetic field were not cancelled during alternating magnetic field demagnetization, and similarly increases with an increase in the strength of the alternating magnetic field. It must be emphasized that such effects are not well understood, but provide a major constraint on the reliability of alternating magnetic field demagnetization, particularly in high fields, and constitute a 'noise level' that will normally be much greater than that of the magnetometers being used to measure the residual magnetization. It is conceivable that apparently magnetically stable directions could be obtained by this method if the specimens and samples from the same site had a similar orientation within the demagnetizer, although such stable directions would be spurious in terms of their geomagnetic, geological or archaeological significance (Cox, 1961; Burmester and Bates, 1979; Dankers and Zijderveld, 1981).

It seems likely that such gyromagnetic effects, when better understood, may provide useful rock magnetic information as the effects are not always

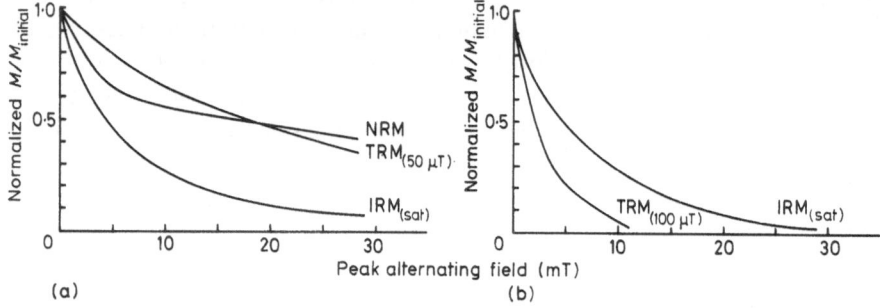

Figure 2.10 Single and multidomain properties. Lowrie and Fuller (1971) have specified six characteristics that enable a distinction to be made between single domain and multidomain behaviour in magnetite. These include the observations that (a) the stability to alternating magnetic field demagnetization of the thermal remanence of a multidomain magnetite is greater than that of the stability of its isothermal remanence if both were acquired in a field of the same strength. Conversely, (b) single domain magnetite has the opposite property.

present and hence reflect differences in the magnetic properties of the magnetic grains within the different samples. In a similar way, Lowrie and Fuller (1971) have made use of the different magnetic properties of single and multidomain particles in order to determine the grain sizes present. The simplest test is based on the fact that single domain particles that have acquired a thermal remanence in a field of less than 1 mT should show a greater stability to alternating magnetic field demagnetization than if they had been saturated by a large magnetic field at room temperature (Fig. 2.10). Similarly, they proposed that comparison of the single or multidomain composition could be evaluated, as single domain particles exhibit a greater stability to demagnetization after being thermally magnetized in a weak magnetic field, while multidomain particles show a decreased stability if the field in which they are cooled is decreased. In practice, such tests are difficult to evaluate, although this sometimes arises because of the chemical changes induced by the heating, with consequent changes in the magnetic properties.

Chapter Three

Magnetic mineralogy and magnetic identification of minerals

3.1 INTRODUCTION

Iron is a common accessory component of all rocks, but usually occurs as an oxide or hydroxide in terrestrial rocks. In meteorites and lunar samples, it occurs predominantly in an elemental form and native iron has been reported in some oceanic igneous rocks (Deutsch *et al.*, 1977), although this is very exceptional. While in a molten state within the Earth (magma), the iron oxides are amongst the earliest crystals to solidify, having solidus temperatures of the order of 1400–1600°C, although most magmatic temperatures are less than 1200°C. At such temperatures, and down to temperatures of about 900°C, the iron oxides commonly contain titanium and form solid-solution series, the composition of which can be plotted on a ternary diagram (Fig. 3.1). If the igneous magma is quench-cooled, then the iron oxides tend to retain their high-temperature solid-solution composition. This normally corresponds to a titanomagnetite composition ($Fe_{3-x}Ti_xO_4$) in the range $0.75 > x > 0.45$, but lies between the titanomagnetite and ilmenohaematite solid-solution series, depending on the magma oxidation conditions (Fig. 3.1). In basic and ultrabasic rocks (silica poor), this means that the composition is normally closer to the titanomagnetite series, while more acidic (silica rich) rocks tend to have compositions approaching that of the ilmeno-haematite series. As quench-cooling is fairly rare under most geological conditions – even basalts erupted under water only tend to have quench-cooled surface layers – so the high-temperature solid solutions are not very common, particularly in old rocks. More normal cooling rates result in the high-temperature solid solutions exsolving to form intergrowths between the titanomagnetite (cubic) series and the ilmenohaematite (rhombohedral) series. This deuteric exsolution tends to result in the preferential formation of the two end members of each series, i.e. magnetite and ilmenite. For even slower cooling, as in rocks intruded and cooled at depths within the Earth, the two solid-solution series both tend to exsolve further, giving rise to magnetite and ulvospinel, and haematite and ilmenite. If such igneous rocks are oxidized at low temperatures, less than 200–250°C, the composition of the titanomagnetites moves towards that of the ilmenohaematites (Fig. 3.1), but

35

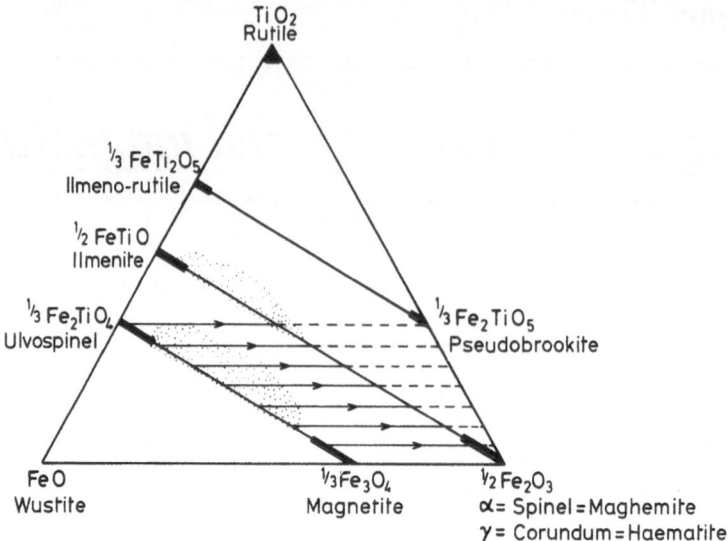

Figure 3.1 Ternary diagram for iron–titanium oxides. The natural solid-solution compositions are shown as shaded, but these normally exsolve into the end members of the solid solutions, together with other oxides, so that the common compositional ranges of the actual minerals within a rock differ from those indicated by the average composition. The commonly occurring minerals are shown by a solid bar. Low-temperature oxidation of the titanomagnetites causes their composition to move in the direction of the arrows, but eventually invert to ilmenohaematite.

such oxidized titanomagnetites are metastable and gradually invert to ilmeno-haematite, especially if heated to some 300°C. These processes may be extremely slow, but will eventually take place over geological time. Ultimately, oxidation leads to the formation of haematite, with varying amounts of rutile or anatase (TiO_2) and pseudobrookite (Fe_2TiO_5). Of this variety of iron–titanium oxides, in most unquenched rocks it is only the fairly pure forms of magnetite and haematite that have magnetic properties at room temperature (Section 3.2). The other magnetic minerals present in some terrestrial igneous rocks, especially if added as a result of predominantly mechanical magnetic over specific compositional ranges (Section 3.2).

Sedimentary rocks may include any of the magnetic minerals found in igneous rocks, especially if added as a result of predominantly mechanical erosion processes (Section 4.3). However, some oxidation will normally occur during erosion, deposition or subsequent burial, with the consequent formation of haematite as the major magnetic constituent in most consolidated sedimentary rocks, although all intermediate oxidation products may occur, of which maghemite is important as having a similar high intensity of magnetization to that of magnetite itself. The presence of oxygen and water will also result in the formation of a series of hydroxides, of which goethite is also magnetic at room temperature. Such minerals are also important as later reduction of the hydroxides may lead to the production of haematite, while metamorphic processes (Section 4.4) can ultimately reverse the oxidation processes, forming molten magma. In contrast, extraterrestrial magnetic materials (Section 9.8) are almost completely dominated by the presence of native iron, with only a little contribution from nickel.

The main characteristics of magnetic minerals found in terrestrial and extraterrestrial materials are given below – more detailed discussions, particularly of crystal structures, are given by Nagata (1961), Stacey and Banerjee (1974) and O'Reilly (1976).

3.2 MAGNETIC MINERALS

3.2.1 Titanomagnetites ($Fe_{3-x}Ti_xO_4$ for $0 \geq x \geq 1$)

Minerals of this composition have a cubic (spinel) structure at room temperature and form a solid-solution series at temperatures above 800°C (Fig. 3.2). If quench-cooled, the composition of terrestrial rocks lies mostly in the range $0.75 > x > 0.45$, with the magnetic properties depending on the actual composition; the unit cell size increasing with increased Ti content and the Curie temperature decreasing (Fig. 3.2). In practice, most naturally occurring solid solutions will therefore have Curie temperatures of less than 200°C so that any thermal remanence is likely to have a low stability as the maximum associated relaxation times will be only some 10^2–10^5 years. This time-scale is comparable to that over which low-temperature exsolution may take place, with an associated chemical remanence being acquired as the blocking volumes are reached (Sections 2.4 and 9.5).

Under normal cooling conditions, the high-temperature solid solution exsolves (deuteric alteration) into fairly pure magnetite with either ilmenite or ulvospinel, of which only magnetite is magnetic at room temperature. Pure *magnetite* (Fe_3O_4) contains two magnetic lattices. In one lattice, A, eight Fe^{3+} cations occur in a tetrahedral configuration, while in the B lattice, sixteen Fe^{2+} and Fe^{3+} cations occur in antiparallel octahedral configuration. Magnetite is thus ferrimagnetic with a spontaneous magnetization of 92 Am^2 kg^{-1} and a Curie temperature of 575°C. The coercivity of magnetite is dependent on grain size, with the maximum coercivity of 2×10^{-1} T (Banerjee and O'Reilly, 1967; Ishikawa, 1967) occurring for single-domain sized grains (c. 0.7 μm diameter). Being a cubic mineral, the 'easy' directions of magnetization are in the [111] directions, but are weak compared with any anisotropy associated with asymmetry in the shape of magnetite grains (Section 2.2). Many of the properties are temperature dependent as the cubic structure changes to orthorhombic on cooling to temperatures below −155°C (≤118 K). At this transition, the magnetization associated with multidomain particles

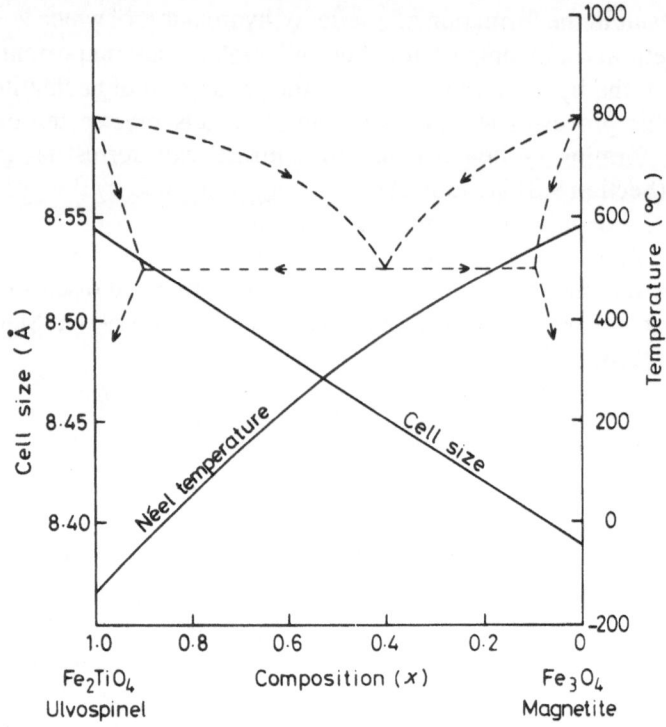

Figure 3.2 The titanomagnetite series. Unless rapidly cooled, the titanomagnetites will normally exsolve into their end members, relatively pure magnetite and ulvospinel, following the phase boundaries indicated. For solid solutions, the changes in Curie (Néel) temperature and cell diameter are shown as a function of composition.

appears to be destroyed (Merrill, 1970), although single and pseudo-single domain magnetizations appear to be unaffected (Ozima and Ozima, 1964). At these temperatures, the anisotropy also changes, although the crystalline anisotropy disappears at temperatures slightly higher (c. $-140°C$) than the structural changes. Most natural magnetites contain impurities (commonly Ti^{4+}, Al^{3+}, Mg^{2+} and smaller quantities of Cr^{3+} and Mn^{2+}, usually with Ti^{4+} and Al^{3+} entering the octahedral (B lattice) sites so that, for high Ti contents, the two lattices become essentially equally and oppositely magnetized and so ulvospinel tends to behave as an antiferromagnetic mineral, although its Néel temperature is so low that it behaves paramagnetically at room temperature.

Low-temperature ($<200°C$) oxidation of magnetite, such as during subaerial (Mullins, 1977) and subaqueous weathering, will often produce *maghemite* (γFe_2O_3), which is a mineral with the composition of haematite, but the cubic structure of magnetite. The formation of this mineral is not fully understood (Freer and O'Reilly, 1980; Moskowitz and Banerjee, 1981;

Özdemir and O'Reilly, 1981) but appears to occur by a diffusion of Fe^{2+} towards the surface of individual grains, leaving lattice vacancies, and the surface Fe^{2+} then oxidizes to Fe^{3+}. Having an essentially similar structure to magnetite, its magnetic properties are similar, with a spontaneous magnetization greater than 70 A m^2 kg^{-1} (Aharoni et al., 1962), possibly up to 85 A m^2 kg^{-1} at room temperature (Stacey and Banerjee, 1974). It is, however, destroyed on heating to some 350°C, when it inverts to haematite, with very large decrease (~99.6%) in spontaneous magnetization.

3.2.2 Ilmenohaematites ($Fe_{2-x}Ti_xO_3$ for $0 \leq x \leq 1$)

As with the titanomagnetites, the ilmenohaematites form solid solutions at temperatures above 900°C (Fig. 3.3), but this series has a rhombohedral (corundum) structure, although sometimes indexed as being hexagonal. Intermediate compositions can only occur naturally in quench-cooled rocks and their cell size and Curie (Néel) temperatures vary systematically with composition (Fig. 3.3). However, their magnetic behaviour differs drastic-

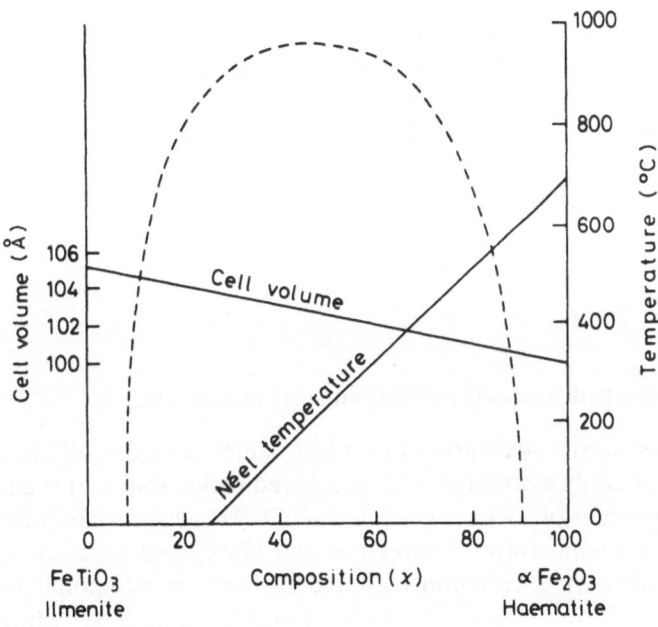

Figure 3.3 The ilmenohaematite series. As with the titanomagnetites (Fig. 3.2), ilmenohaematites will normally exsolve into their end members, haematite and ilmenite (the phase boundary is shown dashed). The variation of Curie (Néel) temperature and cell volume are also shown as a function of composition.

ally, being antiferromagnetic for ilmenite ($x > 0.9$), ferrimagnetic for $0.9 > x > 0.45$, and showing weak parasitic ferromagnetism for haematite with $x < 0.4$. For specific grain sizes and rapid cooling, order–disorder regions can be formed for composition ranges $0.4 < x < 0.55$ between which magnetic interactions may take place causing spontaneous self-reversal (Ishikawa and Syono, 1963) of the direction of remanence when being cooled in a magnetic field (Section 8.4). Solid-solution compositions for individual grains are, however, rare and normally the series exsolves on cooling to form intergrowths of its two end members, haematite and ilmenite, of which ilmenite is antiferromagnetic but behaves paramagnetically at room temperature as its Néel temperature is about $-180°C$.

Haematite (αFe_2O_3)   This is a common constituent of most acidic igneous and sedimentary rocks, and has its Fe cations distributed in two equal lattices, but these are not exactly antiparallel to each other (Morin, 1950; Dzyaloshinski, 1958), giving rise to a spontaneous magnetization of the order of 0.4–0.5 A m^2 kg^{-1} in the basal plane of the crystal, but there is a strong crystalline anisotropy as the spontaneous magnetization along the c axes is very low. The specific magnetic properties are strongly dependent on grain size, impurities and also temperature. On cooling to $-10°C$, haematite becomes a near perfect antiferromagnetic mineral with no spontaneous magnetization. The Néel temperature of pure haematite is generally reported as 675°C, with a somewhat lower value for impure haematite, but Néel temperatures up to 725°C are occasionally observed and generally attributed to defect ferromagnetism. The magnetic coercivity is strongly dependent on grain size, with peak values of 3.5–6.5 T for single-domain sized particles, i.e. around 1–3 µm in diameter, but haematite is virtually non-magnetic when in particle sizes of less than 0.2 µm (Hedley, 1968).

3.2.3 Iron hydroxides (FeOOH) and carbonates, etc.

Four basic crystal structures of iron hydroxides can exist, all loosely termed limonite, mainly associated with weathered rocks, soils and sediments. The most important of these is *goethite* ($\alpha FeOOH$) which is antiferromagnetic, with a Néel temperature between 60 and 170°C, and generally reported as about 110°C. It is commonly quoted to have an additional weak ferromagnetism (Strangway *et al.*, 1968b; Forsyth *et al.*, 1968) which is highly anisotropic, being confined to some 10^{-3}–1 A m^2 kg^{-1} along the c axis. This is generally regarded as arising from lattice distortions creating spin imbalances in the two antiparallel lattices associated with the preferential location of vacancies, Si^{4+} or Al^{3+}. However, goethite is unstable on heating in air and converts to a poorly crystalline form of haematite (protohaematite) at temperatures of 200–290°C (Hedley, 1968) which then converts to true haematite on heating to higher temperatures (Yariv *et al.*, 1979). It is, in fact, conceivable that the apparent ferromagnetism of goethite arises from the common inclusion, within the crystal lattice, of protohaematite at room temperature. The other iron hydroxides – lepidocrocite (γFeOOH), akageníte (βFeOOH) and feroxyhyte (δFeOOH) – can occur naturally, especially lepidocrocite, but are not commonly found in old rocks and are thought to have converted, on ageing, to goethite (Henshaw and Merrill, 1980). They are thus themselves paramagnetic at room temperature, but important sources for the later generation of magnetic minerals. Limonite is strictly a complex iron hydroxide mineral consisting of a mixture of goethite and lepidocrocite, with goethite predominating. It is thus magnetic due to the goethite component. It is this mineral that accounts for much of the red and brown staining in weathered rocks and soils. The magnetic properties of these hydroxides have received little attention, with the notable exception of Hedley (1970).

Iron is a common component of many other minerals, such as many of the carbonates, for example siderite ($FeCO_3$) and ankerite ($(Ca, Mg, Fe)CO_3$) are common constituents of sedimentary rocks. All such minerals are paramagnetic at room temperature and do not, therefore, contribute to the magnetic properties of rocks, but they provide an important source of iron oxides, such as haematite, if they are broken down by either weathering or other diagenetic or metamorphic processes.

3.2.4 Pyrrhotite (FeS_{1+x} for $0 \leq x \leq 1$)

Iron sulphides can vary in composition from troilite (FeS) to pyrite (FeS_2), but troilite is very rare in terrestrial rocks, although common in meteoritic material. Pyrite is common in sedimentary rocks, especially associated with fossilized organic material (which was the probable source for the sulphur), but intermediate compositions are not common, except in some basic igneous rocks. Greigite, Fe_3S_4, occurs in some sediments as a result of microbial activity and is reported to be magnetic, but its magnetic properties have not been investigated because of its chemical instability. Troilite has a monoclinic (pseudohexagonal) structure and is a perfect antiferromagnetic material with a Néel temperature of 320°C. Pyrite has a cubic structure and is paramagnetic at room temperature. The intermediate compositions are dominated by the presence of ordered and disordered regions within the individual lattices, but are mostly slightly ferrimagnetic, except for compositions near $x = 0.14$ where ferromagnetism occurs with a spontaneous value of 13.5 A m^2 kg^{-1} and a Curie temperature of 320°C. Pyrrhotite of a composition Fe_7S_8 has a very high degree of anisotropy, with typical values of 6.3×10^{-4} A m^2 kg^{-1}

along its c axis and 1.3×10^{-4} A m^2 kg^{-1} in its basal plane (Schwarz, 1974).

3.2.5 Iron and nickel (Fe, Ni)

These elements only occur naturally in extraterrestrial materials, with only very rare exceptions. They are, of course, ferromagnetic in the strictest sense. Iron dominates the magnetic properties in most extraterrestrial materials (Section 9.8) and has a saturation magnetization of 218 A m^2 kg^{-1} and a Curie temperature of 770°C. In multidomain form, it is of low magnetic stability, but may have coercivities up to 10–10^2 mT in single domain sizes. *Nickel* has a Curie temperature of 357°C and a saturation magnetization of 57 A m^2 kg^{-1}.

3.2.6 General comments

It must be emphasized that the properties of these magnetic minerals may differ in nature from those determined from their synthetic laboratory forms. The properties of most commonly occurring natural minerals have only been investigated because of their importance to palaeomagnetic studies or for academic interest. This is because their high electrical conductivity or chemical instability renders them unsuitable for use in an industrial context. Furthermore, the properties of naturally occurring magnetic minerals are greatly affected by the presence of impurities and vacancies, as well as by grain size, cooling rates and the like. Similarly, the sequence of the formation (or removal) of such magnetic minerals from pre-existing iron-bearing silicates has been a subject of only very limited investigation in igneous, sedimentary and metamorphic rocks. Many of these processes are also time-dependent, with many important reactions only taking place on a geological time-scale. Similarly, laboratory conditions can only rarely reproduce the physical and chemical conditions that occur naturally, such as within a molten magma. Conversely, of course, the study of the behaviour of magnetic minerals can be a major indication of the natural conditions to which they have been subjected.

3.3 IDENTIFICATION OF MAGNETIC MINERALS

Critical to all palaeomagnetic studies, and to a lesser extent, archaeomagnetic work, is an identification of the minerals carrying the remanence as this can provide direct evidence of the likely age of the magnetization and also of the mechanism by which such remanence was acquired. Minerals associated with weathering, for example, are likely to carry a secondary magnetization of chemical origin, the stability of which will depend in part on the composition of the minerals and in part on their characteristic grain size. Similarly in sedimentary rocks, different minerals tend to form under certain diagenetic processes. The magnetic mineralogy can thus be used to define particular geological conditions which may be relevant to the palaeomagnetic study itself, or of interest in their own right. Furthermore, the instability of certain minerals means that their presence can be identified from the magnetic decay products produced at certain temperatures. The fact that many such identifications can be made remotely, i.e. without grinding and polishing slides (which may require impregnation) or the exposure to air with consequent oxidation of the original minerals, also makes this technique particularly useful in certain conditions.

While magnetic techniques have a clear advantage in being cheap and often effective, they can, of course, only indicate the presence of particular minerals. Optical study is still required to establish their relationship to each other and to any other minerals that are present. Bright haematite can, for example, be seen to be pseudomorphing cubic magnetite and is thus secondary. An additional constraint is that all systems of identification fail to distinguish between the minerals present and those that are carrying the natural remanence. This can often be inferred from the way in which the remanence decays during demagnetization procedures, but the only direct identification is by means of colloidal solutions, the magnetic particles of which, in the absence of an external field, form Bitter patterns (Soffel, 1967) over those parts of the rock that actually carry a remanent magnetization, although these will also be shown by multidomain grains even when forming closure patterns (Section 2.2). Most other methods, such as magnetic separations, isolate those components that acquire a remanence, i.e. have a magnetic susceptibility, in addition to those carrying a natural remanence. Separation can, in any case, be difficult as magnetic minerals are often incorporated within silicates and are adhered to other mineral surfaces, particularly clays, that restrict the efficiency of density and magnetic separation. The need for crushing, in many cases, also raises problems as to the extent to which the observed grain sizes correspond to those occurring in the uncrushed rock. In general, however, most standard techniques for evaluating grain sizes are of little practical use in palaeomagnetic studies because of the importance of single-domain sized particles. As these are generally of the order of 1 μm, they are at or beyond the limits of adequate resolution by both optical and electron scanning systems (mainly because of edge effects). (Electron probe analyses also have the problem of being unable to distinguish Fe^{2+} from Fe^{3+}.) The presence of small grains can thus only be inferred either from the magnetic properties or by assessing whether the tails of any observed grain-size distribution is likely to extend significantly into the single domain range. However, visual examination can often be of direct relevance as the 'redness' of haematite is strongly dependent on grain sizes, with bright clear reds, such as occur at the surface of crudely fired clay pots, corresponding to very small, probably superparamagnetic grain sizes, less than 30 Å (30×10^{-4} μm) in diameter, while deeper reds tend to be characteristic of larger haematite grains.

Palaeomagnetism

3.3.1 Isothermal remanence

The difference in the magnetic susceptibility and coercivity of magnetite and haematite makes this method of analysis the most effective for distinguishing the presence of Ti-poor titanomagnetites and Ti-poor ilmenohaematites – the two commonest magnetic minerals in most terrestrial rocks. The saturation magnetization of haematite is less than 1% of that of magnetite, so that when placed in a strong magnetic field, the induced remanence is, weight for weight, two orders of magnitude greater for magnetite than for haematite. In addition, the titanomagnetites (and maghemite) saturate in fields of 0.01–0.1 T, while haematite does not saturate until fields of over 1–3 T. If d.c. magnetic fields are applied in incremental steps, therefore, the saturation moment becomes constant before 1 T for magnetite, but continues to increase at higher fields for haematite (Fig. 3.4). The study of the isothermal remanence acquisition after heating different specimens can thus indicate the changing ratio of haematite and magnetite as chemical changes occur. The main advantage of this method is that it does not involve any chemical changes and, as the material is being strongly magnetized, it can usually be undertaken on small whole rock chippings, with the minimum of preparation. However, the method can only provide information on the specific composition range within the titanomagnetites and ilmenohaematites if a pure concentration can be obtained of known weight.

3.3.2 Magnetic hysteresis and coercivity spectra

The fact that the saturation moment and coercivity of magnetic minerals depend on their composition means that the area bounded by a hysteresis loop is directly related to the specific composition of the magnetic mineral present and can thus be used to identify the mineral (Wasilewski, 1973b; Schmidbauer, 1975; Özdemir and O'Reilly, 1981, 1982). The problem is that the hysteresis loop is also affected by the presence or absence of single domain particles. This means that, in practice, this technique is mainly of value in rock magnetic studies in which, for example, the mineral composition is known and the proportion of single to multidomain components can then be assessed after different physical treatments.

As the coercivity spectrum of haematite may extend as high as 6.5 T (Section 3.2.2), while the peak coercivities of magnetite are around 0.1 T (Section 3.2.1), the presence of haematite will normally be indicated by the presence of a high coercivity component during alternating magnetic field demagnetization (Section 5.4.2). While the presence of such a component can be taken as clear evidence for the presence of haematite, the absence of such high coercivity behaviour does not exclude the presence of haematite as only low coercivity grains may be present. In a similar way, magnetite has a high

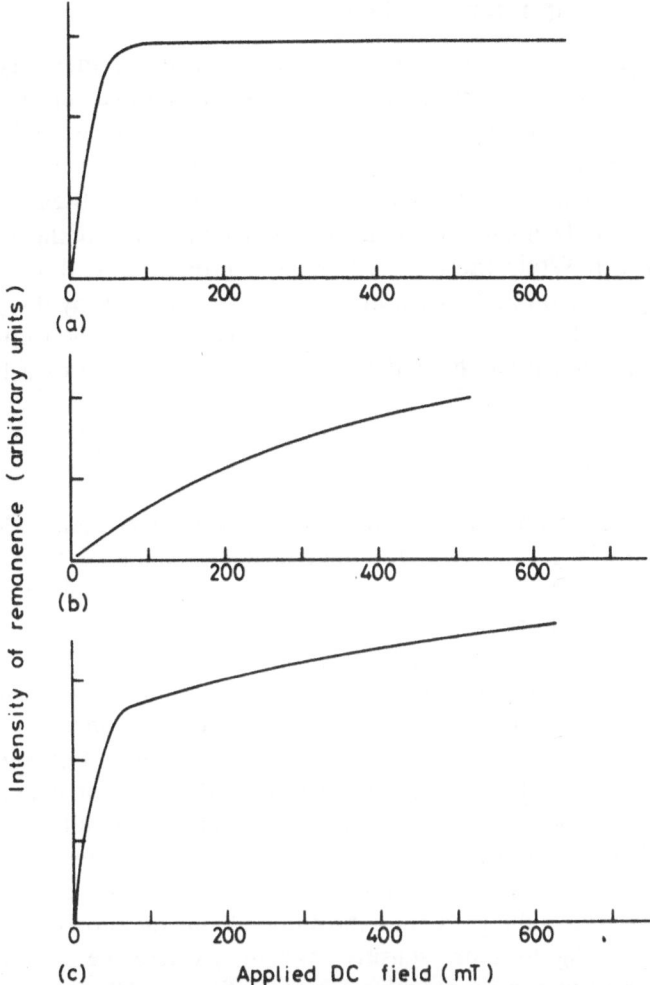

Figure 3.4 Isothermal remanence curves for magnetite and haematite. (a) Magnetite saturates in much weaker fields than (b) haematite, which is often not saturated in conventionally available fields. (c) When both minerals are present, the observed curve is a summation of the two single curves, but magnetite, with a much higher saturation moment, will normally dominate the curve even when only present as a few percent of the total.

saturation moment and this is often reflected in higher initial intensities of magnetization that decrease as the coercivity is exceeded. Such behaviour is obviously not a reliable indicator as similar behaviour could result from the presence of large quantities of lower coercivity minerals of other compositions. Nonetheless, the behaviour of samples during demagnetization can be indicative of the mineralogy involved, particularly if this is taken into consideration with the geological or archaeological features of the samples.

3.3.3 Low-temperature transitions

Both magnetite and haematite undergo changes in their crystal lattice structure on cooling, with associated changes in their magnetic properties. In haematite, this (Morin) transition occurs at some $-10°C$, while in magnetite, it occurs around $-154°C$ (Section 3.2). The presence or absence of these transitions can thus be used to determine the presence or absence of these two minerals, and the magnitude of the transition is indicative of the percentage of that mineral. While these transitions also provide a method of magnetic 'cleaning' (Section 5.4), the transitions can be inhibited by the presence of impurities and lattice defects, which can make the method unreliable for studying natural minerals, although clearly the presence of a transition is diagnostic.

3.3.4 Curie temperature

Each specific composition of magnetic mineral has a unique Curie (Néel) temperature (Figs 3.2 and 3.3), so that the determination of this temperature is diagnostic of such minerals and can be used, for example, to distinguish precise compositions within the titanomagnetites (Özdemir and O'Reilly, 1981; Keefer and Shive, 1981) and ilmenohaematites (Duff, 1979). The procedure is to heat a specimen of a rock while holding it in a strong magnetic field. The magnetization is then monitored as the specimen is heated. Two main types of Curie balance exist, although several other types are available (Collinson, 1983). The simplest is the horizontal translation balance in which the specimen holder is on a cradle that can swing horizontally. The specimen itself is then in a strong magnetic gradient so that any changes in its intensity of induced magnetization, i.e. changes in its susceptibility, cause it to move either up or down the gradient – the motion being monitored by a variety of methods. A slightly more sensitive system is where the specimen is held vertically on a spring, again within a field gradient, and the spring is extended or retracts as the position of the specimen changes within the gradient. However, although more sensitive, the vertical system suffers from changes due to weight loss, such as devolatization, that also give rise to changes in the extension of the spring. Nonetheless, both systems give rise to, in effect, a graph of the change in the saturation moment of the specimen during heating and cooling cycle (Fig. 3.5). In practice, the fields applied are rarely sufficient to saturate haematite, although titanomagnetites are saturated, but this factor makes little or no difference to the Curie point estimations. The presence of more than one magnetic mineral raises some problems in determining the lower Curie temperature. This is often, incorrectly, obtained by simply extra-polating the curve down to the temperature axis for zero intensity of magnet-ization (Fig. 3.5(c)), but the curve in fact represents the summation of the magnetization due to the two minerals and the lower Curie point therefore corresponds to the inflection point on the curve, i.e. the temperature at which

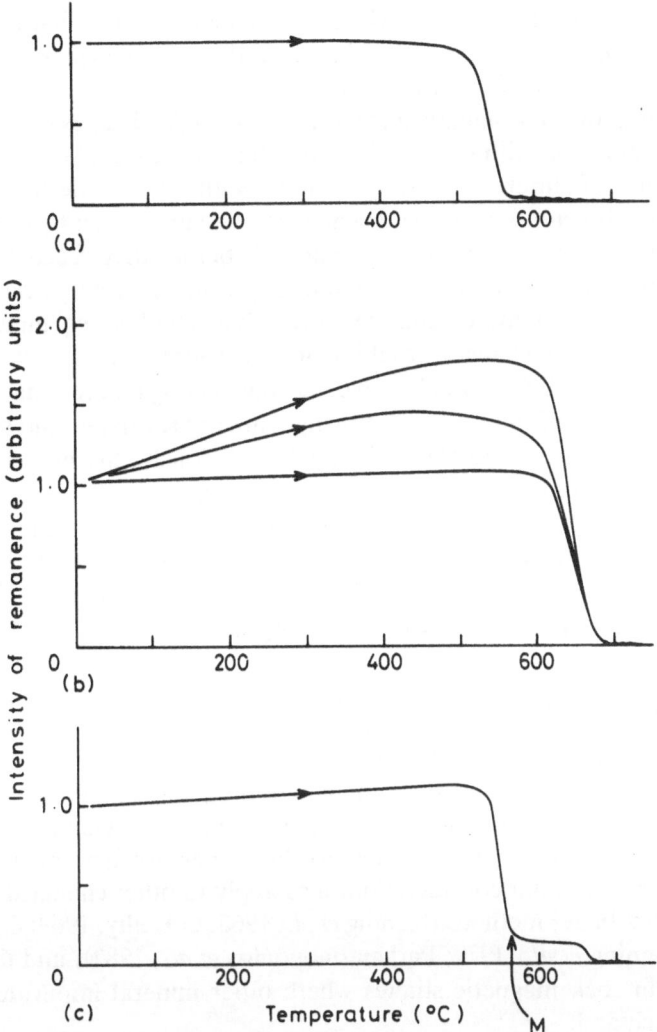

Figure 3.5 Thermomagnetic curves for one and two mineral systems. The intensity of magnetization is conventionally in arbitrary units, the actual concentration of the ferromagnetic minerals not usually being known. (a) A typical thermomagnetic curve for natural magnetite showing a Curie (Néel) temperature of 575°; almost pure magnetite. (b) The curves for haematite are somewhat variable, mainly because the fields applied are not capable of saturation magnetization of the specimen. The curves tend to be flatter for higher applied fields (Duff, 1979). Such increases can be diagnostic of the presence of haematite, although such curves could also indicate the growths of new magnetic minerals. (c) As magnetite has a much higher saturation magnetization than haematite, a thermomagnetic curve for specimens containing both minerals is generally dominated by the magnetite content. The inflection point, M, defines the Curie temperature of the magnetite (Grommé *et al.*, 1979; Moskowitz, 1981).

the contribution from the lower Curie temperature minerals becomes zero. This point can be determined mathematically (Grommé *et al.*, 1979; Moskowitz, 1980), but allowance usually needs to be made for any thermal lag as the surface of the specimen will normally be at a somewhat higher temperature than its centre. Fast heating, to reduce thermochemical changes, may thus show a greater thermal inertia.

Some minerals, particularly haematite, show slight changes in their saturation magnetization with increased temperature, which can be confused with the presence of minute amounts of minerals with other Curie temperatures. Much more difficulty occurs in that chemical changes frequently occur during the heating which may destroy the minerals before they reach their Curie temperature, e.g. goethite and pyrrhotite, giving apparent Curie temperatures as they decompose, and similarly other iron-bearing minerals may undergo chemical changes resulting in the formation of new magnetic minerals (Section 3.6.4). In all such cases, the heating and cooling curves are different as most of these chemical changes are not reversible and the saturation curves are therefore also irreversible. Obviously, the heating curve is most diagnostic of the original mineralogy, especially if the heating is fast and thus inhibits the chemical change, or if undertaken in a vacuum or inert atmosphere. Unfortunately these methods only slightly inhibit the oxidation as the sample itself will normally contain air and volatiles that result in chemical changes, irrespective of the atmosphere.

3.3.5 Chemical analyses

Both wet and dry methods of chemical analysis can be used to determine the total iron and the ferric:ferrous ratio within the sample (Wilson, 1960). The problem with such methods is generally to isolate the magnetic fraction of interest as obviously any iron present in other minerals will be incorporated in such analyses. Similar considerations also apply to other chemical analyses, such as Mossbauer methods (Kundig *et al.*, 1966; O'Reilly, 1969; Coey *et al.*, 1971; Manning *et al.*, 1979; Paramarinopoulos *et al.*, 1982), and their main uses are in rock magnetic studies where other mineral impurities can be largely excluded.

Chemical demagnetization (Section 5.5.4) mainly depends on the fact that small grains are dissolved more rapidly than larger grains as they have a higher surface area per unit volume. However, for grains of identical size and equally good crystallization, the rate of dissolution depends on the composition of the grains. For a weak acid (35% acetic + 25% hydrochloric) the solubility (Henshaw and Merrill, 1980) proceeds in order:

magnetite > ilmenohaematite > haemoilmenite > rutile–sphene

and

magnetite > goethite > haematite.

Chemical demagnetization can therefore attempt to isolate the remanence associated with specific minerals and, conversely, their solubility can be used to estimate the mineralogy present. However, the effect of ranges in grain sizes and generally poor permeability of rocks (which restricts the dissolution to surface layers) severely restricts the use of this technique for mineral identification.

3.3.6 Magnetic study of chemical changes during heating

The heating of a sample will induce chemical changes in its magnetic and also its non-magnetic minerals, resulting in the creation and destruction of magnetic minerals at certain temperatures, some of which can be diagnostic of the changing mineralogy involved. Some of these changes can also indicate that the sample has never previously been heated to that temperature as the

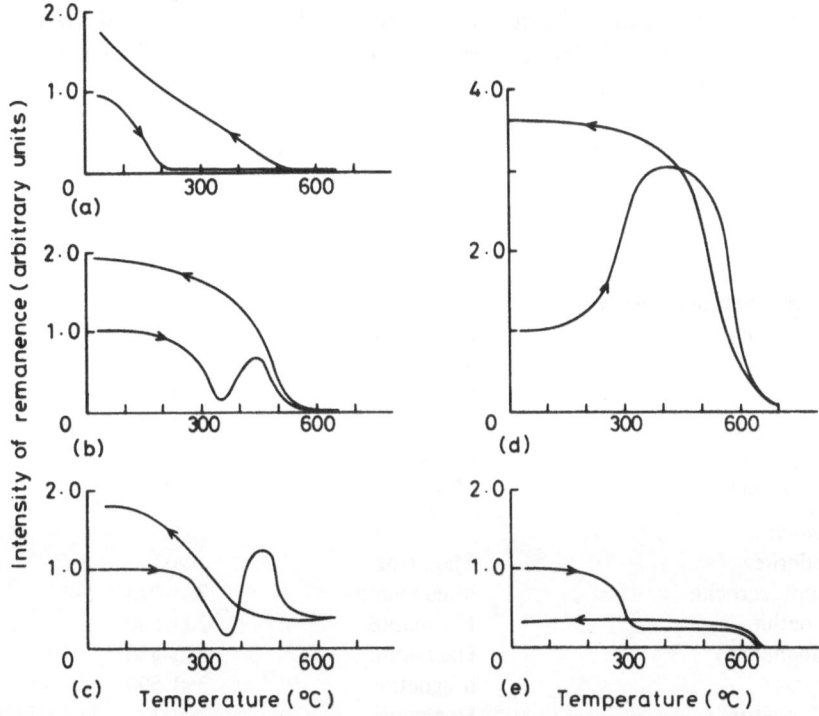

Figure 3.6 Thermomagnetic curves showing thermochemical reactions involving magnetic minerals. The range of possible reactions is great (Table 3.1) and dependent on a variety of factors (see text). In the examples given, only arbitrary units are used for the intensity of remanence so that the total value of, for example, (e) is very much less than that for (a). The specific examples are (a) Ti-rich titanomagnetite exsolving to form relatively pure magnetite, (b) Ti-poor titanomagnetite oxidizing to form maghaemite, (c) Ti-poor titanomagnetite forming titanohaematite and haematite, (d) siderite in sediments converting to magnetite, and (e) maghemite converting to haematite.

49

observed chemical interaction would have already taken place, but these changes can also occur in fresh igneous rocks, reflecting the radically different environment of laboratory heating when compared with natural conditions, particularly in the redox conditions that cannot readily be duplicated in the laboratory.

By far the simplest method of making such analyses is by means of studies of the saturation magnetization as a function of temperature, i.e. the curves used for Curie temperature determination, as these show increases or decreases in the induced magnetic moment as chemical changes alter the magnetic mineralogy (Fig. 3.6). The variety of possible changes is extensive, and only a few common thermochemical changes are listed in Table 3.1; the precise temperature at which such changes occur is also variable, depending on grain sizes, the rate of heating, the size of the sample, the degree of crystallization and so forth – the indicated temperatures are therefore broad generalizations. Nonetheless, the nature of the rock is likely to limit the mineralogy and thus the probable thermochemical reactions. Similarly, the composition of the magnetic minerals can be further assessed, using other methods such as isothermal remanence, after heating to specific temperatures.

Table 3.1 Thermochemical reactions involving magnetic minerals

Initial mineral	Alteration product at temperature (°C)	
Igneous		
Ti-rich titanomagnetite	Magnetite	>300
Magnetite	Maghemite	150–250
Olivines	Magnetite	>300
Pyrite	Magnetite	350–500
Maghemite	Haematite	350–450
Magnetite	Haematite	>500
Pyroxenes	Magnetite	>600
Sediments		
Siderite	Magnetite	>200
Lepidocrocite	Maghematite	220–270
Goethite	Haematite	200–400
Maghemite	Haematite	350–450
Pyrite	Magnetite	350–500
Magnetite	Haematite	>500
Haematite	Magnetite	>550

The temperature range and reaction (e.g. magnetite to haematite or haematite to magnetite) depend critically on the grain sizes and shapes, the presence of impurities (especially water), the atmosphere, rate of heating, etc. They are thus only indicative and poorly defined for natural minerals. (After Eustance, 1981; see also Fig. 3.6.)

Chapter Four

The magnetization of
natural materials

4.1 INTRODUCTION

It is very rare for the magnetization of complex materials, such as rocks, to
have any one simple origin as the composition and grain sizes of their constituent ferromagnetic minerals are generally variable, even within a single
specimen. Furthermore, the magnetic properties of such materials may well
change with time as different minerals equilibrate to the conditions at the
Earth's surface, particularly if they begin to hydrate and oxidize (weather) in
the Earth's atmosphere. Unfortunately even the identification of the magnetic minerals present (Section 3.6) may be only of partial assistance in
understanding the nature of the remanence as generally less than 5% (even
<0.5%) of the total magnetic mineral content actually carries the remanent
magnetization. This means that even rocks with abundant ferromagnetic
minerals should still be physically regarded as a weak solution of ferromagnetic particles within a predominantly paramagnetic matrix. All
remanent magnetizations (Table 4.1) are termed *natural remanent magnetization* (NRM) when they occur naturally, i.e. prior to any laboratory
influences. In most materials, this natural remanence is divisible into at least
two components, one of which was acquired at a specific time, for example at
the time of formation of a rock or when a pot was fired. This component is
termed a *primary magnetization* to distinguish it from any later acquired
magnetizations, such as viscous remanences (Section 2.3.3), that are termed
secondary magnetizations. These components are commonly isolated by
means of partial demagnetization of the natural remanence (Section 5.4) as
the primary component will normally have a differing stability to demagnetization procedures than the secondary components. Unfortunately, while
such a situation is usually true, the ways in which most natural materials
acquire their primary remanence is by no means simple and, ideally, each
sample needs to be examined in detail to determine the nature and origin of its
remanence prior to any interpretations in terms of the behaviour of the past
geomagnetic field or past geological conditions. In the following sections,
some of these intricacies are considered for the different rock types (igneous,
sedimentary and metamorphic) and for archaeological materials, while

Table 4.1 Types of magnetization

Remanent type of magnetization	Abbreviation	Characteristic or definition
Naturally occurring magnetizations		
Natural	NRM	Summation of all components of specimen remanence acquired by natural processes.
Thermal	TRM	Acquired by cooling over a range of temperatures starting at or below the Curie temperature.
Depositional/detrital	DRM	Acquired by the physical rotation of magnetic particles during deposition as a sediment.
Post-depositional	PDRM	Acquired by sediments after deposition but prior to metamorphism or weathering; usually a combination of physical rotation of interstitial particles and chemical changes as sediments consolidate.
Chemical	CRM	Acquired as a magnetic mineral nucleates and grows in a magnetic field.
Shear	SRM	Acquired mostly by unconsolidated sediments when subjected to shear.
Laboratory-induced magnetization		
Isothermal	IRM	Acquired by magnetic particles in steady magnetic field in a few seconds; the field is generally strong. This can also be acquired naturally in lightning strikes.
Anhysteretic	ARM	Acquired when a ferromagnetic particle subjected simultaneously to alternating and direct magnetic fields.
Rotational	RRM	Acquired by a specimen rotating within an alternating magnetic field.
Gyromagnetic	GRM	Acquired by a specimen in an alternating magnetic field without rotation. This can also occur naturally such as during meteoritic impacts.
Shock	SRM	Acquired by magnetic particles when shocked by impact. This can also occur naturally such as during meteoritic impacts.

RRM and GRM: no d.c. field

All 'natural' magnetizations can be duplicated, to varying extents, in the laboratory, and some 'laboratory' magnetizations may also occur in nature. In general, acronyms can be misleading, e.g. SRM has two meanings and their use is avoided in this book.

methods of dissecting the natural remanence are described in the following chapter (Section 5.5).

4.2 IGNEOUS ROCKS

In the context of the Earth and terrestrial planets, all the constituents of their crusts have been at least once through a molten stage. All crustal materials have therefore, at some time, acquired a thermal remanence (Section 2.3.1) as they have cooled in the magnetic field of the planet concerned, or the solar magnetic field. For practical purposes, it is more convenient to distinguish terrestrial igneous rocks as those that have been molten and solidified at some time during the last 4.0 billion years. Such rocks are generally classified on two main criteria, their composition (mostly on the feldspar and quartz content) and grain size. Basic rocks are those with no free silica and are characterized by the presence of olivines, pyroxenes and calcium-rich feldspars. The compositions then grade through intermediate types to acidic igneous rocks, with free silica and sodium-rich feldspars (Turner and Verhoogen, 1960; Cox *et al.*, 1979). Obviously very narrow classifications are possible on the composition, but separation is also made into fine, intermediate and coarse-grained varieties. The composition is generally interpreted as reflecting the degree of differentiation and mixing of the molten rock (magma) from a primitive mantle source while the grain size is considered to indicate the rate of cooling and hence the crustal depth at which solidification took place (Fig. 4.1).

As major igneous bodies form, their composition can evolve as the crystallizing minerals preferentially extract particular ionic groups from the melt (magma), although recharges of the magma can result in changing compositions of the crystallizing fraction, giving rise to zoned crystals. If the rocks were originally completely molten then olivine generally crystallizes first, commencing at about 1890°C, followed by pyroxenes at 1557°C and high-temperature quartz at 1543°C. The position of magnetite within this crystallization sequence is not obvious as it has a solidification temperature of 1540°C but has more commonly formed at much lower temperatures when iron becomes released from other (subsolidus equilibration) reactions (Sections 3.1 to 3.3). Olivines, for example, react with the remaining magma and, if the right oxidation conditions are present, may result in pyroxenes or amphiboles, with magnetite as a by-product. Although titanomagnetites are generally considered to be high-temperature products, there are many examples of granites, for example, where they have only crystallized during the late stages of crystallization. Nonetheless, magma solidification always takes place at temperatures greater than some 800°C, i.e. well in excess of the highest Curie temperature of any magnetic constituents. On this basis, all igneous rocks would normally be expected to be characterized by a thermal remanence (Section 2.3.1) acquired as their constituent magnetic minerals passed through their Curie (Néel) temperatures (a function of their com-

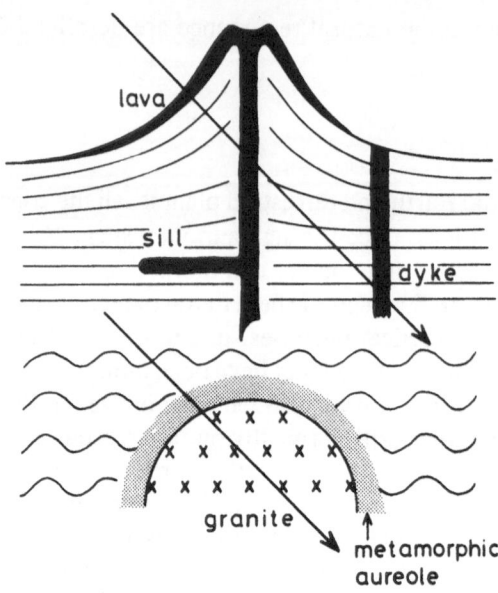

Figure 4.1   The magnetization of igneous rocks. Molten rocks may reach the Earth's surface through pipes and fissures from which they may flow out as lavas. Such rocks are normally erupted at temperatures of 1000–1100°C. They may cool rapidly after eruption, especially if extruded into water, or may take several years to cool to room temperature under subaerial conditions. Not all the volcanic materials reach the surface and so some cool within the upper surface of the crust, heating the surrounding rocks, and only gradually cool to ambient temperatures. At depths of some 10–12 km, large volumes of rock may be molten and cool to form granitic batholiths which eventually become exposed at the surface. Such rocks extensively heat the surrounding country rocks, forming a wide metamorphic aureole. In all these situations, the igneous rocks, and the rocks heated by them, acquire a thermal remanence as they cool. The rate of cooling may, however, vary from a few minutes to several thousand years. Many of these rocks are also associated with the migration of hot waters, often carrying minerals in solution, giving rise to chemical changes in the metamorphic zones that may have associated chemical remanences as new magnetic minerals form.

position) and blocking temperatures (a function of their grain size and composition).

In reality, the situation is even more complex. Coarse-grained rocks have cooled slowly, so their minerals have largely equilibrated with each other during cooling. However, such minerals are also kept at quite high temperatures for long periods so that their remanence will therefore be continually re-set, if geomagnetic changes occur during their cooling (e.g. geomagnetic polarity reversals), by high-temperature viscous magnetizations. In such rocks, the natural remanence is thus likely to represent moderate-temperature viscous remanences acquired during prolonged cooling from temperatures of some 300°C or so, rather than over the full temperature range. Conversely, fine-grained materials may well have acquired a 'simple' thermal remanence while cooling from their Curie temperatures. Fine-grained lavas are likely to have cooled rapidly to a few hundred °C, but it may be some 10 or 12 years before they have entirely cooled. (Submarine lavas may well be quench-cooled and thus reach ambient temperatures very shortly after eruption – Section 9.5.) Although few detailed studies have yet been made, it seems likely that the equilibration of primary titanomagnetites mostly takes place between 750 and 550°C (Haggerty, 1976a), although also at much slower rates at lower temperatures. Rapid cooling through this range may thus inhibit equilibration, but any formation of new magnetic minerals during cooling (deuteric alterations) will give rise to a chemical remanence as the crystallites grow through their critical blocking volumes (Section 2.3.2). For palaeomagnetic purposes, such chemical remanence will have been acquired at a specific time, i.e. during the cooling of the rock, and is thus considered primary. However, the formation of minerals that are not fully equilibrated is likely to induce very slow, prolonged chemical changes over the geological history of the rock. Thus deuteric alterations essentially grade into post-deuteric oxidation, exsolution and devitrification processes.

The distinction between deuteric and post-deuteric alteration is not easy, yet fundamental to palaeomagnetic interpretations. The study of the composition of the different oxides and their phase relationships will normally allow such evaluation (Haggerty, 1976a,b), but the precise oxidation and impurity conditions can vary over short distances within even a single specimen, and obviously different conditions may well exist within the same outcrop, such as between a chilled margin and the more slowly cooled central parts of a lava. However, quite subtle changes may also occur as other iron-bearing minerals, such as olivines, pyroxenes, micas, etc., gradually equilibrate to the low temperatures and pressures at the surface of the Earth. Such changes, possibly taking billions of years, will gradually produce magnetic crystallites that eventually pass through their blocking volume to obtain a stable chemical remanence. It is quite probable that, in many igneous rocks, the remanence is not actually carried by the visible iron oxides. Of the few studies undertaken, Evans and McElhinny (1966) and Evans et al. (1968) have shown that the 2600-million-years-old Modipe Gabbro (Botswana) has a remanence that is carried almost entirely by single domain magnetite exsolved in pyroxenes, while Hargraves and Young (1969) have shown that the stable remanence in the Triassic Lambertsville diabase is carried mainly by magnetite exsolved within feldspars, but cannot be identified microscopically. Even recently magnetized rocks may also have a remanence related to magnetite in the form of rods and spherules exsolved within plagioclases (Davis, 1981). At the extreme range of these and other post-deuteric alterations are, of course, the effects of present-day weathering, or even some earlier weathering event during the geological history of the rock (Section 4.6).

4.3 UNCONSOLIDATED SEDIMENTS AND SEDIMENTARY ROCKS

Sediments can be broadly classified into two types:

1. Those originating by deposition following the mass transport of eroded (detrital) particles by air, water or gravity flow, and
2. Those formed by chemical precipitation, with or without the involvement of biological processes.

In terms of palaeomagnetic studies, most chemical deposits also incorporate some detrital particles, so the physical processes of deposition are of major importance in all sediments, and these will be discussed first. After deposition, unconsolidated sediments gradually convert to consolidated rocks by varying degrees of compaction and chemical interaction between the constituent minerals and migrating fluids. Chemical deposits will therefore be discussed in the context of the chemical changes (diagenesis) affecting all deposited materials.

While an understanding of the processes of sedimentation is fundamental to any interpretation of their magnetization in terms of geomagnetic or geological features, there is an increasing application of palaeomagnetic techniques, particularly the magnetic fabric method, to the study of such processes (Taira and Lienert, 1979). Specific examples of palaeomagnetic features in different sedimentary environments are discussed in Chapters 7 and 8 (mainly unconsolidated sediments) and Chapter 9 (mainly consolidated sedimentary rocks).

4.3.1 Physical orientation processes and unconsolidated sediments

Most detrital (allochthonous) sediments, i.e. those that have formed following mass transport, have been deposited in an aqueous environment. However, even when deposition is subaerial, similar physical processes usually operate as the individual particles generally deposit grain by grain as they are surrounded by air or gases. The detrital particles involved in such deposits will normally have acquired a magnetization by either thermal or chemical processes prior to their deposition. The stability of the magnetization of such grains will therefore be characteristic of the grain size and the composition of the individual particles (Section 2.3) – not by the process of deposition. The shapes of the particles will also be variable, reflecting the mode and duration of erosion and transport. (Deep-sea sediments frequently contain extraterrestrial magnetic grains – Yamakoshi *et al.*, 1981.)

During deposition, all magnetic particles will be acted on by the aligning force of the geomagnetic field, in addition to the gravitational and dynamic forces operating on all detrital particles during deposition (Fig. 4.3(a)). For naturally occurring magnetic particles, Collinson (1965a) has shown that, in the absence of other forces, such a magnetic alignment would take place very rapidly, within 15 seconds. Brownian forces tend to randomize this alignment, particularly for particles less than some 0.1 μm in diameter. All particles are also affected by gravitation forces and water flow. Although calm water conditions can occur, it is more usual that water currents are present, even in deep oceanic environments, but such current motions are damped near the sediment–water interface, resulting in a water layer some 2–3 cm thick in which the flow becomes laminar (Granar, 1958), irrespective of the flow patterns at higher levels (Fig. 4.2). This means that spherical particles, certainly greater than 30–50 μm in diameter, and elongate particles will be systematically rotated out of magnetic alignment so that their long axes become parallel to the bottom and perpendicular to the current (Granar,

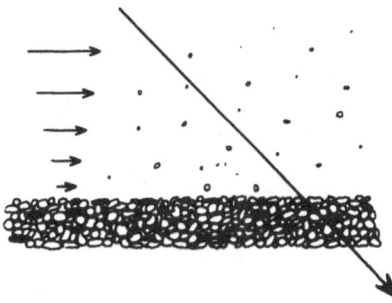

Figure 4.2 The depositional magnetization of sedimentary rocks. As sediment is deposited from flowing water, the grains pass through a zone of laminar flow immediately above the sediments that have already been deposited. Within this layer, magnetic particles are operated on by hydrodynamic, gravity and magnetic forces. Small grains tend to be most readily aligned with the ambient geomagnetic field, while larger grains are more affected by the gravitational and hydrodynamic forces.

1958) if the stream velocities are greater than about 1 cm/s (King, 1955). On reaching the bottom of the laminar layer, particles become subjected to radically different forces, in particular each particle attempts to achieve a position of lowest gravitational energy by rolling into hollows. Asymmetrically shaped particles will also tend to lie with their long and intermediate axes parallel to the sediment interface, i.e. approximately horizontal, and currents will tend to roll elongate particles so that they become aligned with their long axes either perpendicular or parallel to the current flow, thereby imparting a fabric to the sediments (Rusnak, 1957; Granar, 1958; Bagnold, 1966; Allen, 1969; Rees, 1979). Such movements are aided by the stream motions and by each particle still being essentially surrounded by water. (The fabric can now be measured rapidly and precisely, allowing determination of the primary fabric in different types of sediment, as discussed further in Sections 5.3 and 9.4.)

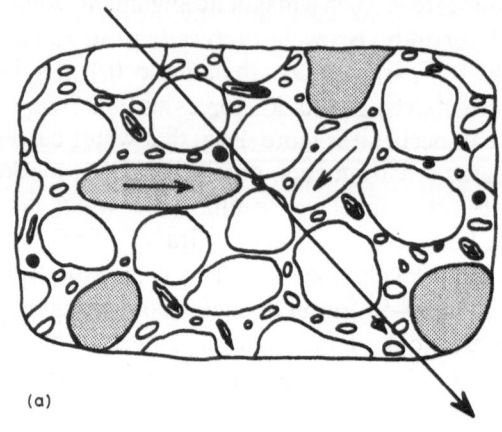

(a)

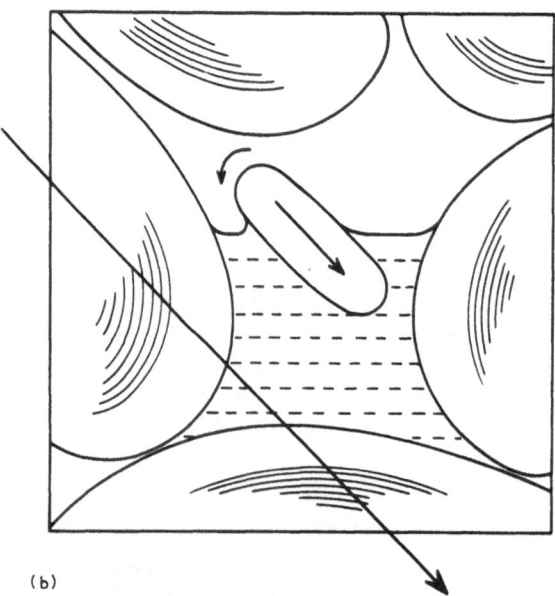

(b)

Figure 4.3 The post-depositional magnetization of sedimentary rocks. (a) Immediately after deposition, many grains are still surrounded by water and are able to rotate into alignment with the geomagnetic field, although the large grains are generally locked in the position determined by the gravity and hydrodynamic forces operating during their deposition. (b) As the sediment consolidates, the water level may change, giving rise to strong surface tension effects, the net result of which is to rotate elongate grains into a shallower orientation than that of the geomagnetic field. Associated with such changes in the water content there are also likely to be major changes in the mineralogy of the detrital grains in the sediment, giving rise to chemical remanences in some cases, and also releasing cements into the intergranular spaces within which new magnetic minerals may grow with an associated chemical remanence in the direction of the ambient geomagnetic field at that time.

The deposition and consolidation of elongate particles (Fig. 4.3(b)), usually magnetized along their longer axes, is palaeomagnetically important because it causes a shallowing of the net magnetization of the sediment towards the horizontal plane. This *inclination 'error'* is common in laboratory experiments, resulting in a shallowing of up to 20° compared with the ambient field inclination (King, 1955), but seems to be rarer in natural sediments. If deposition takes place on a sloping surface, then all particles will have a tendency to roll down slope, further increasing the 'error'. This additional *bedding 'error'* may be as high as 25° in laboratory experiments (King, 1955; Rees, 1965), but both 'errors' appear to be much less in natural sediment samples than in deposition experiments in the laboratory. In natural sediments, the 'error' rarely seems to exceed 5–10° (Granar, 1958) and is often completely absent (Johnson *et al.*, 1948; King, 1955; Griffiths *et al.*, 1960; Thompson and Turner, 1979). The difference appears to reflect the ability of magnetic particles within pore spaces to be re-oriented by the geomagnetic field in natural sediments after deposition. Irving (1957a) found consistent directions of remanence in slumps within now consolidated sandstones and proposed that post-depositional rotations of magnetic particles may occur. In laboratory-deposited sands, Irving and Major (1964) confirmed that such post-depositional remanence could be acquired and Tucker (1980) showed that some 10% of the grains in a mixture of magnetic and non-magnetic particles 1–32 μm in diameter were able to re-align post-depositionally. In natural conditions, fine-grained clays tend to show these 'errors' more than somewhat coarser fractions, such as silts. However, silts and sands often show a greater randomization of their magnetic directions, possibly reflecting the difference in porosity, but also reflecting the greater sphericity of large particles when compared to the platelets of clay particles which often contain magnetite or goethite within their structures. The 'errors' are, in fact, only partially controlled by the strength of the geomagnetic field (Hamano, 1980), and mainly by the size and shape of pore spaces, and the size and shape of the magnetic particles.

One reason for the apparent difference in the frequence of occurrence of inclination 'errors' in laboratory-deposited materials, compared with natural sediments, is probably related to the much slower depositional rates under natural conditions. However, any study of either laboratory or natural sediments, immediately after deposition, is hampered by the difficulty of actually sampling the materials in a way that avoids disturbing the magnetization either physically or magnetically. Such changes can even occur during the collection or handling of apparently stiff clays which may behave thixotropically if subjected to vibration (Symons *et al.*, 1980). (See also Section 4.5.)

Barton and McElhinny (1979) and Barton *et al.* (1980) found that there was less than two days between laboratory deposition and the acquisition of a stable remanence with no significant inclination 'error' for laboratory-deposited natural muds which were accumulated in magnetic fields of 50 μT at a rate of some 1.5 m in 9 months. Nonetheless, this is extremely fast compared with natural deposition rates of some millimetres per year. Very rapidly deposited natural materials also appear to show similar inclination 'errors', such as the 1979 Mt St Helen's volcanic ash deposits (Steele, 1981). Other detailed studies of recent sediments indicate that an inclination 'error' may be present in the uppermost parts, possibly taking up to 100 years before becoming aligned with the geomagnetic field (Suttill, 1980). Most such studies, however, indicate that very short periods of time are involved. Nöel (1976) found that slumping within Swedish varved clays would only affect varves deposited during the previous 9 years, and Verosub (1975) considered North American varves acquired their magnetization less than 3 years after deposition. Addison (1982a), studying the Pleistocene Lisan Formation in Israel, found that intraformational slumping still showed a magnetic fabric parallel to the bedding distortions, but the remanence showed the ambient geomagnetic field direction. It seems clear, therefore, that the rate of post-depositional re-alignment is very strongly controlled by the water content of the sediments immediately following deposition. In natural conditions of very slow sedimentation, with a high water content just below the water–sediment interface, post-depositional re-alignment of ferromagnetic particles less than a few microns in diameter appears to take place very rapidly, but geomagnetic field strengths are too weak to re-align larger particles. However, conditions for partial re-alignment in natural sediments occur, not only at the very top of the sedimentary column, but also as water is expelled from clays, in particular during the gradual consolidation of sediments (Ellwood, 1979a). Such physical re-alignment could take place several thousand years after the original deposition, depending on the rate of sedimentary accumulation and the grain sizes involved. In general, the compaction of clays causes a decrease in their initial high porosity (50–80%) to approximately half in the upper 1 km, with virtual complete loss of the initial porosity at depths of about 2 km. Sands, on the other hand, start with a lower porosity (20–60%) but this only decreases to half at about 4 km depth (Selley, 1976). The dewatering of the clay fractions is not simple in these conditions, as the increasing pressure is only one factor involved in their conversion from clays to claystone. Clay dewatering near the sediment–water interface also occurs spontaneously, forming syneresis cracks (White, 1961), while clay transformations, for example illite to montmoril-lonite, also release water but such conversions are temperature–time dependent and generally occur at depths greater than some 2 km, although this is a function of the local geothermal gradient. Clearly, the porosity is also affected by cementation processes (discussed later).

Where sediments have accumulated in an area of changing water level, it is likely that they will be subjected to alternating drying and wetting. Similar conditions may occur if organic material is present as this is likely to give rise to insoluble gases that may have a comparable effect in that surface tensional forces operate at any gas–water interface and these may physically rotate interstitial grains (Nöel, 1980). It is also important to stress that such processes are also likely to occur if samples, taken from sediments, are allowed to dry (Henshaw and Merrill, 1980; Verosub *et al.*, 1979). The effect of water menisci (Fig. 4.3) is generally to shallow the orientation of the long axes of grains and, hence, to shallow the net magnetic inclination away from the geomagnetic field (Nöel, 1980).

Previously deposited sediments can clearly be disturbed by water currents (or winds for subaerial deposits). As the speed of the current increases, it will change, for a given particle size, from deposition to erosion. Initially this will cause a winnowing of the finer grains, such as clays, but eventually lifts smaller particles which then saltate and eventually become fully suspended in higher velocity currents. Such changes in velocity can arise from a variety of reasons and may cause complete removal of all unconsolidated sediments. It seems unlikely that they should affect the magnetization below the actual sediment–water interface at which the erosion is taking place, but they would remove the layer in which the water content of the sediment is highest (and thus the layer within which post-depositional re-orientation is facilitated). The currents would also result in an orientation of elongate particles at the sediment–water interface. Similarly, ocean waves, which decrease rapidly with depth, still have a major rolling effect on grains being deposited in sub-tidal beach environments – the to-and-fro motion generally causing elongate particles to become aligned with their long axes parallel to the shore-lines. Levels at which such erosion has occurred are thus less likely to preserve an accurate record of the geomagnetic field at or shortly after the time of deposition. In this context, it is important that erosion levels are common, even within deep ocean sediments (Thiede, 1981).

The activity of biological organisms, particularly in the upper few centimetres of the sediment, may also be important. Study of oxygen isotope stratigraphies (Hutson, 1980) indicate mixing extending over some 8 cm in some deep-sea drilling project cores. Similar degrees of bioturbation are also indicated by studies of other isotopes (e.g. ^{14}C, Erlenkeuser, 1980). Clearly such activity physically disturbs the original depositional magnetic alignments, but such effects may well be largely beneficial in that the disturbances may well facilitate post-depositional re-alignments, although this could occur several thousand years after deposition, depending on the sedimentation rate.

In summary, it appears probable that some sediments will have a natural remanence that reflects the direction of the geomagnetic field at the time of their deposition, but rapidly deposited or low porosity sediments are likely to have associated inclination and bedding 'errors' that make the observed remanence shallower than that of the ambient geomagnetic field. Most sediments will also show significant post-depositional alignment, mostly within a few days or months of deposition, but occasionally several years or even decades after their deposition. Older unconsolidated sediments, or younger sediments in areas of high sedimentation, may also show much later re-alignments associated with dewatering of the clay content. All particles less than 0.1 µm in size are likely to have been randomized by Brownian motions, but slightly larger grain sizes are likely to show post-depositional alignments with the ambient geomagnetic field, the precise size range involved depending on the ratio of the ferromagnetic grains to the pore spaces in which they lie at different times during the accumulation of the sediment. Larger grains, mostly those greater than 10–20 µm in diameter, are likely to retain any alignment imparted by the physical processes operating during the actual process of deposition. Conveniently for palaeomagnetic studies, this means that the smaller grain sizes are most likely to reflect accurately the geo-magnetic field direction and these are also the most magnetically stable grain sizes (Sections 2.3, 5.4 and 5.5). It is, nonetheless, likely that most sediments, even excluding chemical changes, are likely to comprise magnetizations acquired at different times during their accumulation. However, chemical changes may frequently dominate even these physical orientations, even in still completely unconsolidated sediments.

4.3.2 Chemical processes and consolidated sediments

As sediments accumulate, their lower layers are subjected to increasing pressure, reducing the pore spacing, expelling water and thus forming more compact rocks. These physical changes are almost always accompanied by chemical interactions as detrital particles adjust to their new physico-chemical environment and as cements are deposited from migrating fluids within the pore spaces. These chemical interactions are termed diagenesis (Pettijohn, 1957), and the cementation part of these processes is generally much more effective in producing consolidated rocks than the processes of physical compaction (Fig. 4.4). Under certain conditions, precipitation from circu-lating fluids may directly form rocks, e.g. stalagmites, which may have an associated remanence (Latham *et al.*, 1979).

The break-down of detrital grains can be seen as an extension of normal weathering conditions (Parron and Nahon, 1980), but the upper layers in sediments may be either reducing or oxidizing, depending on their composi-tion and environment (Curtis, 1977, 1980; Campos and Hallam, 1979). Iron-bearing minerals, such as olivines, pyroxenes and amphiboles, generally continue to disintegrate, producing iron-rich chlorite clays, while feldspars tend to decompose to form illite and kaolinite clays. Even these clays may still be chemically metastable (Lippmann, 1979). In this process the iron ions become very mobile and take part in complex and poorly understood organic–inorganic series of reactions, mostly in the upper few metres of the sediments. The iron compounds rapidly hydrate to form iron hydroxides, such as goethite, lepidocrocite and akagenite (Section 3.2.4), while some titano-

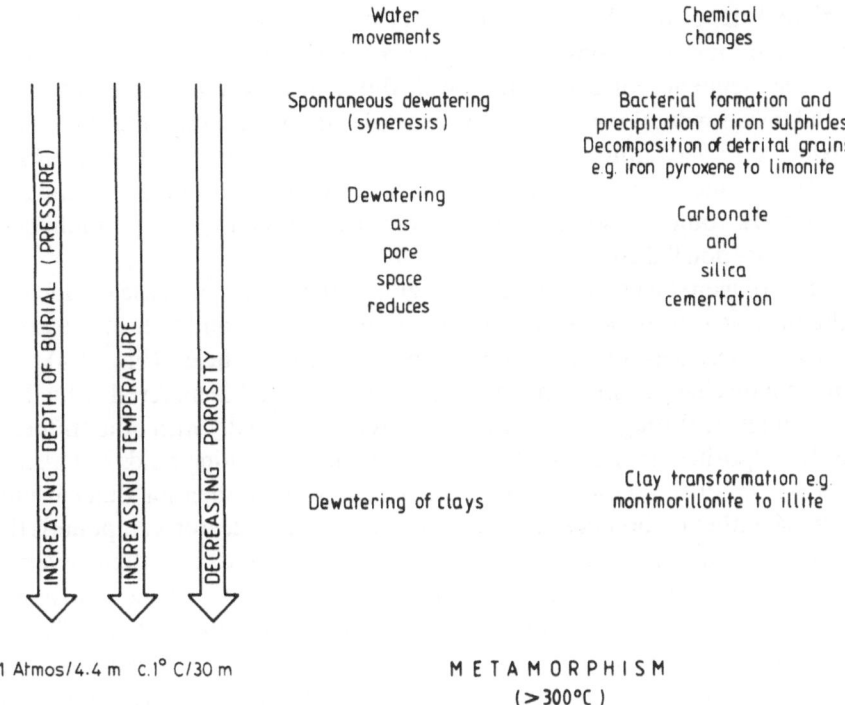

Water
movements

Chemical
changes

Spontaneous dewatering
(syneresis)

Bacterial formation and
precipitation of iron sulphides
Decomposition of detrital grains
e.g. iron pyroxene to limonite

Dewatering

as

pore

space

reduces

Carbonate

and

silica

cementation

Dewatering of clays

Clay transformation e.g
montmorillonite to illite

INCREASING DEPTH OF BURIAL (PRESSURE)

INCREASING TEMPERATURE

DECREASING POROSITY

c.1 Atmos/4.4 m   c.1° C/30 m

METAMORPHISM
(>300°C)

Figure 4.4 Diagenetic processes in sediments. This diagram summarizes the physical and chemical changes that commonly occur as unconsolidated sediments gradually become consolidated into hard, compact rocks, with an associated increase in pressure and temperature as the sediments are buried beneath later sediments. The precise nature of these changes, particularly chemical, depends on the composition of the materials forming the sediments and of any circulating fluids.

magnetites also transform to maghemite (Section 3.2.1), again by poorly understood inorganic–organic processes. These iron hydroxides and oxides then tend to react with any hydrogen sulphide (H_2S) being released by microbial activity (such as *Desulfovibrio* in anoxic environments), resulting in the formation of iron sulphides, such as greigite (Fe_3S_4) and mackinawite ($Fe_{1+x}S$) – of which greigite is magnetic (Section 3.2.5). These sulphides then form pyrite (Fe_2S), that frequently replaces the organic material and is non-magnetic, but in conditions of very high pH and low pE, magnetic pyrrhotite may form (Henshaw and Merrill, 1980). The total time for such pyritization to take place clearly depends on the organic material deposited, the precise oxidation conditions on and shortly after deposition, etc., but is thought to be 90% completed within some 500 years of deposition (Manning *et al.*, 1979), with most reactions taking place within 20 years of deposition, i.e. within the upper metre of sediment (Sweeney and Kaplan, 1973). Obviously, the disintegration of pre-existing magnetic minerals destroys their original depositional remanence and post-depositional rotations in the top few centimetres are similarly likely to be erased by such reactions, but the precise sequence of events may well differ markedly for different rocks. Under continental conditions, for example, the hydroxides are likely to be oxidized to haematite, usually in specular form. All such reactions are likely to be enhanced or retarded by the local conditions so that such oxidation processes could, in some circumstances, be retarded for many millions of years (Section 9.2.3).

The diagenetic changes of detrital grains may either increase or decrease the pore spaces, depending on the minerals involved, and thus may assist or prevent post-depositional rotations of magnetic particles (Fig. 4.4). The diagenetic changes are also the major source for soluble materials which can then migrate through the pore spaces to react eventually with other minerals or be deposited as cements (Collinson, 1974). (Cementing fluids can also, of course, enter the sediments from other beds.) The cementing fluids usually deposit either carbonates or silica, but will often include iron compounds that also become deposited as hydroxides or oxides, that may then revert to haematite. Cementation, with the possible creation of new magnetic minerals, can take place over a range of time (depth), although usually after the redoxomorphic changes of the detrital grains are largely complete. The new magnetic grains will acquire a chemical remanence as they grow, reflecting the direction of the geomagnetic field at the time of diagenesis, which may well extend over several thousand or even several million years (Section 9.4). The growth of the crystals will be influenced by the size and shape of the pore spaces in which they grow. In some cases, it seems possible that such growths would therefore mimic the fabric of the pore spaces. In general, it seems more likely that the initial growth of minute crypto-crystals would not reflect the shape of the larger pore spaces, so that any magnetic fabric resulting is more likely to be attributable to the final shapes of the pore-filling crystals than to crystalline effects (Section 9.3).

In summary, most consolidated sediments are more likely to be character-ized by a magnetic remanence that was acquired during their diagenesis and compaction than during their original deposition. Unfortunately the time of diagenesis is variable, even for similar rock types. In red sandstones, for example, it can apparently occur extremely rapidly after deposition, as in the Triassic Chugwater Formation of the USA (Steiner and Helsley, 1974a,b) and in others it can be delayed for possibly 100 million years after deposition (Turner et al., 1976). Indeed, in permeable rocks, the passage of ground-waters may well lead to the continual re-setting of chemical remanence as magnetic minerals are removed and new ones crystallize. Clearly studies of the remanence of sediments, and particularly consolidated sediments, must be combined with petrological investigation of the relationships of their magnetic minerals. Even then, the interpretation of the time of acquisition of the remanence may be unclear. Conveniently, however, palaeomagnetic measurements can often be applied to date the time of such magnetic acquisitions and to determine the composition of the magnetic minerals and hence the redox state of the sediments at specific times (Sections 9.2 and 9.4).

4.4 METAMORPHIC ROCKS

As rocks become increasingly buried by later sediments and igneous rocks, they are obviously subjected to increasing pressure, initially some 0.3 kb km^{-1}, and also to increasing temperature. The increase in temperature depends on the local tectonic setting, but averages 27°C km^{-1} in continental areas. The increased pressure, but particularly increased temperature, causes some minerals to become increasingly unstable, and these eventually form new minerals at particular pressure–temperature conditions. The recognition of these minerals (Winkler, 1976; Turner, 1968) can therefore be used to determine the metamorphic grade of the rocks being studied (Fig. 4.5). Clearly, such metamorphic grades are initially difficult to distinguish from advanced diagenetic effects (Section 4.3) and there is no clear distinction between low-grade metamorphism and high-grade diagenesis. The conventional distinction is normally taken around 300°C, i.e. within the chlorite grade of metamorphism. One way in which such regional metamorphism is commonly distinguished from diagenesis is that the metamorphic formation of micas, in particular, usually takes place under some degree of tectonic stress (in addition to the hydrostatic pressure). Crystals growing under such conditions then tend to grow with their longer axes perpendicular to the stress. Micas, therefore, tend to form sheets that are aligned at right-angles to the stress field and the rocks therefore become foliated. Shales eventually tend to form slates, and harder rocks develop fracture cleavages as a result of these alignments. At increasing pressure–temperatures, the formation of new minerals makes the rock granular, while at even higher temperatures the rocks show signs of increasing plastic deformation, or even flowage, of the 'softer' minerals, forming a gneissose structure; and at even higher temperatures the rock becomes totally molten.

Under certain conditions, rocks can become heated, with little rise in pressure, or subjected to high pressures with little temperature increase. The intrusion of doleritic igneous rocks, for example, can give localized temperatures of 1000°C or more in surface rocks. Conversely, rocks caught between two moving blocks, such as along a fault or thrust, can be subjected to high pressures and only minor heating. There is, however, a gradation between such superficial effects in that major intrusions of granites, for example, take place at depths of several kilometres and therefore the country rocks have been subjected to regional metamorphism, upon which is superimposed localized heating. The source of the heating, in this example, is large and creates a large peripheral zone around the granite (Fig. 4.1), a metamorphic

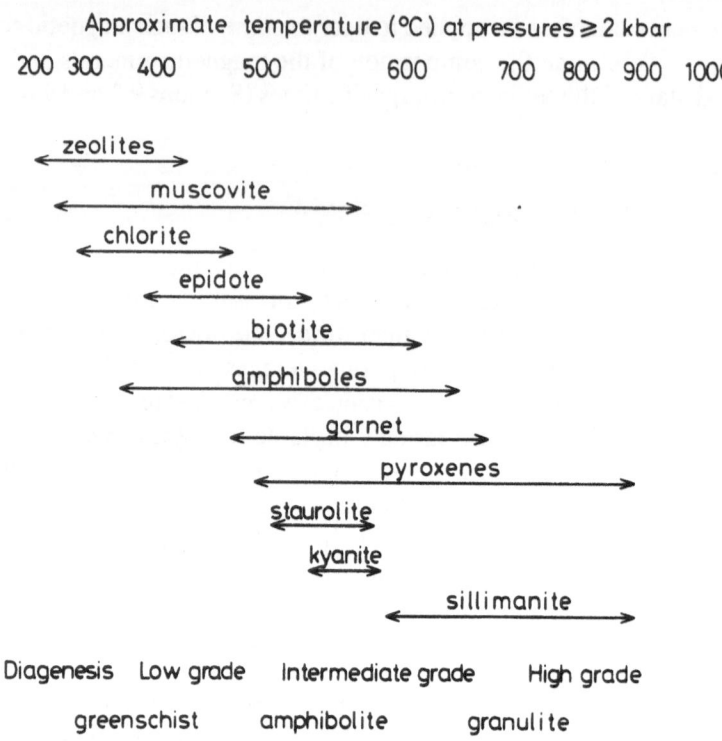

Figure 4.5 Metamorphic grades of rocks. Different metamorphic grades are identified by the formation of different minerals, mostly originating from clays, as the pressure and temperature are increased. In general, temperature increases are more important than pressure and approximate temperatures (NOT to scale) are indicated for pressures corresponding to a depth of burial of some 5–6 km.

aureole, within which the mineralogy (Fig. 4.5), corresponding to increasing regional metamorphism, tends to be similar, but compressed in scale.

Metamorphic rocks are thus characterized by changes in mineralogy, many of which involve iron-bearing minerals. The formation of magnetite is quite common within metamorphic aureoles, although higher oxidation levels or higher temperatures may result in its conversion to haematite. The fact that the crystals also grow, in most cases, in a regional stress field also means that they are likely to possess a shape anisotropy that will also affect the acquisition of remanence (Section 2.2). Possibly more importantly, the fact that temperatures up to and even exceeding the Curie point of the contained minerals will mean that thermal remanence will be acquired as the rocks cool to surface temperatures. However, this cooling can be very slow – possibly tens of millions of years – so that high-temperature viscous remanence will also be acquired at the same time as mineralogical changes are likely to be occurring as the minerals adjust to the changing pressure–temperature conditions. It is thus difficult to exaggerate the complexities of magnetization that are likely to be acquired in most metamorphic rocks. Even in contact metamorphism, where other rocks are locally baked by a shallow igneous intrusion, strong chemical changes are likely to be involved.

4.5 ARCHAEOLOGICAL MATERIALS

Fundamentally the only physical difference between archaeological and geological materials lies in the short time-scale of archaeology, generally less than 10000 years, in contrast with millions or billions of years for geological studies. Conversely, however, archaeomagnetic studies, such as dating, generally require greater precision than for geological studies in which geomagnetic changes of a few thousand years are deliberately averaged out of the analysis (Section 6.3). As the physical processes are identical in both types of material, archaeological materials can be considered in terms of their having a magnetic remanence associated with thermal, depositional or chemical processes (Aitken, 1974; Tarling, 1975a).

4.5.1 **Heated materials**

Obviously volcanic rocks erupted during archaeological times will acquire a thermochemical remanence at the time of their cooling in the same way as older volcanic materials (Section 4.2) and these, or older volcanic materials, may also have archaeomagnetic properties that can be used for a variety of purposes (Section 7.3). In addition, most archaeological materials include magnetic particles if only at impurity levels of less than 0.1%. If these have been heated at any time, then they will also have acquired a remanence on cooling that can be related to the direction and intensity of the geomagnetic field at the time of cooling. Fired clays were, in fact, amongst the earliest of materials that were investigated magnetically (Section 1.2). Until recently, most archaeomagnetic studies have been based on either fired materials, such as pottery, bricks, tiles, etc., or the areas where such firing took place, kilns (Fig. 4.6), hearths, etc. (Thellier, 1937b; Aitken and Hawley, 1967). There are, however, a wide range of archaeological materials that can be utilized in such studies as most metallic objects have been heated during their manufacture and also contain iron or iron oxide impurities. Indeed better results are generally obtained from materials with only very small iron or iron oxide impurities as large-grained iron or iron oxides tend to be less magnetically stable, i.e. they are unable to retain their remanence for long periods. Nonetheless, the time-scale for such retention is often comparable with the archaeological time-scale, although few studies have yet been undertaken in this field, e.g. the use of coins (Tanner *et al.*, 1979; Tarling, 1982a).

Objects that have subsequently been heated, but to temperatures still below the Curie point of their magnetic minerals, will acquire a new magnetization as they cool which will replace the component of remanence previously acquired over the same temperature range. It may thus still be possible to isolate both components of magnetization (Section 5.4). In some cases, the time of the second firing may be of greater importance than the original firing (Section 7.3).

An important feature of archaeological materials is that most of them are man-made and thus the conditions of their firing can be more readily duplicated in the laboratory. Conversely, the magnetic properties can be used to determine the possible conditions of firing (Section 7.3), but, as with most archaeomagnetic studies, little research has yet been undertaken in this field. Certainly many are already known to possess stable remanence – kilns, hearths, bricks, tiles, shards, coins, metal casts, ash layers, slag heaps, etc. This remanence can then be used in a wide variety of ways (Chapter 7).

4.5.2 Deposited materials

As in the geological context (Section 4.3), sediments may accumulate in an archaeological environment and hence preserve a record of the geomagnetic field directions, and possibly strengths, at the time of, or very shortly after, deposition. Despite such materials being common and being of little intrinsic value to archaeologists, there has been remarkably little study of such materials. This reflects, in part, the uncertainty about the exactness with which any sediment reflects the ambient geomagnetic field at the time of deposition (Section 4.3). Nonetheless, various sediments, such as basal sediments in a ditch or post-hole, cave sediments, pond infills, and so forth have provided successful results (Creer and Kopper, 1976; Nöel, Homonko and Bull, 1979; Latham et al., 1979; Hammo-Yassi and Andrews, 1981). Old soils, in an archaeological context, are also likely to be of particular use as their magnetic properties have been enhanced by local firing (Tite and Mullins, 1971; Mullins, 1977). Initial studies must still be considered preliminary, but seem to suggest that sand-sized particles retain the original field direction better than clay-sized sediments. The most serious difficulties are caused by inhomogeneities. Stones, and particularly pottery shards, are often very

---

Figure 4.6 Some archaeological contexts in which magnetizations may be acquired. (a) A wide variety of materials within an archaeological site may have acquired a magnetization at a specific time. Hearths will have acquired a magnetization during their last firing, while bricks, tiles and so forth will have been magnetized when they were fired. Fragments of pottery will also have a magnetization associated with the time of manufacture. Some materials may have a chemical remanence and solids may have acquired a depositional remanence. (b) Kilns provide strongly fired materials with a magnetization that was acquired during their last firing. Problems can arise if the kiln walls have moved or, for large kilns, the local geomagnetic field may have been distorted by the kiln itself by the time it had cooled to near-ambient temperatures.

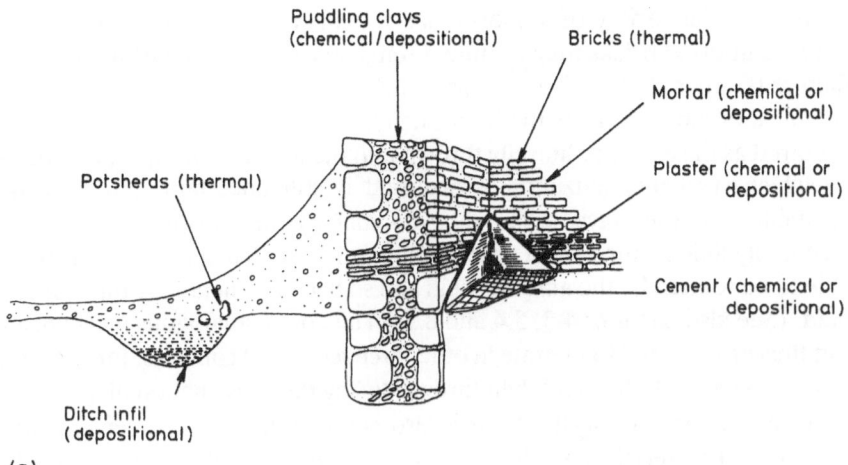

Puddling clays
(chemical/depositional)

Bricks (thermal)

Mortar (chemical or
depositional)

Potsherds (thermal)

Plaster (chemical or
depositional)

Cement (chemical or
depositional)

Ditch infil
(depositional)

(a)

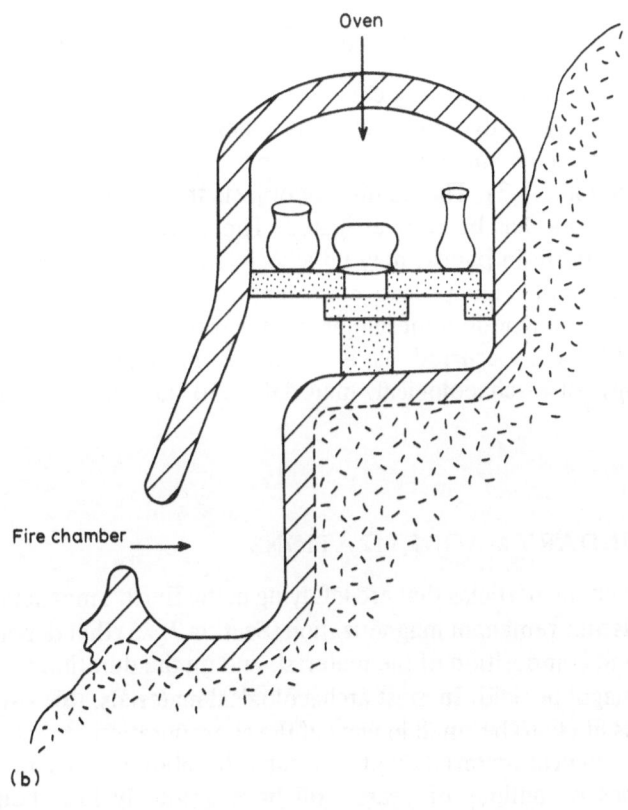

Oven

Fire chamber

(b)

strongly magnetic and their presence, within a single sample, can dominate the remanence of the remaining sediment.

As a special variety of 'depositional' remanence, it is known that hand-made, sun-dried bricks may acquire a remanence at the time of their manu-facture (Games, 1977). This was originally suspected to be predominantly of chemical origin, arising as the bricks dehydrated, but has been shown to be acquired as the clays are literally thrown into the former. The specific process by which such magnetization is acquired is still uncertain, but it seems probable that the clays, when shocked, behave thixotropically, i.e. they essentially fluidize for a short time, during which small magnetic particles are able to be rotated by the ambient field and so become aligned by the ambient field. (See also Sections 4.3, 5.4 and 8.3.) This, or an analogous mechanism, can thus be expected to operate in other archaeological contexts, for example when clays are puddled and then thrown during the construction of a dam, or to form the core of a wall, etc. It is also possible that plaster, cements and similar such materials may also acquire a remanence in this way. It must be emphasized that few studies have yet been made of this phenomenon.

4.5.3 Chemically changed materials

As in geological conditions, any chemical changes in archaeological materials that result in the growth of new magnetic minerals will result in their acquisi-tion of a chemical remanence as the new minerals grow through their blocking volume (Section 2.3.3). The number of objects that come into this category are few and none has been investigated. The dehydration of iron hydrox-ides, such as those in paints, may be associated with the formation of new oxides, for example. The process of cementation may also involve chemical changes in the magnetic mineralogy. In general, however, most chemical changes will have occurred as a result of the gradual decomposition (weathering) of archaeological materials and are hence regarded as secondary.

4.6 SECONDARY MAGNETIZATIONS

All ferromagnetic particles that are left lying in the Earth's magnetic field will acquire a viscous remanent magnetization (Section 2.3.3) that depends on the grain size and composition of the material, and on the duration and strength of the geomagnetic field. In most archaeological materials, this viscous mag-netization is likely to be small in view of the short duration of archaeological time but geological materials that are stable to laboratory treatments will, over millions or billions of years, still have gradually lost their original primary remanence and acquired new viscous magnetizations. Such new components will often have re-set as the geomagnetic field changed, so that most viscous remanences are usually aligned close to the average geomagnetic field direction at the sampling site for the last million years or so.

The direct effect of pressure on remanence is completely dominated by the associated temperature changes in most geological conditions and even on intraplanetary scales (Section 9.8). The pressure effects on the Curie temperature, for example, are still virtually nil at depths corresponding to temperatures of 600°C. Hydrostatic pressure of a few hundred bars (upper crustal pressures) appears to induce a low stability remanence, comparable to an isothermal remanence (Fuller, 1970; Stacey and Banerjee, 1974; Pearce and Karson, 1981) although the saturation moment of magnetite may be decreased by 30% at depths of about 5 km (Kim and Carmichael, 1977) and recycled stress also has some effects (Zlotniki et al., 1981). All such hydrostatic effects are, however, completely dominated by the temperature increase with depth of burial. The effect of uniaxial stresses has been more intensively studied (Stott and Stacey, 1959, 1960, 1961; Kern, 1961a,b,c) as changes in the magnetization of rocks near to an active fault or volcanic activity may provide an indication of stress accumulations prior to failure or eruption. In isotropic rocks, the effects appear to be small and reversible, from theoretical and experimental study, and the observed magnetic changes with both earthquakes (Shamsi and Stacey, 1969; Martin, 1980; Johnston et al., 1980) and volcanoes (Stacey et al., 1965; Johnston and Stacey, 1969a,b) are small, a few nanotesla, and possibly simply related to the closure of cracks (Martin, 1980) rather than a true physical change of the magnetization of individual grains.

Rapidly applied stresses, such as those associated with meteoritic impacts, appear to be associated with a demagnetization effect so that rocks near an impact centre generally have a weaker intensity of magnetization (Cisowski and Fuller, 1978; Robertson and Roy, 1979; Pearce and Karson, 1981). Similar demagnetizations have also been observed associated with blasting, but only low-stability components are affected. Such observations are consistent with laboratory experiments (Shapiro and Ivanov, 1967; Nagata, 1971; Pohl et al., 1975) but are always small and often difficult to detect (Hargraves and Perkins, 1969; Hargraves and Roy, 1974). Nuclear explosions and very high velocity impact studies have indicated the possibility of creating an extremely stable shock-induced remanence (Wasilewski, 1973a; Srnka et al., 1980), but these may well reflect associated thermal heating as much as specific pressure effects.

As time proceeds, almost all rocks gradually chemically equilibrate to the conditions close to the Earth's surface. Even without weathering, therefore, chemical changes usually occur within rocks that may destroy or create new magnetic minerals. Newly created magnetic minerals will acquire a chemical remanence as they grow through the blocking volume (Section 2.3) while the destruction of older minerals will obviously destroy their magnetizations. Such equilibration processes are clearly most extreme in the case of weathering (Section 9.5), which is why it is usually vital to avoid the collection of weathered samples (Section 5.2). In such materials, the new magnetic materials may have a similar or even greater stability to demagnetization procedures (Section 5.4) than the pre-existing minerals. Under such circumstances, the isolation of the different components of remanence can be difficult, or even impossible.

Other sources of secondary magnetization can be numerous. Lightning can have struck the sampling site and have imparted a complex magnetization, predominantly of isothermal origin (Sections 2.3 and 5.4). Lightning strikes are, however, complex in that they involve plasma temperatures of some 15–20 000°C, but the temperature changes are insignificant compared with the discharge of several thousand ampères within some 40–50 μs (Malan, 1963). The effects of lightning tend to be superficial as the current is usually dissipated in the wet upper layers and the magnetic effects usually decrease rapidly with depth, to be generally negligible below some 20 m. Nonetheless, there are clearly good reasons for avoiding areas, such as projecting rocks near mountain tops, that are most likely to have been struck by lightning.

Laboratory-induced magnetizations can, in some sense, be regarded as secondary. Accidentally introduced laboratory magnetizations can be serious. For example, anhysteretic, rotational and gyromagnetic magnetizations may be introduced during alternating magnetic field demagnetization procedures (Section 5.4). While care is taken to reduce the size of these components, they may well be mistaken for other natural components – especially if they are acquired in a consistent manner.

4.7 MAGNETIC INHOMOGENEITY AND ANISOTROPY

4.7.1 Inhomogeneity

Uneven magnetization can occur at all possible levels. On the smallest scale, crystals can only grow if lattice dislocations are present, so magnetic grains will always have internal inhomogeneities reflecting the presence of impurities, vacancies, defects and the like. Archaeological and geological samples will usually contain many millions of individual magnetic grains so that such atomic-scale inhomogeneities are averaged out within a single specimen. However, the grains may themselves have an inhomogeneous distribution. In geological conditions, for example, rocks are commonly banded on scales that may range from microscopic to several kilometres. In archaeological situations, shards often have haematite in their oxidized surface layers but magnetite in the less oxidized interior. Sediments are particularly likely to show such inhomogeneity owing to the presence of larger stones or shards that may have their own magnetization that dominates that carried by their matrix.

The effects of such inhomogeneities depend on their scale. If the scale is small relative to the size of individual samples, then these are generally reduced by spinning during measurement (Section 5.3), although clearly increasing the measurement 'error' if the inhomogeneity is on a scale comparable to the sample size. Larger scale inhomogeneities obviously require adequate sampling to average such effects over the location. However, additional complications can arise if there are two or more components, each with a different distribution. Magnetic grain size distributions may, for example, be bimodal and result in homogeneity for the small grain sizes but inhomogeneity for the larger size distribution.

4.7.2 Anisotropy

Differences in susceptibility in different directions, magnetic anisotropy (Section 2.2), can be very important in causing the observed direction of remanence to depart from that of the applied field (Hrouda, 1982). Anisotropy in geological and archaeological materials arises primarily from crystalline alignments in samples containing haematite (Section 3.2.3) and from shape alignments in samples with magnetite (Section 3.2.1). Additional anisotropy can also occur if magnetic grains, even if individually isotropic, occur in rows or planes (Bathal, 1971), although such grains must be very close to each other, preferably in contact, if they are to create a significant anisotropy. Shape and crystal alignments can occur in a variety of ways and the resultant magnetic anisotropy can be used to determine the nature and direction of the forces causing them (Section 9.3).

Sediments within both archaeological and geological contexts are subjected to a range of stresses during deposition of which the gravity and hydrodynamic forces normally result in elongated and platey minerals being oriented within the bedding plane, and elongated minerals being further constrained into directions either parallel or perpendicular to the flow direction (Section 4.3). Most sediments therefore exhibit an oblate magnetic fabric, slightly imbricated in the direction of flow, that is little affected by burial and dewatering as the larger grains are less free to rotate than the more isotropic, smaller grains. Loading by later sediments, however, tends to reinforce the alignment within the bedding planes. Diagenetic changes often involve the formation of new magnetic minerals, especially from the more platey minerals, such as micas that are usually already aligned within the bedding planes. Diagenetic changes may thus further reinforce the original fabric pattern as the growth of new minerals tends to be controlled largely by the orientations of cleavage planes within the pre-existing minerals, and therefore they often mimic their orientations.

In molten rocks, similar alignments of already crystallized minerals may occur by flowage, although such orientations are generally weak unless the magma has a high viscosity, as is common in the more acidic varieties. Flow-banded lavas may thus have a high degree of anisotropy within the plane of flowage, especially in trachytes, but the magnetic anisotropy of basaltic flows can also be significant (Section 9.3.2). As with sediments, later chemical changes may result in the growth of magnetic minerals that pseudomorph pre-existing minerals and hence mimic their fabric. Clearly similar considerations apply to archaeological materials (Section 4.5) that were once molten, such as metallic casts and the like. In other fired ware, alignment of different minerals also took place during the moulding processes, such as striae formed during the turning of clays on a wheel, potter's marks, or the physical manipulation and rolling of clays as pots are shaped.

Magnetic anisotropies developed in such sediments and heated materials are generally small, less than 10% except in acidic igneous rocks. In the geological context, it is metamorphic rocks that generally show the highest anisotropy. When subjected to uniaxial pressures, most magnetic effects are reversible (Section 2.3), but such stresses are usually accompanied by an increase in temperature that enhances the probability of chemical change and changes in grain shapes. The actual mechanism by which such anisotropy is attained is not clear, although magnetic fabric studies may provide important constraints in such studies (Section 9.3). In some cases there may be a physical rotation of the grains themselves, possibly by crystal plane gliding. In others recrystallization may occur in a plane perpendicular to the stress field as a result of ionic migration or melting at pressure points between minerals that lie with their larger axes along the principal stress direction, etc. (Ramsay, 1967). Irrespective of the mechanism, a very strong crystalline and shape alignment can take place in metamorphic rocks subjected to such axial pressures. Temperature rises almost invariably accompany such uniaxial pressures, enhancing the probability of chemical changes and recrystallization. The growth of later magnetic minerals may therefore occur while the rocks are still under a stress or by later growth of crystallographic axes and shapes that largely pseudomorph the minerals, particularly micas, that are already reflecting the stress reorientations.

4.8 SUMMARY

When account is taken of the complexities in the composition and grain sizes of magnetic particles in rocks and archaeological materials, it is surprising how commonly they have a fairly simple natural remanent magnetization. In general, most igneous rocks and fired materials have a strong primary remanence that was acquired at the time of cooling. This may be of predominantly thermal or thermochemical origin, but such distinctions are frequently not important as both were acquired at the same time and thermal and chemical remanences behave similarly in the laboratory (Section 5.4). Similarly, many sediments have an essentially simple primary magnetization acquired on deposition or within a few years of deposition. However, sediments become more complex as they are compacted and undergo major chemical changes during lithification. These diagenetic changes will normally erase the pre-existing depositional features and they then acquire a chemical remanence relating to the time of lithification. In many sedimentary rocks, this time is not long after deposition on a geological time-scale, that is within a few hundred thousand years or so, but these changes may be much more prolonged in some instances.

As not all rock types or archaeological materials necessarily behave identically, it is obviously essential to evaluate the history of the acquisition of primary remanence prior to any subsequent analysis or interpretation. However, it is also essential that the primary remanence be isolated from all secondary magnetizations. All materials, especially rocks, that have been lying in the geomagnetic field for long periods will have acquired some viscous remanence. Conveniently, these are necessarily of low magnetic stability (Section 2.4) and can thus be removed by partial demagnetization (Section 5.4). More serious are secondary chemical changes that may destroy pre-existing magnetic particles and result in the growth of new ones, thereby producing a chemical remanence that may be more resistant to demagnetization procedures than the primary component, particularly if the primary magnetization is carried by magnetite and the secondary magnetization is connected with haematite formation. Under such circumstances, very careful analyses of the different components of natural remanence are required. The surprise is frequently how well rocks have acquired their primary magnetizations, and preserved them, even after complex geological histories.

Chapter Five

Sampling, measurement and procedures

5.1 INTRODUCTION

In order to dissect the natural remanence into primary and secondary components, partial demagnetization techniques must be used. These methods (Section 5.4) require precise control that is difficult to achieve successfully in the field. On this basis, it is normal to collect samples and to undertake measurements on them in the laboratory. Some field instruments (portable susceptibility meters and magnetometers) can be useful indicators of the general magnetic content and strength of magnetization, and even a simple compass can be useful for determining whether strongly magnetized rocks, such as Tertiary volcanics, attract or repel the compass needle (Doell and Cox, 1967a; Abrahamsen, 1967) and hence see whether they are magnetized parallel or antiparallel to the present geomagnetic field (Sections 8.4 and 9.2). Most palaeomagnetic and archaeomagnetic studies therefore involve the collection and orientation of samples (Section 5.2), followed by laboratory measurements of their susceptibility and remanence (Section 5.3). The remanence must then be examined for its stability (Section 5.4) and the possible age of its components (Section 5.5), thus defining the directions and intensity (Section 5.6) of the geomagnetic field at some specific time. Some

Table 5.1 Definition of localities, sites, samples and specimens

Locality	An area of less than 1 km² from which more than one site has been obtained.
Site	A volume of material that can be considered to have been magnetized at the same time, e.g. sediments from the same stratigraphic level or thin lavas or dykes.
Sample	A single piece of material that has been individually oriented in the field.
Specimen	A single item of material that is measured intact.

A site will therefore comprise several separately oriented samples, from each of which several specimens may be cut. In some cases, only one specimen may be obtained from one sample, but the same terminology is retained to avoid confusion.

standard instruments used in palaeomagnetic research are briefly outlined in this chapter and the reader is referred to Collinson (1983) for detailed considerations of these and additional instruments.

5.2 SAMPLING AND ORIENTATION

Separately oriented samples, of convenient size for measuring, are usually collected from a volume of material in a site, and can therefore be assumed to have acquired their primary magnetization at the same time. Specimens which are of convenient size for fitting into magnetometers, demagnetizers, etc., are then cut from the individual samples (Table 5.1 and Fig. 5.1). Such specimens are commonly cylinders 2.5 cm in diameter, 2.1 cm high, cubes of side 2.1, 2.5 or 5 cm, or cylinders of diameter 5.5 cm, height 4.8 cm. (Generally rock specimens are of the smaller dimensions, but the controlling factor is their insertion into magnetometers and demagnetizers – next section.) Theoretically, the primary remanence for all specimens cut from samples from the same site should thus have uniform magnetic parameters. This may occur, rarely, in archaeological studies, but generally statistical analyses are required to determine the mean directions and intensities within a site (Chapter 6). This means that statistical considerations must be taken into account when determining the optimum number of samples that should be collected from a site. In geological–geophysical studies, this optimum number is 6 or 7 (Section 6.3), with as many sites being sampled as is practicable. Archaeomagnetic investigations usually require greater precision and so 6 or 7 samples per site is often considered to be the minimum number of samples. If such sampling numbers can be obtained, then generally only one specimen need be obtained from each sample (Section 6.3).

The actual shape of the specimen may be critical in many measurements, particularly that of the magnetic anisotropy (Section 5.3.3). For virtually all instruments, the remanence (or the induced magnetization) is taken to correspond to a simple dipole magnet at the centre of the specimen. Such an assumption is invalidated if the specimen is inhomogeneously magnetized (Section 4.7) or is not an exact sphere. Quite serious errors can be introduced in this way, particularly if the specimens are strongly magnetized or have a high susceptibility. It is impracticable to grind spherical specimens and extremely difficult to orient them if ground. The most common shapes are therefore cubes or cylinders, of which cubes reasonably closely approach the ideal spherical shape. For cylinders, the horizontal cross-section is clearly identical to that of a sphere, but the height-to-diameter ratio is critical. Theoretical and practical tests give somewhat different optimum height/diameter ratios, varying between 0.92 and 0.85, for different instruments. In most cases, the optimum ratio appears to be where the height of the cylinder is 86% of the diameter (Porath et al., 1966; Noltimier, 1971; Scriba and Heller, 1978; Addison, 1982b). The importance of the shape effect, however, also depends on the distance of the sensors from the specimen – small specimens, although misshapen, may have little shape effect if they are small relative to the sensor distance, as long as they are correctly centred.

5.2.1 Sampling consolidated materials

Hard substances, such as consolidated rocks or fired materials, can usually be sampled in blocks, *hand samples*, which are subsequently drilled in the laboratory to provide cores, from which specimens can then be sliced. These blocks are generally about the size of small house bricks, i.e. some $15 \times 10 \times 5$

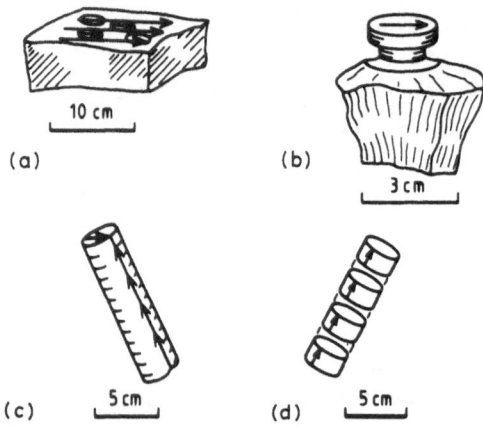

Figure 5.1 Samples and specimens. Samples may consist of individually oriented hand samples (a, b) or drill cores from which individual specimens are cut. In the case of hand samples, individual cores are usually drilled from the hand sample in the laboratory. The cores (c) are then sliced to give cylinders (d) with diameter-to-height ratios of 0.86, the dimensions most closely simulating a spherical sample during measurement.

cm, and can be removed intact from the site by chisels, hammers, etc. If suitable access magnetometers and demagnetizers are available, it is common to collect archaeological materials or fissile geological samples as small hand samples, of side less than 5 cm, that can then be fitted, without further cutting, into suitable magnetometers (see also orientation, below, and Section 5.3). Fissile and fragile samples collected in this way can also be coated or impregnated with non-magnetic resins or paint as long as thermal treatment will not be required.

Alternatively, field cores can be drilled using modified chain saw motors to trepan cores some 8 to 10 cm long, from which specimens can be sliced later. In general, this method is faster, each core taking some 2 to 3 minutes to drill, and more accurate as orientation marks do not have to be transferred from the surface of the hand sample (see orientation). Unfortunately, the system is also more expensive as the wear on field drills is much greater than when drilling under laboratory conditions. The cost of drill bits may thus dictate the sampling techniques employed.

5.2.2 Sampling unconsolidated materials

Present-day sediments are usually sampled by conventional coring techniques such as the gravity corer in which a weighted core barrel is dropped into the sediments from a ship at the surface. Penetration may be up to 10 m, although generally of the order of 1 to 2 m, and it is generally assumed that the corers penetrate vertically and do not twist during penetration. In practice such cores of sediment usually show only minor disturbances during penetration and retrieval, but hydraulic corers are now available (Prell *et al.*, 1980) that provide deep penetration, over 10 m, with a deformation thickness of only 1–2 mm along the core surface. In lakes, most studies have been based on modified versions of the Mackereth corer (Mackereth, 1958, 1969; Barton and Burden, 1979) in which a plastic core is slowly pumped into the sediments, providing continuous cores 5.4 cm in diameter of length 1, 2, 6 or 12 m. Recent versions of this corer have also been designed to prevent twisting of the core during penetration as this can give rise to spurious interpretations in terms of the long-time behaviour of the geomagnetic field. Such lake and marine sediment coring systems have the advantage that they provide continuous sections which can also be analysed from the horizontal components of remanence and susceptibility without removal from the plastic core barrel, but the determination of the vertical component of remanence (Section 5.3) requires the core to be subsampled. This is normally done by slicing vertically through the core and then hand sampling the exposed sediments by pushing in plastic boxes or cylinders that are suitably shaped for use in standard magnetometers. Obviously, such subsampling must be undertaken carefully to avoid disturbance and to retain any orientations.

Subaerially exposed sediments, such as in archaeological ditches or drained pro-glacial lakes, can be sampled with coring systems, but such sediments are generally harder than present-day lake sediments and may be more disturbed by the core penetration. Soft clays can easily be sampled by pressing plastic or copper cylinders into them, the top and bottom of the cylinders then being sealed with plastic disks to reduce subsequent drying out. Stiffer clays, or somewhat more consolidated sediments, can sometimes be carved, with non-magnetic tools, to form *in situ* pillars, over which plastic cylinders can be placed. The sample can then be fixed inside the plastic tub using glues, non-magnetic plaster and other packing. The plastic tub can then be removed and sealed with the sample securely fixed inside it.

5.2.3 Orientation of samples

Although orientation is not necessary for studies concerned with the past intensity of the geomagnetic field (Sections 7.2 and 8.3), it is often advisable that some control over the field position of the samples is obtained whenever possible as this still allows more reliable evaluation of the nature of components of their remanence (Section 5.5). Two standard methods are usually employed (Fig. 5.2), although any standard geological methods of orientation can be employed:

1. *Hand samples* may be oriented by finding a flat surface on the sample, prior to removal. A horizontal line (strike) can then be marked on it, and converted into an arrow (conventionally with the maximum slope of the surface 90° clockwise from the arrow). The direction of the strike arrow, relative to true north, is then determined and the maximum amount of dip of the surface measured from horizontal. Where no flat surface is present, then a flat surface can be provided by gluing a plastic disk to the sample, and orienting the surface of the disk, or a tripod can be placed on the surface (Fig. 5.2(a)), the legs of which are marked on the sample and the strike and dip of the surface of the tripod are then measured.
2. *Field cores* require an orientation table to be used, with a pivotable surface that allows the direction of drilling to be measured relative to true north and the angle of drilling, from horizontal, to be determined (Fig. 5.2(b)).

The angle of dip from horizontal can usually be measured very precisely using inclinometers, but the determination of the direction of the strike arrow (or drilling direction) can be difficult. Conventional magnetic compasses can be used if the sampling site is not strongly magnetic and the local magnetic declination is known. However, sun compasses (Creer and Sanver, 1967) provide a more accurate, faster method of orientation if the location is known within 0.25° (latitude and longitude), the time is known within 2 minutes – and the sun is shining! Alternatively, topographic sightings provide a reliable method of orientation in volcanic terrain if accurate maps are available.

Hand samples of unconsolidated sediments may be oriented in an identical way to other hand samples when sampling cylinders are pushed into them. In these cases, the plastic covering disk can be oriented. Plastic boxes or tubs can be similarly oriented after they have been pushed into long cores of sediments. The cores themselves, however, need to have been oriented. Most long sediment cores, obtained by the hydraulic or Mackereth corers, are assumed to have penetrated vertically – an assumption that is generally true within a few degrees. In most studies, no other orientation data are collected and it has been conventional to assume that the average horizontal magnetic component in the sediments corresponds to true north. While such an

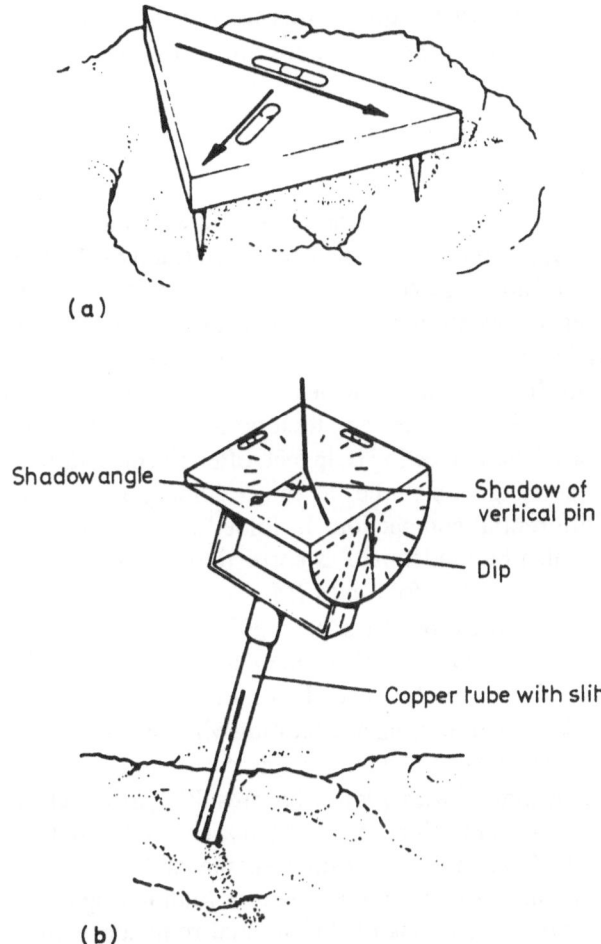

Figure 5.2 Orientation of hand samples and drill cores. Hand samples can be oriented by finding a smooth surface (or gluing on a smooth disc (Fig. 5.1(b)) and marking a horizontal (strike) arrow on it. The direction of this arrow and the dip, from horizontal, of the direction perpendicular to the arrow are then measured. More efficiently, a tripod can be placed on the rock and its surface oriented in a similar way, the position of the legs being marked to allow later re-orientation in the laboratory. Field cores need to be oriented with a table, although again the orientation consists of a directional arrow, usually in the direction of drilling, and the inclination of the drill core from horizontal.

assumption is probably approximately true, the actual declinations of remanence in such cores can only be treated as relative declinations (i.e. relative to each other). More recently, some cores have been oriented by photographing a compass and inclinometer on top of the core barrel, prior to extraction. Other devices have also been tried, such as heating the uppermost few centimetres so that they acquire a thermal remanence in the present field as they cool. Errors in all these methods must, however, be considered to amount to a few degrees.

5.2.4 Tectonic corrections

If the sampling site is no longer in the attitude in which it originally acquired its magnetization, i.e. the rocks have since been tilted or the kiln walls have fallen outwards (Harold, 1960), then orientation corrections are needed for such movements. Conventionally these are taken in the same way as for hand samples. If original horizontal can be defined, for example by determining the bedding plane in shales, then this plane can be measured in terms of its strike and dip (Fig. 5.3). The samples can then be rotated back to their original, pre-tilt, positions. This correction is, in fact, often the most difficult feature to measure in palaeomagnetic studies as the bedding planes are often locally distorted by differential compaction. In folded areas (Fig. 5.3) this simple correction may also be inadequate as correction needs to be made for the effect of the plunge of the fold axes (Macdonald, 1980), although this is insignificant for bedding dips of less than 30° (Fig. 5.3(c)). More seriously, particularly in folded beds, is the probability of intrastratal deformation (Ramsay, 1967; Oertel and Ernst, 1978) during the folding processes. This may take place by fracturing and faulting when the individual units have moved, creating an apparent bedding plane curvature, or it is common for folds to show deformation with 'plastic' flow resulting in a thickening of a bed at the axes of the fold and a thinning of its limbs (Fig. 5.9(c,d)). In the former case, the true bedding may be distinguishable and thus a valid bedding correction can be made in which the apparent folding is ignored, but in the latter case, variable deflections of the original remanence will have taken place, possibly associated with recrystallization and hence remagnetization. A further problem may also arise if the bedding plane had an original dip. This plane will be any value up to the angle of rest, and in subaqueous sediments will rarely exceed a few degrees, but in subaerial deposits, this angle can exceed 20–30%.

Figure 5.3 Tectonic corrections. If the rocks, or other materials, have moved since they were originally magnetized, it is necessary to determine, if possible, their original orientation. If a rock sequence has been simply tilted (a), then the strike and dip of the bedding planes can be measured (Fig. 5.2) and the original position of the rocks can be determined by a simple mathematical rotation. However, if the beds have been folded, this correction is only valid if the folding is about a horizontal axis. If the fold axis plunges (b) then a correction for the dip of the plunge is also necessary (c).

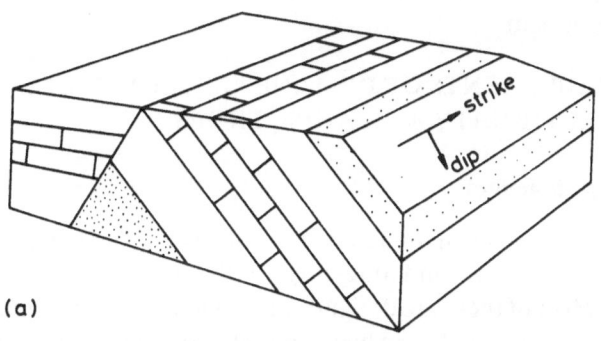

(a)

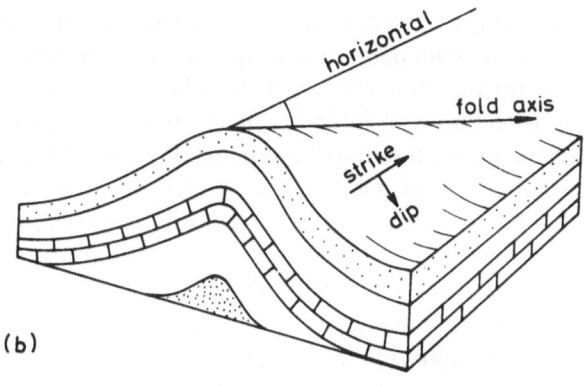

(b)

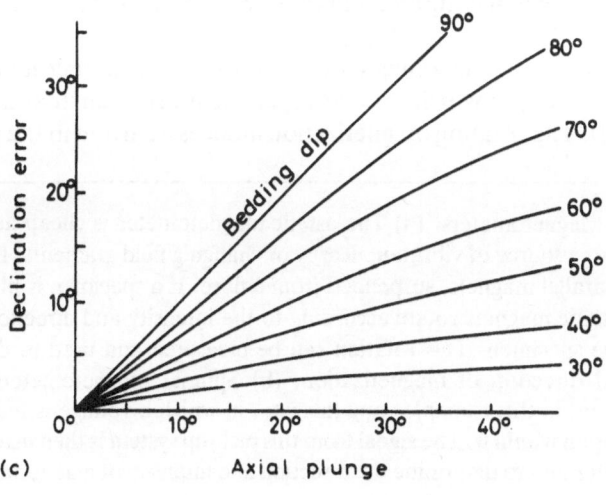

(c)

5.3 MEASUREMENTS OF REMANENCE, LOW-FIELD SUSCEPTIBILITY AND ANISOTROPY

5.3.1 **Magnetometers**

A wide variety of magnetometers are now available (Collinson, 1983) to measure the intensity and direction of the remanent magnetization of specimens. Most of these are designed for 'standard' size specimens (cylinders 2.5 cm in diameter, 2.1–2.2 cm high, and cubes of side 2.1 or 2.5 cm) and most of them have similar sensitivities. The *astatic magnetometer* was one of the earliest instruments (Blackett, 1952) in palaeomagnetic studies (Section 1.2) and is still an important instrument as it is simple yet capable of handling irregularly shaped samples and can also be used to measure low-field suscepti-bility. It comprises a magnet pair that are equally and oppositely magnetized, suspended on a torsion fibre (Fig. 5.4(a)). The magnet system twists in response to the direction and intensity of magnetization of the specimen beneath it. This magnetization may be remanent, if measured in zero magnetic field, or remanent plus induced if measured in the Earth's field. The low-field susceptibility can therefore be determined by comparison of the two measurements.

Other varieties of this instrument are also available, such as the magnetic parastatic system (Thellier, 1938). The sensitivity of these types of instrument is dependent on the strengths of the magnetic system and the torsion fibre and the noise level is largely controlled by the degree of vibration. Measurements usually take between 2 and 20 minutes, depending on the magnetic moment of the specimen and magnetic moments as low as 10^{-9} A m^2 (10^{-4} A m^{-1} intensity for a standard 2.5×2.1 cm cylinder).

More common in palaeomagnetic laboratories are *spinner* magnetometers (Larochelle, 1965) in which standard specimens are spun near a pick-up coil (Fig. 5.4(b)). The rotating magnetization induces a current in the pick-up coil,

---

Figure 5.4 Magnetometers. (a) The astatic magnetometer is cheap and efficient if there is no large source of vibration nearby or changing field gradients. It consists of a pair of antiparallel magnets, suspended from a fibre. If a specimen is placed beneath them, the astatic magnets rotate according to the intensity and direction of magnet-ization of the specimen. This rotation can be measured and used to determine the intensity and direction of magnetization. (b) Spinner magnetometers comprise a pick-up system, in this example, ring fluxgate, in which a voltage is induced when a specimen is spun within it. The signal from this pick-up system is then usually fed into a small computer unit to determine the direction and intensity of magnetization. (c) The fastest and marginally most sensitive magnetometer is the cryogenic type in which a current is induced into a super-conducting ring when a specimen is placed within it. The magnitude of the flux can then be measured and processed by a computer to determine the intensity and direction of magnetization of the specimen. (For further details, see Collinson, 1983.)

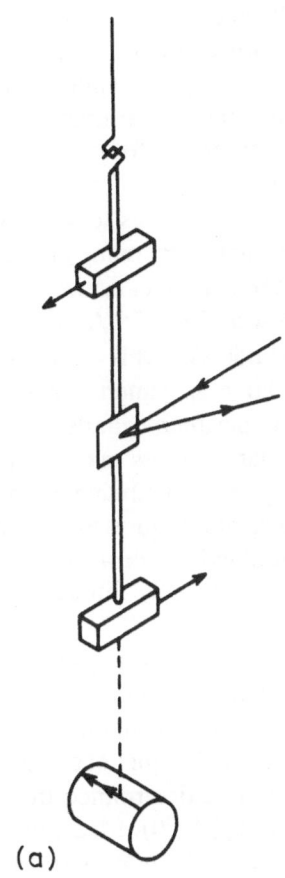

(a)

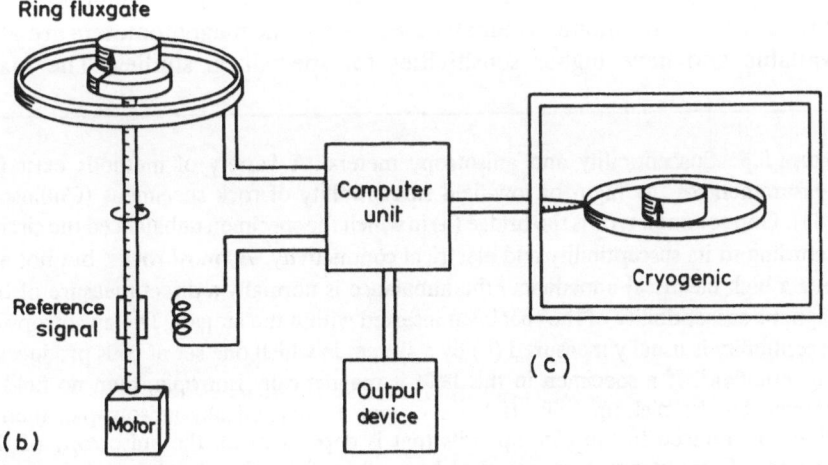

Ring fluxgate

Reference
signal

Motor

(b)

Computer
unit

Output
device

Cryogenic

(c)

the amplitude of which depends on the intensity of magnetization of the specimen in the plane perpendicular to the axis of rotation, while its direction within that plane is given by the phase angle. One of the fastest and most sensitive of these types of magnetometer uses a ring fluxgate pick-up system and is linked to a micro-computer unit for precise determination of the phase angle and amplitude (Molyneux, 1971; Collinson, 1982). This, and similar instruments, takes from 2 to 20 minutes to measure the direction and total intensity of the remanence, again depending on the magnetic moment of the specimen. Adequate measurements can usually be made in some 20 minutes for magnetic moments as low as 5×10^{-10} A m^2 (5×10^{-5} A m^{-1} intensity for standard 2.5×2.1 cm cylinders). Only standard-sized samples can be measured in this way. A large specimen version is available (mostly for archaeomagnetic work) for specimen cylinders 5.5 cm in diameter, 5.1 cm high. Modifications of this large access magnetometer can also be used to measure the horizontal component of remanence in long cores of sediment. It is also possible to arrange fluxgates to form astatic and parastatic systems, the spacing of which can be modified to allow the measurement of even larger specimens (although these are usually difficult to demagnetize partially – Section 5.4).

Cryogenic magnetometers are complex systems that basically comprise a ring that is maintained at liquid helium temperatures (Fig. 5.4(c)). The introduction of a magnetized specimen into the ring causes a persistent current to flow that is proportional to the magnetic moment along the axis of the ring. The current can then be determined by means of SQUID devices (Goree and Fuller, 1977; Petley, 1980). Measurement of the total intensity and direction of magnetization can then be made by inserting the specimen in three different mutually perpendicular orientations, or by the use of three mutually perpendicular rings. These instruments are capable of good sensitivities, 10^{-10} A m^2 moment. Small access cryogenic magnetometers are also available and have higher sensitivities for specialized studies. The main

---

Figure 5.5 Susceptibility and anisotropy meters. A variety of methods exist for measurement of the high- or low-field susceptibility of rock specimens (Collinson, 1983). One common type is the bridge (a) in which the specimen unbalanced the circuit according to its susceptibility and electrical conductivity. As most rocks, but not all, have a high electrical impedance, the imbalance is normally a direct measure of the magnetic susceptibility of the specimen inserted within the air gap. The anisotropy of susceptibility is usually measured (b) by a system in which one set of coils produces a magnetic field. If a specimen in this field is magnetically isotropic, then no field is detected by the pick-up coils. If the specimen is magnetically anisotropic, then a voltage is induced in the pick-up coils that is dependent on the anisotropy of the specimen. Computer systems can then be used to determine the shape of the mean magnetic ellipsoid from three orthogonal rotations of the specimen within the instrument.

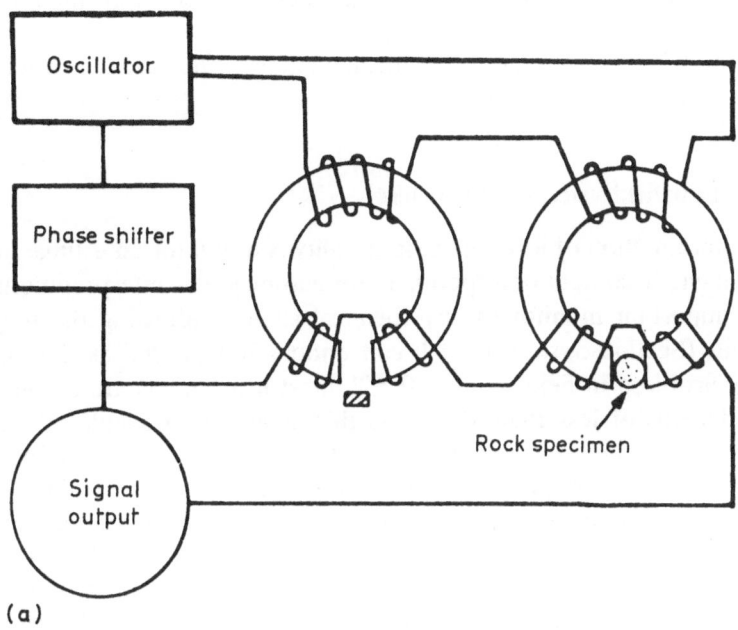

Rock specimen

(a)

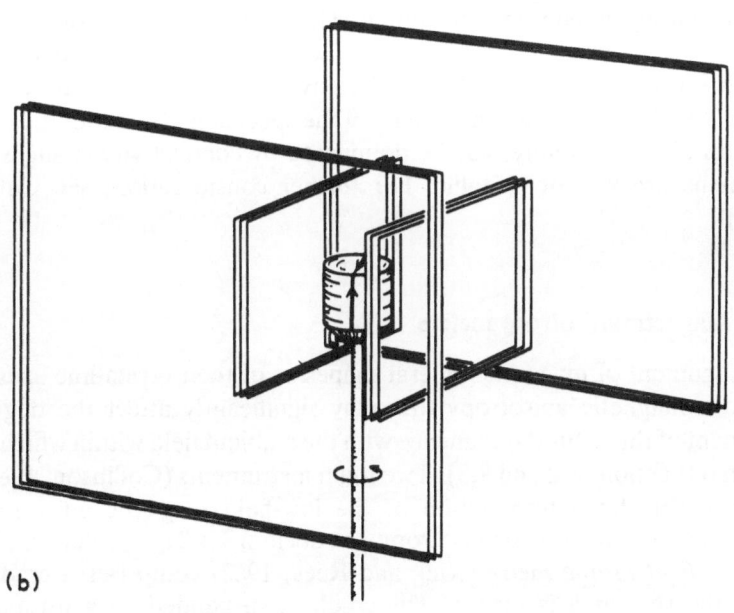

(b)

Palaeomagnetism

advantage of the cryogenic magnetometer over all others is not the somewhat greater sensitivity, but the fact that each measurement can be made in only a few seconds. This means that numerous repeat measurements are practicable, even on very weakly magnetic specimens, thereby allowing improved resolution of the remanence.

For details of these, and other magnetometers, the reader is referred to Collinson (1983).

5.3.2 Low-field susceptibility meters

The determination of low-field susceptibility is useful for an estimate of the total magnetic content of a specimen, for aeromagnetic interpretations, and as an important monitor of any chemical changes affecting the magnetic minerals (Section 5.4). Although conventionally regarded as the magnetization acquired in fields below 10^{-2} T, most instruments are designed for applied fields of less than 10^{-3} T as this avoids the possible addition of isothermal remanences (Section 2.3) to the specimen. One type of susceptibility meter is a bridge (Collinson and Molyneux, 1967) comprising two ferrite rings, with an air gap in each (Fig. 5.5(a)). Primary windings on each ring carry a current that produces an alternating magnetic field across the air gaps of up to 10^{-3} T. A ferrite plug in one gap is used initially to balance the circuit and then the introduction of a specimen into the other air gap unbalances the circuit in proportion to the magnetic susceptibility of the specimen. Another type of instrument measures the change of inductance of a coil when a specimen is inserted into it. These and other types of instrument are fast to use and portable versions are also available. The sensitivity of most of them is rather too low (10^{-3} SI) for weakly magnetic rocks and they are also affected by the electrical conductivity of the specimen. The latter means that readings of susceptibility may be dominated by conductivity changes if the specimens are wet or metallic. For further consideration, see Collinson (1983).

5.3.3 Magnetic anisotropy meters

Any alignment of magnetic mineral shapes or of their crystalline axes gives rise to a magnetic anisotropy that may significantly affect the degree of alignment of the natural remanence with the ambient field within which it was acquired (Sections 2.2 and 9.3). Two main instruments (Collinson, 1983) are available for the determination of the low-field magnetic anisotropy of specimens. (For high-field anisotropy, see Section 3.3.2 and Collinson, 1983.) The *low-field torque meter* (King and Rees, 1962) comprises a cradle, on which the specimen is placed. The cradle is suspended on a torque fibre between the poles of an electromagnet or in a coil and fields of about 10^{-2} T are applied. The suspended specimen then twists, according to its degree of magnetic anisotropy, to attempt to get its axis of maximum anisotropy aligned with the field direction. The torque difference on the specimen in the field is a measure of the susceptibility difference in the plane of measurement. The maximum, minimum and intermediate susceptibility magnitudes and their directions can thus be determined, taking 20 minutes to 1 hour per sample. A very much faster instrument is the *anisotropy delineator* which determines the magnitudes and directions of the differences in susceptibility in different directions (Collinson, 1983). (These differences can be summed with the average susceptibility, determined by a low-field meter, to obtain the same total values as for the low-field torque system.) The instrument (Fig. 5.5(b)) fundamentally comprises two mutually perpendicular coils. A current is passed through one of the coils to produce an alternating magnetic field of some 0.7 mT over the specimen. If the specimen is magnetically isotropic, then no voltage is induced in the second coil, but if the specimen is anisotropic, the exciting field is distorted and a component, proportional to the degree of anisotropy of the specimen, is picked up by the second coil. The pick-up coil is coupled to a small computer unit so that, when the specimen has been inserted in at least three mutually perpendicular axes, the maximum, minimum and intermediate susceptibility values and their directions are computed. Susceptibility differences between the axes of some 2×10^{-8} SI can be determined within some 2–3 minutes. Early versions of this instrument were very sensitive to the shape of the specimens being measured (Ellwood, 1978a; Urrutia-Fucugauchi, 1980a) but later versions are now less sensitive to this effect than low-field torque systems (Addison, 1982b).

5.4 THE STABILITY OF REMANENCE

If any primary magnetization is to be preserved, then the material, when it acquired that remanence, must contain at least a few magnetic particles that have relaxation times of the same order or longer than the age of the primary remanence. Similarly such magnetic particles must also survive to the present day. Obviously any rock that is, say, 1.6 billion years old, but only contains magnetic grains with a relaxation time of less than 1.0 billion years, cannot retain any memory of its original primary magnetization. The requirements for archaeomagnetic studies are, of course, the same, but the relaxation time of some of the magnetic components need only be of the order of a few thousand years. Fundamental to all palaeomagnetic and archaeomagnetic research is therefore the determination of the relaxation time spectrum of samples, usually by determining their present range of blocking temperatures or coercivity (Section 2.3) using partial thermal or alternating magnetic field demagnetization procedures.

5.4.1 Thermal demagnetization

As a specimen is heated, the relaxation time of all its contained magnetic particles is reduced exponentially. This means that the magnetization of grains with low relaxation times will, when heated, cause them to behave superparamagnetically (Section 2.3). On cooling in zero external magnetic field, such magnetizations will thus tend to lie within the easy axes of the magnetic particles. If these 'easy' axes are randomly oriented, i.e. there is no crystallographic or shape alignment of the magnetic particles, then the magnetization of all such particles will be randomly oriented and therefore will not contribute to the observed remanence of the specimen. However, grains with longer relaxation time will still have relaxation times greater than the duration of the heating–cooling cycle, say 20–30 minutes, and will thus not relax into their 'easy' directions but will retain their original magnetic directions. The blocking temperature spectrum of any specimen can therefore be determined by subjecting it to repeated cycles of heating, cooling and measurement, with the maximum temperature increased for each cycle (Irving *et al.*, 1961b). To monitor the chemical stability of the magnetic minerals during such incremental thermal demagnetization, it is essential to measure the low-field susceptibility prior to each reheating as any change in the susceptibility will indicate the destruction or creation of new magnetic minerals.

During cooling, it is obviously vital that there are no magnetic fields operating over the specimens. The prime concern in such thermal demagnetizers (Fig. 5.6(a)) is thus the control of the external magnetic field, although clearly good temperature control is also important. Different sized furnaces are available (Collinson, 1983) some capable of simultaneously heating up to 50 specimens at a time, although 8-specimen-sized furnaces are more common. In some furnaces it is possible to heat and cool in an inert atmosphere or a vacuum and so reduce oxidation of the specimens during heating. (The composition of the atmosphere often makes little difference to the total degree of oxidation as the specimens themselves usually contain gases and vapours that may cause changes in the minerals.) It is chemical changes that often inhibit the use of thermal demagnetization procedures for, in addition to oxidation, many common minerals, in sedimentary rocks in particular, tend to decompose during heating (Section 3.3.6) with the creation of new magnetic minerals. When such minerals form in extremely close contact with already magnetized minerals, they may acquire a magnetization parallel or antiparallel to the magnetization of the mineral they are replacing, depending on whether they are sufficiently close for super-exchange coupling to take place (Section 2.2) or whether they simply become magnetized in the field of the pre-existing magnetic particle. Although it is thus possible for new magnetic minerals to enhance the pre-existing magnetization, it is quite clear that the magnetic properties, after such chemical changes, should be viewed

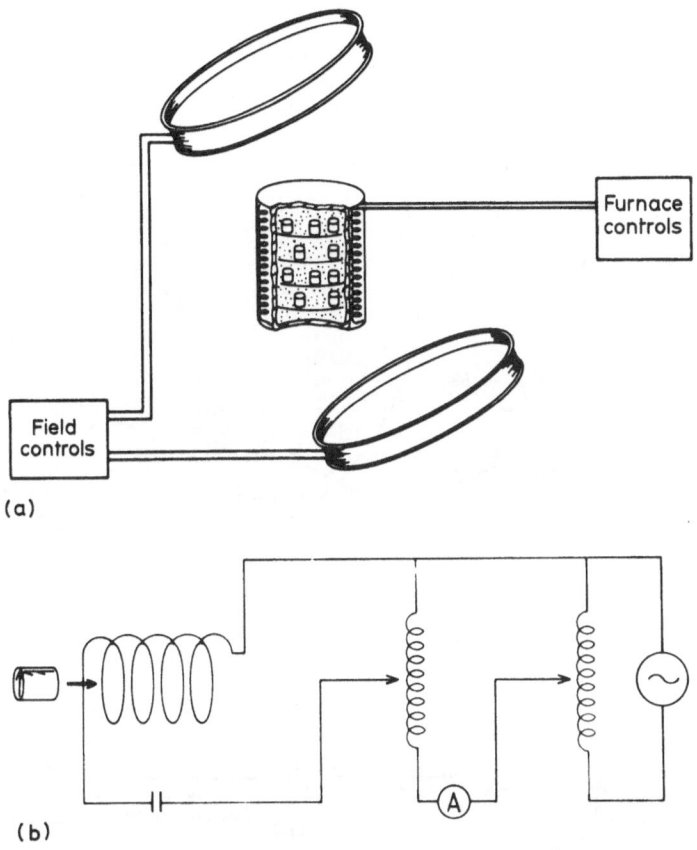

(a)

(b)

Figure 5.6 Demagnetization apparatus. In order to demagnetize specimens, two main systems are used, thermal heating or alternating magnetic fields, both applied in the absence of any direct magnetic field. (a) In the former, the specimens are heated in a furnace, with zero magnetic field being maintained by Helmholtz coils, or their equivalent, or by Mumetal shielding. (b) Alternating magnetic fields are applied by a tuned circuit to give very pure frequency of alternating field along the axis of a coil. The geomagnetic field is again removed by the use of a coil system or Mumetal shielding. (See also Collinson, 1983.)

with extreme caution and should often be discounted from further consideration. (Two examples of thermal demagnetization behaviour are given in Fig. 5.7(a).)

5.4.2 Alternating magnetic field demagnetization

This technique operates in a similar way to thermal demagnetization but dissects the remanence according to the coercivity spectra of the magnetic particles in the specimen (Section 2.3.4). A weak alternating magnetic field is

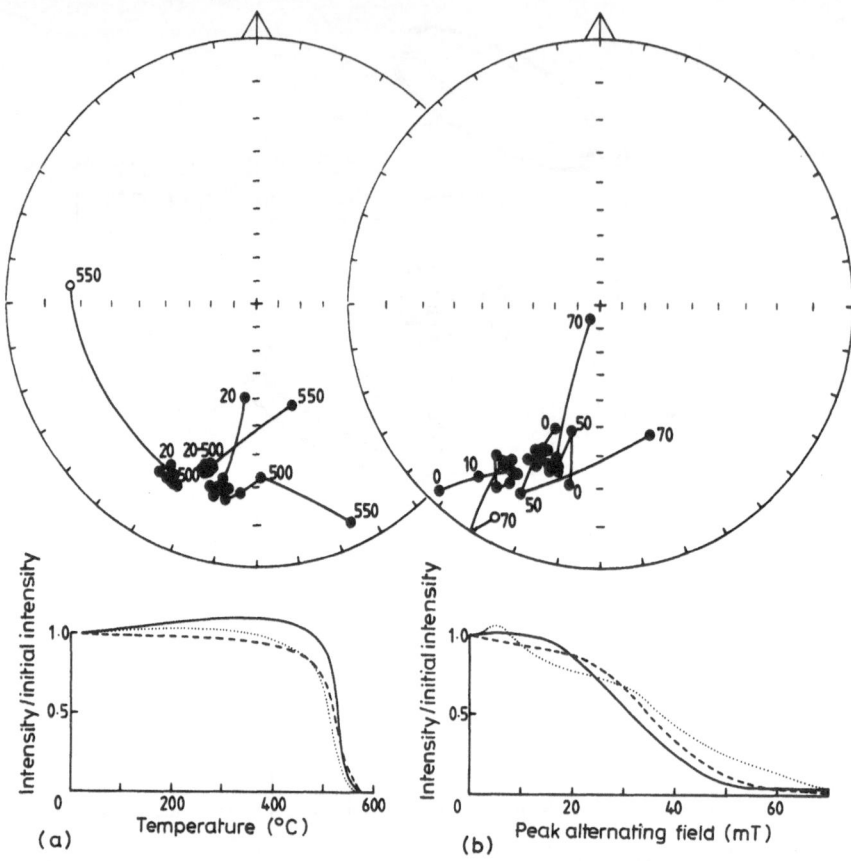

Figure 5.7 Examples of demagnetization behaviour of rock specimens. These examples are based on igneous rocks, some 290 million years old, some of which were thermally demagnetized (a) and others were demagnetized using alternating magnetic fields (b). These samples showed essentially identical stable directions after the initial movement in directions, and became scattered in random directions when their blocking temperatures or coercivities were exceeded.

applied to the specimen, causing the magnetization of grains of low coercivity to follow the applied field. As this is reduced, the magnetization of these particles is left in random positions, i.e. in the direction of the 'easy' axes. As many thousands of grains will be affected, even over a short field range, the magnetization of such particles becomes effectively randomized if the 'easy' axes are randomly oriented, leaving only the remanence of the particles with coercivities greater than that of the applied field (As and Zijderveld, 1958). Successive increments in the applied field, with the remanence of the specimen measured after each increment, allow the coercivity spectra to be analysed very finely up to the maximum applied field of the instrument (Irving *et al.*, 1961a).

Sampling, measurement and procedures

The demagnetizing unit (Fig. 5.6(b)) comprises a coil through which is passed a very pure alternating current to create an alternating magnetic field along the axis of the coil. Specimens are usually rotated within the centre of the coil so that all directions within the specimen are eventually rotated into alignment with the peak field along the coil axis. The current is then gradually reduced while the specimen continues to rotate. Most commercial instruments are designed to reach 100 mT, while individual demagnetizers have been constructed to provide fields up to 600 mT (Collinson, 1982). The rotation system is usually two (Creer, 1959) or three axis (Doell and Cox, 1967b), resulting in either a 95% or 98% presentation of the specimen's directions to the applied field (Hutchings, 1967); in some systems the specimen is not rotated, but inserted into the demagnetizer in three mutually perpendicular positions. It is essential that the geomagnetic field is cancelled over the specimen and that the alternating field has a very pure wave form as otherwise the specimen may acquire an anhysteretic magnetization (Edwards, 1980a,b) that may dominate the remaining natural remanence. To some extent, the process of tumbling the specimen in a two- or three-axis rotator tends to reduce this effect. However, even with such rotations the procedure can give rise to rotational and gyromagnetic magnetizations in certain rock types (Section 2.3.4). The alternating magnetic field demagnetization system therefore tends to become increasingly 'noisy' at fields over some 50 mT and, so far, it is not possible completely to determine the coercivity of specimens in which the remanence is carried by fine-grained haematite as these can have high coercivities (Dunlop and Stirling, 1977; Dankers, 1981), often in excess of 2 T (Section 3.2.3). It is, however, possible finely to dissect the coercivity spectrum of magnetite-bearing specimens and this does not risk the chemical creation or destruction of magnetic minerals – as may happen in thermal demagnetization. In general, therefore, the thermal demagnetization procedure is more generally applied to haematite-bearing materials, such as many sedimentary rocks, and alternating magnetic fields methods are applied to magnetite-bearing materials, such as most igneous rocks. (Examples of alternating magnetic field demagnetization are illustrated in Figs. 5.7 and 5.8.)

5.4.3 Direct current demagnetization

Russian workers frequently use a direct field partially to demagnetize a specimen (Khramov, 1958; Khramov and Andreyeva, 1964). The specimen is oriented so that its direction of remanence is antiparallel to the applied steady field. The specimen is then re-measured, re-oriented (if the remanence has changed direction) and a slightly higher field applied. Stability is indicated when the observed directions of remanence do not change significantly over a range of applied fields, indicating a single component whose coercivity equals or exceeds the fields applied. This system is not widely used in other areas

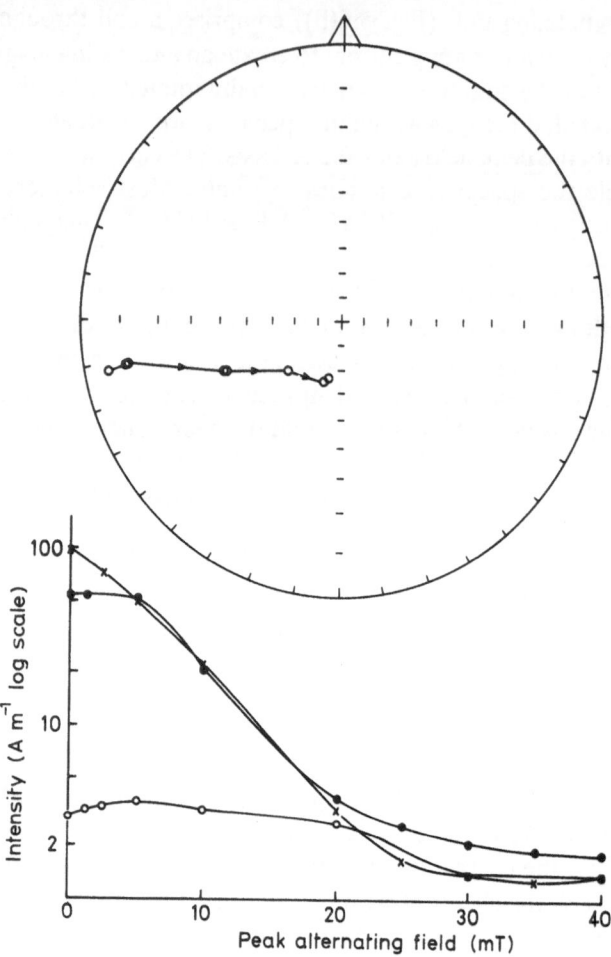

Figure 5.8 Demagnetization of rocks affected by lightning. The magnetization associated with lightning strikes can be extremely complex and overprint all previous magnetizations. It is thus better to avoid such rocks whenever possible, but the predominant effect of lightning is comparable with that of isothermal remanence and tends to be preferentially removed during the initial stages of partial demagnetization. (After Cox, 1961.)

despite its simplicity (other than the difficulty of re-orienting after each step) because of a general distrust of techniques that actually add a remanence to the specimens – such new magnetizations could readily obscure a small stable natural remanence, giving apparent instability even though the natural remanence may, in fact, be highly stable.

5.4.4 Other stability indicators

If the directions of remanence from a collection of rocks of similar ages, but of different compositions and from a wide area, are similar to each other, but different from the geomagnetic field direction in the area, then such *consistency* is itself evidence of stability. (Consistency in the present geomagnetic field direction will normally indicate the dominance of viscous remanences acquired during the last few hundred thousand years, although there have been rare times when the geomagnetic field has had a similar orientation to today.) Where sites have been sampled with differing stabilities, it is often possible to see a '*stringing*' of directions in which the less stable sites are grouped around the present geomagnetic field direction while those of greater stability are displaced nearer to the more stable direction (Fig. 5.9). The presence of such 'stringing' therefore indicates that a stable component is present, to varying degrees, although not necessarily defined (see also Section 5.5). Comparison of directions before and after *storage* for a few months or years can also be informative as directions and intensities of magnetization that change significantly over such short periods necessarily indicate the presence of many magnetic particles with short relaxation times. However, such specimens may also contain a detectable component of much higher stability that can be isolated by thermal or alternating field partial demagnetization. Finally, if *the age of the remanence* can be determined (Section 5.5), then this is, of course, evidence for its stability!

5.5 THE AGE OF THE REMANENCE

While magnetic stability is a necessary prerequisite if the original, primary magnetization is to be retrieved from a sample, secondary magnetizations can readily arise that may have a higher stability than the primary remanence. In particular, any chemical changes that result in the creation of new haematite crystals are likely to be associated with a chemical remanence that may have an extremely high coercivity and high blocking temperature (Section 2.3.4). If the primary remanence is carried by magnetite, of lower coercivity and lower blocking temperature, the secondary component will then have the highest stability. It is therefore essential to establish, if possible, the actual age of the different components of remanence.

5.5.1 Viscous remanence

All materials used in palaeomagnetic and archaeomagnetic research have been lying in an ambient magnetic field, the geomagnetic field, since their formation. They will thus all have varying degrees of viscous magnetizations, depending on the time they have been lying in the field, the strength and constancy of the field, and the relaxation times of the magnetic particles in

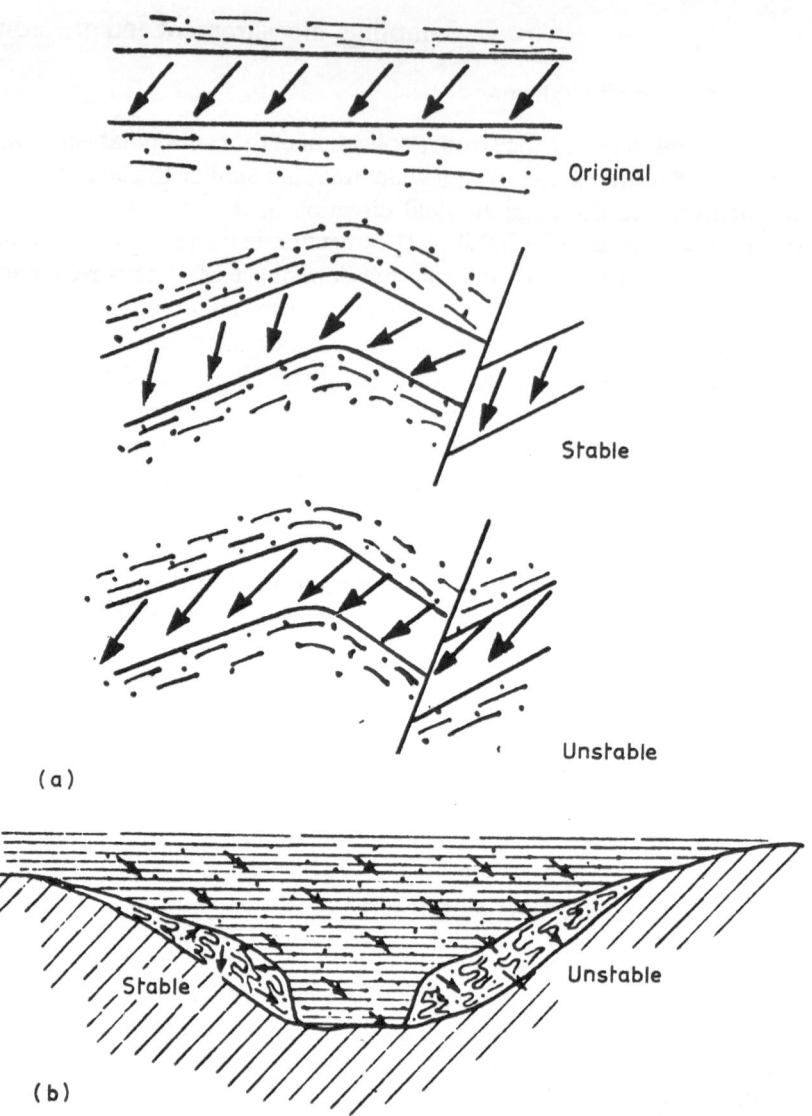

Original

Stable

Unstable

(a)

Stable

Unstable

(b)

Figure 5.9 Fold tests. Where the rock strata, or other structures, have been simply tilted, then standard fold tests can be applied to determine the age of the magnetization relative to the age of the folding by testing the scatter of directions before and after correction for the folding (a, b). Magnetization which was acquired before folding, tilting or slumping, will have retained its original direction and will thus have the same angle to the immediately adjacent bedding plane. Correction for tectonic tilt (Fig. 5.3) will then cause their directions to be less scattered than before correction, i.e. more consistent before correction for the tilting than afterwards. In slumps (b), it may not be possible to define the original bedding so that the directions tend to be scattered if they have a pre-slump origin, while the slump shows consistent directions if the magnet-

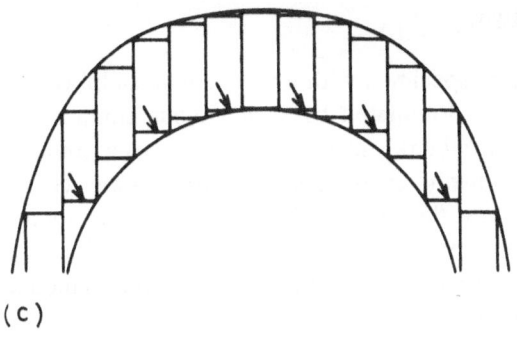

(c)

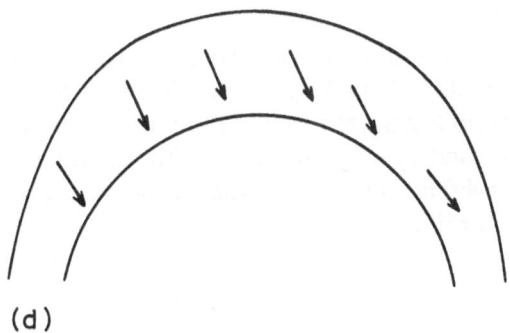

(d)

ization was acquired after the slumping had taken place. (The same considerations apply to conglomerates – the individual clasts would be expected to have random directions relative to each other if the magnetization predates the conglomerate formation.) While these forms of test are ideally suitable for tilted blocks, folds are more complex, requiring correction for any axial plunge (Fig. 5.3), and also for an intrastratal deformation. Folds may form, for example, by the dislocation along numerous slip (cleavage) planes (c) and it may be difficult, or impossible, to distinguish the fact that each individual unit has to be treated separately. In this example the scatter (Section 6.3) is very low relative to the true bedding ($k = 6687$, $\alpha_{95} = 0.8°$), but if the edges of the folded strata are taken as bedding, this correction results in a very large scatter ($k = 4$, $\alpha_{95} = 41°$). Folding, in fact, almost invariably includes some degree of intrastratal deformation. Folds normally show thicker beds on their crests than on their shoulders (d) which means that there is likely to have been plastic deformation within each bed. In this instance, the statistics before and after correction for the bedding planes are $k = 116$, $\alpha_{95} = 6°$ and $k = 3$, $\alpha_{95} = 48°$ respectively, but the mean direction has also been shallowed by some 2° as a result of the plastic flow. The location of the samples within the fold can, however, cause either shallowing or steepening of the mean vector.

97

them. Clearly any specimen with low magnetic stability will have a natural remanence that is dominated by such viscous components, but it is also possible to measure the rate of acquisition of such components by holding the specimen in a known magnetic field for various times (Heller and Markert, 1973), and evaluate the rate at which such components are acquired (Fig. 5.9). Although the extrapolations from such determinations are enormous, they can be a clear indication of the total viscous remanence that is likely to have been acquired.

5.5.2 Consistency

As the process of acquisition of remanence is radically different in igneous and sedimentary rocks, it is unlikely that the directions in these two rock types would be identical unless the remanence observed was primary or completely overprinted by secondary magnetizations. Although the latter can rarely be excluded as a possibility, any such consistency can usually be taken as indicative of a probable primary remanence having been isolated. In the case of baked contacts, where an igneous intrusion has baked sediments, the mineralogy of the two rock types is radically different and directional agreement is most unlikely unless the remanence was acquired simultaneously during the heating episode (Everitt and Clegg, 1962).

An example of a 'negative' consistency test being positive is provided by studies of conglomerates. If the directions of remanence of clasts in a conglomerate are consistent with each other, then the rock type forming the clast is clearly magnetically unstable as such directions would be normally randomized during their deposition. Directions that are stable, but randomly directed between different clasts, clearly indicate that the minerals in the clast are capable of retaining their magnetization from the time prior to deposition, and thus the rocks from which they were derived are likely to have remanences that pre-date the erosion and formation of the conglomerate (Graham, 1949).

5.5.3 Folds, tilts and great circles

Although consistency and the amount of viscous magnetization can be useful indicators of the age of remanence, the only real test is the tilt–fold test (Graham, 1949). In an area where rocks of similar age have been tilted in different directions, or folded, then it is possible to collect sites in which the bedding planes are oriented in different directions (Figs 5.10 and 5.11). In such sites, any magnetizations required after the folding or tilting will be mutually consistent, but components of remanence acquired before the tilting or folding will have been uniform before the tectonic disturbance. Analyses (McElhinny, 1964; McFadden and Jones, 1981) of the grouping of directions before and after correction for the tectonic tilt (Section 5.2.4) will therefore

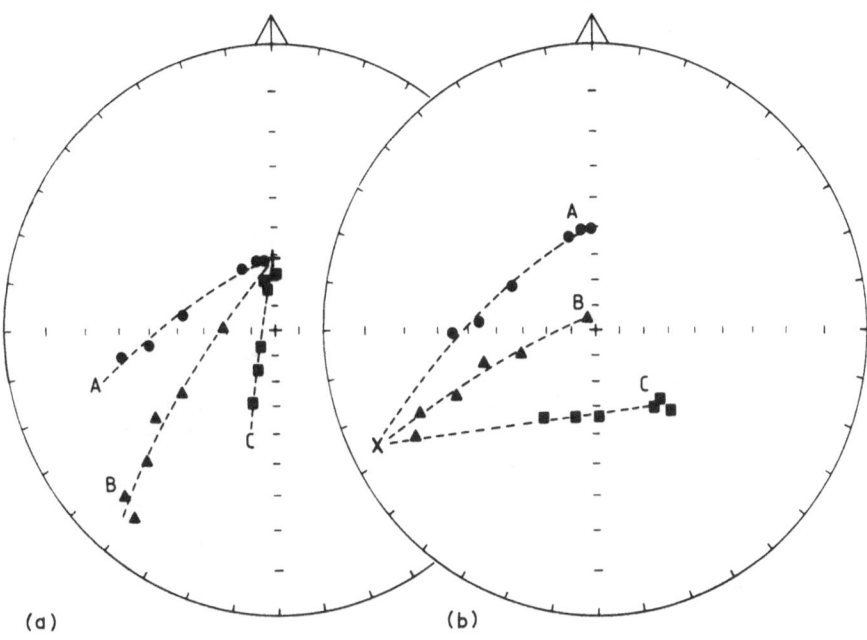

Figure 5.10 'Strung' distributions, before and after tectonic corrections. If two components of remanence are present, but in different ratios in different specimens, then the directions form 'strung' distribution between the two directions of the components. These are also termed 'great circle' distributions as they fall on great circles between the two component directions. In these examples, one of the components is a viscous remanence acquired in the geomagnetic field and the other component was acquired before the beds were tilted. Before correction (a), the directions for different sites radiate away from the present Earth's magnetic field direction, but after correction for tilt (b) they radiate away from the pre-tilting direction. In these circumstances it is possible to determine the direction of the pre-tilt component from where such great circle distributions intersect.

show a closer grouping after tectonic correction if the remanence was acquired before tectonic movements, and a wider scatter if it was acquired after folding (Figs 5.10, 5.11). In cases where the tectonic correction is small, or the scatters are large both before and after correction, the precision of the different scatters needs to be examined statistically (Section 6.3). The test assumes that each side of the fold, or each tilt block, has behaved as an individually rigid unit, i.e. there was no intrastratal deformation (Ramsay, 1967; Oertel and Ernst, 1978 – Section 5.2.4). This test does not, of course, conclusively prove that the remanence is primary, but only that it was acquired before the folding. Nonetheless the stability must clearly be high to survive the disturbances and it is usually reasonable to assume that it can be related to the time of formation of the rocks themselves.

A modified version of this technique can also be used when a stable

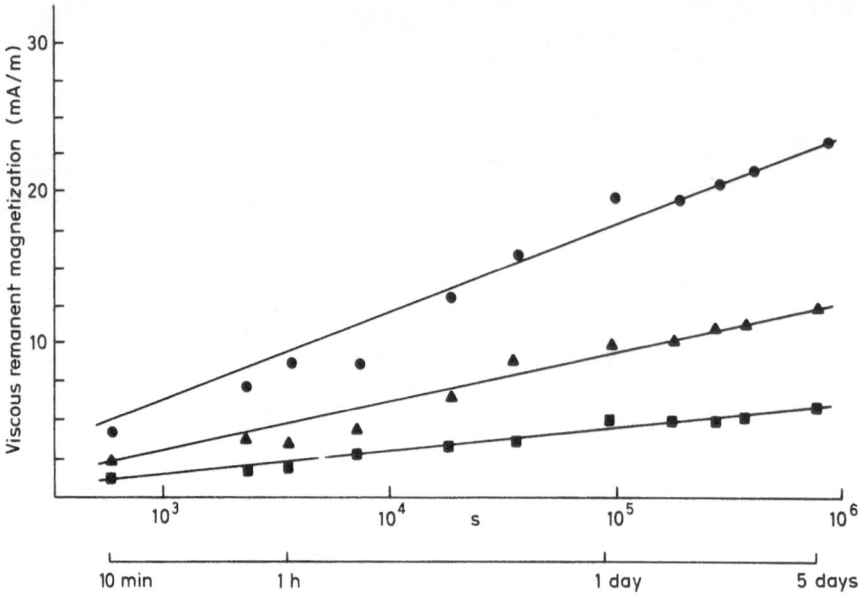

Figure 5.11 Acquisition of viscous remanence. Over long periods of time, the acquisition of viscous remanence, for any given sample, depends on log time, enabling the magnitude of viscous remanence to be estimated if the age of the sample is known, or the viscous remanence can be dated by determining its magnitude and rate of acquisition. Short-term acquisition of viscous remanence can, however, show complications (Fig. 2.8).

component of remanence is known to be present, for example from a 'strung' distribution of directions (Section 5.4.4). When such sites are plotted after correction for tectonic tilt, it is possible to extend the strung distributions along great circles to determine their intersection point (Fig. 5.10) that must correspond to the common stable end point for the collection, although this has not been, or could not be, isolated by partial demagnetization. This 'great circle of remagnetization' technique is commonly used by Russian workers (Khramov, 1958) and is a very useful method of extracting the stable component when other techniques have been ineffective. It does, however, rely on having well-defined strung distributions (Sections 6.3.4, 6.3.5) associated with sites with different tectonic orientations.

5.5.4 Identification of the carriers of remanence

Standard geological investigations will often reveal whether the magnetic minerals of a rock are of primary or secondary origin. The textures of magnetic minerals formed at high temperatures commonly differ from those that have formed by low-temperature oxidation (Haggerty, 1976a,b). Similarly, the textural relationship of the minerals in a sedimentary rock will frequently allow an assessment of the sequence of diagenetic and post-diagenetic changes that have occurred. Identification of the minerals carrying the remanence can then be related to the petrological information and realistic evaluations can then be made of the probable age of any stable magnet-izations associated with specific minerals. The fundamental problem with such investigations is that the stable remanence is usually only carried by a small fraction of the total magnetic grains present and these are likely to be single domain dimensions (Section 2.3) that are on the limits of identification by most optical techniques. Conveniently, the fact that these particles have magnetic properties also means that magnetic techniques can be used to assist in their identification (Section 3.3.6).

(a) *Chemical demagnetization*

In materials that are permeable, it is possible to pass acids through them, usually hydrochloric acid, that will preferentially remove some components of the rock (Collinson, 1967; Henry, 1979; Henshaw and Merrill, 1980). In the case of sediments, for example, carbonate cements can be readily removed, leaving behind the detrital grains. The measurement of the remanence after different amount of acid solution e.g. 35% acetic acid +25% hydroxylamine, may therefore preferentially remove the magnetization associated with the cement, thereby isolating that magnetization associated with the primary grains (Collinson, 1965b; Henshaw and Merrill, 1980). Similar 'cleaning' of igneous rocks has also been attempted in which minerals are selectively etched out of the rock by the action of the acids, thereby isolating the carriers of the remanence (Section 3.3.5) but generally with little success.

The fundamental difficulty with such techniques is the requirement for permeability as otherwise all minerals at the surface of the specimen will be affected, albeit by varying degrees. This means that in low permeability sediments, for example, the magnetization associated with surface detrital grains is lost by chemical action yet the cement inside the specimen may still be largely unaffected. Furthermore, it does not discriminate between detrital grains that have been unchanged since deposition, and those that have been replaced or have oxidized *in situ*, but are still essential detrital grains. For such differentiation, petrological studies are still essential.

(b) *Magnetic carriers and radiometric dating*

The magnetic mineralogy of a specimen can be examined in various ways (Section 3.3). Low-temperature studies (Creer, 1967) are usually more informative as they are still concerned with the natural remanence. High-temperature analyses may also be based on the natural remanence, but may also have associated chemical changes induced by the temperature rise. Saturation magnetization is very informative about the total magnetic

101

minerals present, but can only be used to infer the probable carriers of the natural remanence. Most of these studies therefore need to be carried out in close association with petrological studies.

If a stable remanence had been identified in an igneous rock then the radio-metric age of the rock itself may be of use in assessing the age of remanence if K/Ar methods have been used. The potassium–argon radiometric method, as with all other radiometric methods, assumes that the rocks examined have behaved as a closed chemical system since their formation. In the case of argon, any subsequent heating to temperatures in excess of some 300°C usually affects the argon retentivity (Dodson, 1973). Rocks from which reliable K/Ar ages have been determined are thus unlikely to have been subjected to temperatures in excess of this figure (or to have been held for a very long time at somewhat lower temperatures). Such rocks are therefore unlikely to have acquired a high-temperature viscous component of rema-nence. Chemical changes cannot, of course, be excluded from such rocks as they may well have affected the magnetic minerals, but not those containing K/Ar. It is also difficult to assess exactly what are 'reliable' K/Ar ages! Nonetheless, rocks with such ages are likely to have stable remanences associated with their formation, if subsequent chemical changes can be excluded.

5.6 PALAEOINTENSITIES

Although most palaeomagnetic and archaeomagnetic studies have been con-cerned with directional properties of the geomagnetic field at different times and places, the previous intensity of the geomagnetic field is of importance in many geophysical studies (Chapter 8) and also for archaeomagnetic dating (Section 7.2). In the absence of any secondary components, it is possible to reproduce the field conditions in which the remanence was acquired, and hence determine the strength of the ancient field (F_{anc}). In the case of igneous rocks, or heated archaeological materials, it can then be assumed that their observed intensity of natural remanence (M_{nrm}) is entirely of thermal origin, in which case the following relationship (Thellier, 1937a; Thellier and Thellier, 1942) must hold:

$$\frac{M_{nrm}}{F_{anc}} = \frac{M_{trm}}{F_{lab}}$$

Similarly for deposited materials:

$$\frac{M_{nrm}}{F_{anc}} = \frac{M_{drm}}{F_{lab}}$$

Unfortunately it is virtually impossible to duplicate the natural conditions of deposition under laboratory conditions and also even unconsolidated sedi-ments often carry components of magnetization acquired post-depositionally

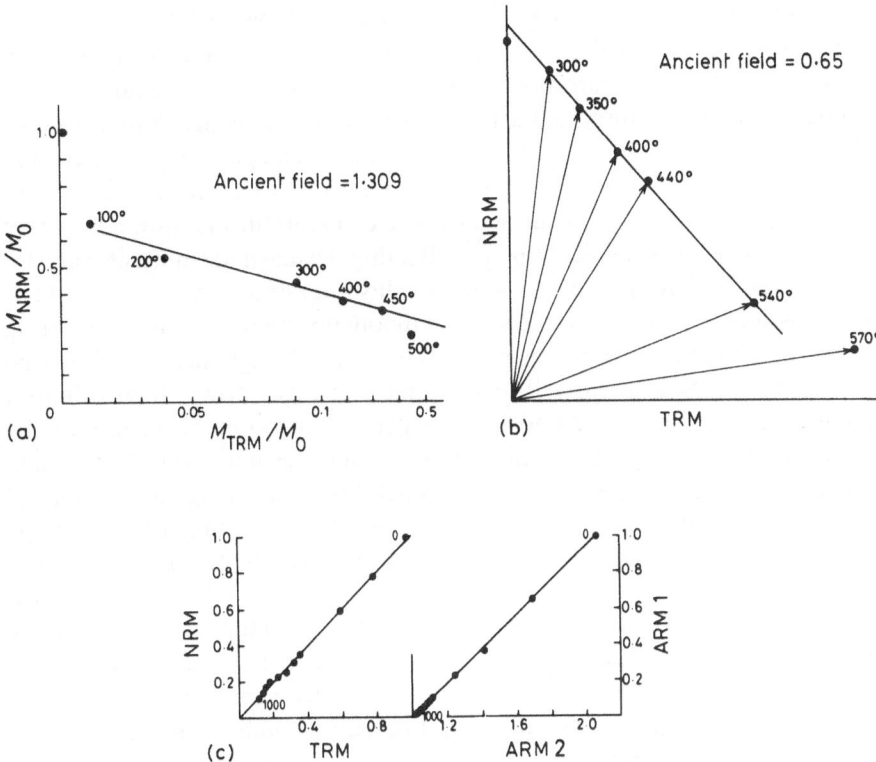

Figure 5.12 Palaeointensity determination. (a) The plot of the normalized intensity of natural remanence against that of a remanence acquired in a known field allows the determination of the ancient magnetic field from the slope of the line. The error in the determination can also be estimated from the goodness of fit (regression). (b) Similarly, a remanence can be applied at right-angles to the observed stable direction of remanence, with the slope of the line on this Ari diagram giving the strength of the ancient field. (c) For materials in which chemical changes may be induced by heating, it is possible to make similar comparisons, but using anhysteretic remanences. In this example, the results are essentially identical as the lines on both graphs are parallel.

(Section 4.3). In practice, therefore, virtually all palaeointensity determinations are based on studies of heated materials.

As even archaeological materials can be expected to contain some secondary magnetic components, it is necessary to ensure that comparisons of the intensity of natural remanence and laboratory-induced thermal remanence in a known field are being made of identical components. The standard procedure (Thellier and Thellier, 1942) is to subject the natural remanence to incremental thermal demagnetization by successive heating and cooling in zero magnetic fields (Section 5.4). The specimen can then be given a thermal remanence by heating above its Curie point and cooling in a known field. This laboratory-induced thermal remanence can then be incrementally demagnetized in the same procedure as in the treatment of the natural remanence. The intensities of the natural and artificial thermal components should then lie parallel to each other when plotted against the demagnetizing temperature (Fig. 5.12(a)). Generally such profiles are not exactly parallel and the lack of parallelity can be used as an assessment of errors in measurement or of chemical changes. Chemical changes affecting the magnetic minerals can also be checked by determination of susceptibility after each heating step and it is also possible to ensure that the same components are present by monitoring the actual directions of both the natural and thermal magnetizations (Urrutia-Fucugauchi, 1980). Clearly if any chemical changes occur, then different components are being analysed and no palaeointensity determinations can be made. Unfortunately, the method outlined involves numerous heating and cooling cycles and therefore a high probability of inducing some chemical changes (Grommé et al., 1979). In order to reduce the number of heating–cooling cycles, Kono (1974) showed that identical results could be obtained if only stable natural remanences were used and each heating step was followed by cooling in a known field perpendicular to the direction of stable remanence. Under this system, the change in intensity of the natural component during successive heatings can be examined simultaneously with the magnetization acquired thermally as this corresponds to the increase in magnetization perpendicular to the natural remanence (Fig. 5.12(b)). When the natural and induced components are plotted against each other, the slope of the line joining successive points corresponds to the ratio F_{anc}/F_{lab}. Clearly any chemical changes again prevent palaeointensity determinations and it is *essential* that the analysis be only undertaken on material of high stability, possibly as evaluated by alternating magnetic field demagnetization prior to thermal analysis. With care, however, this method works as well as the more standard methods (Kono and Ueno, 1977) and has the major advantage of halving the number of heating–cooling cycles.

As in all palaeomagnetic and archaeomagnetic studies, it is important that the materials used are magnetically isotropic (Coe, 1974; Rogers et al., 1979; Aitken et al., 1981), but an additional problem with some archaeological materials can be that the shapes of the objects themselves may affect the intensity and orientation of the geomagnetic field in which they were magnetized and this shape effect will clearly be different for specimens taken from the object for laboratory examination. In both geological and archaeological materials, the rate of laboratory cooling is likely to be much faster than occurred naturally. The natural remanence therefore includes high-temperature viscous components that are not present in the laboratory-cooled samples. These differences in cooling rate can result in the laboratory-calculated ancient intensity being up to 10 or 20% in archaeological materials and probably even greater in geological conditions (Weaver, 1970; Tanguy,

1975; Halgedahl *et al.*, 1980; Walton, 1980; Dodson and McClelland-Brown, 1980), although it is possible to estimate a cooling correction in some instances (Fox and Aitken, 1980). In geological contexts, this cooling rate difference may be extreme, possibly several million years to cool for deeply buried rocks, but may still be small in relation to the effects of chemical changes that have occurred either since the rocks were originally magnetized or during laboratory heating (Coe and Grommé, 1973). Such changes may arise simply because the laboratory furnaces cannot reproduce the oxidation conditions in which the materials were originally cooled. Basalts, for example, are often formed in strongly reducing conditions with consequent changes in the composition of their magnetic minerals in other environments (Kono and Tanaka, 1977) and even the conditions within a kiln are not adequately duplicated in laboratory furnaces. Such chemical changes can be monitored by changes in susceptibility or coercivity (Urrutia-Fucugauchi, 1980), but their occurrence usually inhibits palaeointensity determinations. Such a situation can be particularly difficult with extraterrestrial materials (Section 9.8) in which the iron components readily oxidize (Hoffman *et al.*, 1979; Collinson, 1979), even in an inert atmosphere or *in vacuo*, reflecting the presence of atmospheric contamination well within the specimens.

Laboratory-induced chemical changes can be avoided by the use of alternating magnetic field demagnetization (Shaw, 1974; Banerjee and Mellema, 1974) in which system the natural remanence is incrementally demagnetized first. The specimen is then given an anhysteretic magnetization by demagnetizing in an alternating field while in the presence of a known direct field. This anhysteretic magnetization is then incrementally demagnetized (Fig. 5.12(b)). As the characteristics of thermal and anhysteretic magnetizations are similar, comparison of the demagnetization curves for the original natural remanence and that of the laboratory-induced remanence allows an evaluation of the original field intensity that is free of all chemical changes caused by laboratory experimentation. While good results appear to be achievable by this method (Section 7.2), it is rarely possible completely to demagnetize any specimen by alternating magnetic fields and the alternating magnetic field demagnetization procedure is inherently noisier than thermal methods (Section 5.4). Additional components can therefore be added to the original remanence that could mask or modify the original components of remanence. Spurious palaeointensities may also be obtained, by either thermal or anhysteretic techniques, from materials containing grains that have exsolved (Dunlop, 1969). This possibility may be identified thermally, but would not be evident from the anhysteretic behaviour. Thus this method must be treated with caution, but it has the major advantage that no chemical changes are induced during the experimentation.

One important and intriguing exception to the generalization about palaeointensities being determined largely using fired materials is the use of sundried mud bricks (Games, 1977). These bricks appear to have acquired a

105

remanence when they were thrown into the brick mould. The brick materials can be reconstituted into a wet slurry that can then be stirred or thrown in a known magnetic field. The component of remanence acquired in this way can then be compared directly with the original intensity of remanence to evaluate the previous field intensity. Clearly this system requires a good simulation of the original conditions, particularly of the wetness of the slurry and the possibility that some of the mineralogy may have changed during and subsequent to their initial drying out. Nonetheless, with reservations, this technique gives surprisingly repeatable results for bricks of the same age and the observed variations in the estimated strength of the geomagnetic field for different times appear to be realistic (Games, 1980).

5.7 SUMMARY

There are ways in which most geological and archaeological materials can be oriented and sampled. With care, the orientation errors for any one sample should not exceed the width of the fiducial mark on each of the specimens, i.e. within some 2°. Such errors are comparable with the errors involved in measurement for all magnetometers as long as the intensity of remanence exceeds the noise level and sensitivity of the instruments. The error in the determination of the initial natural remanence of such materials, corrected to their present position in the field, should therefore be no more than 5° for any one specimen and can normally be expected to be no more than 2–3°. The intensity of the remanence and the magnetic susceptibility should also be measured with a repeatability of some 1–2% and this reading should be within some 5% of the true value. In the geological situation, by far the largest source of error is likely to be in the assessment of the tectonic correction by which the observed directions of remanence are converted to their original orientation when their primary remanence was acquired. In the archaeological context, similar errors can also arise if the samples are no longer *in situ*.

The natural remanences can then be dissected, mostly by partial incremental demagnetization by heating or in alternating magnetic fields. Such demagnetization procedures will usually determine the presence of different components and allow their definition. The precision with which such components can be defined will obviously vary according to their magnitude relative to all other components and the noise level of the demagnetization procedures. It can generally be expected that such a directional definition, within any one specimen, will be within a few degrees, c. 2–3°. However, while a stable magnetic component of remanence may be capable of carrying the original primary remanence, it is still necessary to assess the likely time at which it was acquired. None of the tests for this can be considered conclusive on their own and most are negative tests in the sense that they may exclude certain components because they do not fulfil the requirements, but components that fill the requirements are not *ipso facto* necessarily primary. The major test is the fold/tilt test, which allows dating of the remanence relative to the time at which the tectonic movements took place. In most cases, a remanence that has survived tectonic disturbance is likely to have been acquired very early in the rock's history but it could still have been acquired as a result of chemical changes after the rock had originally formed. Similarly, the identification of the carriers of the remanence can often be diagnostic of the likely age of the remanence, but it is often difficult to assess exactly when the carrier was actually magnetized. The major uncertainties in most palaeomagnetic studies are therefore distinguishing whether the magnetic ages of the rocks sampled are, in fact, the same as the age of the rocks themselves. If this can be established, then the next major source of error is usually in the determination of the correction required for tectonic disturbances since the magnetization was acquired.

Chapter Six

Statistical and mathematical analyses

6.1 INTRODUCTION

Statistical analyses are fundamental to all scientific disciplines, but are also the most complex aspect of any such study. This arises only in part from the apparent complexity of the formulae, and is mainly due to the requirement for an understanding of both the statistical and physical assumptions that are implicit in such analyses. For example, it is easy to determine the mean ($\bar{x}$) and standard deviation (s) of a group of n variables (x_i):

$$\bar{x} = \sum_{i=1}^{n} x_i \qquad s^2 = \frac{\sum_{1}^{n}(x_i - \bar{x})^2}{n} \qquad (6.1)$$

This calculation can be applied to any set of numbers and is objective in the sense that anyone else applying the same formulae would obtain the same result, but the answers could be essentially unusable unless the variable being studied had a normal (Gaussian) distribution, i.e. its frequency (f) is given by the formula:

$$f = \frac{1}{s\sqrt{(2\pi)}} \exp \frac{-(x_i - \bar{x})^2}{2s^2} \qquad (6.2)$$

It is not valid, therefore, to undertake this form of averaging in the case of declination, inclination, intensity and susceptibility values as these do not have this form of distribution. The mode of statistical analysis therefore depends on a knowledge, or assumed knowledge, of the distribution function of the property being studied. Conceptually, this means that there is a total distribution, with a true mean and standard deviation. By taking samples from this population, it is then possible to estimate these true values.

It is also obvious that the samples must come from the same population and that is determined by the physical distinction of the variable and the purpose of the analysis. In a study of the intensity of remanent magnetization, for example, extremely high values may be associated with samples from particular sites – sites that have been struck by lightning and therefore have been magnetized by an entirely different mechanism from that at other sites. An analysis of the average intensity of remanence (Section 6.2) would therefore include these sites if the purpose was to establish the average intensity values in the region (for interpretation of aeromagnetic anomalies, for example), but

108

such sites would have to be excluded from an analysis relating to the past strength of the geomagnetic field.

The complexity of statistical analyses therefore arises from the need for both a physical and statistical understanding of the property being examined and of the reasons for the analysis (Davis, 1973). It is thus easy to abuse statistical analyses, particularly now that it is possible to use sophisticated procedures that are available in the form of computer packages which may be completely inappropriate for the purpose intended. In the following sections, the analysis of scalars (intensity and susceptibility) will be discussed prior to an examination of vectors (Section 6.3) for which a distribution function was specifically designed at the behest of palaeomagnetic researchers (Fisher, 1953). The methods of dissection of such vectors and their evaluation for magnetic stability and so forth are discussed in Section 6.4, followed by a consideration of the different levels (specimens, samples, sites) at which analyses can be undertaken. Finally (Section 6.6), the statistical problems associated with magnetic fabric analyses are outlined.

6.2 INTENSITY OF REMANENCE AND SUSCEPTIBILITY

Both intensity and susceptibility depend ultimately on the size and number of magnetic grains and are thus influenced by the grain-size distribution within a rock. Such distributions can be complex as more than one generation of grains may be present. For example, an igneous rock may contain large grains (phenocrysts) that have a distinctly different distribution pattern from that of the fine grains that lie within their matrices. However, virtually all such natural distributions have a log normal (geometric) distribution in which a normal histogram of their frequency is highly skewed with a long 'tail' of high values, but if the frequency of the logarithmic values is graphed, then these distributions have a symmetrical Gaussian distribution (Fig. 6.1(a)) and their mean and standard deviations can be calculated using the log values as the variable. Both low-field susceptibility (Irving 1966) and intensity of remanence (Tarling, 1966) of igneous and sedimentary rocks show this distribution function (Fig. 6.1). (Where more than one distribution is present, the total distribution is still dominantly geometric.)

It is often necessary to compare intensities of magnetization that have been acquired at different latitudes, but the strength (F) of the Earth's field varies according to latitude and it is thus necessary to correct for this effect on the assumption that the geomagnetic field was always dipolar (Chapter 8). For archaeological materials, it is reasonable to assume, to a first approximation, that the geomagnetic field was still essentially an axial geocentric dipole; the normalization can thus be based on the present latitude of the sites. For geological periods, the movement of the continents relative to the average geomagnetic pole means that the normalization must incorporate the ancient palaeolatitude (λ) based on the average inclination of remanence

Palaeomagnetism

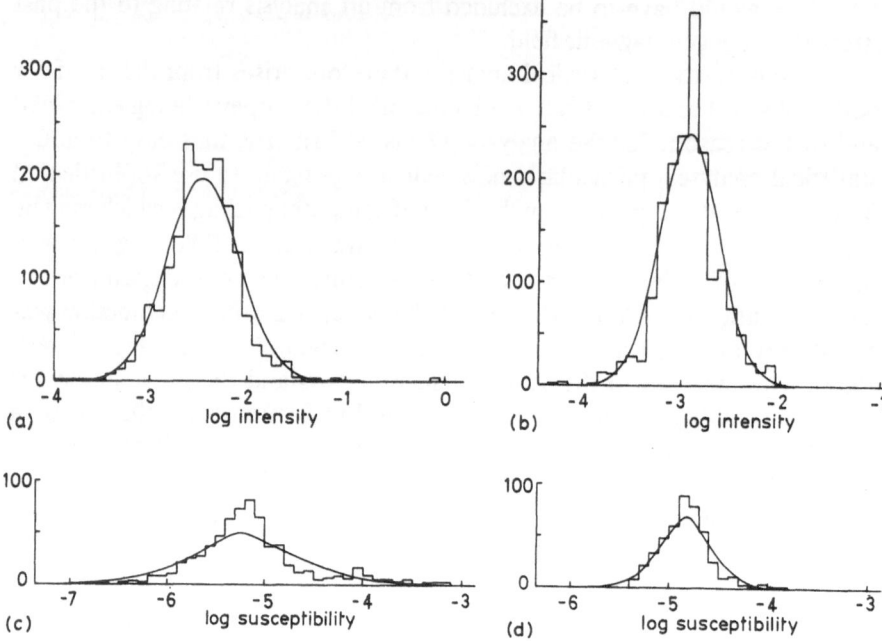

Figure 6.1 The distribution of intensity and susceptibility. If the intensities of remanence of igneous rocks (a) and sedimentary rocks (c) are plotted on a logarithmic scale, then their frequencies have a symmetrical distribution that is defined by a normal distribution, i.e. these scalar values have a log-normal (geometric) distribution. Similarly, the low-field susceptibilities of igneous (b) or sedimentary (d) have a symmetrical log-normal distribution. This shows that the mean of the logarithmic values, and associated standard deviations, etc., should be used, and not the arithmetic mean, in comparing these parameters.

$(\tan \mathrm{Inc} = \frac{1}{2}\tan\lambda)$. Conventionally, such laboratory determinations are expressed as F_{anc}/F_{lab} (ancient/laboratory) (Section 5.6), but, for comparative purposes, it is possible to calculate the equivalent intensity of the field at the equator (F_{equat}), or the geomagnetic dipole moment (M). The equatorial value is given by

$$F_{equat} = F_{anc}(1 + \sin^2\lambda)^{\frac{1}{2}} \qquad (6.3)$$

and the dipole magnetic moment is given by

$$M = F/(1 + 3\sin^2\lambda)^{\frac{1}{2}} \qquad (6.4)$$

where λ is the mean palaeolatitude for the sampling collection. Strictly speaking, these values will have a logarithmic distribution, but the number of observations is generally so low that there is little difference although the mean based on logarithmic values will always be slightly lower than the

110

arithmetic mean and their standard deviations will also differ. An additional uncertainty is whether the palaeolatitude used should be based on the average inclination for the site or for all rocks of similar age. In practice, most authors have used the site mean inclination. As few data are available at the moment, such differences in the models are not yet significant, but it is important that the palaeointensity determined for any one site necessarily incorporates secular variations in intensity that may not have been averaged during the acquisition of remanence. A variation in geomagnetic intensity of 10–20% between sites of similar but not identical age may thus be purely attributable to secular changes rather than to longer term changes in the main dipolar field.

6.3 DIRECTIONAL ANALYSES

6.3.1 Mean directions and poles

Directions of magnetization are usually expressed in polar co-ordinates (declination, inclination and total intensity) or in cartesian co-ordinates (the intensity along the axes of an orthogonal reference system, x, y, z – conventionally representing north, east and vertical – Fig. 1.2). Conversion between these representations is easy (Section 1.3.1). Most calculations are based on the cartesian system, although the results are more usually expressed in polar co-ordinates. (Palaeomagnetic data can, of course, be computed more quickly in a variety of ways, such as matrix products, but the procedures and methods discussed here will be based on the more conventional systems.) It is also convenient, under some circumstances, to represent the directions of magnetization in terms of pole positions, rather than in terms of the directions actually measured. Such calculations are modelled on the observed directions having been acquired in a geocentric dipolar geomagnetic field, i.e. as if the Earth itself were a dipolar magnet. Under this situation, the inclination observed is directly related to the angular distance from the magnetic pole (the magnetic co-latitude, $90 - \lambda$):

$$\tan \text{Inc} = 2\tan\lambda \qquad (6.5)$$

i.e. λ is the magnetic latitude of the sampling location. If the location of the sites (Fig. 6.2) is in normal geographic latitude (λ_s) and longitudes (ψ_s) then the latitude ($\lambda\rho$) and longitude ($\psi\rho$) of the corresponding pole are given by:

$$\lambda\rho = \sin^{-1}(\sin\lambda_s \sin\lambda + \cos\lambda_s \cos\lambda \cos D) \qquad (6.6)$$

$$\psi\rho = \psi_s + [\sin^{-1}(\cos\lambda \sin D/\cos\lambda\rho)]$$

(Standard sign conventions are used, i.e. northern latitudes and eastern longitudes are positive.)

In areas where standard palaeomagnetic analyses are not possible, such as

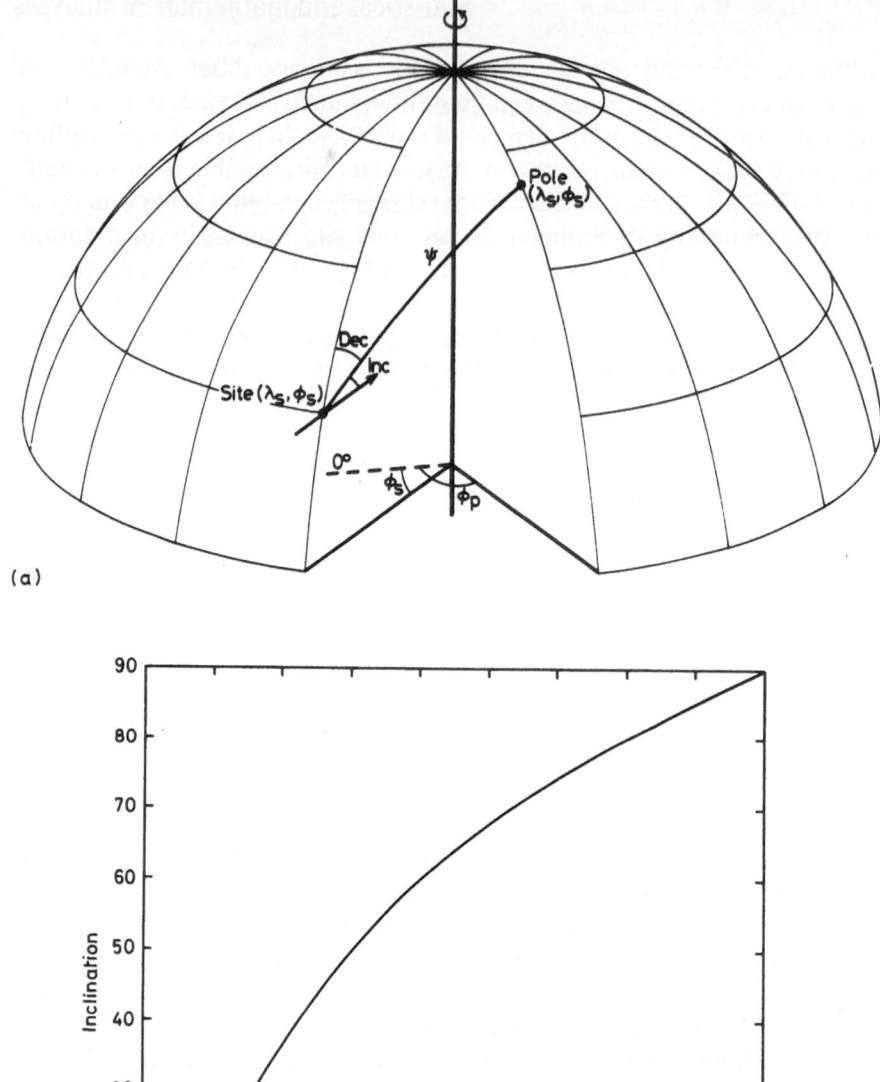

(a)

(b)

$$\tan(\text{latitude}) = \tfrac{1}{2}\tan(\text{inclination})$$

Palaeolatitude

Inclination

oceanic sea-mounts from which no samples can be obtained, it is possible to use the associated magnetic anomaly to evaluate the direction of remanent magnetization (Harrison, 1971; Harrison *et al.*, 1975; Cande, 1976; Gordon and Cox, 1980b). This direction can then be used to calculate the corresponding pole position. The observed magnetic anomaly comprises two components: one induced in the Earth's present magnetic field and one attributable to the remanence acquired by the volcanic core of the sea-mount or other source. Reasonable estimates of the susceptibility are usually possible, and dredged samples can also provide sufficient constraint for the anomaly caused by induced magnetization to be readily calculated and removed from the observed anomaly (Fig. 6.3). The direction and intensity of remanence can then be determined using standard geophysical analyses (Grossling, 1970; Plouff, 1976). Unfortunately, such solutions are not unique and homogeneous magnetization must be assumed. Similarly, it is the total natural remanence that is contributing to the anomaly and secondary components are almost certainly present. Nonetheless such estimates appear to give realistic dates for various sea-mounts in the Pacific (Kodama *et al.*, 1978; Kono and Morgan, 1978; Keating and Sager, 1980) as they are consistent with palaeolatitudes (Section 9.7.3) indicated by associated oceanic sediments (van Andel, 1974). Clearly there are large uncertainties in such estimates mostly resulting from the actual complexity of magnetization of such sea-mounts (Blakely and Christiansen, 1978), but frequently such evaluations could not have been undertaken by any other means and are therefore of major value.

It is important to stress that pole positions, however determined, can be regarded as simply another method of expressing individual directions of magnetization and are often termed virtual poles. In view of the complex nature of the geomagnetic field (Section 8.2), such individual poles will very rarely correspond to the actual geomagnetic pole position. (When sufficient data are available, it is possible that the average of such virtual poles, the palaeomagnetic pole, may correspond to the average geomagnetic pole, but such a further step is an interpretation of the observations and should not be confused with statistical analyses of virtual poles that may, or may not, lead to such an interpretation.) It is convenient in such analyses that the latitudes and

---

Figure 6.2 Calculations of palaeomagnetic poles and palaeolatitudes. (a) The declination of remanence defines the direction of the palaeomagnetic pole at the site and the inclination, assuming the field to be a geocentric dipole, can be used to determine the distance of the site from the palaeomagnetic pole, i.e. (b) the co-latitude $(\psi) = \cot^{-1} (\frac{1}{2}\tan \text{Inclination})$. The pole position can thus be determined using projections or spherical trigonometry. In the latter case, the pole position (latitude $= \lambda_p$, longitude $= \phi_p$), the declination and inclination of remanence at the site (Dec and Inc), and the site latitude and longitude (latitude $= \lambda_s$, longitude $= \phi_s$) are required. (Equation (6.6).)

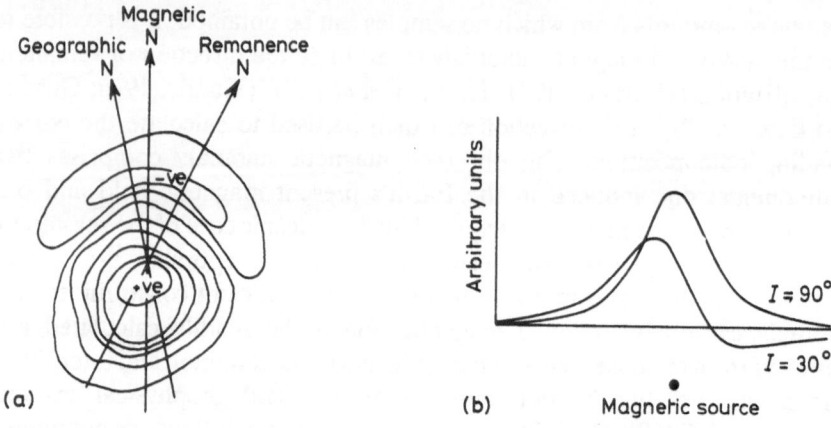

Figure 6.3 Declination and inclination of remanence from magnetic anomalies. Magnetic anomalies associated with, for example, (a) sea-mounts comprise a summation of remanence acquired when the sea-mount originally cooled, any viscous remanence it has since required, and the intensity induced in it by the present geomagnetic field. By estimation or determining the magnetic susceptibility, the component attributable to the present geomagnetic field can be estimated and removed. The shape of the remaining anomaly can then be used to determine the declination and remanence and the inclination (b), the latter causing systematic changes in the symmetry of the anomaly.

longitudes of the virtual poles are exactly analogous to inclination and declination, i.e. in subsequent considerations of directional analyses, identical analyses can be undertaken by replacing inclination by the latitude and declination by the longitude of the virtual pole. (However, polar and directional distributions are necessarily different and the statistics associated with directions are not simply transferable to polar distribution, or vice versa – Section 3.3.2.)

As each magnetic direction is a vector, i.e. it has a direction and an intensity, the average direction of a number N of individual (i) cartesian co-ordinate components can be calculated by standard summation (Fig. 6.4):

$$\mathbf{R}^2 = \sum_1^N x_i^2 + \sum_1^N y_i^2 + \sum_1^N z_i^2 \qquad (6.7)$$

In cartesian co-ordinates:

$$\bar{x} = \frac{\Sigma x_i}{\mathbf{R}} \qquad \bar{y} = \frac{\Sigma y_i}{\mathbf{R}} \qquad \bar{z} = \frac{\Sigma z_i}{\mathbf{R}}$$

and in polar co-ordinates:

$$\overline{\mathrm{Dec}} = \tan^{-1}\frac{\bar{y}}{\bar{x}} \qquad \overline{\mathrm{Inc}} = \sin^{-1}\bar{z}$$

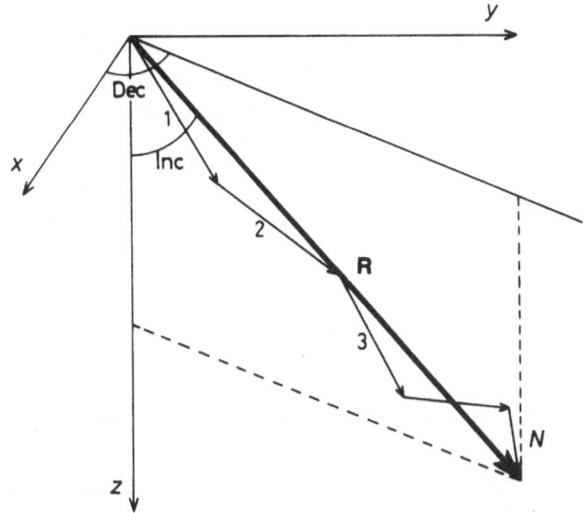

Figure 6.4 The resultant of a number of vectors. When a series of N vectors is summed, the resultant vector **R** has a value that is directly related to the degree of similarity of the individual vectors. If $N = \mathbf{R}$, then all the individual vectors have exactly the same direction. The length of the resultant vector will therefore reduce if the vectors become increasingly dissimilar to each other.

The length of the resultant vector, **R**, is fundamental to all subsequent statistical analyses of precision and scatter and it is assumed, in this analysis, that the magnetic vectors being analysed do come from the same population, i.e. they are samples from the same physical group of magnetic materials. Obviously it is not possible to test whether this assumption is valid for individual samples, but tests are available for determining either whether groups of vectors can be legitimately combined from a statistical point of view, i.e. that they come from the same population, or whether they are sufficiently distinct, i.e. that they cannot be grouped together (Section 6.3.4).

In palaeomagnetic studies, the mean directions are usually calculated giving an equal (unit) weight to each of the magnetic vectors. The reason for this is that there are few meaningful ways of weighting the individual determinations. It is now possible to measure, with essentially equal precision, magnetic vectors with magnetic moments varying from 10^{-12} to 10^{-6} A m^2 (Section 5.3). There is, however, no reason for considering that a more strongly magnetic sample has a more meaningful direction than a weakly magnetized sample, as long as the weak magnetization is approximately an order of magnitude stronger than the noise level of the magnetometer. It is clearly essential to eliminate, i.e. give zero weight, to sample magnetizations that are below the noise level of measurement and clearly less weight should be given to the directions of magnetization intensities that are close to the

115

noise level. In practice, samples falling in this range are usually subjectively evaluated and either rejected or given unit weight. A sensible procedure is to calculate two separate means, with their associated statistics, including and excluding such uncertain intensities of magnetization. Weighting procedures have not been widely used in palaeomagnetic work, although there are many instances where such weighting should be given. For example, different samples show different degrees of magnetic stability and there are a variety of methods for objectively evaluating such stability (Section 6.4), but none has gained widespread acceptance and few are linear. Any weighting system based on such non-linear stability indices could, therefore, give excessive weighting to relatively few observations and hence give a misleading mean direction and an invalid estimate of its precision.

'Geographic' corrections are mostly required in archaeological studies. If the area under investigation is small, 10–100 km², then the variation in the geomagnetic field is sufficiently small to be ignored (excluding local magnetic anomalies), but for larger regions, it is convenient to correct the observed value to what it would be at some central point, thus enabling the definition and comparison of the direction of the geomagnetic field for any one time to be made from studies over an area of several thousand km². At the lowest level, 'corrections' are made to the measured inclination for the dipolar difference (Equation (6.5)) expected between the latitude of the site and that of the central location (Aitken, 1974; Thellier, 1981). This partial correction does not affect declination and thus results in anomalous declination values at the central location. A more effective correction is to use individual site declination and inclination values to determine the location of the palaeomagnetic pole (Equation (6.6)), and then to reverse the procedure to determine the declination and inclination that this would produce at some central location (Nöel, 1980, personal communication). Correction of present geomagnetic field directions in an area of more than 600 000 km² shows differences of less than 1° in both declination and inclination at a central location. Clearly, larger errors will arise for extensions over wider areas as the correction assumes that, for any such region, the geomagnetic pole corresponds to a geocentric dipole. Nonetheless, the area for which this assumption appears adequate can extend over at least three or four states in America, or a few countries in Europe. The advantage is that well-dated observations obtained over a very wide area can then be used to better define a secular variation curve that is applicable over a wide region.

There are clearly problems in obtaining the mean directions of even simple vectors, but there are major difficulties in any attempt to define the locus of a moving vector, e.g. a time sequence. For example, studies of archaeological data and lake sediments provide information on the changing direction of the geomagnetic field at particular locations (Section 8.3). Similarly palaeomagnetic pole positions gradually change with time, so that the poles lie on a polar wandering curve (Section 9.2.3). Sometimes such 'average' polar or secular variation curves have been derived by 'running means', i.e averaging the mean position for, say, three successive means. This can be done for running means of various lengths and the mean values are plotted, along the sequence, forming a new averaged curve. Such a procedure partially averages out the errors in the individual observations, but does not take into account the actual errors, i.e. each point is given unit weight. The main problems are that palaeomagnetic and archaeomagnetic data are rarely evenly spaced in time, as required for this procedure, and that the analysis tends to reduce the amplitude of genuine 'swings' in the data, thus giving apparently lower magnitude of secular variation or polar path loops. A common alternative has been the subjective evaluation of the most reliable points and then drawing a best-fitting curve by eye. This procedure is, in fact, highly successful for relatively small quantities of data, but becomes impracticable for larger quantities.

Parker and Denham (1979) proposed that the mean paths of such time sequences of data could be best evaluated using smoothed cubic splines, but Clark and Thompson (1980) and Thompson and Clark (1981) point out that this does not allow for the different precision of individual points. They have proposed a method of obtaining both the means, and their running precision, using a least-squares fit of cubic splines. The method assumes a Fisher distribution for the data used to evaluate each point (Section 6.3.5), but could be modified for other distributions. As with any statistical analysis it is most effective for precisely determined values and operates most reliably if the data are regularly spaced in time. There is thus still a need for subjective evaluation of the data before undertaking such an averaging procedure, with the added factor that not only must the data themselves be quite well defined, but the age sequence must be known.

6.3.2 Precision

All available statistical evaluations of either the precision of a mean direction or the scatter of directions about the mean depend on two fundamental parameters – the total number of observations (N) and the length of the resultant vector ($\mathbf{R}$) obtained by the summation of the individual vectors. At the simplest level, it is possible simply to compare these two (Watson, 1956a; Watson and Irving, 1957) as the mean direction must obviously be extremely well defined if the N and $\mathbf{R}$ have similar values, while the mean direction must be poorly defined if the summation vector is much smaller than N. There is thus, for any given N value, a certain probability that vectors draw from a random population will have a summation $\mathbf{R}$. On this basis, Watson (1956a) and Irving (1964) tabulated tables of $3 \leq N \leq 200$ at 95 and 99% probability levels for $\mathbf{R}$ values (Table 6.1). For a given collection of N values, if the calculated $\mathbf{R}$ exceeds the significance point $\mathbf{R}_0$ corresponding to that N, then

Table 6.1 95% Significance points ($\mathbf{R}_0$) for N. (After Watson (1956a) and Irving (1964))

N	$\mathbf{R}_0(95)$	N	$\mathbf{R}_0(95)$
3	2.62	12	5.52
4	3.10	13	5.75
5	3.50	14	5.98
6	3.85	15	6.19
7	4.18	16	6.40
8	4.48	17	6.60
9	4.76	18	6.79
10	5.03	19	6.98
11	5.28	20	7.17

Values of $\mathbf{R}_0$ for N equal to 20 can be approximated by

$$\mathbf{R}_0(95) = (N \times 2.605)^{\frac{1}{3}} - 0.04$$

there is a greater than 95 or 99% probability that the mean direction has not arisen by chance selection from a random collection of vectors.

In response to the need for three-dimensional statistical analyses in palaeo-magnetic studies, in particular, Fisher (1953) derived a theoretical distri-bution, the Fisher distribution, for vectors when considered as points on a sphere. By analogy with the normal two-dimensional Gaussian distribution (Equation (6.1)) Fisher proposed that vector points on a sphere would have a probability density given by:

$$P = \frac{\kappa}{4\pi \sinh \kappa} \exp\left(\kappa \cos \psi\right) \tag{6.8}$$

in which ψ is the angular distance between an individual point and the true mean of the vector population and κ is the precision parameter, varying from zero if all the vector points are uniformly distributed, to infinity if they are all identical to the mean.

In the Fisher distribution the precision, κ, is for the absolute total popu-lation of vector points on a sphere. As it is never possible to measure the total population (in palaeomagnetic work this would mean measuring individually the magnetization of every individual magnetic grain of an entire rock sequence!) the actual values of the standard deviation and precision must be estimated from the samples taken from that population. Conveniently, Fisher (1953) showed that the precision parameter, κ, could be estimated simply:

$$\kappa \simeq k = \frac{N-1}{N-\mathbf{R}} \tag{6.9}$$

for the values of N greater than 7 and for κ greater than 3. (κ is analogous to the invariance in a Gaussian distribution, i.e. variance^{-1}, and the precise limits for the validity of this estimate can be questioned – McFadden, 1980a,b,c.) In analyses of unit vectors that have a Fisher distribution, values of k greater than 10 indicate that the observed mean is close to the true mean of the total population, i.e. it is a good estimate, but higher values are preferable for a more reliable estimate of the mean direction (Figs. 6.5 and 6.6). k does, however, increase towards infinity and thus differences between high k values (>100) do not indicate very great improvements in the estimation of the mean direction. (Watson (1956b), Mardia (1972) and McFadden (1980a) have shown that an unbiased estimate of κ is, in fact, given by $(N-2)/(N-\mathbf{R})$ – but this has little practical use. However, in testing the actual fit of data to a Fisher distribution – Section 6.3.5 – McFadden (1980a) proposes that the maximum likelihood estimate of κ, $N/(N-\mathbf{R})$ should be used, although the standard Fisher estimate $(N-1)/(N-\mathbf{R})$ is still suitable for other statistical evaluations.)

In view of the range in possible k values, it is generally easier to assess the reliability of the mean direction by determining the probable distance of the observed mean, based on the samples, from the true mean of the total population. This statistic, still based on a Fisher distribution, is the angular radius of the cone of confidence (α) about the observed mean:

$$\alpha_{1-P} = \cos^{-1}\left\{1 - \frac{N-\mathbf{R}}{\mathbf{R}}\left[\left(\frac{1}{P}\right)^{1/N-1} - 1\right]\right\} \tag{6.10}$$

In palaeomagnetic and archaeomagnetic work, it is conventional to use a probability of 0.05, expressed as α_{95}. This means that there is a 95% probability that the true mean direction lies within the cone of confidence around the observed mean. If $k \geq 7$ then α_{95} can be adequately calculated on the basis:

$$\alpha_{95} \simeq 140/(kN)^{\frac{1}{2}}$$

The smaller the value of α_{95}, therefore, the more reliably has the mean been estimated (Fig. 6.6). Both k and α_{95} are therefore measures of the precision with which the true mean direction (or true mean pole position) has been determined from the samples collected. Occasionally, the statistic δ, the angular standard deviation, is used:

$$\delta = \cos^{-1}(\mathbf{R}/N)$$

Although this makes no assumption about the nature of the vector distributions it is basically a two-dimensional statistic, usable as an objective criterion for accepting or rejecting the mean value of repeat measurements of the same specimen, but of little value for analyses at higher levels (Section 6.5).

In order to make some allowances for the offset of low numbers and irregular distributions, Van Alstine (1980) suggested that more reliable mean

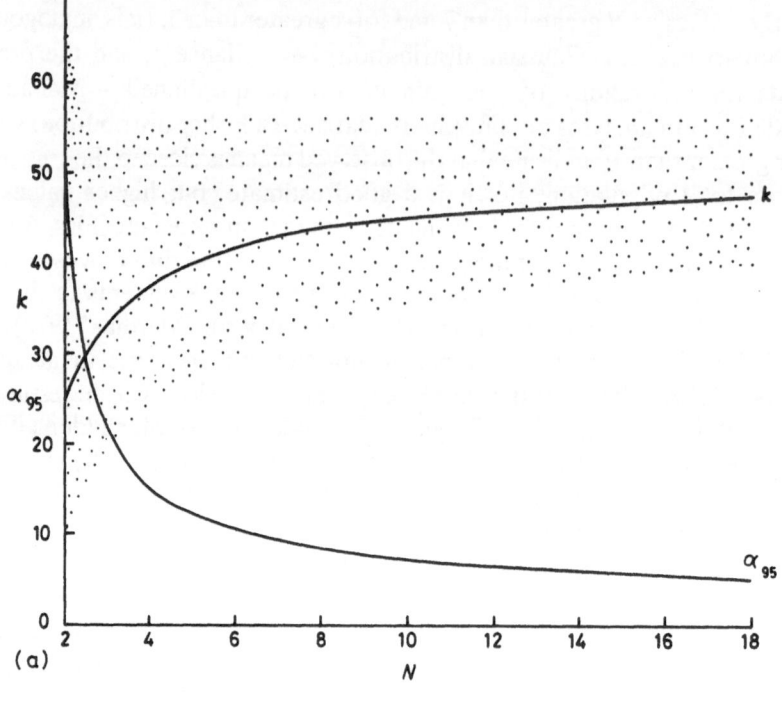

(a)

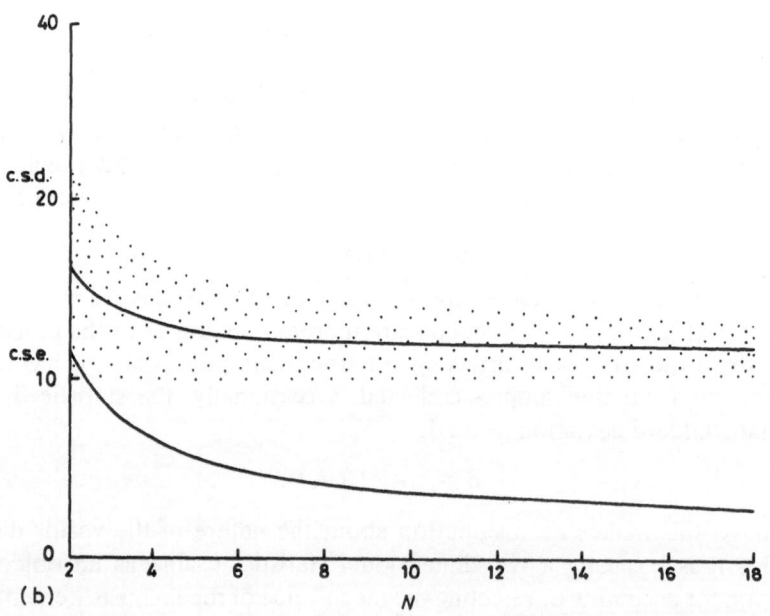

(b)

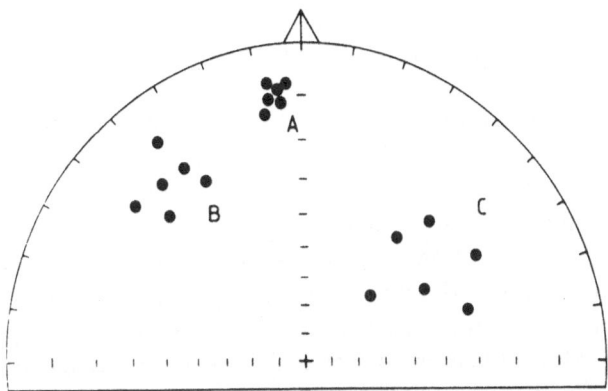

Figure 6.6 Examples of scatter and Fisher statistics. The groups of six vectors are shown here, each individual vector being given equal weight in each group. On this basis, mean directions and associated Fisher statistics are:

Group

	Dec	Inc	N	R	k	α_{95}	θ_{63}
A	353.3	10.3	6	5.99268	683	2.6	3.1
B	311.5	18.8	6	5.94277	87	7.2	8.7
C	54.2	37.6	6	5.80340	25	13.5	16.1

directions and poles could be obtained by means of a contouring system. The modal point, i.e. the point of the maximum density, is given a value of unity, and all other points are weighted according to their probability density function (p), i.e.

$$p = \exp[sk(\cos\theta - 1)] \tag{6.11}$$

where k is the Fisher precision estimate and θ is the angular distance from the modal value. The distribution could then be contoured and a revised value of α_{95} could be evaluated. Although there are some merits for this process, it is not very applicable at the low numbers commonly used in palaeomagnetism as one extra measurement could drastically modify the shape of the distri-

Figure 6.5 Fisher statistics as a function of the number of observations. (a) The measures of precision, k and α_{95}, are very poorly defined for low values of N, with k increasing with increasing N, and α_{95} continuing the decrease. In contrast, (b) the measures of scatter, circular standard deviation and circular standard error, both tend to become constant for N values above about 8. (The 50% probability limits for k and c.s.d. are shown dotted, indicating how the estimates of their value become increasingly reliable for N greater than about 7.) These relationships indicate that considerable caution is required when evaluating the meaning of statistical analyses on small numbers of observations such as are commonly encountered in palaeomagnetic and archaeomagnetic analyses of site directions or poles.

Palaeomagnetism

bution. It is, however, an objective way of determining the most likely mean position from an irregular distribution for which subjective evaluations have often been used in the past.

Under certain circumstances, information is only available on inclination. This can arise particularly when cores that have been obtained from unoriented bore-holes need to be analysed. In some instances, the relative declination can be measured relative to a fiducial mark drawn on the longer sections of core, but often such bore-cores are broken and have rotated relative to each other during the coring operation, thus preventing any relative declination determinations. Where the bore-hole is approximately vertical, the sides of the individual pieces of core can be used to define vertical and so the inclination, relative to this vertical, can be obtained. Such measurements have, on occasions, been simply averaged to give an arithmetic mean and standard deviation, but this is clearly invalid as inclination values form part of a three-dimension distribution. On the assumption that the actual distributions, for any group of samples, are Fisherian, Briden and Ward (1966) derived a mean inclination and associated Fisher statistics which involved a series of tables. This method has been developed by Kono (1980) to provide the mean inclinations (I) of a number (N) of individual inclinations (I_i), and the Fisher estimate of precision, k, by means of two simultaneous equations:

$$\sin \bar{I} \left(\coth k - \frac{1}{k} \right) = \frac{1}{N} \sum_{i=1}^{N} \sin I_i \tag{6.12}$$

$$\sin^2 \bar{I} + \frac{1 - 3\sin^2 \bar{I}}{k} \left(\coth k - \frac{1}{k} \right) = \frac{1}{N} \sum_{i=1}^{N} \sin^2 I_i \tag{6.13}$$

The radius of the cone of 95% confidence, α_{95}, is then given by:

$$\cos^{-1} \left[1 - \frac{N-1}{(k-1)N+1} \left(20^{1/N-1} - 1 \right) \right] \tag{6.14}$$

6.3.3 Scatter estimates

Although scattered vectors will, for any given number of observations (N), have less well-defined mean directions than less scattered vectors (Fig. 6.7), the estimates of the precision of the mean directions (k and α_{95}) are *not* measurements of the actual scatter of the vectors and it is often the magnitude of the scatter that is of interest in, for example, determining the magnitude of past secular variations of the geomagnetic field (Section 8.3). The Fisher distribution (Equation (6.8)), however, involves ψ, the angular distance of any individual point from the true mean. Any given population of vectors with a Fisher distribution, when plotted on a sphere, will thus have a certain percentage lying within a specific distance of the mean. The angular radius (θ)

of a circle about the mean containing either 63 or 50% of the observations is then given by:

$$\theta_{63} = 81K^{-\frac{1}{2}} \tag{6.15}$$

$$\theta_{50} = 6.7K^{-\frac{1}{2}} \tag{6.16}$$

As the true precision parameter, K, can only be estimated, its estimate, k, must be used and is, again, strictly only valid for $N \gtrsim 10$ and $k \gtrsim 7$. (The 63% level is analogous to the standard deviation and θ_{50} to the quartile ranges in two-dimensional Gaussian analyses.)

It is vital to stress the difference between the estimates of precision and those for the magnitude of scatter (Fig. 6.7). It is possible to calculate, for example, a circle containing 95% of the observations, $\theta_{95} = 140k^{-\frac{1}{2}}$, which

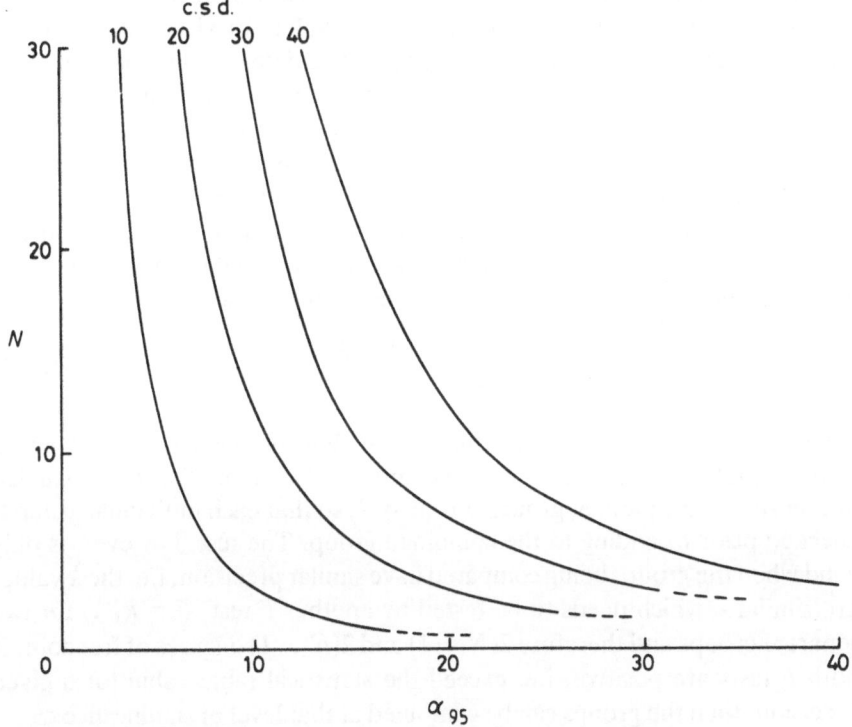

Figure 6.7 The relationship between N, circular standard deviation and α_{95}. There is a close relationship between measures of scatter and precision, that enables an evaluation to be made of the likely precision of many palaeomagnetic techniques. If secular variation of the geomagnetic field is not averaged out, for example, then individual sample directions, or site directions, will have a circular standard deviation which may vary from about 10° to 20°, by analogy with the present geomagnetic field (Fig. 8.7). This means that it would be virtually impossible to define a mean direction or pole with a precision of much less than 5°, even with many hundreds of samples or sites.

can be readily confused with $\alpha_{95} = 140(kN)^{-\frac{1}{2}}$. The former is a measure of scatter and is essentially independent of N, for $N \geq 8$, while the latter is a measure of precision and continues to decrease as the number of observations increases (Fig. 6.5). (To avoid such confusion, it is better if the scatter parameter θ_{95} is not used.)

6.3.4 Combining and comparing groups of vectors

It is frequently necessary to compare two groups of directions in order to establish if there are significant differences, e.g. in the fold tests (Section 5.5.3) which may be interpreted in terms of age differences or tectonic movements (Chapter 9), and also to see whether the data can statistically be combined to obtain an overall mean. The first consideration in such a test for identity is that the data are at the same level of analyses (Section 6.5), e.g. all site mean values are determined at a sample value, or virtual poles are all based on individual sample determinations. The data can then be compared to establish whether they can be combined, i.e. they have statistically been drawn from the same population, or from different populations, in which case they should not be combined. As in most palaeomagnetic studies, a major difficulty in such considerations is the low numbers usually involved, rendering most statistical tests unreliable. The standard method (Watson, 1956b) for comparing B groups of vectors, each group having N_i vectors and a corresponding resultant vector $\mathbf{R}_i$, is by means of the F test:

$$F = \frac{\Sigma N_i - B}{B - 1} \frac{\Sigma \mathbf{R}_i - \mathbf{R}_B}{\Sigma N_i - \Sigma \mathbf{R}_i} \qquad (6.17)$$

where $\mathbf{R}_B$ is the vector sum, $\Sigma_1^n \mathbf{R}_i$; F can then be evaluated in standard statistical tables for 3 and $2(\Sigma N - 2)$ degrees of freedom. This test is carried out on successive pairs of groups, i.e. $B = 2$, so that each individual group is checked prior to adding to the combined group. The test, however, is only valid when the groups being compared have similar precision, i.e. the k values are similar – which needs to be tested by another F test, $F = k_1/k_2$ for two separate groups and therefore $2(N_1 - 1)$ and $2(N_2 - 1)$ degrees of freedom. If both F tests are positive, i.e. exceed the statistical table value for a given precision, then the groups can be combined at that level of significance.

These tests have been extended by McFadden and Jones (1981) and McFadden and Lowes (1981) to make them somewhat more versatile. There are two situations which can arise, one in which the precision estimates of the two groups are statistically similar at an acceptable level of significance, say 95% probability, using their ratio as an F test (McElhinny, 1964):

$$k_1/k_2 = F$$

F is then tested in statistical tables for $2(N_2 - 1)$ and $2(N_1 - 1)$ degrees of freedom. If they pass this null test, then their mean vectors $\mathbf{R}_1$ and $\mathbf{R}_2$ can be compared and tested (McFadden and Jones, 1981):

$$\frac{(\mathbf{R}_1 + \mathbf{R}_2 - \mathbf{R}_v^2)/(\mathbf{R}_1 + \mathbf{R}_2)}{2(N - \mathbf{R}_1 - \mathbf{R}_2)} > \left(\frac{1}{p}\right)^{1/N-2} - 1$$

where $\mathbf{R}_v$ is the vector sum of $\mathbf{R}_1$ and $\mathbf{R}_2$ and p, the probability, is conventionally taken as 0.05, i.e. 95% probability. However, if $k_1 \neq k_2$ when tested, the mean directions can still be compared by defining a correction factor γ whereby the F test of $k_1/\gamma k_2$ is significant at a 95% probability level. The test of the mean directions is then:

$$\frac{\gamma[(\mathbf{R}_1 + \mathbf{R}_2)^2 - \mathbf{R}_v^2]}{2[(N_1 - \mathbf{R}_1) + \gamma(N_2 - \mathbf{R}_2)(\mathbf{R}_1 + \gamma\mathbf{R}_2)]} > \left(\frac{1}{p}\right)^{1/N-2} - 1$$

It is also frequently useful to reduce the effect of scatter at one level of analysis on an analysis at a higher level, e.g. when attempting to determine the magnitude of secular variation, the scatter of site mean values, it is convenient to reduce the 'noise level' attributable to measurement and orientation errors that largely account for the within-site scatter (Section 6.5). Unfortunately this test (Watson and Irving, 1957) also requires similar precision values for each site, which is rare, but the within- (ω) and between-site (β) scatter precisions can then be given by:

$$\omega = \frac{\Sigma(N_i - 1)}{\Sigma(N_i - \mathbf{R}_i)} \qquad \beta = \bar{N}\left(\frac{\Sigma\mathbf{R}_i - \mathbf{R}_B}{B-1} - \frac{1}{\omega}\right)^{-1}$$

where the symbols are as in Equation (6.17), and $\bar{N} = (\Sigma N_i - \Sigma N_i^2)/(B-1)$. The precision estimate, β, is thus the most meaningful parameter for the measure of secular variation scatter as each site represents a spot reading in time of the past geomagnetic field, although this amplitude will be reduced and misleading if the secular variation has influenced the within-site scatter.

6.3.5 Fisher and non-Fisher distributions and error estimates

Although the Fisher distribution was developed because of the requirements for a satisfactory method of analysis of three-dimensional factors, there have been very few tests undertaken to establish whether this model distribution does, in fact, match the observations. One fundamental reason for this is that a large number of observations are required to make rigorous tests.

The test comprises two separate parts (Fig. 6.8). Vectors with a Fisher distribution will have an even azimuthal distribution about the mean, i.e. the same number will occur within each azimuthal segment and the expected frequency of points (E) will follow the Fisher formula, i.e.

$$E = N\{\exp[-k_m(1 - \cos\theta_1)] - \exp[-k_m(1 - \cos\theta_2)]\} \qquad (6.18)$$

125

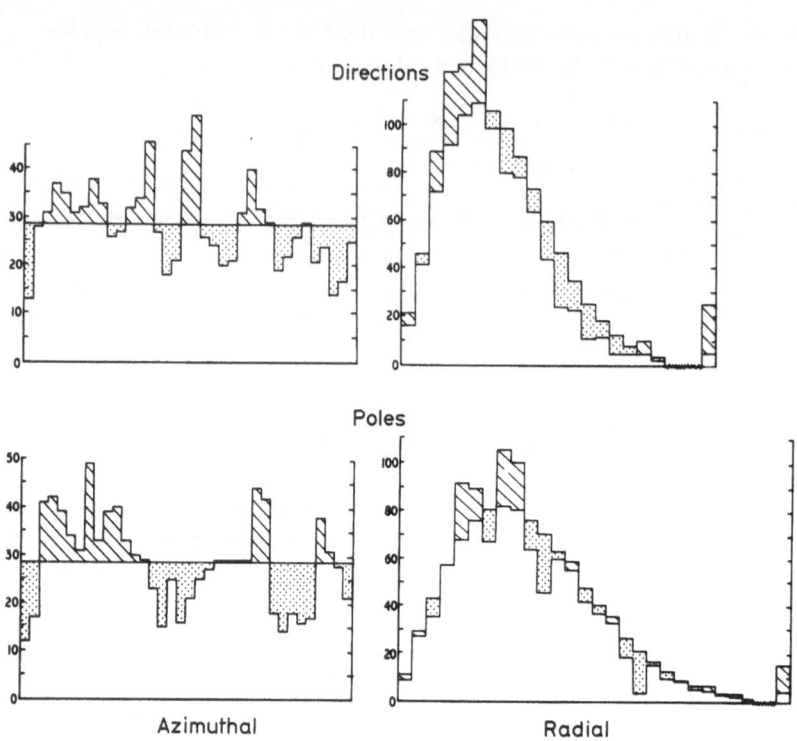

Figure 6.8 Fisher distributions tests for remanence directions and poles. Neither the azimuthal distribution nor the radial distributions of directions and poles show a strictly Fisher distribution, at a 95% probability level, for all collections examined with several hundred determinations. However, the departures are still small in azimuthal components and the radial components are consistently more tightly grouped around the mean direction or pole than they would be for a true Fisher distribution. This suggests that Fisher statistics will tend to give a slightly lower estimate of the precision of the mean than other, as yet untested, distributions. However, it is possible that other distributions may provide more physically meaningful modes of analysis. (The hatched areas are where the observed values exceed the theoretical, and the dotted areas are where the observed numbers are less than predicted by the theory.) Based on Faeroese Lower Tertiary basalts.

for angular distances from the mean of θ_1 and θ_2. (k_m is the maximum likelihood estimate of κ, i.e. $N/(N-R)$ – McFadden, 1980a.) The actual numbers involved in each azimuthal segment or radial band can then be compared with the predicted value using, for example, the standard χ^2 test. Most examinations of the actual distribution of palaeomagnetic directions and poles have been based on less than 70 samples, and so are not very rigorous tests, but most (16 of 20) showed that neither the directions nor the poles had strictly Fisher distributions (Tarling, 1971a; Onstott, 1980), but most tend to show that both forms of data are somewhat more tightly grouped around the mean (leptokurtic) than would be expected for Fisher distributions (Fig. 6.8). This feature suggests that the use and comparison of Fisher statistical parameters, while not strictly correct, is reasonable, and provides slightly conservative estimates of the precision and scatters – assuming the distributions being considered have approximately circular patterns. (Alternative, probably better, but less used distributions are discussed later.)

If directions have a Fisher distribution, then their corresponding poles positions must have elliptical distribution patterns, or vice versa. This arises because of the geometric distortion that is introduced in either deriving pole positions from directions or directions from pole positions. A circle of confidence or circular standard deviation for a group of Fisher distributed directions, for example, will convert to a 2:1 ellipse if the mean inclination is 0° and would only be circular for a vertical inclination (Fig. 6.9). As neither poles nor directions have strictly Fisher distributions it is not clear which is the most physically and statistically meaningful method of approach. As the directions are acquired in a geomagnetic field and the average geomagnetic field approximates to an axial geocentric dipole field (Section 8.5), it is physically more plausible that the virtual poles will have a more Fisherian distribution than the directions as the Fisher distribution exactly describes Brownian motions on a sphere (Roberts and Ursell, 1960) and it seems possible that changing short-term, geomagnetic pole positions can be regarded as either random fluctuations about the mean or random walks with a superimposed bias towards the axis of rotation. On this basis, polar precisions and scatter parameters would probably be better estimated on an assumption that these have a Fisher distribution. Conventionally, however, most workers have determined the statistical parameters on the basis of the directional observations and then extrapolated these to determine the polar error parameters (Fig. 6.9); e.g.

$$\delta m = \alpha_{95} \frac{\cos \lambda}{\cos \text{Inc}} \tag{6.19}$$

$$\delta p = \tfrac{1}{2}\alpha_{95}(1 + 3\cos^2 \lambda) \tag{6.20}$$

where λ is the palaeolatitude (derived from $\tan \text{Inc} = 2\tan\lambda$), δm is the radius of the ellipse of 95% confidence along the direction of the magnetic meridian from the site to the vertical pole, and δp is the same radius, at the pole, perpendicular to the meridian. Within the site itself the errors in declination (δDec) and inclination (δInc) are of use for comparison of components of the vector:

$$\delta \text{Dec} = \frac{\alpha_{95}}{\cos \text{Inc}} \tag{6.21}$$

$$\delta \text{Inc} = \alpha_{95} \tag{6.22}$$

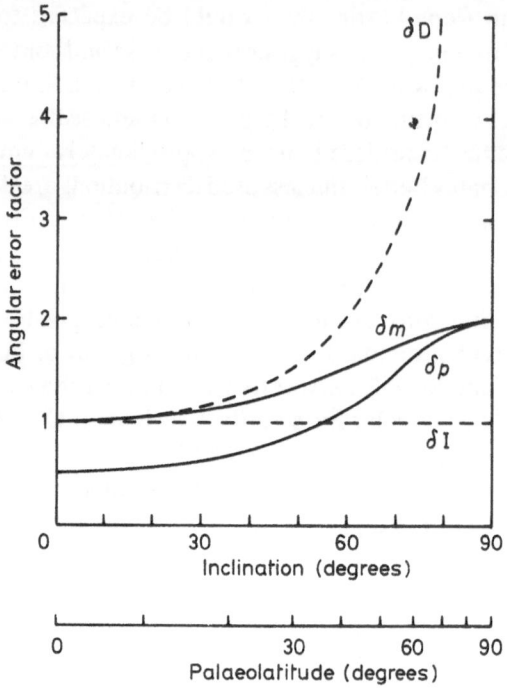

Figure 6.9 Directional and polar errors as functions of inclination. The errors in declination, δD, and inclination, δI, are shown as a function of inclination/ palaeolatitude. A solid angular error of 1° (circular standard deviation or α_{95}) in declination is thus serious at high inclinations, greater than 60°, but inclination errors are unaffected by variation in inclination. The determination of the errors about the corresponding pole, δm (in the direction of the meridian between the site and pole) and δp (perpendicular to the meridian) are both dependent on the inclination at the sampling site, but are less affected by 'solid angle' errors at low inclinations than at high inclinations. These relationships emphasize the fact that declination, or longitudinal values, are only 'solid angles' when measured at 0° inclination and at the equator. This graph allows the determination of the 'errors' by multiplying the observed site parameters (α_{95}, c.s.d., c.s.e. and other solid angle measures) by the factor corresponding to their mean inclination (or palaeolatitude).

Under some circumstances, it is particularly useful to determine whether the angular difference between two groups of vectors is real, i.e. significant at a certain probability level, usually 95%. This type of situation arises, for example, in determining whether directions from two different localities are really discordant from each other or not (Beck, 1980; Magill *et al.*, 1981). The actual differences are, of course, the solid angular separation between the two vectors **R′** and **R″** (or their components of declination or inclination), and the errors on the differences are given by $(\alpha_{95}'^2 + \alpha_{95}''^2)^{\frac{1}{2}}$ – or the appropriate error, δD, δI, etc. All such evaluations assume little or no effect due to regional

128

variations in the average geomagnetic pole, although such effects can be reduced by use of the 'geographic correction' (Section 6.3.1) on the scale of a few 10^3 km.

In view of the fact that most observed distributions of palaeomagnetic data do not strictly conform to a Fisher distribution, Onstott (1980) has proposed the use of the Bingham distribution (Bingham, 1964; Mardia, 1972) in palaeomagnetic analyses as this is suitable for circular distributions, but is somewhat denser nearer the mean direction than the Fisher distribution (and will therefore reduce the leptokurtic feature). Furthermore it is applicable to bipolar data, such as the axes of magnetic susceptibility axes and observations

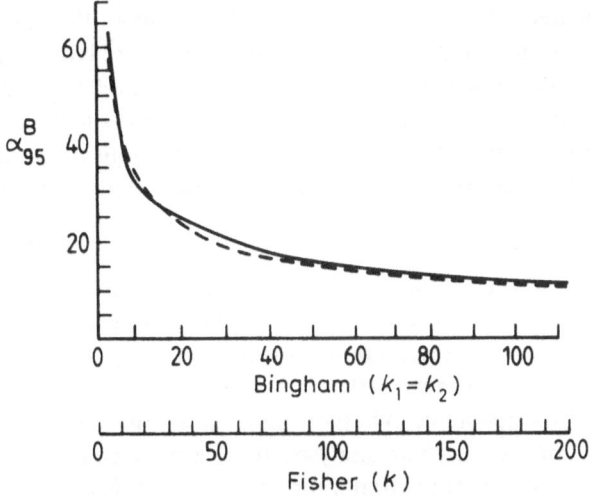

Figure 6.10 Fisher and Bingham statistics. Where the distribution of directions or poles has a circular symmetry, it is possible directly to compare the Bingham precision values, k_1 and k_2 with the Fisher k parameter. The estimates of the angle of the cone of confidence are then almost identical for both forms of distribution. Where the distribution is not symmetrical, such comparisons are not valid. (After Onstott, 1980.)

of rocks that include polarity changes in the geomagnetic field, and is also applicable to elliptical (such as arise from converting assumed circular directional functions to polar, or vice versa) and elongate (strung) distributions that occur when two or more components are present of different relative magnitudes (Section 6.4). Unfortunately, the calculations in evaluating the parameters of this distribution are complex, involving the use of eigenvector and complex integrations and numerical search packages necessarily involving computer facilities, and these are slow and therefore expensive in computer time. These difficulties have so far made Bingham analyses little used although the increasing availability of computing facilities may lead to their increasing use and fuller testing, particularly in view of their application to vector definition and analyses (Section 6.4). For a circular distribution, some of the parameters are directly comparable with those for a Fisher distribution analysis (Fig. 6.10). Bingham statistics provide two estimates of precision, k_1 and k_2, which are identical to each other for a circular distribution and can thus only be compared directly with the Fisher k under those conditions (Fig. 6.10). Similarly, the Bingham statistics which are comparable to Fisher's α_{95} then correspond to the major and minor axes of ellipse of 95% confidence, k_1 or σ_{31} and k_2 or σ_{32}, respectively.

Onstott (1980) has thus made a very strong case for the widespread use of Bingham statistics, although rigorous tests of this distribution are still required so that implicit assumptions in both the statistical distribution and the palaeomagnetic data can be assessed. For example, the Bingham distribution can be applied to palaeomagnetic directions involving approximately antiparallel (bipolar) distributions and intermediate directions, representing polarity changes of the geomagnetic field, with associated intermediate, transitional, directions between the two stable directions of the geomagnetic field. It is normal practice in palaeomagnetic studies to reverse one or other of the polarity directions and then to use Fisher analysis on the combined data. This often raises the problem of the treatment of transitional directions. Palaeomagnetic studies of the transition of the geomagnetic field indicate that such transitional directions are likely to be few compared with the obvious grouping of normal or reversed polarity direction. It can thus be argued that such intermediate values have little effect on either the mean direction or the parameters, irrespective of whether their directions are left or rotated through 180°. For the Bingham analyses, there is no need for any rotations of either fully reversed or transitional directions, which is an apparent advantage. However, in most studies (Chapters 8 and 9), it is the mean direction of the geomagnetic field that is required, irrespective of polarity, and thus transitional directions should be actually eliminated as only corresponding to anomalous behaviour that is only of importance in studies of transitional geomagnetic behaviour.

6.4 ANALYSES OF VECTOR COMPONENTS AND THEIR STABILITY

Fundamental to all geomagnetic, archaeomagnetic and palaeomagnetic interpretations is the definition of a direction of remanence in a collection of samples that can be related to the direction of a past geomagnetic at the collecting site. However, most samples, when collected, will have a remanence that comprises at least two components – one acquired at some specific time, such as when the sample cooled, and a viscous magnetic component that it has acquired while lying in the ambient geomagnetic field. In some archaeological materials and even some geological samples, the secondary, viscous, component may be so small in relation to the primary component that it can be ignored but in many cases this is not the case and needs to be tested by partial demagnetization procedures (Section 5.4) which preferentially remove the lower stability components. Many geological samples, in particular, may also contain other components of remanence acquired at different times during the rock's history, some of which may be useful in determining its evolution (Chapter 9). Two features, within the demagnetization procedure, thus need to be defined – the identification of specific components of remanence and an assessment of their stability – hence leading to their mode of and possible age of origin.

As the viscous remanence is associated with the time effect of the geomagnetic field over the last few 10^1 to 10^5 years it can usually be identified as it is in the direction of either the present geomagnetic field at the site or its average direction (the axial geocentric dipole field) for the last 7×10^5 years. The direction of remanence, during demagnetization, thus moves away from the present field direction and, if only one other component is present, its direction then remains constant during subsequent demagnetization until the coercivity or blocking temperature of its component magnetic grains has been exceeded (Section 5.4). If several components are present, then the observed vector moves successively from the lower stability component towards the summation of all higher components. (Similarly it is possible to examine the properties of the magnetic vectors being removed during progressive demagnetization and thus define their direction and stability.) All forms of analysis therefore ideally require a specific vector component to be defined within at least two successive demagnetization steps, i.e. three measurements of the remaining or removed vector. If a vector is present that is completely dominant over the corresponding blocking temperature or coercivity range, then this component can be identified by a variety of forms of analysis. The main problems arise when different components have overlapping blocking temperature or coercivity properties so that the observed remanences are summations of two or more components. Very closely spaced demagnetization steps will sometimes allow this difficulty to be reduced, but there are practicable limits to the magnitude of steps employed, with 25–50°C or 5–10 mT being the lower practical limits for analyses of large numbers of samples, although clearly finer increments of temperature or field can be used to examine the magnetization of individual specimens. Clearly if all the components have a wide range of coercivities or blocking temperature it may not be possible to differentiate between the individual components.

The main clue to the presence of such components is thus the directional changes in the observed vector during progressive demagnetization, but additional evidence can also be provided by changes in its intensity. The behaviour of intensity alone can sometimes be used to identify different components of remanence. Hoblitt and Kellogg (1979), for example, examined the magnetization of recent volcanic clasts to determine their temperature of emplacement on the basis of intensity changes. In this

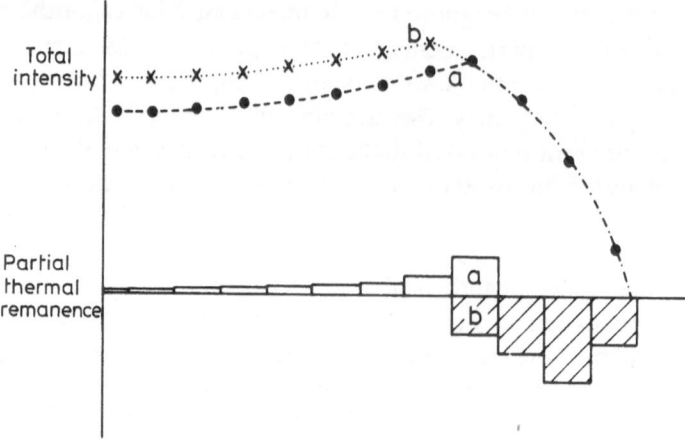

Figure 6.11 Vector isolation using intensity behaviour. As a rock containing two antiparallel components is progressively demagnetized, the total intensity will generally increase slightly as one component is preferentially removed, reaching a peak just before the reduction of the higher temperature, antiparallel component. In such situations, the intensity behaviour can be used to determine the approximate emplacement temperature if the lower temperature component, acquired after emplacement, has an obtuse angle to the higher temperature component (Hoblitt and Kellogg, 1979). (This situation is similar to that involving self-reversal mechanisms (Fig. 8.8).) The definition of the peak temperature, which is always slightly below the emplacement temperature, depends on the blocking temperature spectrum, the degree of obtuseness of the two vectors, and the temperature of emplacement – being generally harder to define the peak temperature for lower emplacement temperatures. For older rocks, the presence of a third component, a viscous remanence, can further complicate this approach.

analysis, it was assumed that the clast has two components of magnetization – a high-temperature one, acquired prior to emplacement, and a lower temperature one acquired as it cooled down after emplacement. By measuring the directions of remanence, clasts could be selected in which these two components were magnetized in approximately opposite directions to each other. On partial demagnetization (Fig. 6.11), the low-temperature (post-emplacement) component is preferentially removed. As this is antiparallel to the high-temperature component, the observed intensity of remanence of the clast rises until the antiparallel component has been removed, after which the observed intensity value of remanence of the clast reduces. The intensity value therefore peaks at the temperature at which the lower temperature component is totally removed, thus marking the emplacement temperature. The disadvantage of this system is, of course, that it is only in clasts in which the components are both discrete and approximately antiparallel that such a procedure works successfully. If the two components have overlapping blocking temperatures (or coercivities), then the peak temperature is obscured and the method is obviously not practicable if the components are not approximately antiparallel, thus requiring considerable analysis to determine the appropriate clasts for study. Indeed, the intensity behaviour is generally a poor indicator of the properties of the components of geophysical interest, although it is clearly of value in rock magnetic evaluations.

One of the earliest statistical methods for both distinguishing specific vectors and establishing their stability, was the stability index, devised on the basis of directional changes alone (Tarling and Symons, 1967). The argument is simple. When the vector has moved from the viscous remanent direction (or other low-stability component) to a higher stability component, the direction of the vector should remain constant for successive increments in demagnetization levels (Fig. 6.12). The directions should thus group tightly when this vector has been defined, i.e. the scatter (circular standard deviation or c.s.d.) of the directions will be small. An indication of stability and the adequacy with which this vector is defined is thus provided by a search for the lowest scatter over the widest range of demagnetizing treatment. The stability was thus defined as c.s.d./range of treatment^2, for both thermal (°C units) and alternating magnetic field (0.1 mT units) treatment. The maximum value for two or more successive demagnetization increments (i.e. three or more vector measurements) was thus defined as the stability index (SI):

$$SI = max(c.s.d./\text{range of treatment}^2) \qquad (6.23)$$

Comparison of stability indices for both thermally and alternating field demagnetized rocks, mostly containing magnetite, indicates similar values for the same rocks, irrespective of treatment, up to 300°C and 30 mT and a good agreement with subjective estimations of stability:

SI <1 unstable; 1–2 metastable; 2–5 stable; >5 very stable

The stability index thus provides a method of vector definition as well as stability that is independent of changes in intensity (Fig. 6.12). This form of analysis, as outlined, only defines the most stably magnetized vector which must, of course, be dominant in at least three successive demagnetization steps, but can be modified to search for other vectors that have been isolated in a similar way but are not as well defined because the directions are somewhat more scattered or they are only defined over a shorter range of coercivity or blocking temperatures.

A variety of other stability indicators (Fig. 6.12) have subsequently been proposed, all based on alternating magnetic field demagnetization of igneous rocks (Wilson *et al.*, 1968; Briden, 1972; Symons and Stupavsky, 1974; and others). The Briden index (Briden, 1972)

$$SI_{i,i+1} = 1 - \frac{\mathbf{R}_i - \mathbf{R}_{i+1}}{\mathbf{R}_i} \text{ for demagnetization steps } i \text{ and } i+1 \qquad (6.24)$$

is useful as it normally ranges between +1 and −1 with the most stable component occurring at the maximum SI value (most other indices range from zero to infinity) but this index needs to be calculated on constant increments of the applied field if the indices are to be really comparable. The Stupavksy and Symons (1978) index, PSI, depends on the rate of change in the vector components as demagnetization progresses:

$$PSI = 57.296 \times 10^4 [(dx/dH)^2 + (dy/dH)^2 + (Dz/dH)^2]^{\frac{1}{2}} \qquad (6.25)$$

in which the rate of change is in radians per 0.1 mT applied field. The PSI is, therefore, a direct measure of the rate of change in the total vector removed in millidegrees per 0.1 mT. Stupavsky and Symons (1978) determined a subjective evaluation of the PSI:

$$\leq 5 \text{ very stable; } 5\text{--}20 \text{ stable;}$$

$$20\text{--}50 \text{ moderately stable; } 50\text{--}200 \text{ low stability; } \geq 200 \text{ unstable}$$

All these different stability indices measure different features of the demagnetization behaviour and must be considered complementary to each other (Fig. 6.12). Most of them probably overemphasize the importance of changes in intensity of the vector – it is more important that the vector is well defined, i.e. does not change direction, rather than whether it remains of constant intensity or not. The Briden index and PSI, for example, would give

---

Figure 6.12 Examples of stability behaviour and analyses. Three examples are illustrated here, involving two or more components. The behaviour of the remanences is illustrated on both stereographic and As–Zijderveld (cartesian) projections. (a) A two-component system, with a well-defined stable remanence and a viscous component directed towards the present geomagnetic field. Thermal demagnetization causes the vector to move systematically away from the present geomagnetic field direction, stabilizing at the original remanent direction at 400°C, and remaining there until it becomes demagnetized by 550°C. (b) A three-component system, with a viscous component, another associated with magnetite and thus of lower coercivity than another component carried by high coercivity haematite. (c) Alternating magnetic field demagnetization does not isolate any clear single component, although the confinement of the vector within a single sector would allow the estimation of a single vector up to applied fields of 50 mT. (d) An essentially unstable sample, although alternating magnetic field demagnetization allows an evaluation of the polarity of the sample, but not to define a direction. The statistical analyses of these examples are:

	Max stability index	Range	Briden index (0.9)
(a)	5.5	450–550°C	450–550°C
(b)	13.5	100–160 (mT)	80–160 (mT)
(c)	1.1	5–20 (mT)	50–20 (mT)
(d)	0.7	5–80 (mT)	–

(Considerable information can also be extracted from similar plots and analyses of the components removed during demagnetization, but these are not illustrated here.)

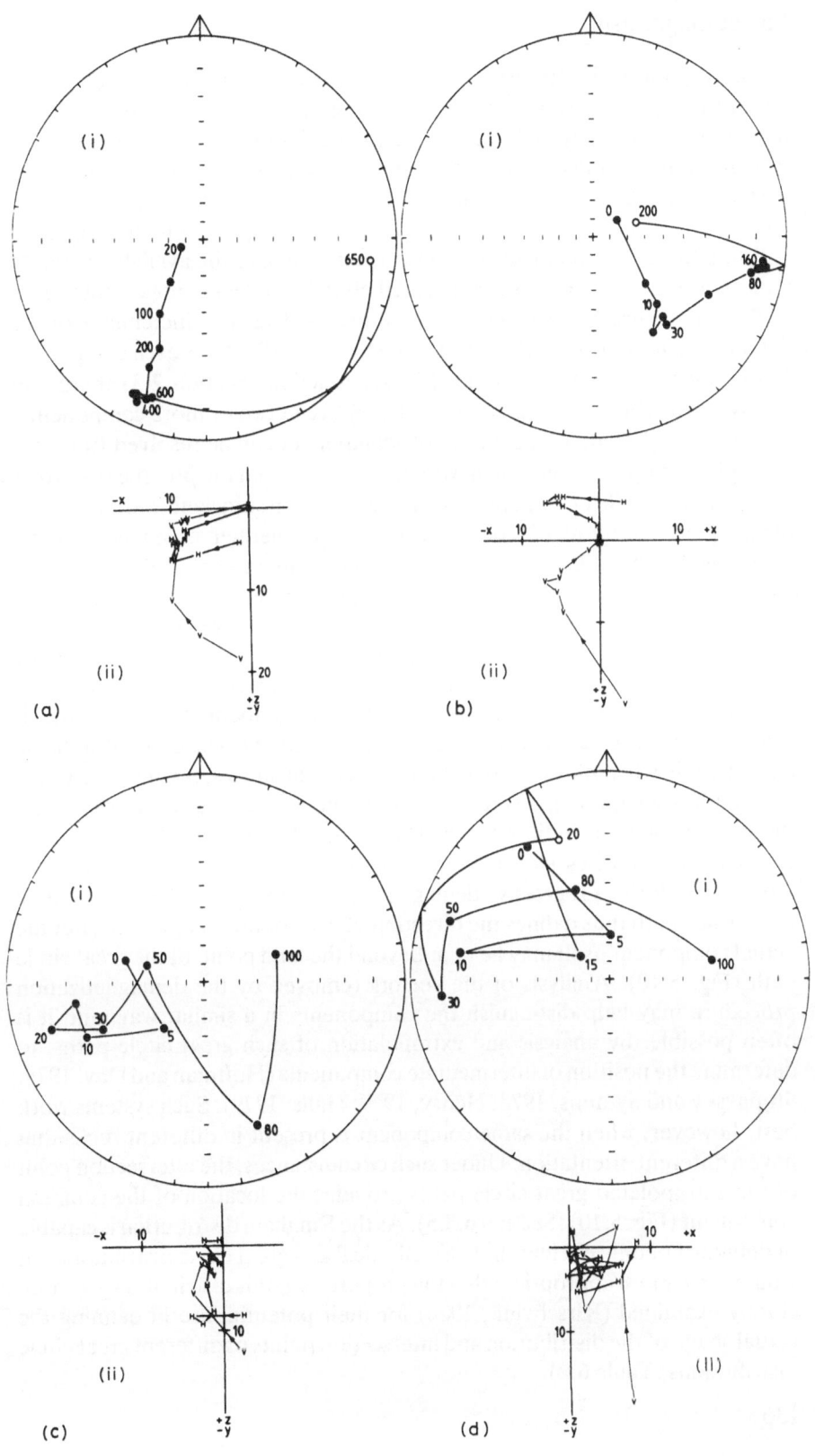

apparently low stabilities for a vector that remained of constant directions, but of rapidly decreasing intensity, while the SI (Tarling and Symons, 1967) would show extremely high stability. In physical terms, the Briden index and PSI are therefore most meaningful, but in most palaeomagnetic applications of such observations, the definition of the direction of the vector is much more important and hence could be regarded as the most important in that context.

While there are arguments for, and against, most of the available stability indices, none of them can be considered effective unless a single component of the remanence has become dominant over at least two increments of the demagnetization process (Stupavsky and Symons, 1978). As each component has its own exponential stability (decay) function (Section 2.3) there will almost always be some degree of overlap between two or more components. In many 'simple' situations, the stable component can be resolved from, for example, the viscous component with relative ease, but the presence of two or more higher stability components will often result in sufficient overlap of their blocking temperature or coercivity ranges for neither component to be adequately defined by these analyses. An effective method of deciphering such overlapping components has been by determining the 'great circle' path of magnetization (Khramov, 1958; Creer, 1959). If two components are present with dissimilar but overlapping ranges of coercivity or blocking temperatures then as one is being preferentially removed, the resultant vector remaining moves towards the higher stability component and the vector removed moves towards the lower stability component. The points thus lie on a great circle when plotted on a sphere. In some instances, the removal of the low-stability component results in the isolation of the higher component, and the standard stability indices can then be applied, but often the higher stability component comprises two or more separate components with therefore a movement, during progressive demagnetization, along a different great circle path. Each path thus defines the direction of each of the components, but the actual component itself may lie at or beyond the 'end point' of the great circle path (Fig. 5.10). Analysis of the vectors removed by the demagnetization procedure may help distinguish the components in a similar way, and it is often possible, by analysis and extrapolation of such great circle paths, to determine the position of intermediate components (Hoffman and Day, 1978; Stupavsky and Symons, 1978; Henry, 1979; Halls, 1979). Such systems work best, however, when the same component is present in different rocks that have a different orientation. Under such circumstances, the intersection point of the extrapolated great circle paths provides the location of the common component (Fig. 5.10) (Section 6.3.5). As the Bingham distribution is capable of defining not only circular, but also elliptical and great circle distributions, it would seem more appropriate that the properties of this distribution are more closely examined (Kirschvink, 1980) for their potential use in defining the actual shape of the distribution and intersection points of different great circle distributions (Table 6.2).

136

Table 6.2

Bingham function	Distribution
$k_1 \simeq k_2 \simeq 0$	Uniform circular
$k_1 < k_2 \leq 0$	Unimodal
$k_1 \simeq k_2 \ll 0$	Bipolar (or large **R**)
$k_1 \ll k_2 < 0$	Symmetric girdle or elliptical
$k \ll k_2 \simeq 0$	Asymmetric girdle

(Onstott, 1980) $k_1 = \sigma_{31}$ $\quad k_2 = \sigma_{32}$

There are two common ways in which such demagnetization behaviours are illustrated – the cartesian and stereographic projections (Section 1.3) – which are, of course, quite independent of the statistical and mathematical treatments. Both systems have advantages and disadvantages and are thus largely complementary. In plotting on cartesian systems (often termed a Zijderveld diagram, but first published by As (1960)), linear sections of the path can be seen clearly if the vector concerned is being reduced in intensity (Fig. 6.12), but shows as a scatter of grouped points if there is little or no change in the magnitude of the vector. On the stereographic (or other spherical projections) only the directions are shown and systematic changes in direction, e.g. the differential decrease of two separately directed components, plot as great circles, while vectors that show only changes in intensity show constant directions, i.e. they plot as a group of directions clustered about the mean vector direction. On the As–Zijderveld diagram (Fig. 1.2(b)), it is more difficult to visualize the total directional motions since the horizontal and vertical components must be plotted separately. All such plots have one graph showing x vs. y, but it is then possible to plot z against x, y or $H = (x^2 + y^2)^{\frac{1}{2}}$ (so that the vector angle from the horizontal plane is the inclination). Zijderveld, in Dunlop (1979), advocates the adoption of the convention of plotting z vs. x, although obviously this would not always be the most informative representation. The definition of the origin, i.e. the units used in such plots, can also lead to confusion. The only 'real' zero intensity is the limit of sensitivity of the available magnetometer, or of the demagnetization treatment, but is often taken as the intensity of remanence remaining after the last demagnetization step. These difficulties make comparison of such diagrams difficult unless they are plotted uniformly, in which case much information becomes hidden if there is a major decrease in intensity. For example, an intensity decrease from 100 to 1 mA m^{-1} that is plotted on axes with units of 10 mA m^{-1} will necessarily decrease towards the origin of the graph, suggestive of the isolation of a single component during the final stages, but identical data, plotted on a scale of 1 mA m^{-1} (or even the sensitivity of the magnetometer, possibly 0.01 mA m^{-1} units) may well not appear to have a vector that will pass through zero, thus indicating that a single component has not yet been isolated. Nonetheless, such a projection makes it easy to illustrate linear segments of the demagnetization behaviour for individual specimens on such diagrams if they are accompanied by changes in intensity. The stereographic projection is thus less informative, especially when components are present with rapidly changing intensities of remanence, but provides a much greater visual appearance for the relationship, in space, of the different vectors and can, for example, be used to examine the great circle distributions of sequences of overlapping components more effectively than on the cartesian projection. Obviously it is advisable that both methods of plotting are used in any analyses, with the cartesian method most effective where large changes in intensity are involved. It is important to stress, however, that the measurements of linearity, coplanarity, etc., while undertaken on cartesian components, do not require the data to be presented in cartesian co-ordinates. The choice of illustration is, therefore, dependent on whether the intensity changes are considered to be of major importance, in which case the cartesian plot is the most visually effective, or whether the directional changes are the most fundamental, in which case the stereographic projection is generally most effective.

6.5 LEVELS OF DIRECTION ANALYSES AND SAMPLING NUMBERS

As the precision parameters, such as k and α_{95} (Section 6.3.2), are dependent on the total number of observations, N, it is obviously true that the more observations that are made, the higher the precision with which the mean can be defined (Fig. 6.5). However, analyses can be made at a range of levels (Table 5.1). Specimens are the lowest level as these are the actual objects that are measured. These are cut from separately oriented samples, so that sample values are necessarily based on the mean of specimen measurements. The samples are taken at sites, so that site mean values can be calculated on either specimen or sample values, and, finally, sites can be grouped into localities or regions – the mean of which can be based on sites, samples or specimens. If the only object was to obtain the highest precision for the final mean, then clearly this can be obtained by combining all specimen values. However, each level represents the averaging, and thus reduction, of the effect of some physical parameter or source of error. It is therefore more important to assess the different physical meanings of averages obtained at each successive level and then determine which is the most physically appropriate statistical analysis.

At a specimen level, the magnetometer tends to average out, within the specimen, inhomogeneity in the distribution of the magnetized grains and also of their anisotropies. The limit to this is, of course, that of the instrumental sensitivity and criteria are normally required to distinguish meaningful

138

measurements from instrumental noise. For any single measurement, the measurement can be accepted or rejected on the basis that it corresponds to a value at least one order of magnitude greater than that of the noise level. Most weakly magnetized specimens are, however, measured repeatedly. (Most specimen magnetizations are, in fact, measured at least twice in all standard instruments, and usually at least four times – Section 5.3.) Criteria are therefore required to determine what degree of repeatability is acceptable. Various parameters for this have been proposed (Harrison, 1980c; Lowrie *et al.*, 1980c; Briden and Arthur, 1981), all based on the solid angle about the mean. In practice, the criteria are then subjectively evaluated on the basis of whether the measurements of different specimens are in agreement or not. Again objective criteria can be used for this, but such criteria depend on the ultimate precision required as well as the time available for repeating further measurements.

The combination of specimen data, to obtain individual sample means, is then normally undertaken giving equal weight to each specimen vector (Section 6.3.1) on the basis that, if having passed the reliability test, no one specimen vector is necessarily more reliable than another. (This assumption is not, of course, always valid but leads to weighting problems – Section 6.3.1.) A sample value therefore represents a summation of specimen values in which random inhomogeneities and anisotropies of magnetization on a scale of ≤ 10 cm^3, and random instrumental noise, have been at least partially averaged out.

Site mean directions will be identical, whether based on specimens or samples, if there are the same number of specimens cut from each sample, but the precision of the vector will appear to be greater when the number of specimens exceeds that of the number of samples. At this level, an averaging, giving unit weight to each sample, will average out random orientation directions made in the field, but if the number of specimens per sample is uneven, a site mean direction, calculated giving equal weight to each specimen, would be biased towards those samples from which most specimens were cut. Any error in the orientation of such samples would thus be less effectively averaged in this procedure. In order to obtain the most error-free site mean, therefore, either exactly the same number of specimens must be taken from each sample or the direction must be calculated giving equal weight to each sample value.

Similar arguments then apply at higher levels. If the same number of specimens are cut from the same number of samples obtained at all sites, the mean directions would be identical whether calculated on specimens, samples or sites. For example, combining sites also has the effect of averaging out bedding (tilt, tectonic) corrections (Section 5.2.4) for areas that have been uniformly tilted, but exactly uniform collecting would allow this to be undertaken at all levels. However, analysis at each separate level allows an evaluation of the sources of error at each stage. It can be important to assess the magnitude of field orientation errors, for example, before proceeding to combine site values as some sites may contain serious errors, while others do not. On this basis, the most conservative error estimates are obtained by successively averaging through at each level, i.e. equal weight to all reliable specimen measurements, then equal weight to samples to obtain site mean values, and then to higher levels. At each stage spurious data can then be examined for human errors, etc., as well as for different physical effects.

Such a procedure can be designed to optimize the amount of useful information that can be obtained for the minimum amount of measurement time. At a sample level, most information provided by the study of a large number of specimens is either already known, i.e. the instrumental noise, or can be evaluated at a site level. For example, major anisotropy or inhomogeneity effects at a specimen level will also be fairly obvious at a site level. On this basis, there seems little extra information to be obtained, under normal circumstances, by measuring a large number of specimens per sample. Indeed, one specimen per sample is frequently sufficient to detect the effects of inhomogeneities, etc., so that doubling the measurement time, i.e. two specimens per sample, will rarely yield twice as much useful information. (Sites showing 'strange' characteristics can, of course, be examined in further detail.)

The optimum sampling programme is thus to obtain 6 to 12 samples per site, as this allows a good estimation of the precision and scatter parameters (Fig. 6.5), and clearly as many sites should be obtained per region as is practicable for the problem being tackled and the local field conditions.

6.6 MAGNETIC FABRIC

The basic measurements available for analyses of an anisotropy ellipsoid are the susceptibility, χ, along its major, intermediate and minimum axes (Section 5.3) and their orientations.

If crystalline anisotropy can be ignored, as for magnetite, then if the three axes are equal, $\chi_{max} = \chi_{int} = \chi_{min}$, i.e. if the sample is isotropic, these correspond to an average spherical shape for all the grains in the sample. Departures from the average spherical shape can therefore be quantified as the magnitude of anisotropy and the ratios between the three axes can be used to determine whether the shape is predominantly prolate (cigar/cylinder shaped) or oblate (pancake/disc shaped). An oblate ellipsoid would be characterized by $\chi_{max} \simeq \chi_{int} > \chi_{min}$, while a prolate ellipsoid would have $\chi_{max} > \chi_{int} \simeq \chi_{min}$. A wide variety of anisotropy parameters is available (Table 6.3) and no one set has been widely adopted. This arises, in part, from there being three basic instruments available, the low-field and high-field torque types and the spinner type (Section 5.3). The spinner types mostly operate with applied fields of 0.4–0.7 mT, the low-field torque meters use up to about 10 mT, while the minimum field in high-field systems is about 10 mT.

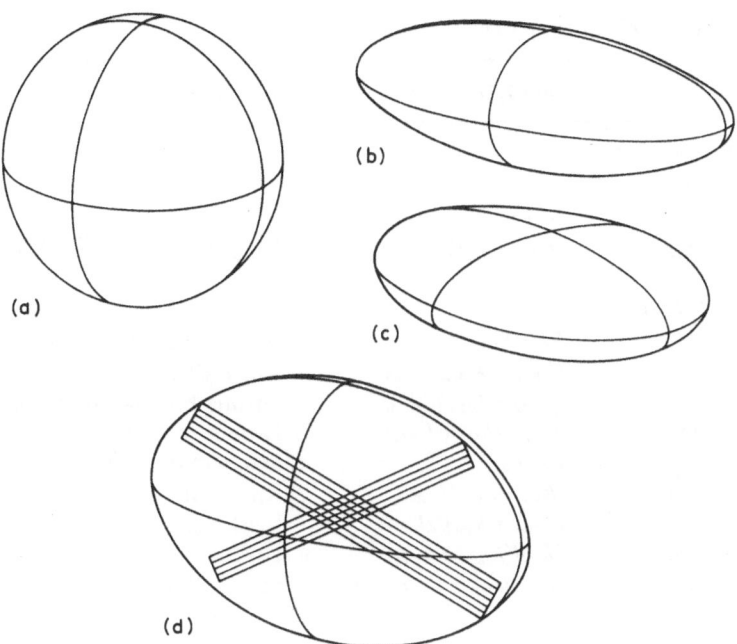

Figure 6.13 Susceptibility ellipsoids and parameters. (a) A sphere has equal suscept-ibility in all directions and hence the axes are undefined and the anisotropy is zero. (b) The susceptibilities along the three axes are 30.1, 10.1 and 10.0 and are thus clearly prolate, i.e. highly lineated. (c) The susceptibilities are 30.1, 25.0 and 10.0, i.e. oblate and with a clear foliation. The parameters (see text) for these two ellipsoids are:

	Mean susceptibility	A	L	F	q	$V°$
(b)	16.73	1.99	1.20	0.01	1.98	4.0
(c)	21.70	0.80	0.24	0.69	0.29	59.7

(d) Each susceptibility ellipsoid is, in fact, the summation of the total of all individual grains and the components cannot be separated magnetically. However, as the components change their magnitude or orientation, the shape and orientation of the susceptibility ellipsoid vary, thus allowing an interpretation of their probable nature and origin from the geological contexts in which they are observed.

The assumption of the linearity of susceptibility with applied field is therefore invalid for comparison between high- and low-fields methods, and it is even unlikely to be linear between 0.7 and 10 mT. This means that each instrument essentially measures a somewhat different property and inter-instrument calibration may only be valid for specific specimens. Three quantified parameters are required: the total magnitude of the anisotropy, the degree of lineation, and the degree of foliation (Fig. 6.13); it is not possible arbitrarily to recommend any three parameters for adoption (Rahman, Gouch and Evans,

Table 6.3 Some anisotropy parameters

Term	Susceptibility axes	References
Magnitudes of anisotropy		
Anisotropy degree	k_{max}/k_{min}	Nagata, 1953
Absolute anisotropy	$(k_{max} - k_{min})/k_{int}$	Rees, 1966
Total anisotropy	$(k_{max} - k_{min})/k_{mean}$	King and Rees, 1962; Owen, 1974
Lineation/foliation		
Lineation	k_{max}/k_{int}	Balsley and Buddington, 1958
Magnetic lineation	$(k_{max} - k_{int})/k_{mean}$	Khan, 1962
Lineation	$(k_{max} + k_{min})/2k_{int}$	Urrutia-Fucugauchi, 1980a
Lineation degree	$2k_{max}/(k_{int} + k_{min})$	Hrouda *et al.*, 1971a
Foliation	k_{int}/k_{min}	Stacey *et al.*, 1961
Magnetic foliation	$(k_{int} - k_{min})/k_{mean}$	Khan, 1962
Foliation	$(k_{max} + k_{int})/2k_{min}$	Balsley and Buddington, 1960
Foliation degree	$2k_{int}/(k_{max} + k_{min})$	Urrutia-Fucugauchi, 1980a
Lineation/foliation	$(k_{max}k_{min} - k_{int}k_{min})/$ $(k_{max}k_{int} - k_{max}k_{min})$	Stacey, 1960
Strain indicator	$(k_{max}k_{min} - k_{int}k_{min})/$ $(k_{int}^2 - k_{max}k_{min})$	Flinn, 1962
Shape, oblateness, prolateness		
Prolateness	$(k_{int} - k_{min})/(k_{int} - k_{min})$	Khan, 1962
Prolateness	$(2k_{max} - k_{int} - k_{min})/$ $(k_{int} - k_{min})$	Urrutia-Fucugauchi, 1980b
Oblateness	$(k_{int} - k_{min})/(k_{max} - k_{int})$	Khan, 1962
Oblateness	$(k_{max} + k_{int} - 2k_{min})/$ $(k_{max} - k_{int})$	Urrutia-Fucugauchi, 1980b
E factor	$k_{int}^2/(k_{max}k_{min})$	Hrouda *et al.*, 1971a
Ellipsoid shape	$(k_{max} - k_{int})(2k_{max} - k_{int}$ $- k_{min})/(k_{int} - k_{min})$ $(k_{max} + k_{int} - 2k_{min})$	Urrutia-Fucugauchi, 1980b
Shape indicator	$(k_{max}k_{min} - k_{int}^2)/$ $(k_{max}k_{int} - k_{max}k_{min})$	Urrutia-Fucugauchi, 1980a
q factor	$(k_{max} - k_{int})/$ $\frac{1}{2}(k_{max} + k_{int}) - k_{min}$	Granar, 1958
Angle V	$\sin^{-1}\left[\dfrac{k_{int} - k_{min}}{k_{max} - k_{min}}\right]$	Graham, 1966

The mean susceptibility, k_{mean}, is given by $(k_{max} + k_{int} + k_{min})/3$. Note that there are other parameters in the literature, such as the emplacement factor (Ellwood, 1979) and the current velocity indicator (Ellwood and Ledbetter, 1977), but all are based on the manipulation of the three basic observations, k_{max}, k_{int}, k_{min}. (See also Hrouda, 1982.)

1975). Furthermore, many of the available parameters are interdependent and most are altered, to some extent, by changes in the bulk (total) susceptibility. This has led to erroneous assessments of changing oblateness or prolateness when it has simply been the total number of grains, i.e. the total susceptibility, that has been changing. Similarly, some formulae simply define a boundary line between two different characteristics and cannot be used to determine the magnitude of the departure from that boundary. The ellipticity, e, for example, simply defines a shape as oblate or prolate and not the degree of oblateness or prolateness, i.e. similar e values can represent radically different shapes (Fig. 6.13).

Whenever possible it seems best to use the simplest expressions available, but normalized for the total susceptibility, i.e.

Magnitude of anisotropy $= A = (k_{max} - k_{min})/k_{mean}$ (Owens, 1974)

Degree of lineation $= L = (k_{max} - k_{int})/k_{mean}$ (Khan, 1962)

Degree of foliation $= F = (k_{int} - k_{min})/k_{mean}$ (Khan, 1962)

in which the mean susceptibility, k_{mean}, is given by $(k_{max} + k_{int} + k_{min})/3$. In addition, the quotients V and q are widely used in studies of unconsolidated sediments (Table 6.3) and are worth retaining. It is emphasized, however, that even these parameters do not adequately parametize the features and can only be used cautiously for quantitative comparisons. The fact that grain sizes and susceptibilities have a log normal distribution (Section 6.2) further suggests that comparisons should be made on the basis of logarithmic rather than arithmetic values.

Under present circumstances, it is difficult to determine physically meaningful average values for the magnitude of the fabric and shape of the ellipsoids, particularly if the total susceptibility varies within a site. In practice, simple arithmetical averages are usually made at a site level and clearly subjective evaluations are also feasible, but it is dangerous, at this stage, to attempt to determine wide-scale average values where the bulk susceptibility or shapes of the ellipsoids change by large amounts.

As each ellipsoidal axis is dipolar, it has two possible directions. In most cases, the maxima (prolate cases) or minima (oblate cases) can be expected to be closely grouped and so axial directions can be reversed to give a minimum grouping for that axis. Conventionally, such directions are then analysed using Fisher statistics (Section 6.3). This procedure is reasonably valid (subject to the discussion in Section 6.3) as the close grouping corresponds reasonably well to the Fisher model. Unfortunately, the other two axes may not be so clearly uniform (Fig. 6.13) and can be expected, when of similar magnitudes, to lie within a plane (great circle distribution) at right-angles to the better grouped axis. Such a distribution does not conform to a Fisher distribution and cannot be analysed in this way. This situation is not necessarily important in all cases as the pole to the plane, the well-grouped axial

Palaeomagnetism

directions, can be used to describe the plane and how well it is defined. The main problems arise for triaxial ellipsoids $k_{max} > k_{int} > k_{min}$, if there are changes within a site as each group of sample axes may, for example, be 'smeared' into a great circle. Similarly, problems occur as a prolate ellipsoid changes towards an oblate shape, or vice versa, as the maximum axes are well grouped for the one case, but the minimum axes for the other – intermediate triaxial stages are thus difficult to analyse. In such circumstances, it seems clear that Bingham statistics (Section 6.4) are more applicable to the analyses of magnetic fabric data.

Chapter Seven

Archaeological applications

7.1 INTRODUCTION

As most archaeological materials contain magnetic particles, if only as impurities (Section 4.5), they will have magnetic properties that can be used in a variety of ways (Thellier, 1938; Aitken, 1974; Tarling, 1975a). One of the main properties is that they may acquire a remanence at some specific time. As the geomagnetic field changes direction and intensity in both time and space (Sections 8.2 and 8.3), the time of this acquisition can be determined by comparison of these magnetic parameters with known records of the past geomagnetic field at that locality. Such archaeomagnetic dating (Section 7.2) can be undertaken with unoriented samples, using intensity, or with oriented samples, using both the intensity and direction of remanence. Where the past variations of the geomagnetic field have already been established, archaeo-magnetic dating can be as accurate as the original dating of the record, but even when these records are absent, relative dating between different samples can be undertaken by the identification of differences in the magnetic parameters. If these differ and are related to the past geomagnetic field then the samples are of different age, while similar parameters would strongly indicate a similar age for their acquisition of a remanence. This means that relative dating is generally possible, even for times when absolute dating cannot yet be undertaken.

The actual magnetic properties can also be used in a variety of ways. The rate at which a sample acquires a viscous magnetization can, for example, be used as another way of archaeomagnetic dating (Section 7.2.4). However, the fact that most magnetic measurements can be made quickly, and without any other apparent effect on the sample, means that certain properties can be used to determine the sources for some materials, such as obsidians (Section 7.3), while the directions of remanence can be used to determine previous orientations of the artefacts when they were manufactured. Such observations therefore provide cheap, effective, non-destructive techniques that have an application in a wide variety of archaeological problems.

7.2 ARCHAEOMAGNETIC DATING

Observatory records indicate that the Earth's magnetic field gradually changes in direction and intensity by amounts that differ from one part of the

145

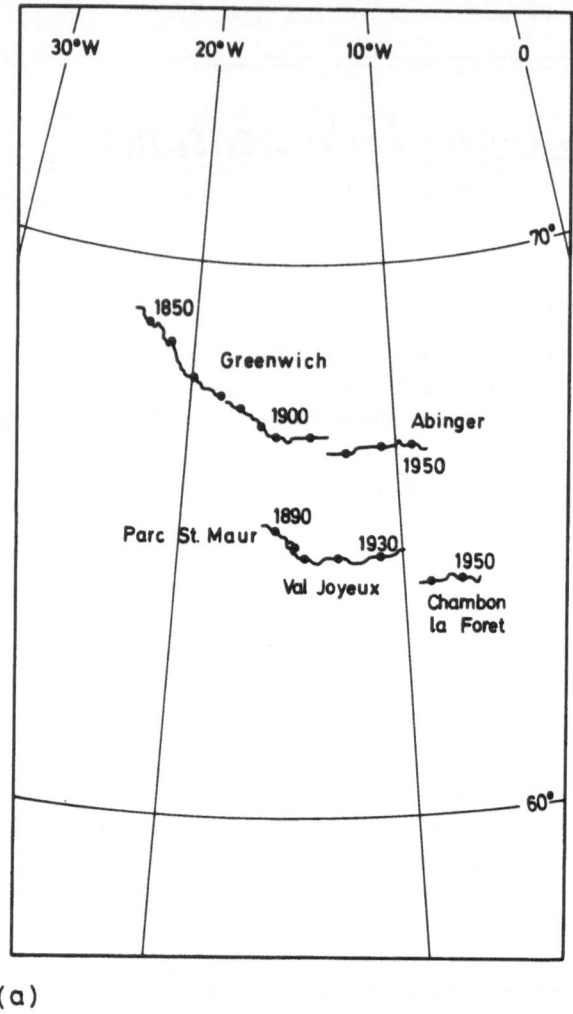

(a)

Figure 7.1 Secular variations at London and Paris, 1850–1950. (a) A stereographic projection of declination and inclination values. The upper curves refer to observatories in and near London, while the lower curves are based on changing locations for the Paris observatories. (b) The changes in intensity, μT, for the two locations are combined and also shown as percentage changes from the average value.

globe to another and also differ in time at the same locality (Sections 8.2 and 8.3). These long-term changes, secular variations, correspond, in London and Paris, to changes of some 0.25° year^{-1} in direction and 0.05% year^{-1} in intensity during the last 150 years or so (Fig. 7.1). If such changes are regarded as typical of the rate of secular change, then this would imply that archaeomagnetic dating, using direction changes, should be possible to within about

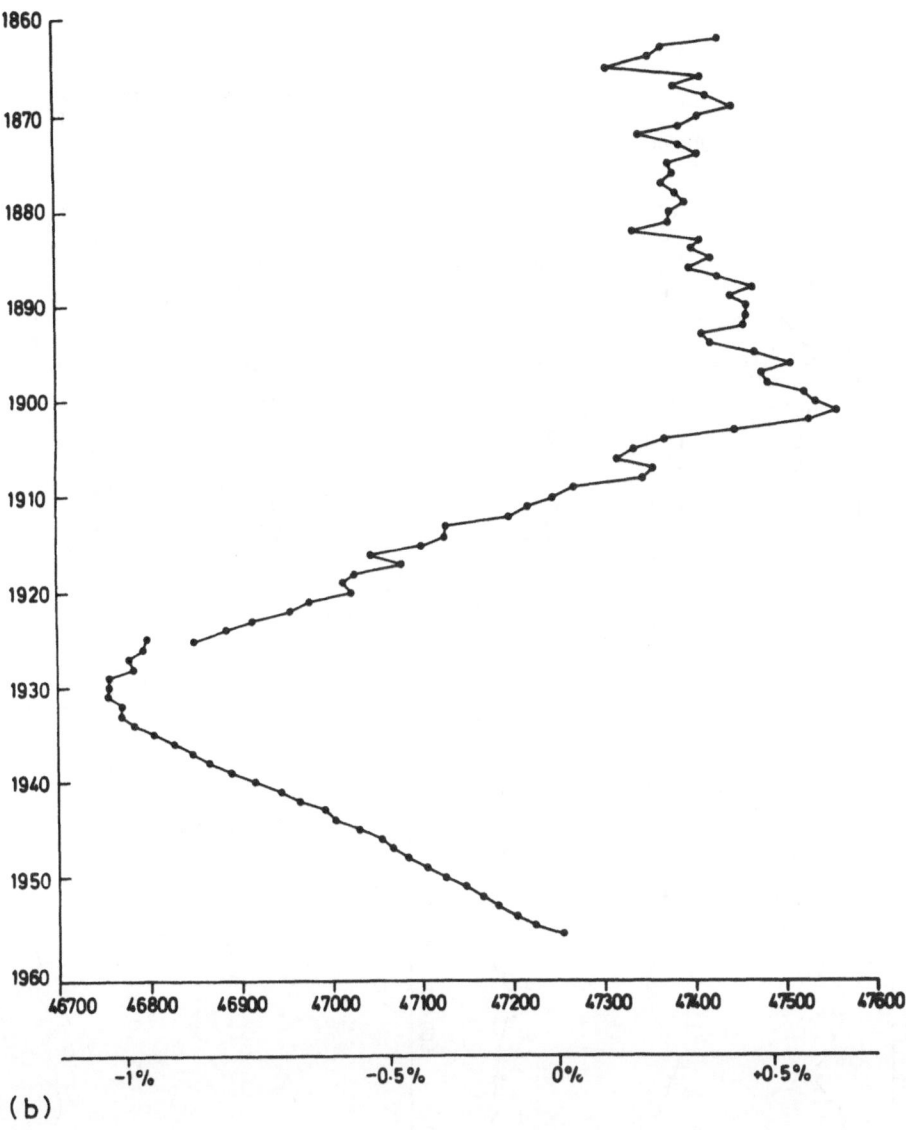

(b)

5–20 years, while intensity changes would give a potential dating accuracy of some ±20–50 years. However, studies of observatory records elsewhere indicate that the rate of change in direction can be both faster and slower, correspondingly increasing or decreasing the accuracy of the method. Changes in intensity also differ in their rate of change, in different areas at different times, but are generally slower than directional changes. Obser-

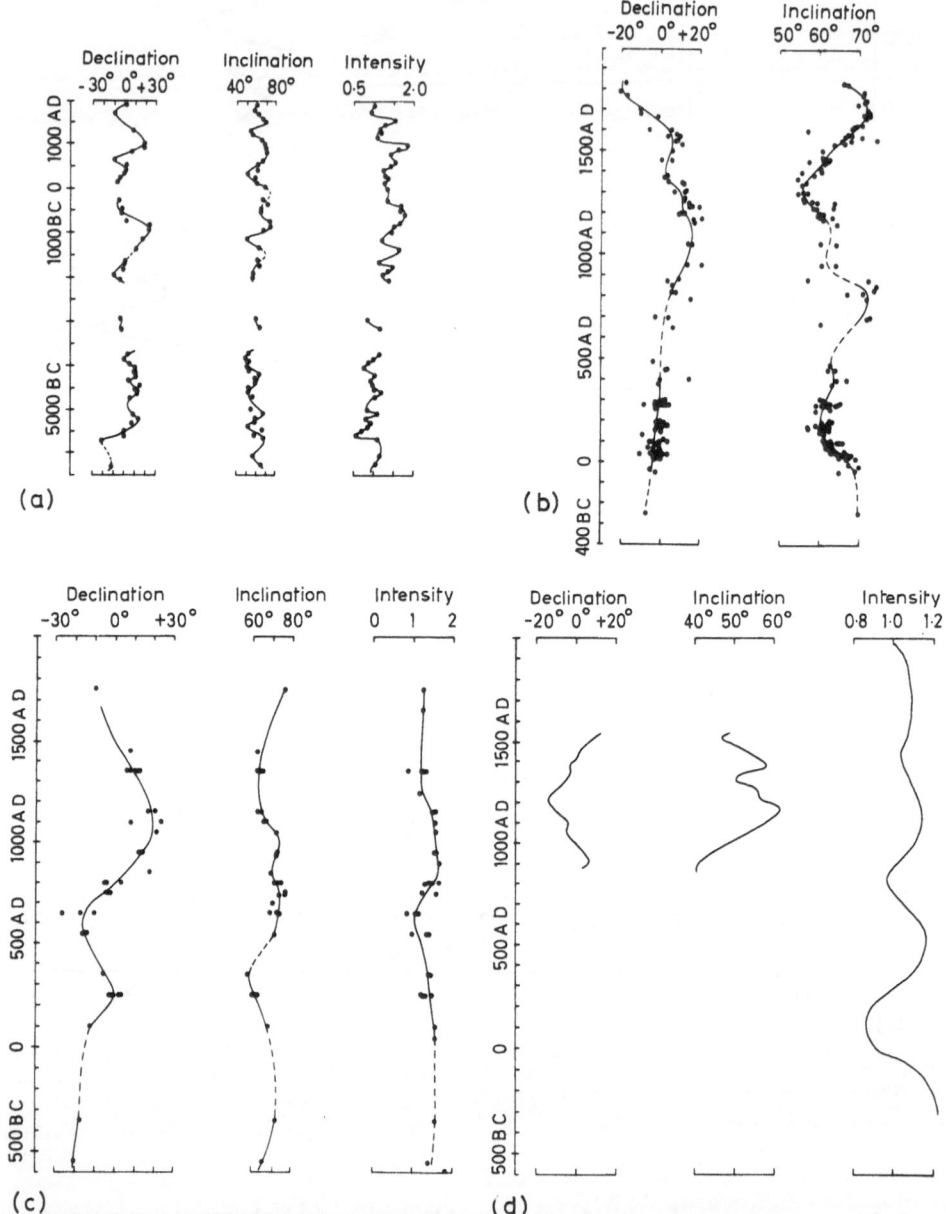

Figure 7.2 Archaeomagnetic master curves for declination, inclination and intensity. (a) Bulgaria (after Kovacheva, 1980). (b) Paris (after Thellier, 1981). (c) Ukraine (after Rusakov and Zagniy, 1973a,b). (d) Tucson, Southwestern USA (after DuBois, 1975). (See also Fig. 8.4.)

vatory records further indicate that the nature of such secular changes tends to be spatially similar on a scale of several 10 000 km². This spatial coherence is particularly important in archaeomagnetic dating studies as observatory records of secular variations are of very specific location and limited duration, mostly less than 100 years, and only a few extend back to around 1600 AD. Thus, 'master curves' of dated secular changes need to be constructed from archaeomagnetic studies of materials that have already been dated by other means (Section 7.2.3 and Fig. 7.2).

7.2.1 **Intensity**

The main advantage of dating based on the changing intensity of the geomagnetic field is that unoriented material can be used (Fig. 7.3). This means that there is an abundance of suitable materials in most archaeological sites as fired clays are common. Furthermore, such materials can often be well dated archaeologically for construction of master curves as pottery types are a common method for stratigraphic dating in an archaeological context. Until recently, the number of results has been limited because more measurements are required to establish the strength of field in which the magnetization was acquired (Sections 5.6 and 8.3; Fig. 7.3) and there are more sources of error than for directional studies – mostly arising from the need for repeated heating and cooling. Attempts to reduce these errors, such as by use of alternating magnetic fields (Shaw, 1974; Bagina and Petrova, 1976; Gunn and Murray, 1980), have been of varying success, making the reliability of any one observation somewhat uncertain. The development of systems for rapid heating and cooling, using small samples measured with a cryogenic magneto-meter, has vastly increased the speed of such investigations (Walton, 1979), but serious errors can still arise from

1. Inhomogeneity
2. Anisotropy (due to mineral alignments, such as may occur during pot manufacture, as well as the shape of the pot itself)
3. Differences between the original and laboratory firing atmosphere and
4. The original and laboratory heating and cooling rates (Section 5.6).

Nonetheless, the speed of this method and its ability to use samples a few millimetres in dimension means that it can now be readily applied to a wide range of problems. While there are uncertainties about the magnitude of many of the corrections (anisotropy, shape, cooling rates, etc. – Section 5.6), the similarity of experimental runs suggests that many of these errors are about constant for most archaeological materials. For example, the original cooling rate for pottery is likely to be very similar to avoid contractional cracking and the cooling rates used in the same laboratory are generally the same for each run. On this basis, the absolute palaeointensity values may be in error by a few per cent, but the relative intensity differences are likely to be valid. Clearly some problems may arise in comparing results from different laboratories and also in attempts to determine the absolute geomagnetic field intensity (Chapter 8) or comparing over large distances (several 10^3 km – Section 6.2).

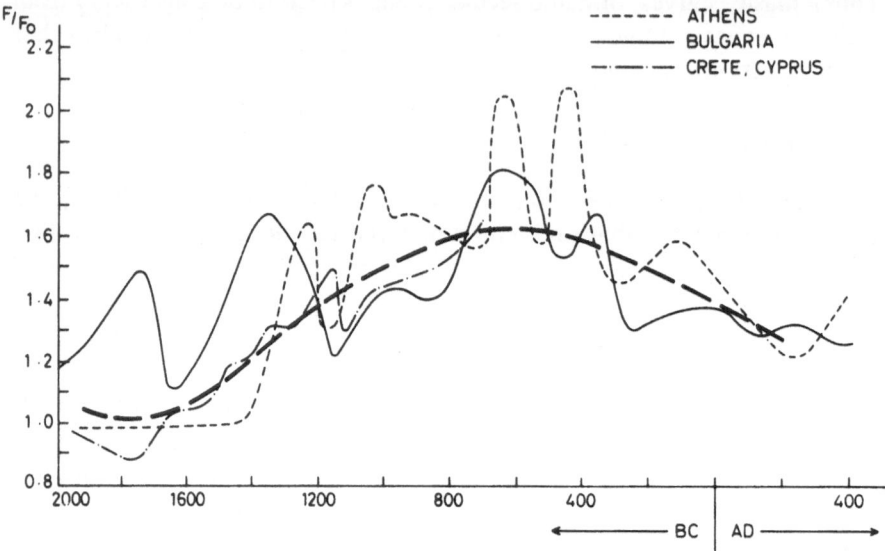

Figure 7.3 Palaeointensity variations in Southeastern Europe. The thicker dashed line represents an average curve for palaeointensity observations in the Aegean region, mostly Athens and Crete (Walton, 1979; Liritzis and Thomas, 1980; Shaw, 1979). The continuous solid line is for palaeointensity variations in Bulgaria (Kovacheva, 1980).

So far, most palaeointensity studies have been localized (Smith, 1967a,b; Burlatskaya and Nachasova, 1977; Kovacheva, 1982; Liritzis and Thomas, 1980), often being concerned with local problems in, for example, Japan, the USSR, southeastern Europe (Fig. 7.3) and the southwestern United States. There is some evidence for very rapid changes in the geomagnetic field at specific times (Walton, 1979) which would allow very precise dating by this method under such circumstances. It is not certain, however, whether such differences have arisen by, for example, the use of materials that have had different original cooling histories.

7.2.2 Directions

As with intensities, age determination using directions of remanence requires a dated master curve of the directional changes of the geomagnetic field (Section 7.2.3 and Fig. 7.2). This may comprise the horizontal (declination) or vertical component (inclination), or preferably both. Unlike intensity studies, the samples must have acquired their magnetization in a known position. Ideally, therefore, this method of dating is best applied to archaeological materials that are still *in situ*, i.e. they have not moved since they acquired their original magnetization. Fired materials, such as hearths and small kilns, thus provide excellent material for archaeomagnetic directional dating (Section 4.5), and geological materials can also be used in some situations (Hoye, 1981). Sediments, within an archaeological context, may well have a magnetization associated with the time of their deposition (Section 4.3) and the direction of this remanence can be measured in oriented *in situ* samples. It is occasionally possible to use materials that are no longer *in situ* if they can be at least partially oriented. Some pots, for example, had their final firing with either their rims or bases in a horizontal position, as can be judged from glaze runs. In such materials, it is possible to measure the inclination of their magnetization and relate it to records of the past inclination of the geomagnetic field. Unfortunately, while decorative, non-utilitarian pottery may frequently have been fired in horizontal positions, most domestic pottery, after its initial biscuit firing, was usually stacked in quasi-random orientation within the kiln. (Conversely the magnetization of shards can be used to assess the method of stacking in the past or for reconstructing pots – Section 7.3.)

Most fired archaeological materials are quite strongly magnetic and their magnetization can be measured rapidly and accurately (Section 5.3). Most of the errors in this method of dating therefore arise from orientation determinations (Section 5.2) and particularly from the possible presence of undetected tilts. Kiln walls, for example, frequently fall outwards (Harold, 1960) and differential subsidence, causing tilting, may well occur in many archaeological structures such as hearths, burnt walls, etc. Where such tilts exceed a few degrees, they will usually be visible, but movements up to 5° or so may well occur without recognition (Hoye, 1982). Under such circumstances, unsuspected consistent errors could be present, the magnitude of which depends on the relationship between the direction of remanence and the direction of tilting (Section 5.2.4). Without such errors and with adequate sampling (Section 6.3), an accuracy of some 1–2° can be expected for stably magnetized materials, corresponding to some ±5 years for times when directional changes were at similar rates to those observed in London and Paris during the last 200 years or so. A more realistic estimate of the likely errors in any one stably magnetized site yields a potential precision of some ±20 years, but higher or lower accuracy may occur when the rate of secular variations was greater or lesser than that observed in Western Europe during the last two centuries.

Records of the geomagnetic field show occasional repetitions of declination or inclination values (Fig. 7.2). Clearly materials magnetized at such times will have similar directional properties, and thus could be assigned to two or more possible ages. In many instances, however, the archaeological context is sufficiently restrictive to eliminate most 'cross-over' points. It is, however, rare for all three geomagnetic parameters – declination, inclination and intensity – to be the same, so the determination of both directional and intensity parameters is thus usually diagnostic of a specific time.

So far, most archaeomagnetic studies have been concerned to establish the validity of the technique and to obtain master curves for dating purpose (next section). In other studies, the samples have been of parochial interest. However, in one particular case, the archaeomagnetic dating has been of special interest. It has often been considered that the sudden end of the Minoan civilization in the Eastern Mediterranean, c. 1450 BC, was associated with the eruption of the island of Santorini (Thira) in which the ashes destroyed the crops and tsunami destroyed the fleets. This event is also frequently linked to the legend of Atlantis. The archaeological evidence suggests a short time gap between the eruption and the destruction of the

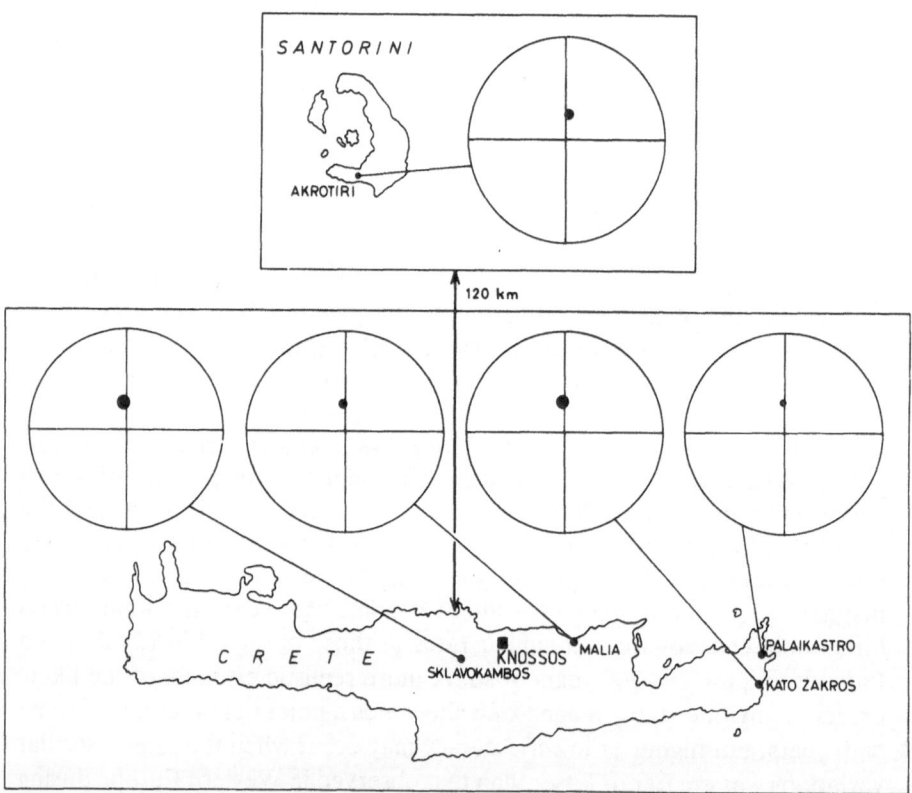

Figure 7.4 Late Minoan directions, Aegean, Mediterranean. The directions for the fired materials from Akrotiri, Santorini, differ from those of the destruction levels of Minoan Palaces in Crete. This confirms the archaeological evidence that the former took place in Late Minoan IA times, and the Cretan Palaces destructions were later, Late Minoan IB. (Work in progress; Downey, 1981, 1982, 1983.)

Minoan thalassocracy as Late Minoan IA pottery is found on Santorini, but the destructions on Crete appear to be associated with the later Late Minoan IB styles – although the magnitude of the time gap is unclear. It is even uncertain, for example, whether the destructions on Crete were themselves simultaneous, or occurred over several years. Standard dating methods, such as ^{14}C, have errors associated with them that far exceed the possible age differences. Magnetic studies of the igneous ashes and archaeological materials in the region (Fig. 7.4) showed that the main explosive eruption of Santorini was preceded, with a clear time gap, by a more standard eruption (Plinean) that was contemporaneous with the destructions on central Crete. The explosive eruption, when an 80 km^2 caldera was formed, was later and this was approximately contemporaneous with the destructions in Eastern Crete (Tarling, 1978c; Downey, 1981, 1983). The consequences of this relative dating have still to be considered, but provide the first scientific evidence for such age differences and similarities.

Sediments may also preserve records of the geomagnetic field direction at the time of deposition (Fig. 7.5), but such records may often be affected by currents operating during deposition and by post-depositional compaction, chemical changes and water movements (Section 4.3). Whilst clearly less reliable, such materials are quite common in archaeological environments (Hassan, 1978), occurring within ditches, post-holes and so forth, as well as in associated geological deposits such as lake and pond sediments, cave deposits (Creer and Kopper, 1976; Nöel et al., 1979; Hammo-Yassi et al., 1982) and puddled clays (Games, 1977). Such materials may well provide an indication of the age of the sediments in the absence of better methods, even though some smoothing may be necessary to extract the geomagnetic 'signal' (Fig. 7.5).

7.2.3 Master curves and assessment of standard methods

As observatory studies of the geomagnetic field only extend back for, at most, 400 years, it is only relatively recent material that can be dated by direct comparison with observatory records. The extension of such records over archaeological time therefore requires archaeomagnetic observations based on well-dated materials for earlier times. Obviously, such records must depend on the adequacy of both the archaeomagnetic observations and the dating methods employed. Precise dating, sometimes in calendar years, can often be obtained, for any given site, from either historical records or, more commonly, from the stratigraphic information available for that site. Shards, associated with the site, may, for example, provide a very precise control on the age of any particular level. Studies in Europe, in particular (Kovacheva, 1980; Thellier, 1981; Clark et al., 1983), and to some extent in the south-western United States (DuBois, 1975), the Ukraine (Rusakov and Zagniy, 1973a,b), and Japan (Kinoshita, 1970; Hirooka, 1971), have established reasonably precise curves for declination, inclination and intensity changes over time-scales of 2 to 8000 years for these areas (Fig. 7.2).

Conveniently, geomagnetic changes tend to be regional (Section 8.2), which means that data from a wide area can be incorporated into constructing such master curves (Section 6.3.1). Thellier (1981), for example, uses data from most of France, partially corrected to Paris (Fig. 7.2(b)); Clark *et al.* (1983) corrected British data to a central location, Meriden, and DuBois (1975) corrected southwestern United States data to Tucson (Fig. 7.2(d)). The area of Bulgaria was considered sufficiently small (Kovacheva, 1980) for no corrections to be necessary (Fig. 7.2(a)). Obviously such corrections could introduce uncertainties as the actual spatial variations of secular change have not yet been adequately established, but it seems likely that the errors arising from such 'corrections' are likely to be less than 1° in direction (Section 6.3.1) and less than 1% in intensity (Section 6.2). It is important to emphasize, however, that the master curves already available should not be regarded as final. Techniques continue to be improved, thereby isolating the stable component more precisely and improvements in sampling and measurement also allow more samples per site to be measured and thus the direction and intensity to be better defined (Sections 6.2 and 6.3). Similarly, an understanding of the effects of anisotropy, rates of heating and cooling, the sources of orientation errors and so forth, are now better appreciated. However, the main sources of error in the existing curves do not arise from the quality of the archaeomagnetic determinations used so much as from the reliability of the original dating.

Where historical dating of the master curves is available, high precision can be obtained, but many archaeological sites that have been used have been dated from their context, the evaluation of which may change. More serious dating errors can arise from other dating methods. ^{14}C dating, for example, may be slightly in error when uncorrected for the dendrochronological correlation (Suess, 1970b; Ralph, 1972; Stuiver, 1978), but may be drastically in error where old carbon is present, particularly in sediments that may contain geological carbon, such as may occur in lake sediments derived from limestone terrains. As with most techniques, the errors tend to increase with age so the errors in ^{14}C become more significant on a scale of several hundred years for materials several thousand years old. As better dated materials

---

Figure 7.5 Comparison of archaeomagnetic curves based on fired materials and sediments. (a) Comparison of lake sediment observations, fine line, with the standard archaeomagnetic curve for Britain, thicker line. The sedimentary data are based on work on Lake Windermere and Loch Lomond (Thompson and Turner, 1979; Turner and Thompson, 1979). (b) Comparison of observations of pond sediments in South Yorkshire, Britain, with the archaeomagnetic curve. The pond data are based on a three point running mean, possibly accounting for the reduction in the amplitude in the swing of declination. The apparent shallowing of the direction in the sediments is very clear.

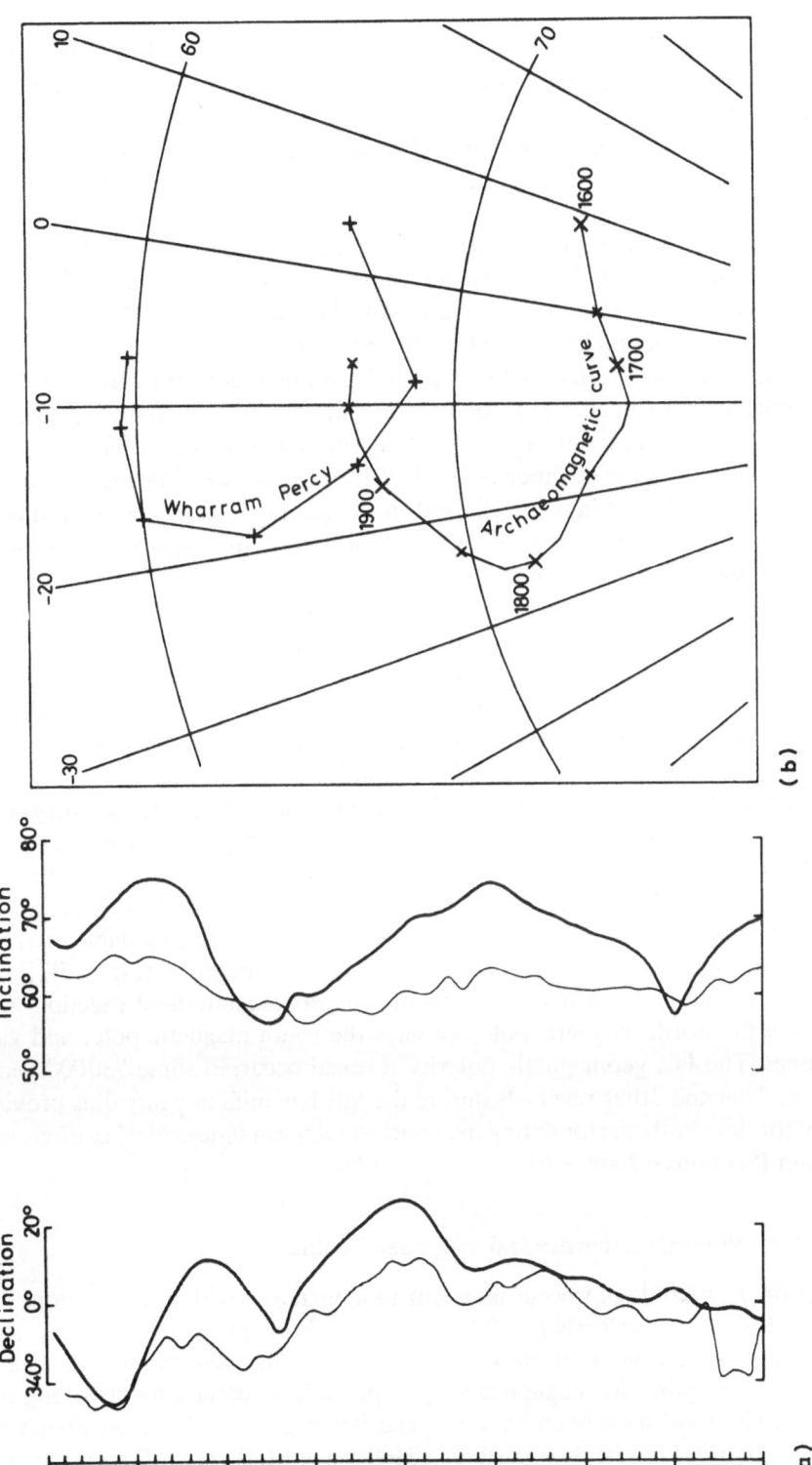

(a)

(b)

become available, it is essential that they are incorporated into the extant master curves, thus leading to a greater accuracy in archaeomagnetic dating techniques.

The magnetization of long cores of present-day lake sediment (Section 5.2) provides a rapid method of determining the overall character of directional changes in the geomagnetic field at precise locations for the last 10000 to 15000 years. Studies in Britain, for example, show that swings in declination and inclination (Fig. 7.5) can be correlated with the archaeomagnetic record established on fired materials (Turner and Thompson, 1979). The maximum westerly declination in observatory records, around 1825 AD, can be readily identified at approximately the expected level in a sediment core (Fig. 7.5). Similar swings can also be recognized in lake sediments in many parts of the world (Section 8.3 and Fig. 8.5) such as the Great Lakes (Banerjee *et al.*, 1979), Lake Tahoe (Palmer *et al.*, 1979), western New England (Verosub, 1979a,b) in the United States, and in Australia (Barton and McElhinny, 1981). The reliability of such records is not good in terms of the absolute directions of the geomagnetic field as such swings can vary in amplitude by several tens of degrees, even for sediments from the same lake of the same age (Section 8.3). Ultimately, precisely dated sediments, such as the glacial varved clays in Scandinavia and North America, may provide the best dated record of secular variations in directions (Section 8.3) for the last 15000 years or so, but such studies are in their infancy, despite such materials being amongst the earliest to be studied (Section 1.2).

Much older and more dramatic changes in the direction of the geomagnetic field may also prove useful for archaeomagnetic dating (Sections 8.4 and 9.2). The reality of most major geomagnetic excursions has not yet been well established and it is also unclear whether they occur on a local or world-wide scale. If they do occur on a large scale, they could provide invaluable markers for correlation purposes between different areas of the world (Fig. 9.7). Even more dramatic are actual reversals of the geomagnetic field (Section 8.4) when the north magnetic pole becomes the south magnetic pole, and vice versa. The last geomagnetic polarity reversal occurred some 730000 years ago. This and other reversals during the last few million years thus provide major time horizons for dating over anthropological time-scales, as discussed later (Sections 9.2 and 9.6).

7.2.4 Viscous remanence and 'alignment' dating

As the magnitude of viscous magnetization increases with time (Sections 2.3 and 4.6), and it is clearly possible to measure the total viscous magnetization present in a specimen (Section 5.5.1) it should, after determining the rate at which it acquires its magnetization, be possible to determine how long the specimen must have been lying in a specific orientation in the geomagnetic field. In order to test this possibility, Heller and Markert (1973) investigated the viscous remanence in three (igneous) stones from the Roman Wall, northern England, that had been constructed in the 1st century AD. The agreement between the samples was very poor, although the results could be interpreted as being consistent with an age of almost 2000 years. The uncertainties seem to arise predominantly from assessing the rate of acquisition of viscous remanence (in view of the internal discrepancies) but other sources of error are likely to arise from the fact that the strength of the geomagnetic field changes with time and the observed natural viscous remanence will represent an integration of such changes. Nonetheless, while this technique will not provide an accurate, absolute dating method, it has a potential for distinguishing between, for example, objects buried during the last few tens or several hundred years, for example, the case of the Glozel materials (McKerrell *et al.*, 1974).

Yet another possible method of archaeomagnetic dating is based on the fact that it was not realized in Europe until the end of the 16th century that the direction of magnetic north differed from that of true north (Section 1.2). Many buildings constructed before this time were therefore constructed using magnetic compass orientations assumed to correspond to true North (i.e. they were thought to point to the pole about which the universe was thought to rotate). Such structures were therefore aligned along the direction of magnetic declination at that time (Searle, 1974) and so it should be possible to date these constructions, particularly those associated with religious purposes, by comparison of their actual orientation with the archaeomagnetic records for changes in declination that are now becoming available (Section 7.2.3).

7.3 OTHER APPLICATIONS

Although archaeomagnetic dating has been undertaken for many years (Thellier, 1938a,b), the number of detailed studies is still few. Even fewer are the considerations of how the magnetization of archaeological materials can be utilized in other ways. Until recently, the main application of this property has been in terms of geophysical exploration, i.e. using portable magnetometers to detect the presence of materials with a different susceptibility or remanence compared with the surrounding materials. Such techniques (Aitken, 1974) are clearly of continuing importance, particularly as faster and more sensitive magnetometers have been developed allowing surveys to be carried out more rapidly and more efficiently than previously (Kuznir and Tarling, in press). Such exploration systems will not be discussed here, but more speculative applications will be considered. These are intended as indications of ways in which archaeo- and palaeomagnetic techniques and analyses have been used for unusual purposes, with the hope that these examples may stimulate consideration of other archaeological problems for which magnetization may offer at least partial solutions.

7.3.1 Sourcing (provenance) of archaeological materials

The determination of the source of various materials, provenancing, is of major importance in establishing the existence and extent of trading contacts between different regions. In some instances, it seems probable that the magnetic properties may provide a rapid, non-destructive, cheap method of either establishing the provenance, or limiting the number of potential sources. As most archaeological materials contain magnetic minerals (Section 4.5), the composition and proportion of these minerals may be diagnostic of the location of manufacture or derivation. Metallic objects, for instance, were characterized by quite high impurity contents until the advent of electrolytic smelting. Most analyses of such objects require the destruction of at least parts of the object in order to undertake a determination of the composition of such impurities, from which the original source of the ore can be determined, e.g. lead impurities in coins (Brill, 1970; Gale, 1978). Coins and other objects appear to carry a readily determined proportion of magnetic impurity (Section 4.5) that may be of value in such studies – but has not yet been investigated.

An example of the potential of such sourcing is that of obsidian artefacts from the Eastern Mediterranean region, but it is emphasized that similar principles may well apply to a wide variety of archaeological materials, both natural and man-made. Obsidian is a form of volcanic glass that has a conchoidal fracture, making it very suitable for the manufacture of arrow heads and the like. Such artefacts occur in many prehistoric sites in Europe and are generally abundant and well preserved, but there are only about 20 potential geological localities from which these can have been derived. The provenance of such artefacts therefore provides an ideal method of establishing trading patterns in prehistoric times and, if they could be provenanced in sufficient numbers, it would be possible to quantify the magnitude of such trade routes (Cann and Renfrew, 1964; Renfrew et al., 1966; Wright, 1969; Cann et al., 1969). Most available techniques, such as major element analysis, petrological studies and so forth, are not very effective discriminants (Renfrew et al., 1965), with the exception of minor elements analyses, such as those based on neutron activation and X-ray fluorescence analysis (Gordus et al., 1968; Aspinall et al., 1972; Hallam et al., 1976) which require skilled operators and are slow, destructive and expensive ($>£100 \approx \$200$). As obsidians are rapidly cooled igneous rocks, they contain magnetic impurities, the composition of which is likely to be quite distinct for different obsidian flows as the solid-solution series of the titanomagnetites and ilmenohaematites (Sections 3.2 and 3.3) are unlikely to have had time to exsolve into their end members. On this basis, many of the magnetic properties of different flows should differ, reflecting differences in the composition, grain size and concentration of such minerals. Furthermore, obsidians cooling at different times will become magnetized in geomagnetic fields of different strength. It thus seems probable that obsidians may have magnetic properties that are sufficiently restricted to specific flows to allow these properties to be used for characteristic artefacts derived from them. McDougall *et al*. (1983) found that many known obsidian sources in the Eastern Mediterranean could be identified simply on

1. Their initial intensity of remanence
2. Their low-field susceptibility and
3. The magnetization acquired when placed in a direct magnetic field of 350 mT (Fig. 7.6).

It also appeared that determination of the magnetic stability characteristics when subjected to alternating magnetic field demagnetization was likely to be a further 'finger print', although the three simply measured parameters were often sufficient for source identification. Furthermore these parameters can be determined in the field should laboratory analysis prove impracticable. In this particular case, therefore, it appears that non-destructive, rapid (about 5–10 minutes per sample) magnetic 'sourcing' can be undertaken on obsidian

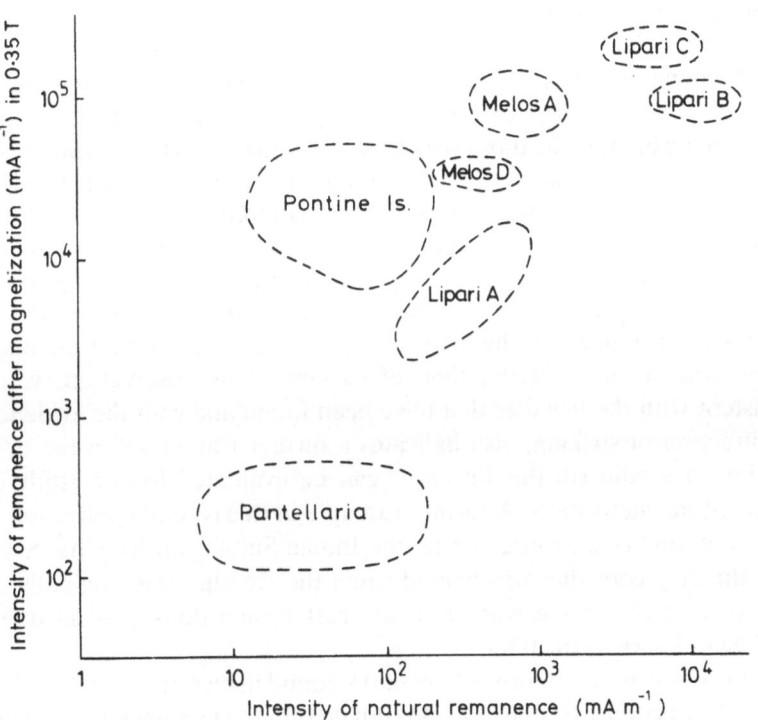

Figure 7.6 Obsidian sourcing using magnetic properties. The magnetic properties of obsidians that are easiest to measure also provide quite effective discrimination between different obsidian sources.

artefacts greater than 0.7 mg in weight (a few millimetres in dimensions), with most known sources being identifiable. (Even smaller samples may be usable with cryogenic magnetometers.) Where such magnetic provenancing is difficult because of similar characteristics at different sites, as occurs with all sourcing methods, other magnetic properties may still prove effective and, as a minimum, the magnetic method appears to provide a rapid way of sorting artefacts into those readily sourced and those that require more expensive sourcing methods.

7.3.2 Technological and other potential applications

As with provenancing, there are few studies of the ways in which magnetic properties of archaeological materials can be utilized. The iron impurity content of cast metals may, for example, reflect the technological state of smelting, as well as the types of ore being used and hence provenance. Coins made since the widespread use of electrolytic smelting (19th–20th century) have virtually no magnetic content (Tarling, 1982a) but studies of older coins may prove of interest, especially when most magnetic properties of, for example, a coin can be examined quickly and returned with no change in anything other than its magnetization. Again, coins cast by centrifugal processes are likely to have a different remanence from those that have been cooled when struck horizontally. Similarly, the composition of the die, or of iron or steel bands supporting it, is likely to distort the direction of the geomagnetic field, while those struck with non-magnetic tools alone may well reflect the geomagnetic field at the time that they were struck. Unsilvered Roman coins, for example, appear to have acquired a remanence at the time that they were struck that is steeper than that expected for the locality of the mint and may be attributable to the presence of an iron supporting collar for the bronze dies (Tarling, 1982a), but the fact that the emperor's head always appears to have been on the lower die appears to confirm that the magnetization was acquired at the time of striking. This observation, which is consistent with the few dies that have been found and with the evidence for the direction of striking, also indicates a further way in which the original position of a coin (in this instance) can be evaluated from a study of its remanent magnetization. A further example of this type of application is for the orientation of a bronze statue, the Indian Sultanganj Buddha. Samples from the clay core that was heated when the Buddha was originally made indicate that the statue was originally cast upside down, facing due east (Tarling in Learmouth, 1983).

As the Curie temperature of haematite, found in many fired archaeological materials (Section 4.5), is 675°C, it is not possible to use magnetic techniques to assess temperatures of heating above this value. However, firing at lower temperatures will mean that a pot, for example, originally fired at say 800°C, and then later heated to only 300°C, will have two components of remanence.

The lower temperature component will normally have a different direction and thus the temperature of the second heating can be determined from the temperature at which it is destroyed during incremental thermal demagnetization (Sections 5.4 and 9.5.2). Similar considerations also apply to the assessment of the emplacement temperature of volcanic tephra (Section 9.5) covering archaeological sites, such as those of Pompeii, Italy, and Akrotiri, Greece, and the temperature of natural or deliberate firing of buildings (Incoronato, 1981; Downey, 1982b).

Most pottery, bricks and so forth are reddened by the firing process. This is largely in response to the reduction of other iron-bearing minerals and produces very fine grained haematite, often less than 100 Å in diameter. However, the redox conditions during firing are often such that it is only the outer few millimetres that show this change. The presence of a darker layer of unoxidized material between the reddened layers can readily be determined magnetically as the outer red layer usually contributes little or nothing to the total remanence of the material since haematite of this grain size is superparamagnetic (Section 2.3). It seems possible that studies of the magnetic mineralogy, when combined with other evidence, may be important indicators of the redox conditions in furnaces and kilns. Apart from a technological interest, such features may be important in archaeomagnetic determinations of, for example, palaeointensity (Section 7.2.1) for which it is important to reproduce, as far as practicable, the original conditions of firing and cooling (Section 5.6).

Pottery stacking in a kiln is also likely to have varied from time to time. Previous random stacking of pots, tiles, bricks, and the like during firing can be determined from the random magnetization of each unit, while systematic packing would result in systematic directions of magnetization. Pottery reconstruction can similarly be attempted as each shard from an approximately spherical pot will have a unique direction of remanence relative to its outer surface. As these directions were acquired parallel to each other at the time of firing of the original pot, each shard can be relocated relative to other shards from the same pot (Burnham and Tarling, 1975). Such a technique is clearly not as efficient as skilled fitting of shards, but has a clear potential in positioning isolated shards that cannot be fitted, thus determining the probable shape of the vessel.

Chapter Eight

Geomagnetic applications

8.1 INTRODUCTION

The geomagnetic field is one of the most intriguing aspects of the Earth and is also one of its earliest geophysical properties to be studied scientifically (Gilbert, 1600) yet its origin can still only be loosely 'explained' in terms of magnetohydrodynamic motions in the Earth's outer core. An understanding of the origin and behaviour of the geomagnetic field is therefore crucial to any understanding of the core itself, the interaction between the core and mantle, and the origin of the Earth itself. Conversely, much of the information on the properties of the core and lower mantle is based on geomagnetic observations.

A major constraint on any investigation into the cause of the field, based on present-day observations, is that many core processes can be expected to have time-scales of the order of a few thousand years. Conductivity considerations, for example, indicate that any electrical currents, unless regenerated, would decay with a half life of some 2–4000 years. Evaluating such processes ideally requires observations carried out over somewhat longer time-scales. Palaeomagnetic observations are thus crucial in making such assessments.

Apart from the intrinsic value of such palaeo-geomagnetic studies, the hypothetical models for the nature of the past geomagnetic field are fundamental to most of the geological interpretations that have revolutionized the understanding of the tectonic processes that have and continue to operate on the rocks of the Earth's surface. It is therefore not possible to separate the geological (Chapter 9) and archaeological (Chapter 7) applications from discussion of the validity of such geomagnetic models. In this chapter, therefore, the relevant properties of the present geomagnetic field are considered first, mainly determined from present-day and direct historical measurements (Section 8.2). The properties of the field, as indicated by the properties of the field over the last few thousand years, from evidence of archaeomagnetic and palaeomagnetic studies, are then considered (Section 8.3) before those of even longer time-scales (Section 8.4). The average nature of the geomagnetic field throughout geological time is then outlined (Section 8.5).

162

8.2 THE PRESENT GEOMAGNETIC FIELD AND HISTORICAL OBSERVATIONS

Information on the strength and direction of the geomagnetic field comes from three main sources – magnetic observatories, surveys on land and sea, and satellite measurements. The most detailed records are from observatories that record changes in the geomagnetic parameters on a range of time-scales. Regional surveys cannot be carried out instantaneously but are mostly repeated every 5 to 10 years in the more habited regions. Satellite observations are made continuously, but have generally lower precision because of their greater distances from the source. Examination of observatory records shows many forms of short-period changes, some of which are regular, such as the 12 and 24 hour cycle, monthly changes and annual cycles. Others are irregular, such as magnetic storms associated with sun-spot activity on the Sun. These forms of short-term changes are termed *transient variations* and are of the order of 0.1% of the total field, although magnetic storms may be of the order of 1% of the normal field.

Information from observatories and surveys has been combined to determine the spatial variations of the geomagnetic field at 10 year intervals, for inclination and intensity, and at 5 year intervals for declination. Such maps (Fig. 8.1) involve smoothing to remove variations going from land to sea, the local effect of oceanic islands, the equatorial jet stream and so forth. (These are frequently smoothed by the use of second or third order polynomials over areas of some 20–40°.) More recently, satellite observations of the intensity of the field have been analysed to produce maps based on spherical harmonics up to the eighth order. (The coefficients of the International Geomagnetic Reference Field are now published annually.) Since about 1965, therefore, there has been adequate global coverage to provide reliable spatial coverage of the geomagnetic field (Barraclough *et al.*, 1975; Barraclough, 1982), but the data for Antarctica and many parts of the oceans must be regarded with caution for times before 1955 and much of the data are unreliable, except at observatories (Institute of Geological Sciences, 1978) and for specific areas, such as India, for times before 1915. Local data are, however, available for earlier times, particularly at observatories, such as London (Malin and Bullard, 1981), Paris and Batavia. Observations, of variable reliability, were also made by many of the early navigators, many of whom took regular declination readings, with both declination and inclination readings when in harbour. Thus many of these earlier data, although less reliable than observations made with modern instruments, nonetheless provide major sources of information on the longer term variations in the geomagnetic field (Braginskii, 1972; Barraclough, 1974; Thompson, 1982).

The present geomagnetic field (Fig. 8.1) corresponds closely to that of a uniformly magnetized sphere, as originally postulated by Gilbert (1600), in that it has two magnetic poles, although the northern pole at 73°N 100°W is

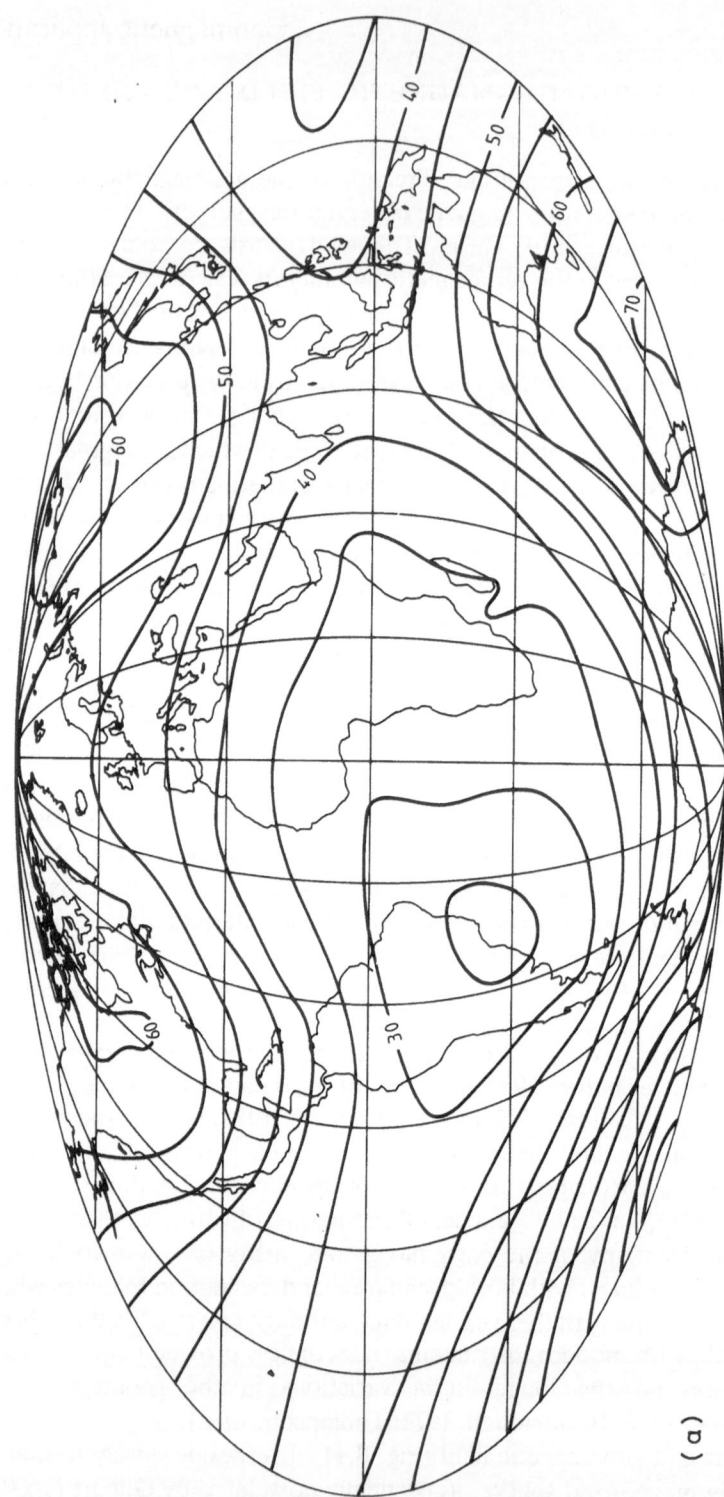

(a)

Figure 8.1 The present geomagnetic field. (a) Total intensity (*F*) of the field in mT.

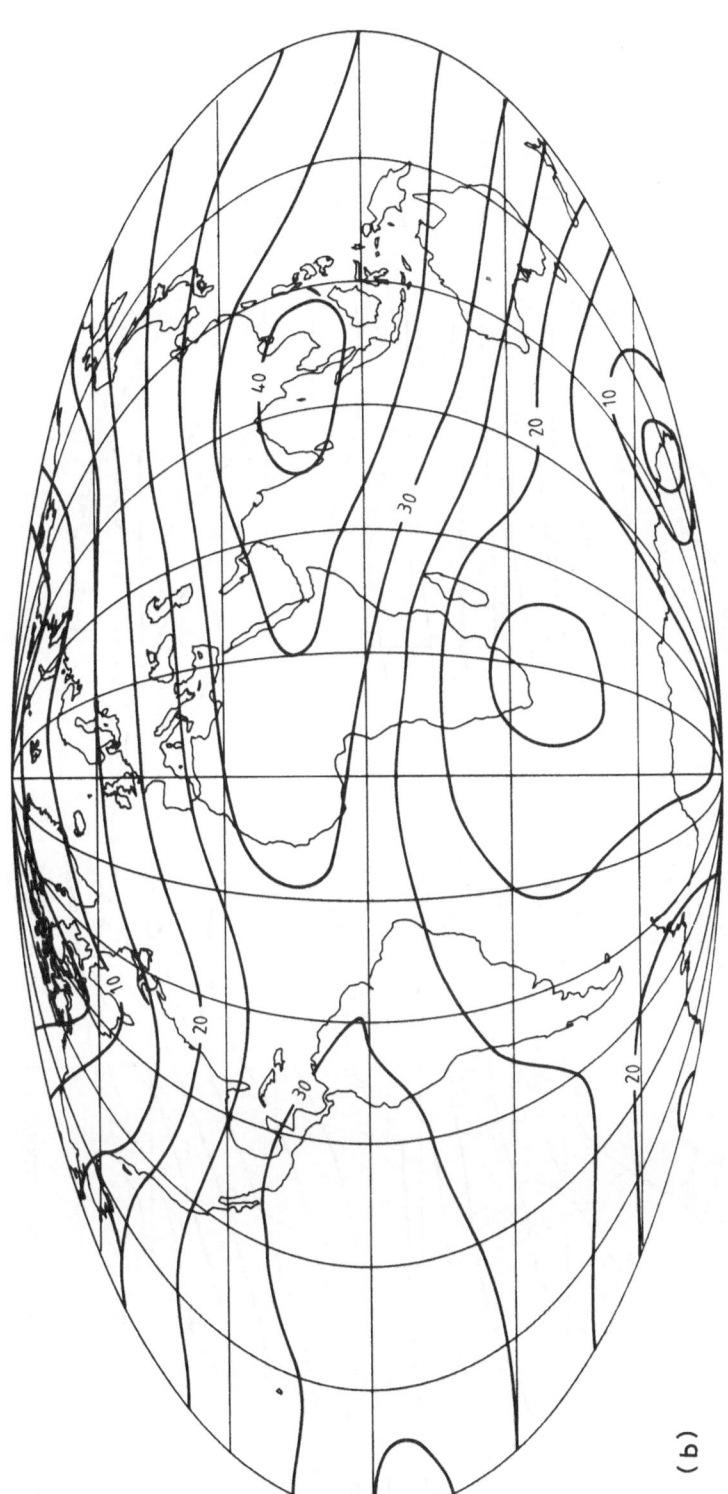

Figure 8.1 (b) The intensity of the horizontal component (*H*) in mT.

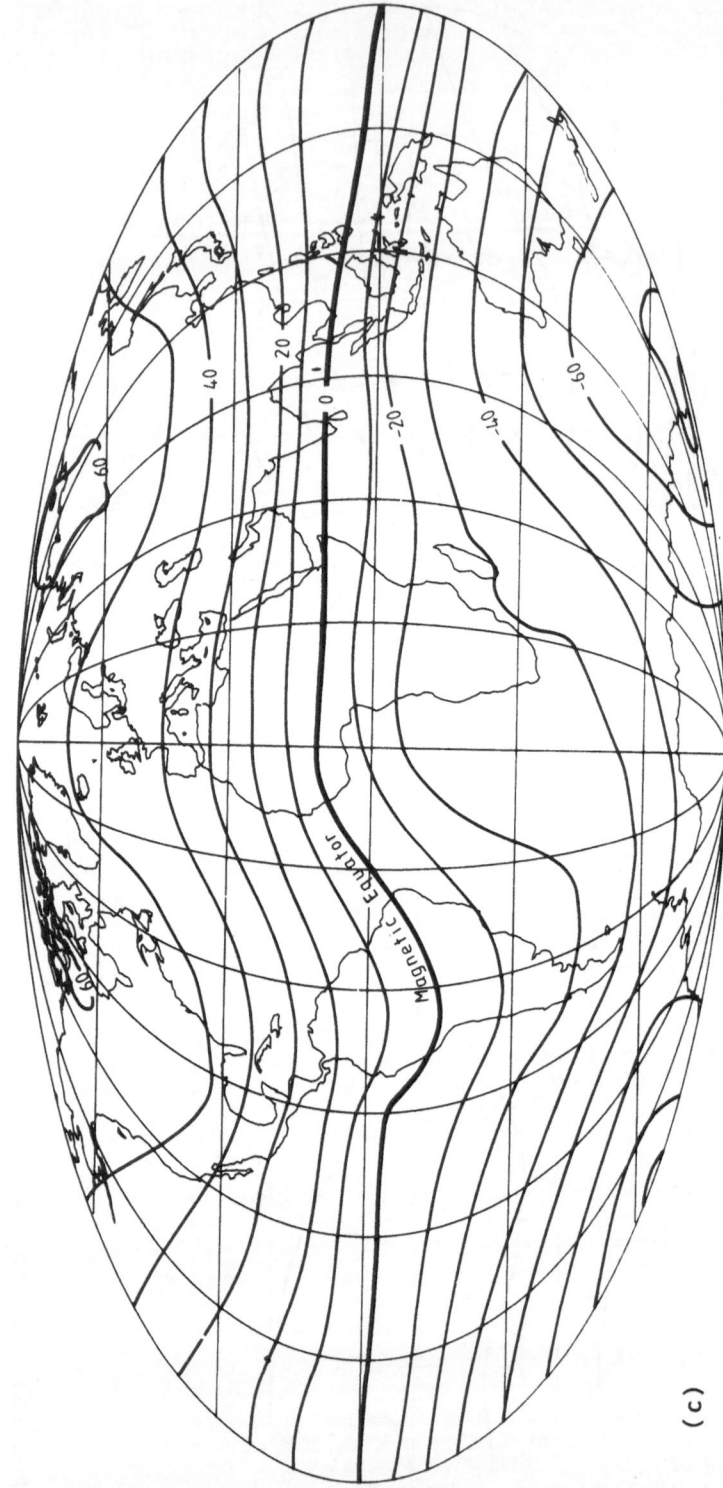

Figure 8.1 (c) The vertical component of the field (V) in mT.

(c)

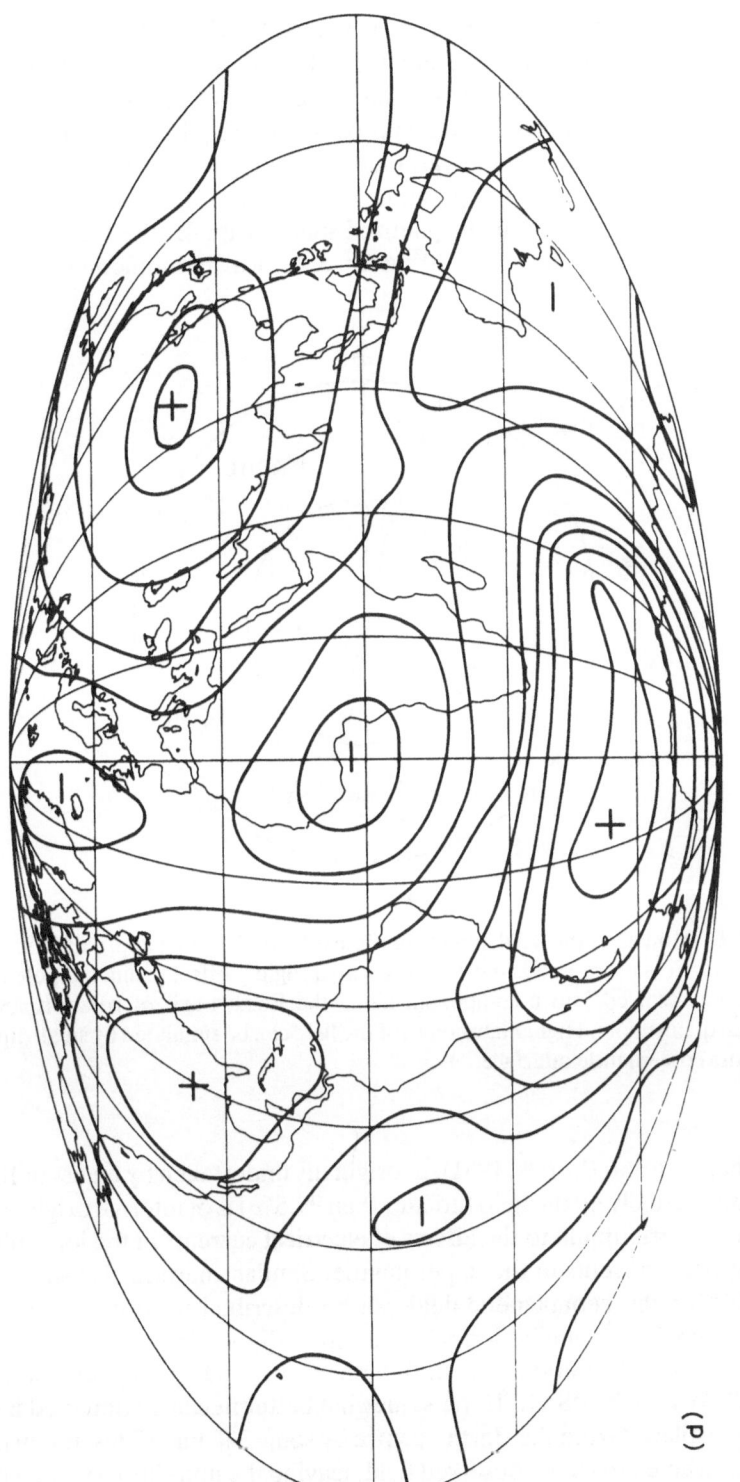

Figure 8.1 (d) The z component of the non-dipole field, contoured in 4 μT intervals.

not exactly antipodal to the southern pole at 68°S 143°E. The distribution of inclination does, however, correspond closely to that of a uniformly magnetized sphere, i.e. tan inclination = 2 tan(latitude), and so does the variation in total intensity which varies from approximately 30 μT near the equator to 60 μT near the poles, with the variation, according to latitude, being given closely by $F = (M/r^3)(1 + 3\sin(\text{latitude}))$ – the formula for the magnetization on the surface of a uniformly magnetized sphere with M, the total magnetic moment, being at present 8.01×10^{22} A m^2. Spherical harmonic analysis of the

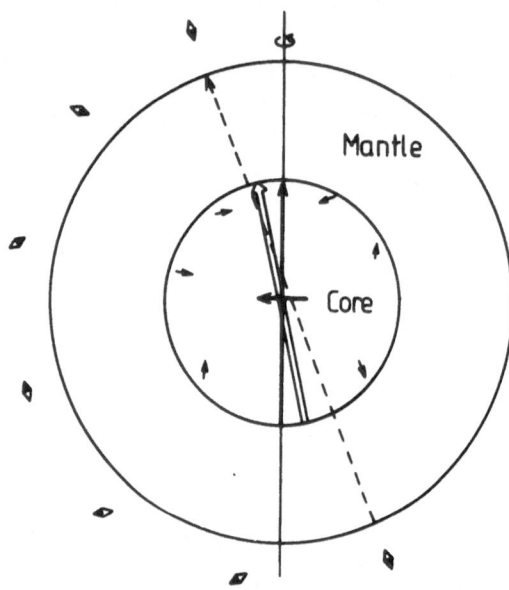

Figure 8.2 Models of the axial and offset dipolar field. To a first approximation, the geomagnetic field corresponds most closely with a slightly offset inclined dipole field. This can be resolved into a component along the Earth's axis of rotation and an equatorial quadrupole. Higher harmonics of the field can be simulated by small dipoles close to the core–mantle interface.

field (Chapman and Bartels, 1951), as originally undertaken by Gauss in 1839, shows that virtually all the field (greater than 99.5%) is of internal origin, with the balance attributable to the effects of electrical currents in the ionosphere inducing eddy currents in the upper mantle. Similar analyses also show that some 80% of the geomagnetic field can be described in terms of the first harmonic, corresponding to a single geocentric dipole, inclined at 11½° from the Earth's axis of rotation (Fig. 8.2), with corresponding magnetic poles at 78.5°N 70°W and 78.5°S 110°E. (A somewhat better fit can be obtained if this dipole is displaced from the Earth's centre by some 340 km.) This dipole field can be subtracted from the observed field, leaving the non-dipole field, which shows a surface pattern (Fig. 8.1(d)) comprising some 12 regions of approximately continental scale, varying in intensity between some ±1.5 μT. These areas can be simulated by the fields of 8 dipoles at depths of about 4800 km, i.e. at the base of the outer core, or by some 12 dipoles close to the core–mantle boundary (Lowes and Runcorn, 1951).

Observatory and survey records (Figs. 7.1 and 8.4) show that the geomagnetic field undergoes long-term changes, *secular variations* (Gellibrand, 1635; Halley, 1692; Vestine *et al.*, 1947; Institute of Geological Sciences, 1978; Madden and Le Movël, 1982) on time-scales longer than 1 year (excluding the 11 year sun-spot cycle), and these variations differ in magnitude from one observatory to another, although there has been an average tendency for the total magnetic field strength to decrease. Such a decrease

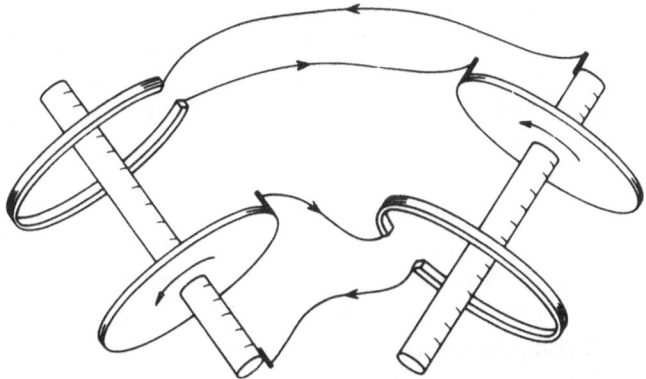

Figure 8.3 Dynamo models of the geomagnetic field. Coupled self-reversing dynamos provide one mechanism by which the geomagnetic field could show random changes in its polarity and also retain an average axial symmetry.

appears to be predominantly associated with a decrease in the magnetic moment of the dipole field of some 18 nT year^{-1} (Petersons, 1974) during at least the last century. The spherical harmonic analyses of earlier geomagnetic fields indicate that a similar rate of decrease has been occurring since observations have been made, but the reliability of these analyses is poor. (The continuation of such a decrease would, of course, result in the complete loss of the geomagnetic field in some 2000 years' time.) The direction changes do not, however, show very much coherence, with different components of the field increasing, decreasing, or remaining stationary at different observatories. Nonetheless, when analyses are made of the rate of change of the field, on a global scale, there is evidence for a distinct westward drift, during at least this century, of the non-dipole components at some 0.2° year^{-1} (Bullard, 1950; Nagata, 1965), although there are local variations, even an eastward drift in some locations (for example at Sitka, Alaska (Skiles, 1970)). If such a

169

westward drift were persistent, then it has implications for the internal properties of the Earth. One possible explanation is that the core is rotating at a slower rate than the mantle (Halley, 1692) – the tidal drag of the Moon, however, is operating to slow the mantle–crust rather than the core (Munk, 1968; Lagus and Anderson, 1968; Jacobs and Alldredge, 1975; Rosenberg and Runcorn, 1975; Tarling, 1975b; Lambeck, 1975, 1980). This explanation therefore raises many crucial questions concerning the changing rotation rate of the Earth and the nature of the core–mantle coupling. When did the Earth–Moon distance correspond to some 20000 km, i.e. the Roche limit at which distance gravitational forces should disrupt the two planetary bodies? Although the core–mantle coupling must be magnetohydrodynamic (Hide, 1966; Roberts, 1972; Yukutake, 1973), what changes may occur with time? Is it physically plausible that there could be times when the core was rotating faster than the mantle, corresponding to an eastward drift? Fortunately, the sense of motion of the field pattern past a single site can be determined from an examination of the direction and shape of the vector loop at that location. Downward vectors, such as those observed in the northern hemisphere, show a clockwise looping with time as the field, positive or negative, passes from east to west below the site (Runcorn, 1959; Skiles, 1970).

It is not clear to what extent the entire present non-dipole field partakes in the observed westward drift (Dodson, 1979; Yukutake, 1979) and Hide and Malin (1970) suspected that the observed westward drift had only existed for the last 500 years. Yukutake and Tachinaka (1968) have proposed that the non-dipole field comprises two components, one of which partakes in the westward drift and one that remains stationary, a standing component. The definition of such components is hampered by the shortness of the observed secular variation record and the fact that the non-dipole field is changing in magnitude on a semi-continental scale (Vestine *et al.*, 1947; Bullard *et al.*, 1950). Observations indicate that the time-scale of overall non-dipolar changes appears to be of the order of a few hundred to a thousand years or so (Thompson, 1982) possibly corresponding to the time-scale of the decay of eddy currents in the core. On the other hand, changes in the core–mantle coupling appear to be able to take place over only a few months (Ducruix *et al.*, 1980) – although only two such drastic changes have occurred this century.

Models for the origin and maintenance of the geomagnetic field have ranged from the Earth being a uniformly magnetized sphere (Gilbert, 1600) to the concept of electrical currents being maintained as a result of convective motions in the outer core. The fact that the Earth is not a uniformly magnetized sphere is known because the only rocks capable of carrying a remanence are those at temperatures below their Curie temperatures i.e. the upper 25–30 km of the Earth, yet these have maximum observed intensities of magnetization that are at least 1000 times too weak to account for the geomagnetic field. Even if some unknown process created a uniform magnetization in the inner (solid) core, it is impossible to see how polarity reversals could occur

(Section 8.4). Similar difficulties are also encountered by models that have the magnetization as a fundamental property of rotating matter (Blackett, 1947). It is therefore generally accepted that the field is generated by electrical currents (Larmor, 1919) circulating in the liquid part of the core (Elsasser, 1939). These currents would decay spontaneously in some 2000 years unless they were continually regenerated by some form of dynamo action (Bullard, 1949a,b; Elsasser, 1955, 1956a,b). The strength of the field in the core is weak, about 10^{-2} T (Hide and Roberts, 1979), and there appears to be little difficulty in creating convective motions within the outer, liquid, core of the Earth as sufficient energy is available from a range of possible sources, such as the latent heat produced as the inner core gradually solidifies (Loper and Roberts, 1979), disturbances caused by irregularities at the core–mantle interface (Hide and Horai, 1968) or heat generated by small concentrations of radioactive elements (Ringwood, 1977) introduced with sulphides as they segregated to form the Earth's core very early in the history of the Earth (Section 9.8). The dynamo action has an axially related symmetry caused by the Earth's rotation (Krause and Radler, 1979) and dynamos must be coupled in such a way that spontaneous reversals of polarity (Section 8.4) can take place. The feasibility of such coupled dynamo action (Fig. 8.3) was demonstrated in models by Lowes and Wilkinson (1963, 1968), but there are many uncertainties about the true nature of the generating mechanism (Jacobs, 1963; Rikitake, 1966; Busse, 1979; Hide, 1982; Gubbins, 1982; Krause, 1982). Critical to any evaluation of these and other magnetohydrodynamic mechanisms is an evaluation of the time-scales of many geomagnetic properties. It is generally considered that the non-dipole field arises from convective disturbances near the core–mantle boundary and, until recently, that the westward drift of the field arises from the difference in rotation rates between the Earth's core and mantle. Any fuller understanding requires a much closer definition of the nature of secular variations and polarity changes over much longer time-scales than are available from observatory records. The available palaeomagnetic data already provide major constraints on many postulated mechanisms (Sections 8.3, 8.4 and 9.8), although considerably more observations, with much higher precision, are still required.

8.3 SECULAR VARIATIONS AND THE DRIFT OF THE NON-DIPOLE FIELD

As most geomagnetic observatories have only been established in the 20th century and the oldest records of geomagnetic changes extend only over the last 400 years, it is clearly important to establish the amplitude and periodicity of such variations in order to constrain models for their generation. Their precise delineation would also provide valuable references for the dating of archaeological materials (Section 7.2) and of several geological processes (Section 9.2). Over archaeological time-scales it is possible to use archaeomagnetic measurements of the geomagnetic intensity and direction at known locations for times which can, ideally, be defined historically or, from their archaeological context, within 5–10 years. In many archaeological situations, the dating may not be as precise and may be only definable in terms of radio-carbon (^{14}C) age. Such archaeomagnetic observations (Chapter 7) do not normally provide a continuous record of the geomagnetic changes at any one location as suitable dated materials only occur at specific times in specific locations. Continuous, or generally continuous, records of geomagnetic changes can be obtained from sediments that have been accumulating during the period of interest. Similarly, but much more irregularly, there is the possibility of using a sequence of lava flows that provide a series of spot-readings of the geomagnetic field as each lava cooled. Thus a wide range of palaeomagnetic observations are available that allow definition of the properties of the geomagnetic field over archaeological and geological periods of time. Such records are best defined for the last 10–15 000 years (Yukutake, 1979), for which dating techniques are generally sufficiently precise to allow the periodicity of such phenomena to be studied.

The establishment of master curves for archaeomagnetic dating has been discussed earlier (Section 7.2.3). At the moment, the available records only allow a temporal analysis of the periodicity of such changes in specific areas (Figs 7.2 and 8.4), such as Europe (Kovacheva, 1980; Thellier, 1981; Clark *et al.*, 1983), Japan (Kinoshita, 1970), the Ukraine (Rusakov and Zagniy, 1973a, b), and the southwestern United States (DuBois, 1975). The longest record of archaeological data, 8000 years, is for Bulgaria (Kovacheva, 1980, 1982). Observations for France and Britain are more detailed but for shorter periods (Thellier, 1981; Hammo-Yassi and Clark, 1982). Fourier analyses of each record are clearly required to determine the precise harmonic content, but it is clear from the available data that there is considerable variability, with swings in declination varying in duration from a few hundred years to over one thousand, with inclinations possibly having somewhat shorter period oscillations (Figs 7.2 and 8.4). The data for intensity variations (Sections 5.6 and 7.3) are possibly less reliable but do not appear to be related to the periodicity of either declination or inclination (Figs. 7.2 and 8.4). It is still dangerous to draw too many inferences from such data bases. There are, for example, some archaeomagnetic indications for very rapid changes in the local geomagnetic intensity (Section 7.2.1 and Fig. 7.3) such as in the Mediterranean and France between 100 BC and 300 AD (Shaw, 1979; Walton, 1979). Such rapid changes appear, at the moment, to be real but it is still uncertain whether cooling corrections need to be applied (Section 5.6) and, if so, whether they should be applied uniformly, in which case the rapidity of the changes remains constant, or different corrections need to be applied to different materials, in which case both the total intensity determinations and the rate of change will be altered.

A major contribution from archaeomagnetic studies is the clear evidence

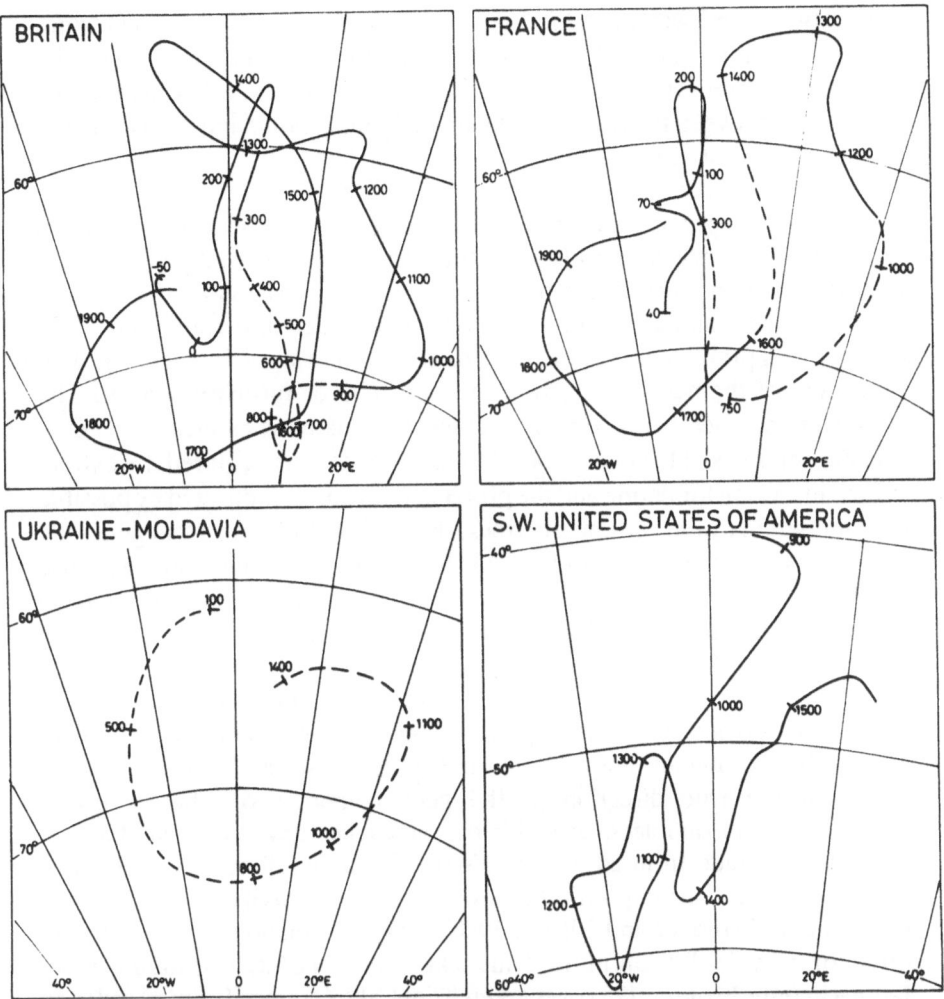

Figure 8.4 Secular variations of directions on an archaeological time-scale. The British data are based on Clark *et al.* (1983); French on Thellier (1981); Ukraine–Moldavia on Rusakov and Zagniy (1973a,b); and the United States data are after DuBois (1975). (See also Fig. 7.2.)

that both eastward and westward drift of the field have occurred during archaeological time (Section 7.2) as the curves for various areas show both clockwise and anticlockwise movements of the geomagnetic vector. Even during historical times, the geomagnetic field in the southwestern United States appears to indicate an eastward drift (Fig. 8.4) and examination of European records indicates both westward and eastward motions of non-dipolar components when comparing Franco-British, Bulgarian and Ukrainian results (Kovacheva, 1982). This observation therefore places less constraints on the nature of the core–mantle coupling and it is probable that

173

the eastward and westward motions of the geomagnetic non-dipolar components are not clearly related to differential rates of rotation between the core and mantle. However, some relationship must exist because of the correlation between the rate of westward drift this century and changes in the Earth's rate of rotation.

Present-day lake sediments provide an invaluable, nearly continuous record of geomagnetic changes at one particular location, usually for periods of 10–15 000 years, and mostly since the end of the last Ice Age (Fig. 8.5). Unfortunately the records are not always continuous as breaks in sedimentation can occur and sediments can have been eroded at certain times. The most serious problem is that the rates of deposition vary, with the additional possibility of the occasional erosion level, slump or turbidity deposit. The main way in which cores of such sediments are dated is by means of ^{14}C at specific intervals. However, considerable sediment is required to extract sufficient carbon for dating and the problem is further complicated by possible contamination by old carbon (old trees, shells, limestone, etc.) that generally results in older 'ages' being obtained. Secondary problems may also arise from the extent of post-depositional rotation, the possibility of chemical changes (particularly organically controlled), and effects of the actual coring process and subsequent transport (Section 5.2.2), of which the previous tendency for core barrels to 'corkscrew' into the sediment (a problem now solved) can obviously produce spurious long-term trends in declination. In view of these problems, few core records can yet be used as absolute measurements of the true direction of the geomagnetic field. The swings in declination, for example, even within the same lake, may be correlatable, but vary in magnitude from 20° to 90°, almost certainly reflecting the effect of currents during their deposition (Verosub, 1977) as observed in laboratory experiments (Bressler and Elston, 1980). Even correlation of the swings in direction may be difficult in the same lake. Creer *et al.* (1975) found correlations in only 39 of the 175 cores taken from Lake Geneva, Switzerland, and Palmer *et al.* (1979) found it impossible to correlate small scale features over 1–2 km in Lake Tahoe, Nevada, although some larger scale features could be identified. Clearly such problems mean that palaeomagnetic data from such sources must be treated cautiously, but the geomagnetic importance of such records is nonetheless extremely high.

Figure 8.5 Lake sediment palaeomagnetic records for North America and Europe. European records (Windermere, Bielersee, De Joux) show peaks in the swings of their declination and inclination, the patterns of which (but not the absolute magnitude) can be matched from one lake to another. Similarly, matching features can be found between the records preserved in some North American lakes (Minnesota, Thunder Bay) – but the correlations between European and North American data are not very clear, and it can often be difficult to obtain correlations even within the same lake (see text). (After Creer, 1981.)

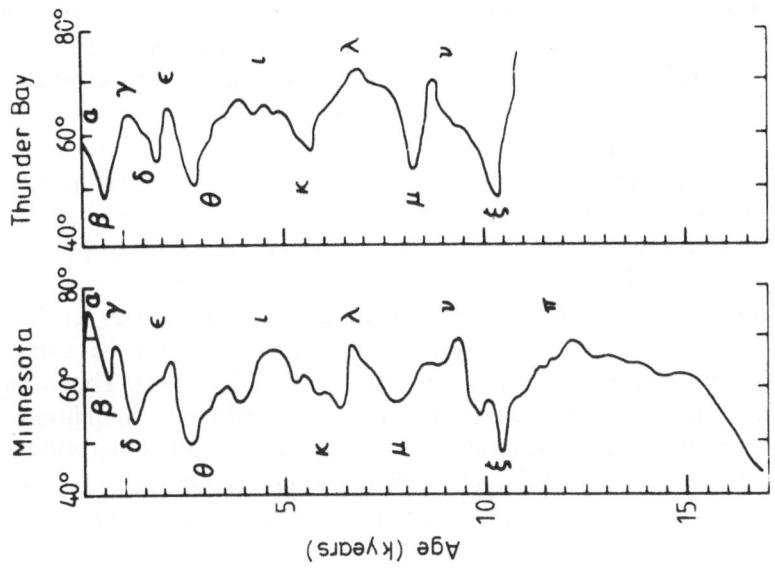

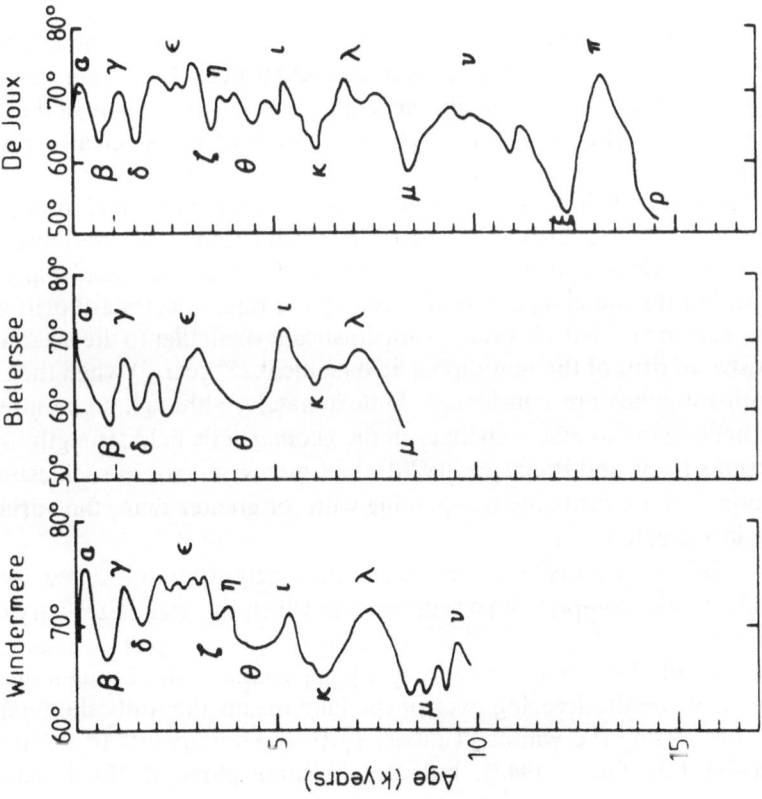

Palaeomagnetism

Detailed records are now available (Fig. 8.5) from many lakes in Europe (Thompson, 1975; Creer et al., 1975; Creer, 1977; Thompson and Turner, 1979; Turner and Thompson, 1979; Tucholka, 1980; Creer et al., 1981) and North America (Banerjee et al., 1979; Palmer et al., 1979; Verosub, 1980; Mothersill, 1981; Creer and Tucholka, 1982), and similar observations are becoming available from other parts of the world (Nakajima and Kawai, 1975; Thompson and Oldfield, 1978; Hyodo and Yaskawa, 1980; Barton and McElhinny, 1981). However, until better spatial coverage is available it is difficult to assess the geomagnetic significance of the available observations, particularly in view of the noise level (especially in dating) of the results. Nonetheless, the data (Fig. 8.5) are moderately consistent with secular variations over the last 10–15 000 years, being comparable with those shown in observatory records for the last 100 years or so (Fig. 7.1). The swings in declination, inclination and intensity appear to be of similar amplitudes, and their patterns, for any one time, appear to be regional rather than global, i.e. on a semi-continental scale. The only common trend in these sedimentary cores appears to be for the inclination of the geomagnetic field to have been shallower in the past than today. It seems almost certain that this trend is not, in fact, geomagnetic but represents the effects of compaction (Sections 4.3 and 9.4). Studies of European lake core data have largely shown comparability with the archaeomagnetic record (Section 7.2) and it has often been possible to correlate the observed swings in declination and inclination from one part of Europe to another (Thompson and Turner, 1979; Creer, 1981; and Fig. 8.5). Similar correlations also appear to exist between some of the North American data (Fig. 8.5), although the agreement is less clear – possibly reflecting the greater distances over which correlation has been attempted. Creer (1981) and Creer and Tucholka (1982) examined both European and North American records, particularly for inclination, and considered that some correlations appear to be possible, at least in the broad pattern of swings, but the matchings were delayed by the rate of westward drift of the field, i.e. some $0.14–0.15°$ year^{-1} – not drastically dissimilar to the present rate of westward drift of the non-dipole field (some $0.25°$ year^{-1}) when the errors in both estimates are considered. Unfortunately, although some attempts have been made to assess changes in the geomagnetic field strength in such sediments (Levi and Banerjee, 1976), the sources of error in such estimates (Section 5.6) are probably comparable with, or greater than, the variations being investigated.

Some lake sediments, however, can be extremely accurately dated. Varved glacial deposits comprise lake sediments in which the grain sizes vary on an annual basis. Each year is thus represented by a couplet of coarse–fine sediment with the spring–summer layer being coarse grains brought in as the ice melts while the freezing over of the land means that only the finer silts settle out during the winter (Tauber, 1970; Fromm, 1970). In Sweden, in particular (De Geer, 1940), but also in some parts of North America

(Verosub, 1979a,b), it has been possible to count the varves back from the present day, thus establishing a chronology that is accurate, in absolute terms, to within absolute limits of 300–475 years even 10000 years ago (Fromm, 1970). Even where such an absolute scale does not exist, it is possible to undertake varve counts within an exposed section and determine the duration of the sedimentation record even though its absolute age is unknown. Clearly there are likely to be problems with current effects, particularly in the summer layers, but the winter layers are generally deposited under quiet conditions. Furthermore, evidence for penecontemporaneous slumping and faulting suggests that the magnetization in many of these varved deposits is acquired within 2 or 3 years of their deposition (Section 4.3). Unfortunately detailed studies of dated sequences of varves are still few (McNish and Johnson, 1938; Johnson et al., 1948; Ising, 1943; Granar, 1958; Griffiths et al., 1960; Nöel, 1975; Verosub, 1979a,b) and yet these could provide a major source of information for delineating the secular variations and longer term behaviour of the geomagnetic field (Section 8.4) over the last 14000 years or so.

For earlier times, the accuracy of dating tends to become similar to or much greater than the expected period of secular variation. It thus becomes impracticable to assess the periodicity of secular variations for earlier periods, particularly as such older magnetizations are likely to contain components that are harder to remove by demagnetization techniques. In some instances it seems possible that records of the periodicity may be preserved in certain sediments, such as limestones (Turner, 1975), where the rate of diagenesis has been very rapid so that the primary magnetization, when isolated, appears to reflect the magnetization acquired during a few years only. It is then possible to estimate the rate of deposition and thus the time period covered in any one rock sequence. In the Carboniferous (c.300 million years) limestones of Britain, for example, the variation in declination and inclination may be attributable to secular geomagnetic changes (Turner, 1975). Assuming sedimentation rates as in the Bahamas today, the measured periodicity is similar to that of secular variation today, i.e. a few thousand years. It is unclear, however, whether the observed changes are, in fact, attributable to geomagnetic factors alone, and obviously the assumption of modern Bahaman sedimentation rates must be treated cautiously. Nonetheless, it is clear that this principle of establishing the past periodicity of geomagnetic variations has a considerable potential for different geological periods.

The question of the persistence of westward drift of at least parts of the non-dipole field now appears to be answered from archaeomagnetic studies which show (Fig. 8.4) clear evidence for westward drift during the last 300–400 years for most of the world, although anticlockwise looping (eastward drift) in the southwestern United States (Fig. 8.4). For somewhat older times, eastward and westward drift appear to be equally common. In lake sediments, the directional definitions are somewhat poor, but show the presence of eastward and westward drift, with the latter possibly dominating,

Palaeomagnetism

with clear exceptions such as the eastward drift in Lake Mono (California) sediments some 24 000 years ago (Denham, 1974). Such observations seem to make it unlikely that the sole cause of the present westward drift is a differential rate of motion between the core–mantle. The most likely explanation seems to be that the present dominance of westward drift is somewhat unusual and is related to changes in the dipole sources within the outer core resulting in an apparent differential rotation. The stationary and moving parts of the non-dipole field are thus reflecting changes in the electromagnetic configurations within the core, possibly involving relatively small motions of the material itself.

Although most studies have concentrated on directional changes of the secular variation, intensity changes also occur and determinations are available from many parts of the world (Smith, 1967a,b; Burlatskaya and Nachasova, 1977), but most of these are of uncertain reliability, particularly in view of the difficulties that are now known to occur in the determination of palaeointensity (Section 5.6). Nonetheless, the data can be analysed to attempt to isolate dipolar and non-dipolar components and estimate whether the non-dipolar components also show a westward drift (Cox, 1962; Bucha, 1970a,b). Most people that have carried out this examination have concluded that there is some evidence for such a drift, but the reason for this is mostly that peaks, for example, in Japan, can be predicted to occur at specific times in Europe and North America if a drift of 0.2–0.3°/year is assumed. Such peaks can readily be found as the data are few and the scatter large. The evidence for, or against, a persistent westward drift of the intensity of the non-dipole field must be considered, at this stage, inadequate for making such tests. Nonetheless it is possible to evaluate changes in the total dipole moment if it is assumed that these will give rise to long-term (few 10^3 years) changes, while non-dipole components will only give rise to shorter duration changes (c. 10^2 years). On this basis (Bucha, 1970b), it seems that the main dipole moment was at its highest value just over 2000 years ago (Bucha, 1970) and the change of this dipole moment also coincides with corrections to ^{14}C dating (Fig. 8.6) indicated by comparison with dendrochronology (Section 9.7). This correlation indicates that Bucha's estimates may be correct, but the data base is extremely poor in quality and particularly in age (Barton et al., 1979).

Although dating methods are generally too imprecise for determination of the periodicity and sense of motion of the secular variations of the geomagnetic field for materials older than some 100 000 or so years, palaeomagnetic observations can still be used to assess the magnitude of such variations and therefore to test models for their origin. In a sequence of lavas, for example, it is generally impossible to assess the time interval between successive flows, although erosion and weathering can indicate that the gap was long. Nonetheless, each lava provides a spot reading in time of the direction and intensity of the geomagnetic field at the time that it cooled. A lava sequence thus provides a series of spot readings of the field during the interval when the lavas were

178

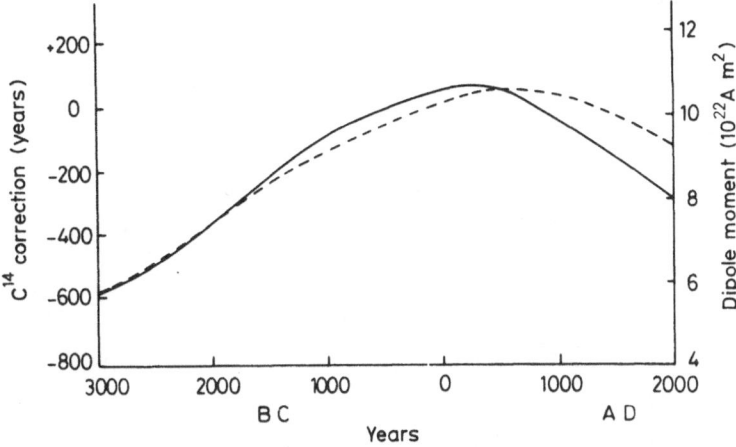

Figure 8.6 The dipole moment and ^{14}C calibration curve. The mean dipole moment, as a function of time (Bucha, 1970b), is the solid line and shows strong resemblances to the curve (dashed) for correcting ^{14}C dates to the dendrochronological scale (Damon *et al.*, 1970). This suggests a probable relationship between the two, possibly reflecting the effect of changes in the height of the ionosphere. Superimposed on this average calibration curve are the 'Suess wiggles' that have very short periods compared with the longer term variations in the geomagnetic field.

erupted. The magnitude of their scatter, measured as the circular standard deviation or angular deviation (Section 6.3.3), can then be compared with those generated by either models of the geomagnetic field (Cox, 1962) or with the present geomagnetic field, assuming that the entire scatter in directions along a line of latitude is due entirely to the westward drift of the non-dipole field (Fig. 8.7).

The present geomagnetic field shows a scatter, along lines of latitude, that is twice as great near the equator as near the poles (Cox, 1962; Creer, 1962; Tarling, 1967a) and an asymmetry between the northern and southern hemispheres. Studies of lava sequences in and bordering the Pacific showed a low scatter of palaeomagnetic directions (Doell and Cox, 1971) that were initially interpreted as indicating a persistence of the low magnitude non-dipole field in this region for at least some 5 million years. With improved dating, it is now clear that these low values of scatter correspond to lava sequences erupted very rapidly over relatively short time intervals (McWilliams *et al.*, 1982). On this basis, each sequence of lavas corresponds, to a first approximation, to a spot reading in time, rather than a series of 'readings' within a secular variation cycle, and there seems little evidence for the long-term persistence of the present suppression of the non-dipole field in the Pacific. Evidence from other areas, such as the Atlantic (Ellwood *et al.*, 1973; Dodson, 1980; Saemundsson *et al.*, 1980), the Indian Ocean (Watkins, 1973; Amerigian *et*

179

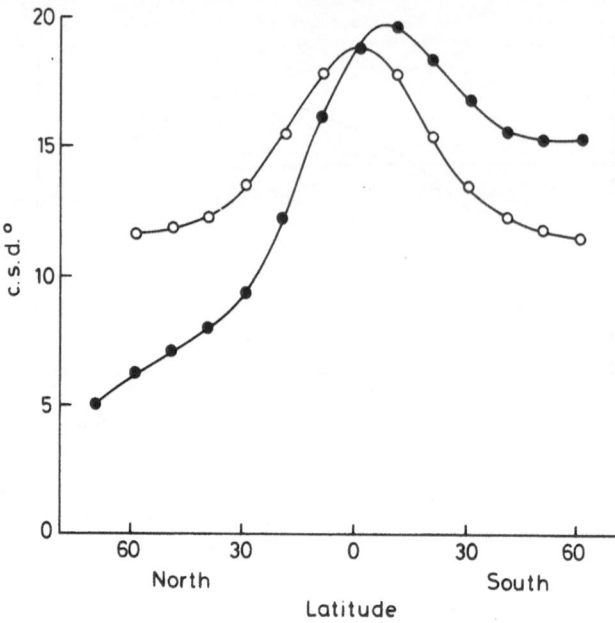

Figure 8.7 The present magnitude of secular variation. If the present field is rotated past the same point, then the variation in the direction of the field exhibits a much greater scatter near the equator than at the poles. There is also a distinct difference between the scatter observed in the different hemispheres. One model for testing against the palaeomagnetic data (Section 8.3) is therefore to assume that secular variations arise simply from the westward drift of the non-dipole field. If the non-dipole field was similar to that today, then a similar latitudinal variation in the circular standard deviation of the palaeomagnetic data would be expected for particular rock sequences, such as lava sequences.

al., 1974) and so forth, does not indicate particularly low magnitudes of scatter, and often exceeds the predicted scatter.

It is not clear whether more subtle changes in the magnitude of secular variation can be defined at this stage. Cox (1975) has pointed out that the present non-dipole field tends to be characterized by a zonal distribution with positive vertical components tending to be in higher latitudes than negative. Coupland and Van der Voo (1980) consider that the available palaeomagnetic data for the last 7 million years or so indicate a persistence of this distribution, although Harrison and Watkins (1979) found consistently lower inclinations in Eastern than in Western Iceland for flows of the same age.

Studies of even older lava sequences have generally been less intensively investigated and problems of remagnetization, decay of primary remanence and so forth make any interpretation of such sequences less reliable than for Cenozoic times (<80 million years). There does, nonetheless, seem to be a somewhat greater scatter for rocks erupted in lower latitudes than for those erupted at higher latitudes (Tarling, 1967a, 1971a) and no obvious differences between rocks erupted in northern as compared with southern latitudes. In general, therefore, the properties of the past secular variation appear to be broadly consistent with the patterns observed today, but with no particular asymmetry between hemispheres.

8.4 POLARITY REVERSALS, TRANSITIONS AND EXCURSIONS

8.4.1 **Polarity reversals**

The most important restriction of almost all models for the geomagnetic field is that they must allow the Earth's field to change polarity on a series of time-scales. No one theory is yet capable of fully explaining such reversals or their periodicity, but many earlier models have had to be discarded on this observation alone (Section 8.2). The solar magnetic dipole, although only a small part of the total solar field, shows a 22-year cycle in its polarity, but there is no obvious connection between stellar processes and those operating within the Earth and other planets. The evidence for terrestrial polarity changes is thus based on palaeomagnetic observations alone and it is obviously critical to establish that the observed polarity changes are attributable to geomagnetic changes and not to some physical, self-reversal mechanism occurring in certain rocks.

Néel (1951) proposed four possible mechanisms by which self-reversal could occur (Fig. 8.8) by means of magnetostatic or superexchange reactions between either different minerals or different compositional volumes within the same mineral grain, although the magnetostatic interactions are now thought to be too weak (Uyeda, 1955; Stacey, 1963). One convincing example of self-reversal is known, the Haruna dacite (Nagata *et al.*, 1951; Ishikawa and Syono, 1963). This is, in fact, a tuff that has been quench-cooled when it was erupted into water. It has thus retained, in part, ilmenohaematite in its solid solution composition of $Fe_{1.52}Ti_{0.48}O_3$, which would normally have exsolved towards its end members (Section 3.2.2). The self-reversal property, however, is not only restricted to this composition, but also requires the rock to contain both ordered and partially ordered phases; completely ordered or completely disordered phases are not able to self-reverse. Similarly, the actual sizes of the grains in this rock are also critical (Westcott-Lewis and Parry, 1971). Verhoogen (1956, 1962) suggested that a further possible self-reversal mechanism could occur as a result of the ionic migration between different lattices, thus changing the relative magnetic strength of lattices of opposite polarity (Fig. 8.8) in antiferromagnetic minerals (Section 2.2). However, while such behaviour is technically possible, the composition range over which such mechanisms might operate is small and corresponds to a very high degree of oxidation that is metastable under most geological conditions (O'Reilly and Banerjee, 1967). It is not possible to discount this mechanism in

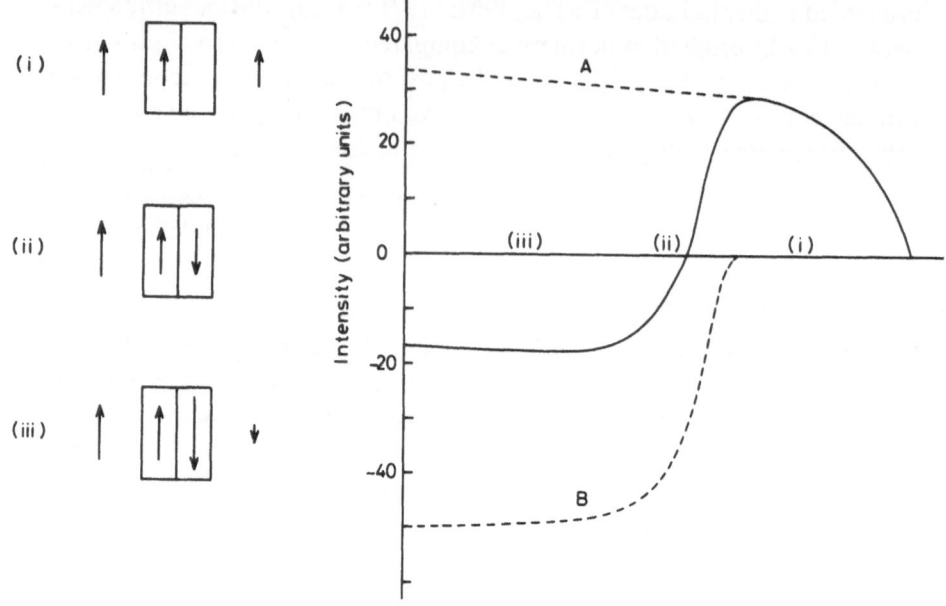

Figure 8.8 Self-reversal mechanisms. Although self-reversals are extremely rare in natural rocks, they can occur where two different magnetic minerals are present or in magnetic materials with different volumes having different structures and hence different magnetic properties within the same grain. As the grains cool, the higher Curie point material becomes magnetized in the direction of the ambient field. On reaching the Curie temperature of the other mineral, this cools in the combined effects of the geomagnetic field and that of the already magnetized component. Under some circumstances, this lower Curie point material can have a magnetization stronger than that of the higher Curie temperature material, so reversal of polarity eventually occurs. This process is reproducible in the laboratory. If, however, the higher Curie point material is removed or becomes non-magnetic (e.g. due to ionic migration or chemical instability), then only the reversed component will be observed and no self-reversal will be observed in the laboratory. (Similar non-geomagnetic reversals can also occur if the hot material is in motion during cooling – see Fig. 6.11.)

all cases, but it would seem that this could only occur very rarely under natural conditions. Occasional self-reversals have been reported in a variety of igneous rocks, but usually only after repeated heating–cooling cycles and almost certainly associated with chemical instability. However, apparent self-reversals can readily occur if later chemical remanences have been acquired. The process is, of course, not reproducible in the same sample as, on cooling after heating, both components become magnetized in the same direction. Oceanic basalts, for example, may become magnetized as they cool and initially have the polarity of the ambient field, but chemical changes, such as serpentinization, maghematization, and so forth (Section 9.5), may well occur at a slightly later time, possibly when the field is of a different polarity.

The chemical remanence associated with these later processes will thus have an opposite polarity to that of the unoxidized magnetic grains. On heating, the polarity could reverse, but this is not repeatable because

1. On cooling both minerals acquire the same polarity, and
2. Maghemite, in particular, is unstable on heating (Section 3.3).

Thus, on our present knowledge of self-reversal mechanisms, they can only occur in minerals that have, or have had, specific compositions. This means that there would be expected to be a general correlation between the polarity observed and the mineralogy. Such polarity:petrology correlations have been found in metamorphic rocks (Balsley and Buddington, 1958) and in a few basaltic sequences (Ade-Hall and Wilson, 1963; Ade-Hall, 1964; Wilson and Watkins, 1967; Watkins and Haggerty, 1968; Ade-Hall and Wilson, 1969). In both cases, the more highly oxidized minerals were more commonly associated with reverse polarity than with normal polarity, and vice versa. There is, however, no causal relationship for such a relationship between the oxygen fugacity and polarity. Identical examination of many other sequences has not found the correlation (Larson and Strangway, 1966; Watkins and Haggerty, 1968; Ade-Hall and Watkins, 1970) and the general consensus is that the observed correlations are fortuitous (Larson and Strangway, 1968; Watkins and Haggerty, 1968). Nonetheless, self-reversals can occur, albeit under very specific rare conditions, and apparent self-reversals can readily occur if two or more components of remanence are of different ages.

The evidence for the fact that the geomagnetic field does change polarity is, in fact, overwhelming. The same polarity change can be dated at exactly the same time in both sediments and igneous rocks (Section 9.2.2). Ocean sediments, for example, have an upper layer that is normally magnetized that overlies a reversely magnetized layer. The boundary occurs at different depths in different places (Fig. 8.9) but at exactly the same time 730 000 years ago, whether dated radiometrically or stratigraphically (Opdyke et al., 1966; Ninkovich et al., 1966; Opdyke, 1972). The same polarity change is also recorded at the same time in subaerial volcanic rocks (Cox et al., 1965; Cox and Dalrymple, 1967a,b; Cox, 1969; Dalrymple, 1972). No mechanism is known for causing such complete reversals of magnetization in sediments (although depositional factors can cause major changes in the apparent geomagnetic field direction) and both rock types acquire their remanence by entirely different means. Yet both processes would have to cause self-reversals at the same time on a world scale. Similar considerations also apply in areas where haematite-rich sediments have been baked by a magnetite-rich intrusion as, in this instance, the mineralogy is quite different (Fig. 5.10). Virtually all cases of such baked contacts that have been examined have an identical polarity, with only 3 of 157 cases having somewhat different directions (Wilson, 1962; Irving, 1964; McElhinny, 1973a), and the exceptions probably reflect exsolution phenomena which have formed a second component of remanence at

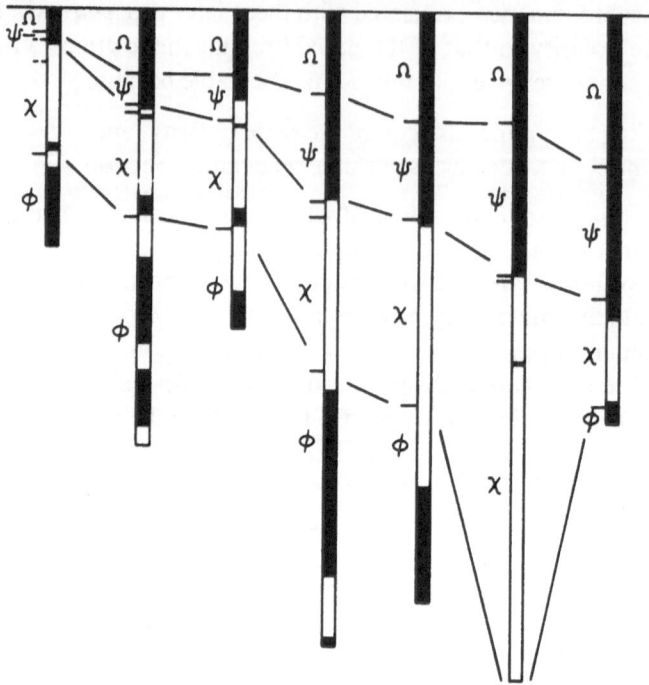

Figure 8.9 Correlation of reversals in deep ocean sediments. Oceanic sediments are usually dated by their microfossil content, although some radioactive isotopes are also suitable. The dating on microfossil zones (Greek letters) shows an excellent correlation with polarity changes in the different sedimentary cores. (After Opdyke *et al.*, 1966.)

a somewhat later time, rather than reflecting self-reversal features. Self-reversal mechanisms would also result in a reduction of intensity as the components become equally strongly magnetized, and the change of polarity would be instantaneous as the magnitude of the opposing components changed. Yet intermediate directions of magnetization occur during a polarity change (Section 8.4.2). If, in fact, the geomagnetic field reverses polarity, and self-reversals were common, then approximately half of the observed baked contacts would show antiparallel polarities and neither could there be a correlation between polarity and petrology. It seems reasonable to conclude, therefore, that self-reversals are very rare and, except for the Haruna dacite, most reported cases more probably reflect magnetization at two different times, rather than genuine self-reversals. However, in terms of geomagnetic theory, only one reversal of the geomagnetic field needs to have taken place to impose major constraints on theories for the origin of the field (Section 8.2).

For the last 150 million years or so, the most detailed record of geomagnetic polarity changes is provided by analyses of the magnetic anomaly patterns observed in the ocean basin. The model for their origin is that new oceanic crust is being continually generated along the crests of the oceanic ridges. As these rocks cool, they become magnetized in the polarity of the ambient field and are then carried away from the ridge by the action of convection currents within the mantle. The oceanic crust therefore acts as a tape-recorder of the polarity changes of the geomagnetic field. Unfortunately, the oldest known oceanic rocks, in the northwestern Pacific, are only some 180 million years old and the total length of the well-defined record is some 150 million years. The dating of this record is not simple (Section 9.2) but is now considered to be accurate, ±2%, for much of the last 80 million years. Analysis of the frequency of such polarity changes over the last 45 million years shows that, on average, about three polarity changes occur per million years but the frequency of such changes is random, i.e. they have a Poisson probability function:

$$P = \frac{(k\lambda)^{k\lambda} t^{k-1}}{k} \exp - k\lambda t$$

where λ is the reversal frequency, t is time and $k \equiv 1$ for a Poisson function (Naidu, 1971; Cox, 1975, 1981; Phillips, 1977; Laj et al., 1979; Fuller et al., 1979; Reyment, 1980). In more detail the change from reversed to normal appears to be somewhat more stable than the change from normal to reversed (Phillips, 1977). The sequence older than 45 million years was also initially thought to be somewhat more systematic in frequency than the younger, but recent versions of the time-scale have tended to result in a more random pattern. It seems probable, therefore, that the polarity frequency during the last 80 million years will all be random, thus indicating that the core retains no memory of its previous polarity, but changes simply as a result of some randomly occurring, but as yet unknown, process.

Polarity reversals for older periods (Fig. 8.10) are necessarily less well dated (Section 9.2), but periods of rapid polarity change appear to character-

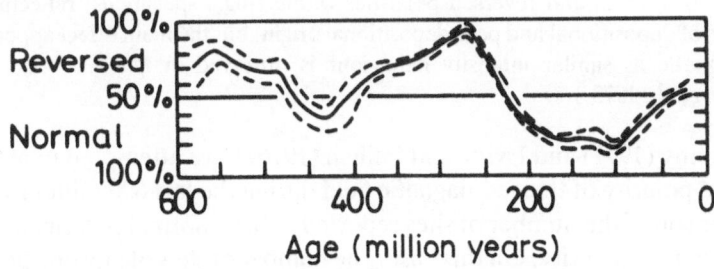

Figure 8.10 The frequency of polarity changes during the Phanerozoic (0–600 million years BP). Analysis of the percentage of observations of normal and reversed polarity allows an evaluation of the relative duration of normal or reversed polarities, although it does not necessarily indicate the number of polarity changes. (After McElhinny, 1971 and Irving and Pulliah, 1976.)

ize the Lower Triassic (220–230 million years) and possibly the Devonian (390–410 million years) but there are also prolonged periods of constant polarity. In the Cretaceous the geomagnetic field appears to have been almost entirely reversed between 70 and 120 million years ago, while mostly during the Permian, the field was almost entirely reversed between 245 and 295 million years ago (Section 9.2). The polarity sequence in rocks much older than 300 million years old is extremely difficult to determine (Section 9.2), reflecting dating difficulties, problems of magnetic overprints and so on.

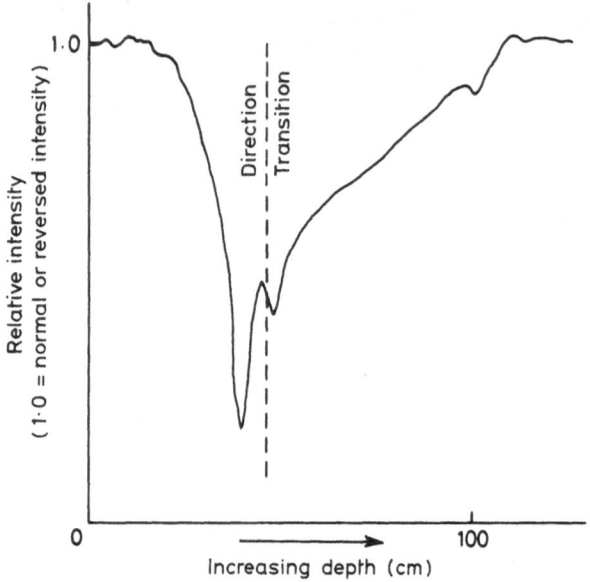

Figure 8.11 Intensity reduction during a polarity transition. In rapidly deposited oceanic sediments, the transition in direction takes place over only a few centimetres, but the decrease in intensity of remanence before and after the transition in direction occurs over much longer periods. Some of the reduction in intensity may be due to a mixture of normal and reversed polarities within single specimens, reflecting components of depositional and post-depositional origin, but the main effect appears to be geomagnetic as similar intensity behaviour is observed in igneous rocks. (After Opdyke *et al.*, 1973.)

McElhinny (1971) and Irving and Pulliah (1976) have attempted to assess the average polarity of the geomagnetic field during the last 600 million years by comparison of the number of sites reported to have normal polarity and those with reversed polarity. On this basis, the number of sites of any one polarity is likely to be few if there were few polarity changes, although obviously it could reflect intensive sampling that happened to be within a possibly short interval of one polarity or the other. With sufficient number of observations, however, this sampling effect is likely to reduce, and there appears to be some tendency for an approximately 300 million years cycle in the polarity bias (Fig. 8.10). It is not clear what such observations imply for geomagnetic dynamo processes and the observations for rocks older than 300 million years are increasingly suspect in terms of their age of magnetization (Section 9.2), but it is clear that there are processes within the Earth's core that can suppress or enhance geomagnetic polarity changes over very long time scales. The fact that reversals are also recorded throughout the geological record (Section 9.2) also suggests that the core processes generating the field have always had this property to reverse. Some of the older reversals, 1100–1250 million years ago, for example, do not appear to be exactly antiparallel (Piper, 1977; Pesonen and Halls, 1979; Pesonen and Nevanlinna, 1981) and tests appear to indicate that this is not due to the presence of other components. Such features, if substantiated, could be of major importance for understanding the earlier nature of the geomagnetic field. The complexity of the magnetizations in such old rocks, however, necessitates caution before such behaviour is widely interpreted as being solely of geomagnetic origin – similar detailed studies of even older reversals would also be particularly interesting.

8.4.2 Transitions of polarity

The duration and behaviour of the geomagnetic field during a transition from one polarity to the next are of major importance in defining the time-scale and nature of the physical parameters controlling the geomagnetic field. Specific characteristics may also provide useful time-markers for identifying the same reversal in different regions (Section 9.2). However, the frequency of polarity transitions, in relation to the precision of radiometric dating, means that it is often difficult to be certain which transition is being studied. Most of the more precise data have therefore been obtained (Fuller et al., 1979; Liddicoat, 1982) from studies of the last polarity transition, 730000 years ago (Section 9.2). Furthermore, most of the information is from oceanic sediments as these provide nearly continuous records of the changing properties of the geomagnetic field but individual specimens tend to average out changes on a scale of less than 1000 years (Section 5.2).

It is not practicable to determine the absolute intensity of the geomagnetic field directly from sediments but the intensity of remanence almost invariably decreases (Figure 8.11), with little change in susceptibility, in sediments on and slightly to each side of the transition (Ninkovich et al., 1966). This also characterizes the few older transitions studied in oceanic sediments and appears to be geomagnetically caused, although the decrease could also be generated by the occurrence of mixed polarities within the sediments, resulting from a combination of depositional remanence, post-depositional rotations and bioturbation effects (Section 4.3). It is, however, confirmed by observations on lava extruded during a polarity change (Ito and Fuller, 1967; Watkins, 1969; Larson et al., 1971; Dagley and Lawley, 1974; Burakov et al., 1976) and appears to correspond to a decrease to some 10–20% of the usual field during periods of constant polarity. In a few studies, however, there is evidence for anomalously high intensities of the geomagnetic field occurring approximately half way through the transition (Shaw, 1975, 1977; Prévot, 1977). The interpretation of such palaeointensities is difficult (Section 5.6) and some of the apparent variations could reflect different rates of cooling and hence different acquisition abilities, rather than actual geomagnetic changes.

The actual transition movements of directions, within any one site, tend to be predominantly in inclination so that the corresponding palaeomagnetic poles tend to follow a great circle path between the two polarities, but Fuller *et al.* (1979) point out that there is always some declination shift and, in a few cases, consistent changes have been observed that seem to simulate secular variation changes during a transition (Watkins, 1965b; Goldstein *et al.*, 1969). It is difficult to assess the significance of some of these changes in view of changes in the rates of eruption or sedimentation. There is, for the last reversal, no clear evidence for a common path (Hoffman and Fuller, 1978; Fuller *et al.*, 1979), thus indicating that it is not caused by a 'toppling' of the main dipole, but is more consistent with a gradual loss of the main dipole, leaving the non-dipole field, followed by a gradual reappearance of the dipole field in the opposite direction. However, common transitional polar paths have been suggested, mostly passing through either the site location or its antipodal point (Hoffman, 1977, 1979, 1982) and identical paths for the same transition have also been proposed (Freed, 1977), but there was no evidence for this in a detailed analysis of transitional directions occurring during the last 20 million years in Iceland (Kristjansson and McDougall, 1982).

The duration of directional transitions is difficult to estimate as radiometric and even palaeontological dating methods are much too imprecise, and depositional rates cannot be assumed constant for oceanic sediments. Nonetheless, most estimates on sedimentation rates are consistent, suggesting some 4000 to 5000 years (Harrison and Somayajulu, 1966; Niitsuma, 1971; Opdyke *et al.*, 1973), although durations of 8–10 000 years cannot be excluded (Clement *et al.*, 1982). Similar time-scales are indicated by statistical studies of the frequency of transitional directions, generally defined as more than 40° away from the usual direction (Wilson *et al.*, 1972), observed in igneous rocks over specific periods of time (Cox and Dalrymple, 1967a,b; Cox, 1969). Changes in intensity clearly extend over longer periods, the decrease and subsequent increase each taking some 2–3000 years so that the total transition takes some 10 000 years. Deep tow magnetometer traverses across oceanic magnetic anomalies show that the zones between normal and reversely magnetized zones of the ocean floor are very narrow (Larson and Spiess, 1969; Klitgord *et al.*, 1975; Macdonald *et al.*, 1980) corresponding to a duration of some 10 000 years for the intensity change, with somewhat shorter duration possibly occurring over even shorter periods. Older reversals have been less intensively studied, but generally indicate similar properties for transitions some 200 million years ago (Van Zijl *et al.*, 1962a,b) and 500 million years ago (McElhinny, 1970). At the moment, however, no data are available from Pre-Cambrian rocks that can be regarded as recording secular changes that have been isolated successfully from remagnetization effects. The scatter of directions for such early times is, however, broadly similar to that observed during much younger periods and there seems no reason for not considering that secular variations have occurred of similar magnitude, and possibly similar periodicity, throughout the geological history of the geomagnetic field.

8.4.3 Geomagnetic excursions

The ability of the geomagnetic field to remain of one polarity for periods of at least 10^5 years is well established, but the evidence for changes on a shorter

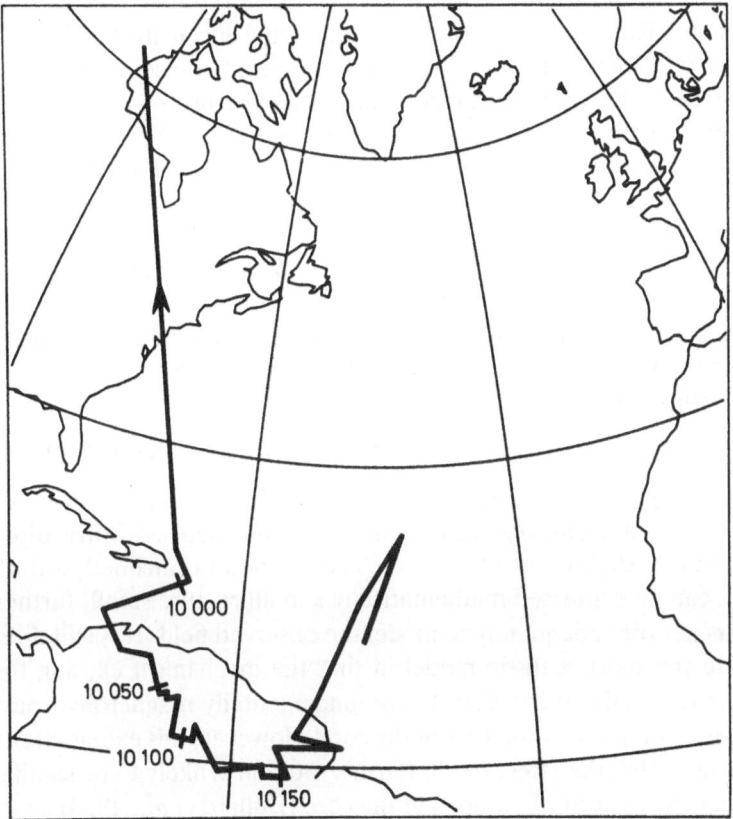

Figure 8.12 An example of a possible geomagnetic excursion. During a geomagnetic excursion, the virtual pole may move well away from an approximately axial geocentric dipole location, to possible reverse or to lie in very shallow latitudes. This example was recorded in varved sediments at Starno, southern Sweden, and may, as with many other putative excursions, be explained in terms of sedimentological effects. Ages are in BP. (After Nöel, 1976.)

time-scale is much less reliable (Verosub and Banerjee, 1977; Harrison, 1980a; Verosub, 1982). This partially reflects the fact that such 'excursions' (previously also known as events and flips – Watkins, 1976) are too brief to be recorded in oceanic magnetic anomalies or most oceanic sediments. Even in other types of rock, it is difficult to be sure whether anomalous directions are genuine observations or have arisen from some error in collection, handling, or local sedimentological (Fig. 8.12) or tectonic effects. Excursions do not necessarily involve reversals of polarity, but generally reflect the position of the corresponding palaeomagnetic pole in intermediate to low latitudes, often passing into the opposite hemisphere for durations of some 10^2 to 10^3 years. There are additional problems in attempting to establish if such excursions are a global phenomenon or whether the effects causing them are merely regional.

Over a dozen different excursions have been reported (Table 9.1), with many in the Russian Loess deposits, as occurring during the last 300 000 years (Section 9.2), of which at least two or three seem to be genuine geomagnetic excursions – the Blake excursion, in particular, probably occurring on a world-wide scale. Kristjansson and Gudmundsson (1980) reported at least eight excursions in Icelandic basalt sequences in the last 120 000 years. As the existence of some of the excursions recorded in sediments is in doubt, it is clearly premature to develop too many theories on the basis of indicated vectoral and intensity changes. There does appear to be some evidence for a reduction in the field at such times, and it seems reasonable to attribute them to abortive attempts of the geomagnetic field to reverse (Cox *et al.*, 1965), although none of the existing theories satisfactorily accounts for the observations (Harrison, 1980a).

8.5 THE GENERAL NATURE OF THE GEOMAGNETIC FIELD

It is possible to view the present Earth's magnetic field in a variety of ways (Section 8.2): as an axial geocentric dipole, an inclined geocentric dipole or a dipole that is slightly offset from the centre and also inclined. All of these models can be expressed mathematically and all require small, further components in order adequately to model the observed field. It is still difficult to evaluate the most realistic model in that the mechanism causing the field remains uncertain, other than being fundamentally magnetohydrodynamic processes within the liquid part of the core. However, this explanation would also suggest that the effect of the Earth's rotation is likely to be significant in affecting the pattern of motions in the core (Bullard *et al.*, 1950) so that the simplest, most physically plausible model would be an axial geocentric dipolar model, with small-scale flow patterns giving rise to the higher harmonics which form the non-dipole field. The possibility that some components of the non-dipole field may be permanent features (Section 8.3) would clearly be a modification of the simple model, possibly requiring long-term topographic features on the core–mantle interface (Hide and Horai, 1968; Malin and Hide,

1982), but even these features, if verified, would be expected to be changing over time-scales of a million years or so. The time-averaged geomagnetic field should, therefore, correspond with that of an axial geocentric dipole and all other features should be effectively averaged out if the geomagnetic field is

Table 8.1 Mean directions of selected archaeomagnetic observations

Location	N	Declin- ation	Inclin- ation	α_{95}	Dipole	Angular difference	References
Paris (France)	137	2.4	63.6	0.8	66.2	2.8	Thellier (1981)
Meriden (UK)	42	3.6	67.4	1.8	68.8	1.9	Unpublished
Bulgaria (all)	44	1.8	58.8	2.1	61.4	2.7	Kovacheva (1980)
Bulgaria (500 years)	16	356.8	58.5	3.8	61.4	3.3	Kovacheva (1980)
Tucson (USA)	15	359.2	51.5	3.5	51.5	0.5	DuBois (1975)
Ukraine (USSR)	40	0.5	68.3	2.0	63.5	4.8	Rusakov and Zagniy (1973a)

N is the number of observations used: In the case of the Meriden and Tucson data, these are 50 year intervals, and 500 year intervals for the Bulgarian data. Other calculations are based on the total published data with probable overemphasis on Roman times. The French, British and United States data are each corrected to a central datum, but the Ukrainian and Bulgarian data are uncorrected. The declination and inclination columns give the mean declination and inclination values, and α_{95} is their precision. The dipole field direction would correspond to $0°$ in declination and the inclination to the figure tabulated under dipole. The angular difference between the mean archaeomagnetic direction and the dipole direction is given in the final column.

sampled over periods of a few thousand to a million years or so. Such a model is clearly testable by palaeomagnetic studies and its validity is fundamental to many geological interpretations (Chapter 9).

Archaeomagnetic observations are now available for periods of time up to 8000 years (Kovacheva, 1980), although are mostly restricted to the last 2000 years (Sections 7.2 and 8.3). The mean values for these (Table 8.1) all lie within $5°$ of the axial geocentric dipole position for that locality, and mostly lie within $3°$. Such statistics are hard to evaluate as the data are irregularly spaced in time. The somewhat lower inclinations in most observations could, for example, reflect a bias due to the large number of Roman materials compared to those of Medieval age. The general, but not complete, agreement with the expected axial geocentric dipole would therefore seem to indicate that secular variations, when averaged over some 2000 years or so, lie close to, but do not coincide exactly with, the axial geocentric dipole model. However, the disparity is small, $2–3°$, compared with the amplitude of secular variation, some

20° (Fig. 8.7), indicating that most non-axial components of the geomagnetic field would be averaged out in periods of some 8–10000 years, although somewhat longer time-averaging may be necessary to remove them completely – 27000 years has been suggested by studies of some oceanic sediments (Opdyke, 1972).

Oceanic sediments accumulate slowly and hence a single specimen may represent several thousand years of deposition, thus partially averaging out secular changes within the measurement of each specimen. Unfortunately

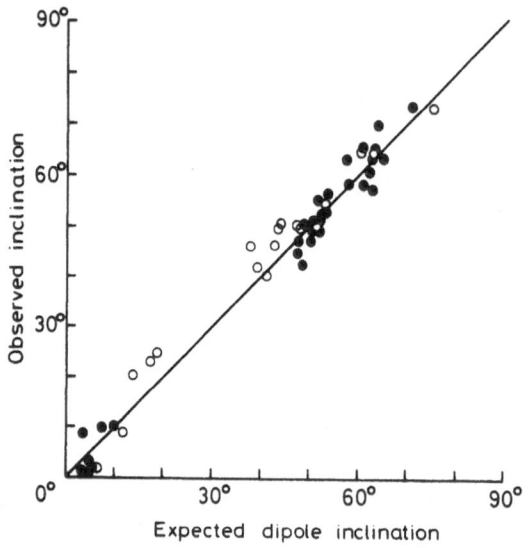

Figure 8.13 Observed inclinations and the axial dipole field. In oceanic sediments deposited during the last million years or so, the slow rate of deposition usually means that secular variations are averaged out within the grains contained in a single sample. Such inclinations, from both hemispheres, are the same (within 1–2°) as those expected for an average axial geocentric dipole field during this time. (After Opdyke and Henry, 1969 and Opdyke, 1972.)

most sample cores are collected unoriented (Section 5.2) so that the declination determinations are arbitrary. However, assuming vertical penetration, inclination and relative declination measurements are possible. These inclinations very closely correspond to those expected for an axial geocentric dipole (Fig. 8.13) for sediments deposited during the last few million years, the best fitting dipole being within 1° of the present rotational pole (Opdyke and Henry, 1969). However, studies of terrestrial igneous rocks, erupted during the last 25 million years, show much greater scatter as each lava or dyke represents a spot reading of the geomagnetic field and secular variations are therefore not averaged out within an individual sample or between samples from the same lava or dyke, unless it is very thick (Section 6.3). When plotted on a global scale, the palaeomagnetic pole positions, calculated assuming an axial geocentric dipole model, are also clearly grouped around the Earth's rotational axis (Fig. 8.14). However, when each collection is considered separately (Fig. 8.14(inset)), there is a tendency for the corresponding poles to occur on the far side of the present rotational pole, and also slightly to the right of it (Wilson and Ade-Hall, 1970; Wilson, 1970, 1971; Wilson and McElhinny, 1974; Saemundsson *et al.*, 1980). The right-handed feature is

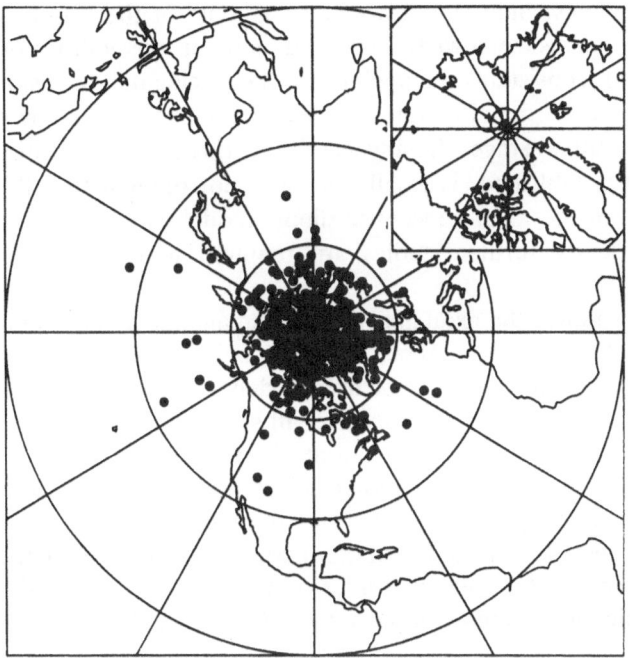

Figure 8.14 Palaeomagnetic pole positions for igneous rocks less than 20 million years old. When individual site poles are plotted, they clearly group around the axial dipole (geographic north) rather than the present geomagnetic axis (Tarling, 1971a). However, if they are plotted with a common site longitude (inset) there is a tendency for the pole positions to be slightly right-handed and distinctly far sided (Wilson, 1970). Note that the inset figure (based on data from McElhinny, 1973a) is on a much larger scale than the main figure.

within the statistical errors, but the far-sided aspect is sometimes 2–3° outside the present statistical limits (McElhinny, 1973a). It may be questioned whether this departure is, in fact, meaningful, but it is unlikely to be caused by plate motions as it can be found in rocks 2–3 million years old as well as 20–30 million years old. Nonetheless, it does only correspond to a net shallowing of the inclination at the locality by some 1–2° relative to the expected dipole inclination. Wilson and others (*op. cit.*) have shown that the far-sided effect can be modelled by a northward displacement of the dipole by between 200–300 km along the Earth's axis of rotation, 4% of the Earth's radius. This displacement would be recorded in a change in the latitudinal relationship to the magnitude of secular variations (Harrison and Watkins, 1979; Harrison, 1980a) and there is some indication for this in Icelandic lavas (Coupland and Van der Voo, 1980). An alternative explanation could lie in a false assumption that all such lavas are magnetically isotropic. In intermediate to high latitudes, any anisotropy in the plane of the lava would tend to cause a net shallowing of the inclination, so that an average anisotropy of some 3–5% in the plane of lava flow could account for most of the observations. This magnitude seems somewhat high and another possibility could be the inclusion of consistent shallowing by the accidental inclusion of some excursion or transitional vectors. These will tend to give a shallow component to any overall average. In any case, the observed departure, although statistically established, is small, and it is not thought that this seriously detracts from the axial geocentric dipole model as a working hypothesis, although clearly further studies are required to ensure a more global coverage.

For older times, the axial geocentric geomagnetic model can be tested in two stages. For a continental block that has behaved as a single tectonic unit, palaeomagnetic directions can be used to define the locus of a single pole for each region (Fig. 8.15). The fact that all such tests show a common intersection point for rocks magnetized at about the same time, and that this corresponds with the pole position calculated assuming a geocentric dipole, indicates that the geocentric aspect of the model is a very good approximation. To test the axial nature of this dipole is more difficult. Unfortunately, the only evidence for palaeolatitudes comes from palaeoclimatic data, each type of which is imprecise within some 30° or more (Section 9.7). It is only when all available palaeoclimatic evidence is assembled, that the uncertainties for any one type of observation become sufficiently reduced to provide adequate tests of the axial nature of the geocentric dipolar field. Such data (discussed in more detail in Section 9.7.3) indicate a good agreement between palaeomagnetically determined palaeolatitudes, but the imprecision of the palaeoclimatic data means that the axiality can only be estimated to be within some 10–15°. However, there are no reasons for suspecting any significant departure of the geomagnetic field from an average axial geocentric dipole for at least the last 2700 million years, other than during times of polarity transitions or excursions.

Most of the directional properties of the geomagnetic field therefore seem to be remarkably similar throughout geological time. The average field seems to have been an axial geocentric dipole, other than during brief times of polarity changes. Polarity changes have also occurred throughout the last 2700 million years or so, although there are some indications for long-term changes in their frequency, and secular variations appear to have had similar magnitudes and periodicities – although all these generalizations require more extensive and intensive studies. The intensity behaviour of the geo-

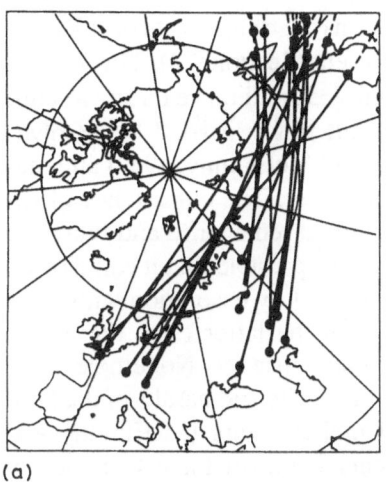
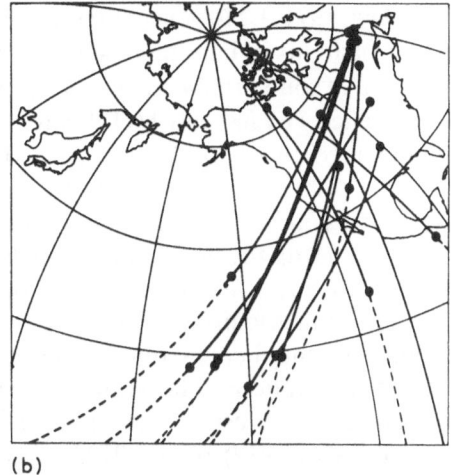

(a)   (b)

Figure 8.15 Continental blocks and the nature of the average geomagnetic field. If the mean directions of the palaeomagnetic vector are plotted on a globe, they define the location of a magnetic pole, with no further assumptions about its nature. The geocentric dipole position corresponding to these directions can also be plotted (dots). The fact that the lines intersect each other would suggest that the simplest hypothesis is for a single pole to lie near such intersections and the fact that the intersections correspond closely with those calculated for a model geocentric dipole field strongly supports the validity of the model. In these two examples (a) Triassic for non-Alpine Europe, and (b) Pre-Cambrian, c. 1400 million years ago, for North America, the observations can also be taken to indicate the tectonic stability of the regions from which they have been taken (see Fig. 9.18). For the Permian of Europe and the Cretaceous of North America, see Tarling (1971).

magnetic field is clearly anomalous during polarity changes, but the few data available do not indicate any differences in the magnitude of the field during normal and reversed polarity states during the last 60 million years or so. However, it has been postulated that there may be prolonged periods of reduced geomagnetic intensity. In the Lower-Middle Jurassic, for example, some 150–190 million years, magnetic anomalies are not readily definable in the oceanic basins creating a 'quiet zone' (Vogt *et al.*, 1971; Larson and Hilde, 1975; Cande *et al.*, 1978) that are often interpreted to indicate a weak geomagnetic field at this time. On the continents, few reliable Jurassic data are available (Section 9.2) and many sedimentary formations appear to be somewhat more weakly magnetized than similar older formations (Steiner, 1980). However, the lack of good stability in igneous formations effectively restricts adequate determinations of palaeointensities (Section 5.6) and identical behaviour could result from very frequent polarity changes, resulting in mixed polarities in many rocks (Steiner, 1980; Hijab, 1982). In general, the few determinations of ancient palaeointensities indicate geomagnetic field strengths of comparable magnitudes to those observed today

(Tarling, 1971a). However, all such observations are fraught with problems of the probability of later remanences being present (Section 9.2). The Modipe Gabbro of southern Africa has an age of some 2350 million years and a stable remanence (Evans and McElhinny, 1966) but this is carried by exsolved titanomagnetites with the pyroxenes (Evans and McElhinny, 1969) and may thus be primary (deuteric) or acquired much later. The consistency of the remanence indicates that this is a genuine thermal remanence and therefore associated with the time of intrusion, but it is also possible that it could result from a chemical remanence having been acquired by exsolution within the pyroxenes at some later, but specific, time. In the latter case, the age of remanence may bear no relationship to the radiometric age. Nonetheless, the oldest observations so far available, such as the dacite and basalts of the Duff Formation in Western Australia of 3450 million years (McElhinny and Senanayake, 1980) or the 2700 million years Poohbah Lake Complex of Ontario (Dunlop, 1980), do not appear to have intensities of magnetization strikingly different from those of much younger formations of similar nature. While it would be clearly desirable to obtain actual palaeointensity determinations for such old materials, there seems no reason for considering that the general intensity of the geomagnetic field has changed significantly for something like the last 3000 million years. Such a conclusion would be consistent with the formation of the Earth's core during a very early stage of planetary development, with no subsequent major changes in its radius ($<5\%$) or composition (Tarling, 1978a).

Chapter Nine

Geological applications

9.1 INTRODUCTION

The occurrence of magnetic minerals in virtually all rocks means that palaeo-magnetic studies can be applied in a very wide range of contexts. In particular, rocks can be magnetically dated on a range of scales, from 1^1 to 10^8 years (Section 9.2). For most other purposes, such as structural studies (Section 9.6), it is essential that sufficient observations are made to allow an adequate definition of the average geomagnetic field at some particular time. Usually this means that secular variations need to be averaged out, thus allowing the location of the average geomagnetic pole to be defined and hence the Earth's rotational pole for some particular period. It is important to stress, therefore, that most geological interpretations are fundamentally based on specific geomagnetic models. The fact that the geological interpretations that have been made on the basis of palaeomagnetic data appear to provide sensible interpretations that are consistent with other geological data indicates that such geomagnetic models are valid. However, it is unlikely that such models are always precisely valid. Observations must, for example, be omitted if relating to a period of a polarity transition when the field may not be dipolar and is certainly not typical. The real precision of all such interpretations is thus only partially indicated by statistical analyses (Chapter 6), as there is the additional uncertainty about the reliability of the geomagnetic model on which most of the interpretations are based.

9.2 MAGNETIC DATING

The dating of geological materials can be undertaken over a range of time-scales using the intensity and direction of remanence, and also its various components. As in the archaeological situation (Section 7.2), absolute dating by such methods can only be as good as the information on the past properties of the geomagnetic field. Unfortunately, theories of the cause of the Earth's magnetic field (Chapter 8) are inadequate for making sufficiently reliable estimates of these parameters and so any absolute dating, using such properties, must be predominantly based on previous observations and can, therefore, only be as reliable as their original dating. Nonetheless, the explosion in the quantity of palaeomagnetic data during the last decade or so,

197

Palaeomagnetism

combined with more reliable dating by both stratigraphic and radiometric methods, means that the precision of palaeomagnetic dating has improved considerably, although it is essential that the available records are continually updated and reassessed as more data accumulate. Nonetheless, while good absolute dating requires good 'master' records of geomagnetic properties, it is always possible to attempt relative dating using the properties of the geomagnetic field on the basis that the geomagnetic field rarely has repeated the same direction and intensity at more than one time. However, while opposite polarities of the geomagnetic field isolated in different rock sequences clearly establish that their magnetization was acquired at different times, the identification of an identical polarity only means that they could be, not necessarily are, of identical age. In the following sections, the methods of magnetic dating will be considered briefly in terms of increasing time-scales, with absolute dating methods considered before the potential of using that parameter for relative dating. Particular examples of such applications are given in Sections 9.4 and 9.5. It is important to emphasize, however, that all such dating methods date the age of the remanence rather than the age of the rock. It is therefore essential that the relationship between the age of the rock and the time of acquisition of its remanence is determined (Section 5.5). (All radiometric dates are corrected to the latest decay constants (Steiger and Jager, 1977) and the geological time-scale is given in Table 1.1.)

9.2.1 Secular variations

The occurrence of secular variation cycles means that rocks acquiring their remanence at different times will normally be magnetized in a geomagnetic field of different direction and intensity. Records of these variations in different parts of the world are now becoming available for the last few thousand years and form the basis for archaeomagnetic dating (Section 7.2). Clearly any material, whether in a geological or archaeological context, that acquired its magnetization during recent geological times can, potentially, be dated by the methods discussed in Section 7.2. Indeed, the 'master curves' used for such dating incorporate geological materials, such as historic lava flows, in their construction. This section will thus be primarily concerned with relative dating applications of this technique, mostly in recent sediments, and with pre-Quaternary applications in a geological context. These are, however, also discussed later (Section 9.4) and are only outlined briefly.

The presence of changes in declination and inclination in lake sediments reflect the interaction between geomagnetic secular variations in strength and direction with depositional and post-depositional forces (Sections 4.3 and 8.3). The most recent changes can often be related directly to archaeo-magnetic records, and hence provide an absolute dating method (Sections 7.2 and 8.3), but older swings can often be attributed to geomagnetic changes, although the exact times of such changes may be poorly known. In many

198

cases, it has become possible to establish a radio-carbon age for the peaks of such swings (Thompson and Turner, 1979; Creer, 1981; Section 8.3) although such dates must be treated cautiously in view of the effects of old carbon in a sedimentary environment (Blong and Gillespie, 1978; Megumi, 1978; Wintle and Huntley, 1979). Nonetheless, it is clear that such changes can frequently be used as reliable time-markers, sometimes more effective and certainly much cheaper than radiometric techniques (Thompson, 1977; Oldfield *et al.*, 1978; Bloemendal *et al.*, 1979; Walling *et al.*, 1979; Goodwin, 1980), particularly if depositional effects, mainly changes in the rate of deposition, can be assessed from magnetic fabric (Section 9.3) or other sedimentological techniques. However, correlation of the same directional features of remanence, even within the same lake, is not simple and usually requires a subjective evaluation of all available evidence.

Older sediments rarely preserve the precise parameters of the geomagnetic field at the time of their deposition due to physical changes during compaction and eventual lithification and to chemical changes during diagenesis (Sections 4.3 and 9.4), but it does seem possible that such features are preserved in certain sediments, such as limestones that underwent very rapid diagenesis (Turner, P., 1975). If the observed variations in directions of remanence observed in such sediments are, in fact, genuine records of secular variations during the time of cementation of these limestones, then clearly these may give a first-order approximation of the time-scale of diagenesis in earlier times although clearly any such arguments tend to be circular (Section 8.3) as there are no independent methods of determining the actual periodicity of such secular changes in the remote past.

Under certain circumstances, igneous rocks may also be dated by means of secular variations, even for those formed long before Recent times (Bogue and Coe, 1981). Lava sequences, for example, provide a series of spot readings of the geomagnetic field during the time of their extrusion (Section 8.3). Without other time-markers, their directions and intensities cannot be usefully evaluated, but it is possible to correlate sequences of lavas, displaced relative to each other by faults, by matching the directional properties of individual flows (Hailwood *et al.*, 1983). Another situation in which records of secular variations may be recorded is in igneous layered complexes (Schwarz *et al.*, 1979) and thick baked contacts where cooling may have taken place over time-scales of several thousand years. Unfortunately, metamorphic zones have usually had a very complex history (Sections 4.4 and 9.5) involving both heating and the migration of chemically active fluids which result in extensive chemical remanence being acquired. Furthermore, in areas that were heated at depth, the long times involved during cooling are likely to have resulted in extensive moderate-temperature viscous remanences having been acquired (Sections 2.3 and 9.5), effectively destroying any record of secular variations that may have occurred during the metamorphism. Nonetheless, under special conditions of purely thermal metamorphism, it is possible that records of secular changes may be preserved that will, ultimately, allow an evaluation of the time-scale of cooling. At the moment, however, it is more likely that the absence of systematic secular changes is simply used to confirm the complexity of such metamorphic zones (Section 9.5).

9.2.2 Reversals and excursions

The observation that the geomagnetic field undergoes changes in its polarity (Section 8.4) is, of course, based entirely on palaeomagnetic observations and theories of the behaviour of the geomagnetic field remain constrained by this evidence rather than providing a predictive theory. Furthermore, much of the evidence available indicates that the occurrence of polarity changes, during at least the last 80 million years, has been on a random basis and hence only definable by observation. Nonetheless, it has been the occurrence of such polarity changes that has been the main, most precise, method of dating the separation of continents during the last 200 million years (Section 9.6). This precision mainly arises because such reversals occur on a world-wide basis and are, geologically speaking, instantaneous, taking some 2–8000 years to occur (Section 8.4). This makes it a time-marker that is possibly more instantaneous than that of fossil extinctions intervals in which isolated communities may survive long after the demise of the species elsewhere, thus resulting in strongly diachronous extinction levels when comparing different regions of the Earth. Furthermore, such polarity changes will be recorded in igneous and sedimentary rocks being deposited at that time, providing an opportunity to determine their synchronism in both marine and continental environments.

(a) *Reversals*

Changes in the polarity of the magnetization of rocks are relatively easy to measure, as high accuracy is not required. Deflection of a compass needle can, for example, be used to determine the polarity in a sequence of lavas (Section 5.3), enabling a magnetostratigraphy to be constructed (Einarsson, 1957b; Doell and Cox, 1967a; Harrison *et al.*, 1979). Such a stratigraphy can then be compared with a similar succession elsewhere in order to obtain relative dating between them. If the sequences accumulated at similar rates, then such matching is relatively easy and reliable, but different rates of eruption or sedimentation can create apparent increases or decreases in the lengths of individual polarity intervals, rendering correlation more difficult, unless some other marker horizon is present and the interruptions in the sequences are not so great that entire polarity zones are omitted. The presence of other marker horizons, such as a specific stratigraphic horizon, may also mean that absolute dating is possible, thus evaluating the rates of deposition (Channell and Medizza, 1981) or eruption. Problems arise because, in isolation, any one polarity zone looks essentially similar to another unless its duration can be adequately assessed, but the record of oceanic magnetic anomalies and their dating (Section 8.4) means that specific anomalies can often be identified, of known age, if the approximate age of the measured sequence is known. A simple illustration of such dating is where the rocks are known to be less than 1 million years in age, then any truly reversed rocks must correspond to magnetizations acquired during the Matuyama Epoch, and hence be older than 730 000 years (Section 8.4). Such dating is of major importance in several radically different fields. In the siting of nuclear power stations, it is required, in many parts of the world, that the construction be away from any fault that has shown movement during the last 500 000 years. This age range is, in fact, difficult to date by most radiometric methods, and the presence of a reversal can be of considerable significance in making such an assessment. Another major field is anthropological, with the dating of the sediments, and occasionally igneous rocks, associated with hominid sites during the last 5 million years or so being greatly improved by means of the polarity stratigraphy (Johnson et al., 1975; Opdyke et al., 1977; Ninkovitch and Burckle, 1978; MacFadden et al., 1979; Drake et al., 1980; Liddicoat et al., 1980). Indeed, some of the earliest evidence in Britain for the presence of man occurs in cave sediments deposited during the Brunhes/Matuyama boundary (Hammo-Yassi et al., 1982). This time horizon similarly provides a correlatable horizon in volcanic rocks, for example between different Pacific islands (McDougall and Tarling, 1963), and in deep ocean sediments, it can be used to determine the extinction rates of certain organisms (Erlenkeuser, 1980), checking radiometric methods of dating sediments (Ku et al., 1968), the degree of bioturbation mixing of oxygen and other isotopes (Hutson, 1980) and similar bio-sedimentological processes, the growth of manganese on nodules (Crecelius et al., 1973; Greenslate, 1978; Toth, 1980), and so forth.

The last few polarity reversals are readily determined in oceanic sediments where they are datable by fossils in the same cores. Examination of such records (Figs 9.1 and 9.2) now extends well back into Tertiary times, covering the last 20 million years or so (Opdyke, 1972; Berggren, 1972a,b; Cox, 1975; Berggren and Van Couvering, 1974; Opdyke et al., 1974; Theyer and Hammond, 1974; McDougall, 1979; Dunn and Moore, 1981). The reversals during the last 5 million years can also be radiometrically dated in terrestrial volcanic sequences, such as on Iceland and Hawaii, with an accuracy which is generally less than the duration of a polarity zone. However, such dating still requires statistical analyses (Cox and Dalrymple, 1967a,b; Cox et al., 1968; Mankinen and Dalrymple, 1979) to obtain sufficient precision. For older sequences (Fig. 9.2), considerably greater numbers of radiometric dates are required to obtain comparable precision and, in most terrestrial volcanic sequences, it is not easy to define the actual polarity zone, on a world-wide scale, that is being dated. The primary dating of the polarity time-scale during the Cenozoic (the last 80 million years or so) is thus based on evaluation of the ages of marine magnetic anomalies that have resulted from ocean-floor spreading. (Conversely, the reality of ocean-floor spreading is predominantly based on the dating of the anomalies – Section 9.6.)

Originally (Section 1.2), the most recent oceanic magnetic anomalies were dated (Vine and Mathews, 1963) by matching them with polarity zones in terrestrial volcanic sequences that had been erupted during the last 3.5 million years (McDougall and Tarling, 1963; Cox *et al.*, 1963a,b). Older anomalies were then dated by assuming that the spreading rate determined for the last 3–4 million years remained constant in the South Atlantic throughout its opening (Heirtzler *et al.*, 1968). One particular traverse in the South Atlantic was chosen on an assessment that this area appeared to have the simplest spreading history. This evaluation was, in fact, remarkably astute as it is only within the last few years that a maximum error of some 20% has been determined for the older anomalies (Table 9.2) although corrections of this order were applied by Tarling (1971a). Revisions of these ages have been based, fundamentally, on two approaches. The sediments immediately overlying the oceanic igneous floor have now been sampled and their ages dated from their fossil content. Assuming that sedimentation commenced very shortly after the lavas cooled, then these closely date the time that they acquired their magnetization and hence date the anomaly observed at the surface. The second approach has been to identify the polarities in terrestrial sequences (Lerberkmo *et al.*, 1979; Barghorn, 1981) and to obtain radiometric dates for igneous rocks associated with them. In practice, this means dating polarity zones in an igneous sequence by fossil dating of inter-bedded sediments, and then determining the radiometric age of the fossil zones. Such techniques ideally require each anomaly to be drilled in the ocean basins and its counterparts in a volcanic sequence to be identified and dated. This is impracticable, at this stage, especially as a large number of radiometric dates would still be required to minimize the inaccuracy of radiometric dating (Mussett *et al.*, 1980). The procedure is therefore to attempt to determine the age of specific anomalies, such as Anomalies 5, 6, 13, 21, 24 and 31, and then to interpolate the ages of intermediate anomalies assuming that ocean-floor spreading has remained constant between successive dated anomalies.

An example of such dating is that for Anomaly 24 (Fig. 9.3 and Table 9.1), which is the oldest oceanic magnetic anomaly so far identified between Greenland and Europe (Section 9.6). The sediments overlying igneous rocks which cause this anomaly have been sampled in several locations and all yield a stratigraphic age which is on the Palaeocene–Eocene boundary (Maxwell *et al.*, 1970; Sclater *et al.*, 1974). In the western United States, volcanic tuffs of this age are interbedded with a terrestrial sequence that has been attributed to the same stratigraphic age and the tuffs have been radiometrically dated as being 53–54 million years old (Funnell, 1964). On this basis, the age of Anomaly 24 is now regarded as being 52–53 million years old (Table 9.2),

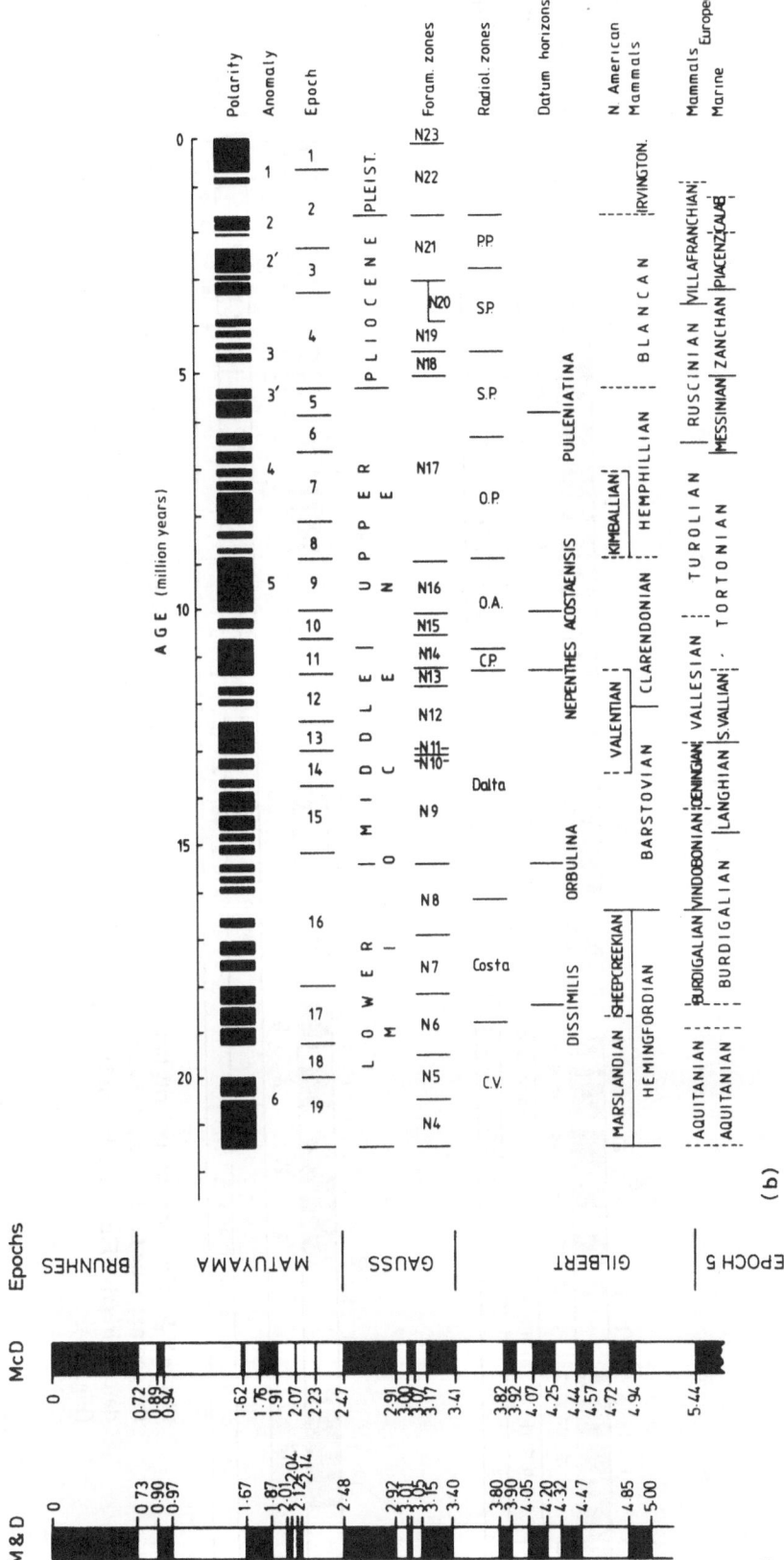

Figure 9.1 The Miocene to Recent polarity time-scale. (a) Considerable detail is now available for the nature and age of reversals of the geomagnetic field during the last 5 million years, and the studies of polarities in oceanic sediments, in particular, have enabled (b) comparison of the radiometric/polarity time-scale with those indicated by their fossil floral and faunal content. M & D is Mankinen and Dalrymple (1979); McD is McDougall (1979).

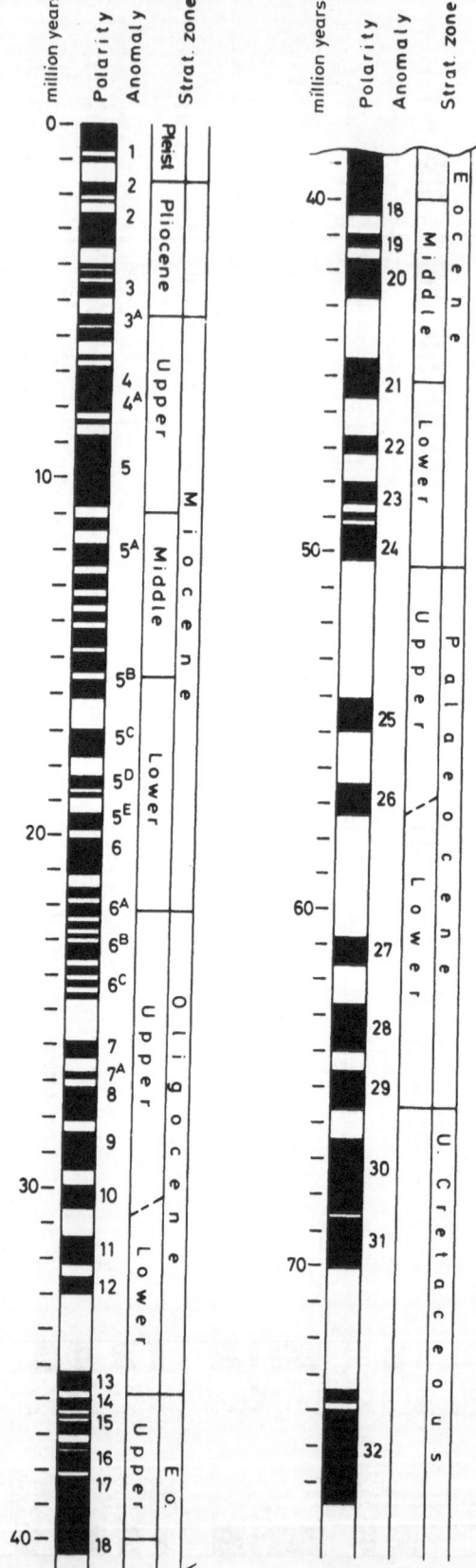

Figure 9.2 The Cenozoic polarity time-scale. The study of oceanic magnetic anomalies has enabled the polarity sequence for the geomagnetic field to become very well established for the last 60–80 million years. The younger sequences can also be dated directly (Fig. 9.1), but older sequences need to be correlated against key, dated anomalies, such as Anomaly 24 (Fig. 9.3).

Table 9.1 Proposed ages for Anomalies 13 and 24

| Method | Age (million years) | | Reference |
	Anomaly 13	Anomaly 24	
Spreading rates	39.0	61.5	Heirtzler *et al.*, 1968
Stratigraphic	38.0	53.5	Tarling, 1971a
Stratigraphic	36.5	58.0	Sclater *et al.*, 1974
Stratigraphic	37.0	58.0	Berggren and Van Couvering, 1974
Stratigraphic + radiometric	35.5	50.5	Tarling and Mitchell, 1976
Stratigraphic	36.5	53.0	Hailwood *et al.*, 1979

The radiometric ages (new decay constants) are given to the nearest 0.5 million years and may thus differ from the published original values using older decay constants.

some 8 million years younger than that originally estimated by Heirtzler *et al.* (1968). However, there are still uncertainties about these correlations. The exact relationship between the terrestrial sequence and the marine sequence is only poorly established in the United States (Schorn, 1971; Savage, 1980), and may also differ from the sediments of similar fossil-determined age in Belgium, France and Britain where glauconites have given a radiometric age of about 53 million years (Odin, 1978; Fitch *et al.*, 1978). To complicate the assessment further, lava sequences in Eastern Greenland have interbedded marine sediments extending across the Palaeocene–Eocene boundary (Soper *et al.*, 1976a,b) and the lavas are entirely reversed (Tarling, 1967b; Faller, 1975; Hailwood *et al.*, 1973), suggesting that they were magnetized during a prolonged polarity zone of opposite polarity to today, such as that immediately preceding Anomaly 24. This suggests that there was a slight time-delay, possibly 1 million years, between the cooling of the ocean floor basalts that give rise to Anomaly 24 and deposition of the sediments now covering them. The East Greenland basalts themselves have radiometric ages between 48 and 53 (average 51) million years (Hailwood *et al.*, 1973; Tarling and Mitchell, 1976). Such dates therefore suggest that the age of the Palaeocene–Eocene boundary is slightly younger than that observed in the United States, or that the tuffs in western America carry excess argon, or the lavas in Greenland have lost argon. In summary, therefore, the age of Anomaly 24 is only known to be around 51–53 million years, but a more precise radiometric age is not yet defined. Such an uncertainty is small, probably ±2%, but could be significant in statistical assessments of the frequency of polarity changes (Section 8.4). There are also clearly dangers in assuming that the extant polarity time-scale is exact to better than 2%. Nonetheless polarity changes provide time-markers

205

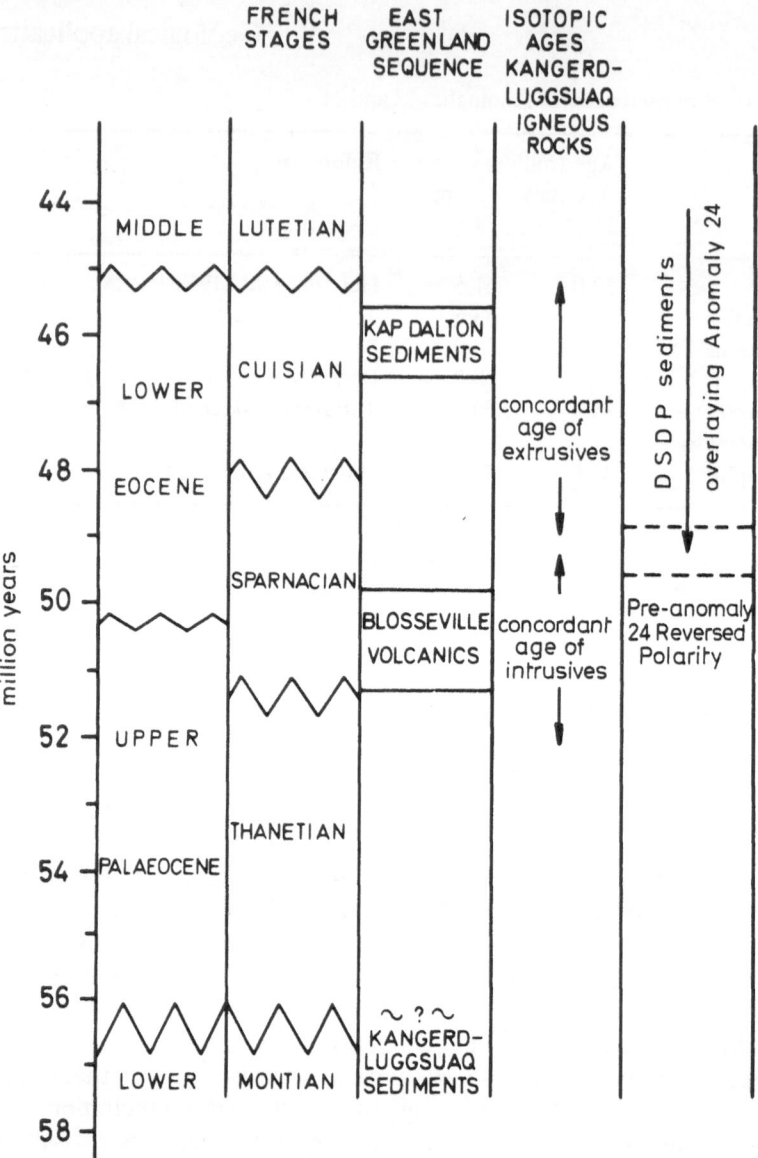

Figure 9.3 The dating of Anomaly 24. The age of the sediments immediately over-lying igneous rocks which have been associated with Anomaly 24 shows that these igneous rocks must have acquired their magnetization close to the Palaeocene–Eocene boundary. The same aged rocks also occur in East Greenland where they can also be radiometrically dated, thus enabling the age of Anomaly 24 to be determined. However, errors must arise in assigning the age of the anomaly to that of the sediments immediately overlying it as there could be a time-gap between the eruption and their covering of sediment. There could also be errors in correlating the stratigraphy established for the oceanic sediments with those of continental sequences, thus giving rise to errors of a million years, or possibly more, due to such correlations. Additionally there are problems of the precision of the radiometric dates that have usually a repeatability error of only 1–2%, but may show systematic errors due to, for example, argon loss or excess old argon, being incorporated in the rocks. On this basis, it is not considered that anomalies much older than 10–20 million years have yet been dated to an accuracy of much better than 1–2 million years.

that are probably only diachronous to less than some 2000 years on a global scale, and their stratigraphic ages are fairly well known – it is their precise ($\pm 2\%$) radiometric ages that remain poorly defined.

Only oceanic magnetic anomalies occurring at the very end of the Cretaceous have so far been drilled in the ocean basins, so that precise stratigraphic correlations, such as illustrated for the age of Anomaly 24, are no longer possible. The dating of Mesozoic reversal sequences (66–235 million years) therefore depends primarily on the matching of anomalies observed in the oceans with those found in sedimentary and igneous sequences, mostly on land, that may be dated by fossil or radiometric methods (Helsley and Steiner, 1969; Ness et al., 1980). The sequence of marine magnetic anomalies (Fig. 9.4(a)) indicates periods of changing polarity at the start and end of the Cretaceous (66–133 million years) separated by a long interval within which there are no distinct linear anomalies (Larson and Pitman, 1972; Larson and Hilde, 1975), the Cretaceous magnetic quiet zone. Studies of overlapping sections of carbonate rocks in Italy, traversing the Cretaceous–Tertiary boundary and extending almost to the base of the Cretaceous (Alvarez, 1977; Channell et al., 1979c; Lowrie et al., 1980a,b), have delineated similar changing polarities in the upper and lower parts of the sequence, separated by a zone extending from the uppermost Aptian into the basal Coniacian (110.5–87 million years) of continuous normal polarity. On the basis of these correlations, and taking into account evidence from cores of marine sediments (Keating and Helsley, 1978a,b,c; Hailwood et al., 1980), Lowrie et al. (1980a,b) have proposed that the younger polarity changes are, in fact, even younger than thought on ocean-floor spreading arguments, but the older polarity changes are older, suggesting, therefore, faster rates of spreading in the Upper Cretaceous and slower in the Lower Cretaceous, although the changes in rate are small, of the order of 7%.

The arguments for the age of polarity changes in the Jurassic (133–184 million years) are essentially similar to those for the Cretaceous, but become increasingly difficult and therefore less precise (Fig. 9.4(b)). Marine magnetic anomaly sequences are present, mainly in the Pacific (Larson and Pitman, 1972; Larson and Hilde, 1975; Larson and Helsley, 1975; Larson, 1976) but any age estimates based on ocean-floor spreading rates must obviously be corrected for proposed changes in the Cretaceous rates (Lowrie et al., 1980b). Jurassic palaeomagnetic studies (Steiner and Helsley, 1972, 1975a,b; Steiner, 1980) are also hampered by the increasing probability of chemical overprints, and possibly a weaker geomagnetic field during this time (Section 8.5), and by uncertainties in dating of some of the terrestrial sandstone sequences because of the sparsity of fossils, most studies having been undertaken in Colorado and Wyoming, Utah and Arizona. Surprisingly, these are mostly char-acterized by mixed polarities, with reversed polarities sometimes appearing dominant, yet most studies of Jurassic anomalies led to predictions of a quiet normal zone for much of the Jurassic. It seems probable that the lack of

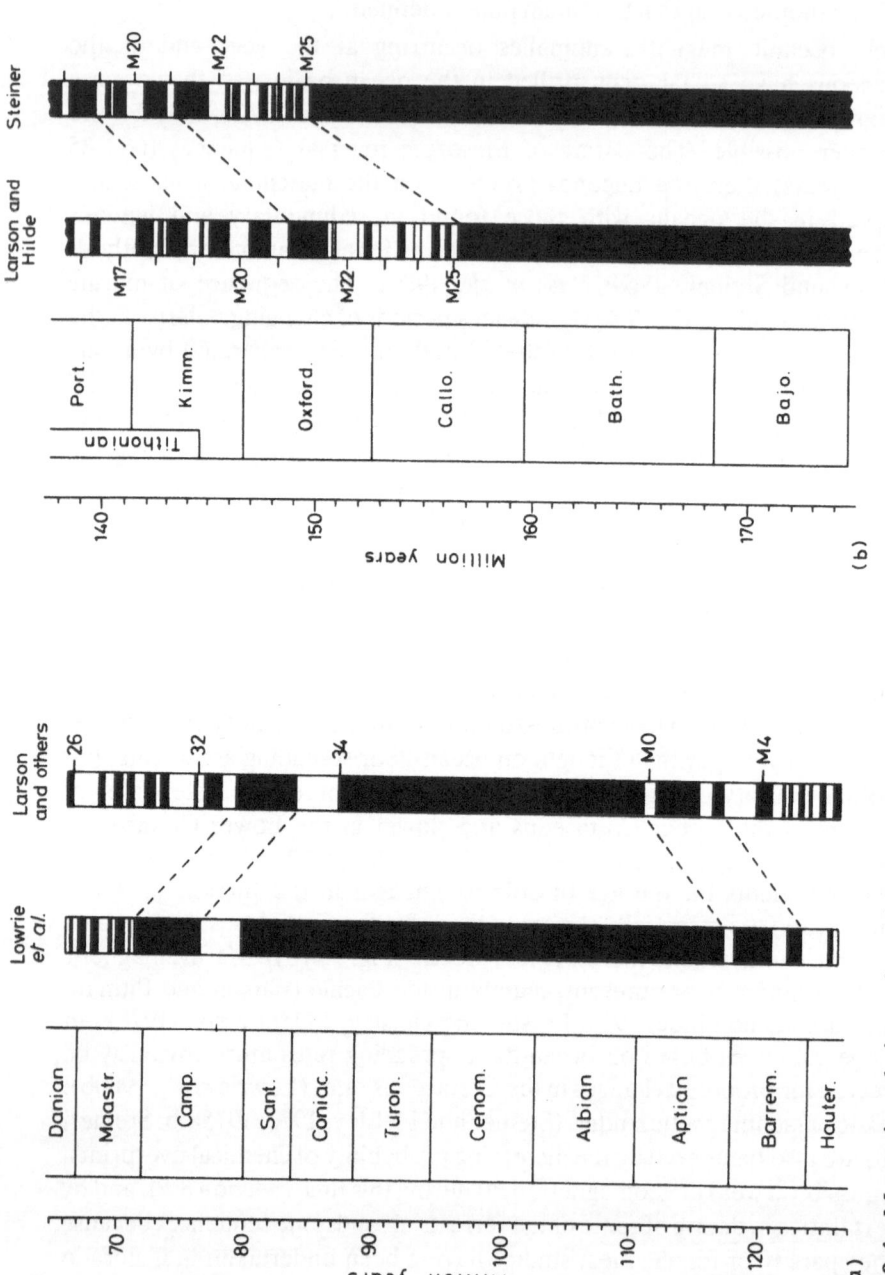

Figure 9.4 Mesozoic polarity scales. Although the polarity scale is becoming increasingly established from studies of continental rocks, much of the information is still derived from oceanic magnetic anomaly studies, with general agreement on the general pattern of the scale, but uncertainty about the dating. (a) Mainly Cretaceous polarities based on oceanic anomalies after Lowrie *et al.* (1980a, b) and Larson and Pitman (1972), Larson and Hilde (1975). (b) Mainly Jurassic anomalies based on Larson and Hilde (1975) and Steiner (1980).

apparent anomalies in Middle Jurassic may therefore reflect

1. Extremely rapid changes of polarity, so that their corresponding anomalies tend to compensate each other or
2. A weak axial field so that the field is both weak and characterized by large-scale secular variations.

Few data are yet available for the Lower Jurassic, but in the Triassic (194–235 million years) there are frequent reversals of polarity indicated, mostly by studies of the Moenkopi and associated formations in the south-western United States (Picard, 1964; Helsley, 1969; Helsley and Steiner, 1974; Steiner and Helsley, 1974a; Elston and Purucker, 1979; Shive et al., 1979) and Europe (Burek, 1964). These sequences have very clearly delineated polarity zones, but there is considerable argument concerning the mode of acquisition of their magnetization (Section 9.4), with most of the sedimentological evidence indicating that it is of diagenetic origin (Walker et al., 1981), although the sharpness of the boundaries and their lateral persist-ence would suggest that such diagenetic changes must have taken place extremely rapidly after deposition and therefore still record changes of the polarity of the field penecontemporaneously with their deposition. Such a situation could readily lead to the magnetization of specific strata being some-what later than the beds on either side, thus resulting in a greater number of apparent polarity changes than occurred in reality. Unfortunately for such periods, there are no oceanic magnetic anomalies that can be used as guide lines as the oldest extant oceanic floor is only some 185 million years old (Fig. 9.5), although such dates are still based on assumed ocean-floor spreading rates and small areas of slightly older ocean floor may yet be defined.

For all older periods (Fig. 9.6), the only evidence for the age of polarity changes is based on direct palaeomagnetic studies. These clearly become less precise with increasing age, reflecting the probability of later magnetizations being present (Section 9.2.3). In particular, many sequences appear to have a dominant overprint magnetization that was acquired during a prolonged period of reversed polarity, the Kiaman, that extended from late Westphalian (Europe) – Desmoinesian (USA) to uppermost Permian, Lower Scythian (USA) – upper Tartarian (USSR), from 305–315 to c. 250 million years ago, with only one or two brief normal episodes (Irving and Parry, 1963; Khramov, 1967; McElhinny, 1969b). The uncertainty in the dating reflects problems of intercontinental stratigraphic correlation for these periods. Despite such uncertainty, the Kiaman is the longest known period of constant polarity (>50 m. years). For earlier times, the problems of overprinting and even relative dating become dominant. Russian studies in Siberia, in particular, appear to be least affected by overprinting (Section 9.2.3), and polarity time-scales have been proposed for much of the last 600 million years (Molostovsky et al., 1976). Such scales still need to be better defined and correlated with sections in other regions. It is clear, however, that

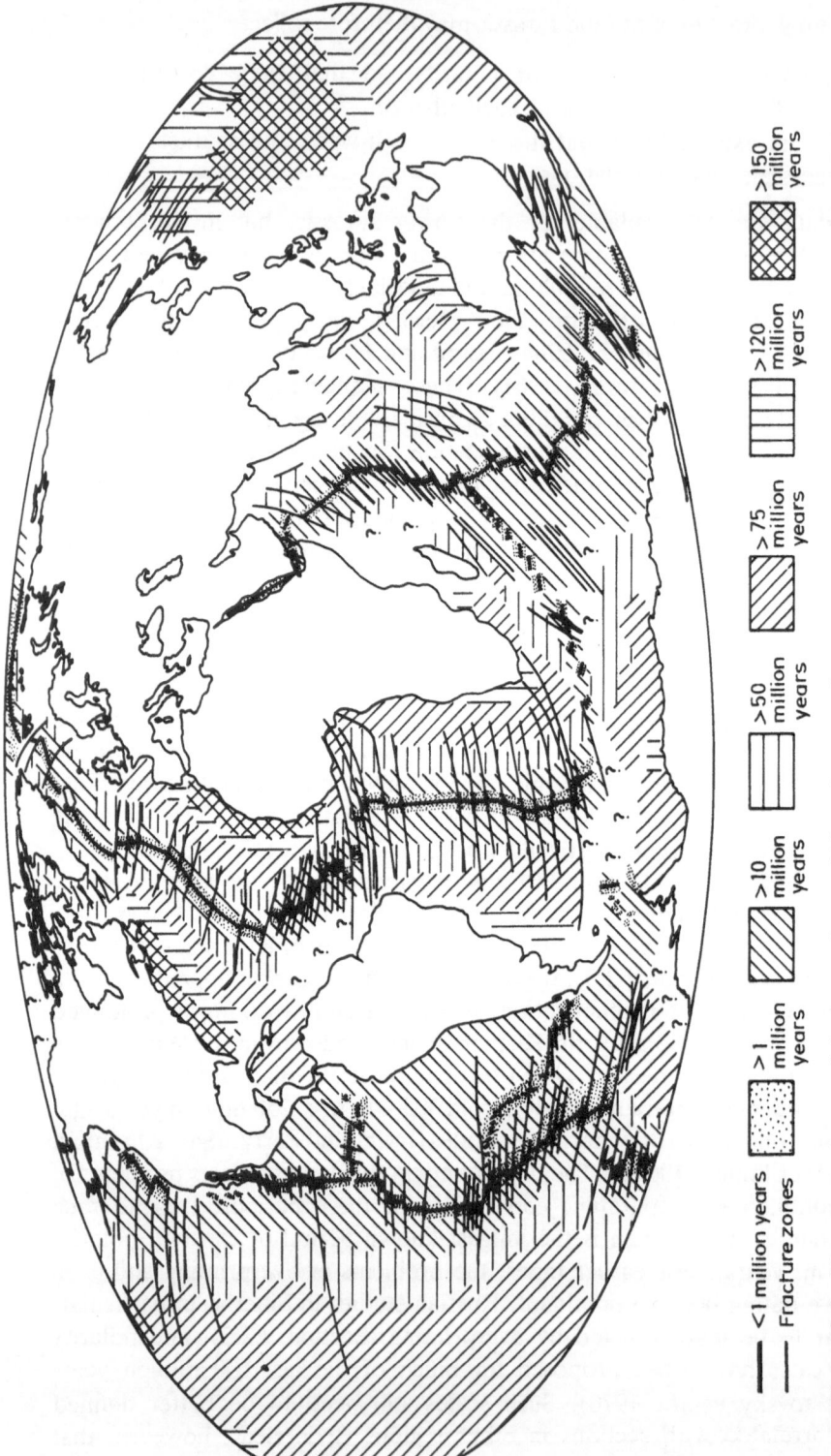

Figure 9.5 The magnetic age of the ocean floors. Although most of the ocean floors have now been dated, many of the ages are based on extrapolation from known areas, with such extrapolations often depending on particular reconstruction models for the continents. Nonetheless, the fact that no ocean floor is currently known that is older than some 200 million years is, in itself, a fundamental constraint on models for the evolution and formation of the ocean basins.

Legend:
- <1 million years
- Fracture zones
- >1 million years
- >10 million years
- >50 million years
- >75 million years
- >120 million years
- >150 million years

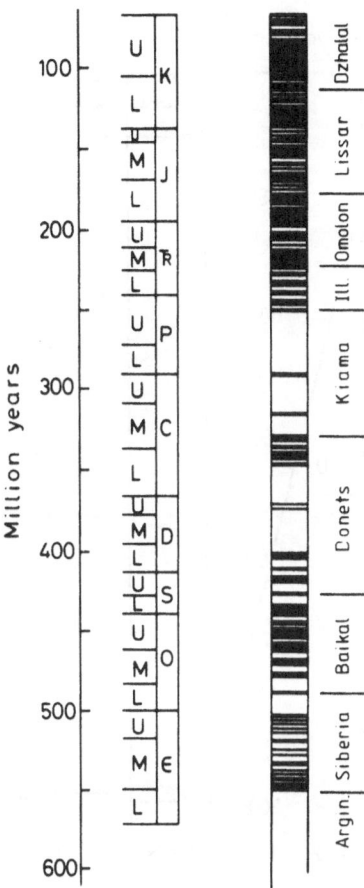

Figure 9.6 The Russian polarity scale for the Phanerozoic. (After Molotovsky *et al.*, 1976.)

geomagnetic polarity changes have occurred during most geological time as reversals of polarity also occur throughout the Pre-Cambrian. At this stage, however, it is impossible to establish realistic correlations between different areas based on such evidence alone.

(b) *Excursions*

Departures of the geomagnetic field direction more than about 40° away from its normal direction, for periods of 10^2–10^4 years, but without actually totally reversing are termed polarity excursions (Section 8.4) of which many have been reported as occurring during the last 300 000 years or so (Table 9.2 and Fig. 9.7). However, the vast majority of these are reported from studies of sediments, comprising wet lake and dry lake deposits, glacial tills and varves, and so forth, although a few anomalous directions are reported from igneous rocks. In many instances, it is thus difficult to be sure that the excursions are geomagnetically controlled and thus of application over a wide area, possibly

211

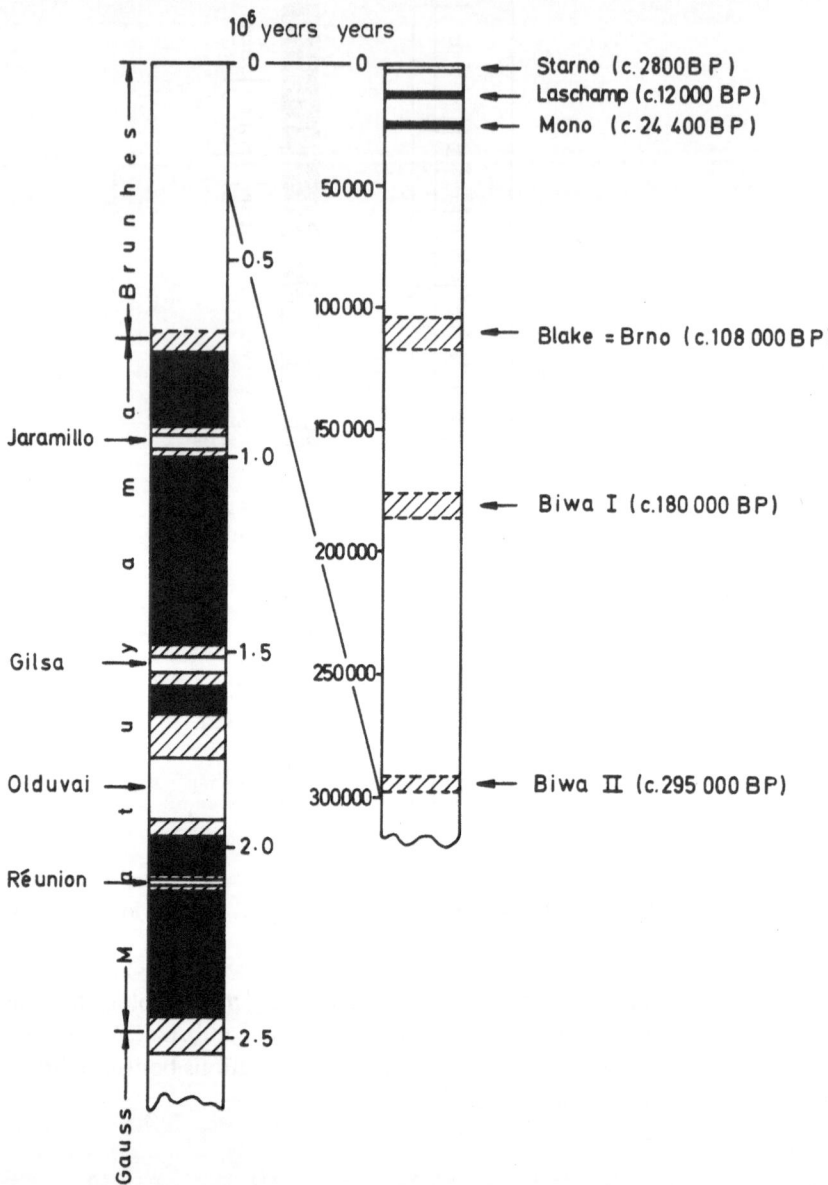

Figure 9.7 Polarity changes and excursions during the last 2.5 million years. The validity of many of these excursions has yet to be firmly established. See also Table 9.2.

world-wide. Only one excursion, the Blake excursion, appears to be found in Atlantic and Pacific sediments (Ninkovich *et al.*, 1966; Smith and Foster, 1969; Yasakawa *et al.*, 1973; Creer *et al.*, 1980). Many other apparent correlations may well reflect the problem that occurrences of disturbances, particularly in sediments, are not uncommon, so that reported findings of one excursion allow the discovery of apparent correlations elsewhere – the 'reinforcement syndrome' (Watkins, 1972). Many of the individual reports have been summarized by Verosub and Banerjee (1977) and will not be duplicated here. It is, however, important to be aware of the problems of both reinforcement and unreliability of most forms of dating during this period of time.

The Laschamp–Olby excursion provides a typical example of the difficulties of dating such excursions in igneous rocks. This was discovered in France by Bonhommet and Babkine (1967) and dated by K/Ar as being less than 20000 years old (Bonhommet and Zahringer, 1969). Thermoluminescence of pebbles then gave 41000 ± 4000 BP and U/Th dating gave 39000 ± 6000 BP (Condomines, 1978). Hall and York (1978) gave a revised K/Ar and Ar/Ar ages of about 47000 BP. Huxtable et al. (1978) determined thermoluminescence ages for the sediments baked by the flows as about 25800 BP. Valladas et al. (1977) and Labeyrie and Gillot (1977) gave ^{14}C, thermoluminescence and K/Ar ages of 33000 ± 4000 BP, while Gillot et al. (1980), using the same three techniques, gave ages of 36000 ± 4000 BP for the Laschamp flow and 42000 ± 5000 BP for the Olby flow. This range of possible dates clearly reflects the great difficulty of using any of the standard radiometric methods of dating over such time-scales. It is now reasonably clear that the probable ages are, in fact, greater than 30000 years BP, in which case they could relate to anomalous directions reported in glacial tills in New England and Ontario (Soloyanis and Brown, 1979; Stupavsky et al., 1979) and also to anomalous directions associated with aboriginal fireplaces in Australia, the Mungo excursion (Barbetti and McElhinny, 1972, 1976), for which an increasing range of possible ages are appearing. They could also correlate with the Maelifel excursion in Iceland (Peirce and Clark, 1978). To complicate this evidence even further it has also been claimed that it is another example of self-reversal (Heller, 1980).

Although it seems clear that excursions exist there are grounds for questioning the existence of some of the proposed excursions (Table 9.2), and there is even greater uncertainty about their age, so it seems premature to use these changes as a dating method. However, if they are geomagnetic and occur on at least a continental scale, they will clearly have great value as stratigraphic markers during a time that is particularly difficult to date by radiometric methods. In particular, it is essential that well-dated sequences, such as the Scandinavian varves (Sections 4.3 and 7.2), are examined in combination with fabric studies, with geomagnetic interpretations restricted to observations from the winter layers of quiet, largely current-free deposition (Sections 4.3 and 7.2).

The significance of reversal stratigraphy, magnetostratigraphy, has been indicated earlier for the last reversal, mainly in the context of sedimentology. Much older rates of diagenesis can be evaluated from the presence of different polarities, sometimes in the same stratum (Channell, 1978). The uncertainties about the boundaries of the Kiaman illustrate how such world-wide

Table 9.2 Some proposed excursions during the last 300 000 years

Excursion	Date*	Location	Material	References
Starno	c. 2800	Sweden	Glacial varves	Nöel and Tarling, 1975; Nöel, 1975
Gothenburg	10–15 000	Sweden	Glacial varves	Mornër et al., 1971; Mornër and Lanser, 1975
	11 950–12 103 +	Sweden	Glacial varves	Nöel and Tarling, 1975
	c. 12 000	Czechoslovakia	Loess	Bucha, 1973
	8–10 000	Lake Michigan	Sediments	Vitorello et al., 1974
Nørre Lyngby	14 000	Denmark	Sediments	Abrahamsen and Readman, 1980
Laschamp/Olby	20 000–* 47 000	France	Volcanics	Bonhommet and Babkine, 1967; Bonhommet and Zahringer, 1969; Candomines, 1978; Hall and York, 1978; Huxtable et al., 1978; Barbetti and Flude, 1979a; Gillot et al., 1980; Heller and Petersen, 1982
Lake Biwa	c. 18 000	Japan	Sediments	Nakajima et al., 1973
Imuruk Lake	c. 18 000	Alaska	Sediments	Noltimier and Colinvaux, 1976
Gulf Mexico	15–18 000	Gulf Mexico	Sediments	Clark and Kennett, 1973; Freed and Healy, 1974
Rubjerg	23 000	Denmark	Sediments	Abrahamsen and Knudsen, 1979; Abrahamsen and Readman, 1980
Lake Mono	25 000	California	Sediments	Denham and Cox, 1971; Denham, 1974; Liddicoat and Coe, 1979
Mungo	17 000–31 000	Australia	Baked sediments	Barbetti and McElhinny, 1972, 1976; Barbetti and Flude, 1979b
Maelifell	28–31 000	Iceland	Volcanic	Peirce and Clark, 1978
	30 500	Ontario	Glacial tills	Stupavsky et al., 1979
	32 000	New England	Glacial tills	Soloyanis and Brown, 1979

214

Olby	42 000*	France	Volcanics	Gillot et al., 1980
	45 000	Washington	Dry sediments	Othberg, 1973
Blake	104–117 000	Atlantic, Pacific Japan, Italy	Sediments	Smith and Foster, 1969; Denham, 1975; Ninkovich et al., 1966; Wollin et al., 1971; Kawai et al., 1972; Yasakawa et al., 1973; Creer et al., 1980
Biwa	176–186 000	Japan	Sediments	Kawai et al., 1972
Biwa	292–298 000	Japan	Sediments	Kawai et al., 1972

instantaneous time-markers can be used to improve stratigraphic correlations. The precise location of the Permian–Triassic boundary is clearly at a different time-level in the USSR compared with Europe–North America (Cain et al., 1979) if normal polarity zones have been correctly identified. Similarly there are major differences between Russian and Western stratozones (Bouroz et al., 1975; Wagner and Higgins, 1979) in the Carboniferous (Mississippian–Westphalian). Such differences are not merely of academic interest as the zonations are often based on different lithological units associated with transgressions and regressions of sea-level during the Carboniferous (Ramsbottom, 1971, 1973, 1979), but also extending throughout the Mesozoic (Hart and Tarling, 1974; Hancock and Kauffman, 1979; Jenkyns, 1980). Many of the world's oil reserves are directly related to such transgressions (Vail et al., 1977; Tissot, 1979; Hallam and Bradshaw, 1979) and would thus define times of interest in other areas of the world should such changes be global, i.e. eustatic. If they are eustatic, then they raise many academic problems of how such eustatic changes are caused. With the exception of increasing or decreasing ice sheets, the only way in which such changes can occur is by changes in the morphology of the ocean basins, for example, the growth or decline of oceanic ridges (Donovan and Jones, 1979). Such changes, however, would take place on a significantly longer time-scale than is observed in changes of sea-levels in North America, Europe and the USSR (Wise, 1972; Whitten, 1976; Ryan and Cita, 1977; Reyment and Mörner, 1977; Hallam, 1977, 1978; Leggett, 1978; McKerrow, 1979). This implies that there is a further control on either the shape of the ocean basins or the water content of the basins – as yet unspecified, yet indicated by the existence of world-wide hiatuses in deposition within the ocean basins (Hays and Pitman, 1973; Harrison, 1980b).

On a more local scale, reversals can provide an excellent method for dating largely unfossiliferous sequences, such as desert sandstones, distinguishing between, for example, Permian desert sands with essentially reversed polarity and Triassic sands with mixed polarities. Even when fossils are abundant, they may be extremely mixed and of little value as stratigraphic markers, as in some of the flysch basins of the southern USSR or East Indies. In such zones, correlation between different polarity zones provides specific time horizons that can then be used to determine the regional structure (Section 9.6). On an even smaller scale, such polarity changes can provide improved correlation between different regions in the same continent, thus relating to fossil migration patterns, such as that of Palaeocene mammals in the San Juan Basin of the United States (Lindsay *et al.*, 1978). Igneous sequences can similarly be correlated on polarity sequences such as in the Miocene lavas of the Columbia Plateau (Watkins, 1965); these few examples merely indicate some of the applications of such polarity stratigraphies that have already been used – their potential uses are even more enormous.

9.2.3 Palaeomagnetic poles

The magnetic vectors in rocks of any specific age can be used to calculate the position of their corresponding palaeomagnetic pole on the standard geocentric dipole model for the average geomagnetic field (Section 6.3). Both the theoretical and observational evidence support the reality of this model (Section 8.5) so all rocks magnetized at the same time should have the same palaeomagnetic pole position. However, different areas of the Earth's surface (tectonic plates) are moving relatively to each other and so, although the Earth's axis of rotation remains fixed in space, the pole position appears to change relatively to each tectonic block. Successive determinations of the position of the poles relative to any one stable tectonic block therefore lie on a curve – *the polar wandering path*. Each tectonic block will have a unique polar wandering path which represents its movement relative to the Earth's axis of rotation. The age of magnetization of rocks of unknown age on the same block can thus be determined by comparing their corresponding palaeomagnetic poles with those of known age. Cursory examination of the polar wandering paths for the main continents during the last 300 million years (Fig. 9.9) indicates that the average motion of a continent relative to the pole is some 0.3°/million years. If a palaeomagnetic pole can be defined within 2–3° for rocks of unknown age, then the age of their magnetization can be determined within some 7–10 million years. However, greater or lesser precision may occur at times of faster or slower movement of the tectonic block relative to the pole.

Fundamental to such dating is, of course, the reliability of the polar wandering path for that particular tectonic block and it must be emphasized, at this stage, that such polar wandering paths are still only poorly defined, even for the major continental blocks. A further difficulty is also created by the need to establish that the rocks being examined do actually belong to the same tectonic unit for which the curve has been derived (Section 9.6). Additionally, the palaeomagnetic data for both the curve and the rocks being dated need to be free from errors due to inhomogeneity and anisotropy, and of sufficient quantity for the short-term fluctuations of the geomagnetic field – secular variations and polarity transitions and excursions – to be either eliminated or averaged out. This generally means that igneous rocks, which often preserve a record of past secular variations, may need to be sampled more extensively than sediments, in which the magnetization within an individual sample may well tend to average out secular variation effects (Sections 4.3 and 6.3). Finally, it must be emphasized that the magnetic dating corresponds to the age of magnetization of the rocks which may be younger than their geological age.

When these criteria have been fulfilled, absolute dating of the age of magnetization can be attempted by comparison with established polar wandering paths. However, where such paths are not yet established, relative dating can be attempted. Comparison of the remanent directions in a suite of iron ores, igneous and sedimentary rocks can establish similarities and dissimilarities in age, thus establishing whether the iron ores are epigenetic, syngenetic, etc., as evaluated by Symons (1967a,b) for Minnesota iron ores. As the polar wandering paths become increasingly well documented, it becomes increasingly possible to attempt absolute dating, although this can, of course, only be as accurate as the definition of the polar wandering path. Under certain circumstances, it is possible to use polar wandering paths established for other tectonic blocks, but the accuracy of such dating is then additionally dependent on the accuracy of the fitting of the two tectonic units. An example of such dangers is provided by the dating of the Dwyka varves in South Africa. The Dwyka varves, when originally sampled, were stratigraphically poorly defined and could be Devonian to Permian in age. On the basis of the Du Toit (1937) and Smith and Hallam (1970) reconstructions for the southern (Gondwanan) continents, it was found that the Dwyka varve pole corresponded closely with the Australian Lower Carboniferous palaeomagnetic pole position and it was thus attributed to this age. Shortly afterwards, but by different people, the Lower Carboniferous agreement between the Australian and Southern African data was being quoted as evidence for the correctness of the Du Toit and Smith and Hallam reconstructions. (The age of the Dwyka varves is now known to be basal Lower Permian – Anderson and Schwyzer, 1977.)

As polar wandering paths are of vital importance in both dating and most structural analyses (Section 9.6), a review of these paths is given below. It is not practicable to comment on, or even to tabulate each of the 8–9000 palaeomagnetic poles that have now been published. (Most of these data have been tabulated in pole lists – Irving, 1960a,b, 1961, 1962a,b; Irving and Stott, 1963; McElhinny, 1968a,b, 1969a, 1972; McElhinny and Cowley, 1977, 1978,

217

1980; Hicken *et al.*, 1972; Irving and Hastie, 1975; McElhinny *et al.*, 1977, 1979.) In general, the pole paths for each major continent (Section 9.2.3(d)) have necessarily been determined using somewhat inadequate criteria and subjective evaluation. In order to compile these paths, the available pole positions have been selected broadly on the following criteria:

Devonian poles and sites of Australia

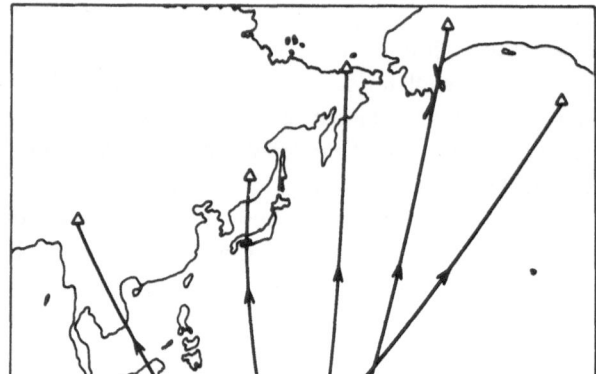

(a)

Figure 9.8 Some problems in establishing polar wandering curves. (a) Devonian of Australia; most sites are from southern Australia and may well have been affected by the Kanimblan orogeny, causing tectonic rotation of the sampled regions, with the exception of Central Australia. (b) The Devonian of northwestern Europe, if simply averaged, gives a mean pole D. It seems probable that many of the data are affected by later magnetization, as indicated by most of the Russian data being grouped closely near to the Permian mean pole positions, P. Some northern Russian data do not appear to have been affected by this 'overprinting' and it is likely that most northwestern European data comprise varying degrees of Permian overprinting, with the true Devonian pole lying near to the equator. (c) The Jurassic pole for North America is still badly defined because of the lack of data. The Cretaceous poles for cratonic North America (+) are well grouped, but the available Jurassic data are very scattered (×); those selected by Vugteveen *et al.* (1981) are shown as solid dots for Cretaceous and triangles for Jurassic. The two curves are (i) dashed for Van der Voo and French (1974), incorporating data from other continents after geometric reconstructions to close the Atlantic Ocean, and (ii) barred for Irving (1979) based on a 'running mean' of the available data. The scatter in the data is shown inset which shows the Irving path with associated error circles. (See text for further discussions.) Ages in million years.

218

Devonian poles of 'extra-alpine' Europe

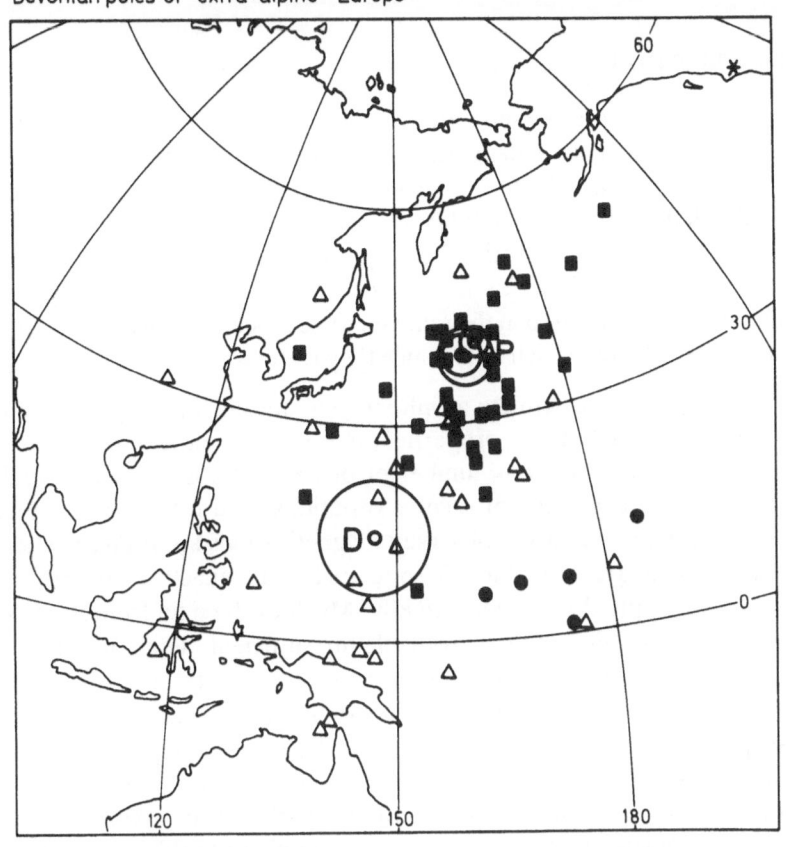

△ N.W. European ● N.Timan and N.Russia

■ S. Russia, Moscow and Urals

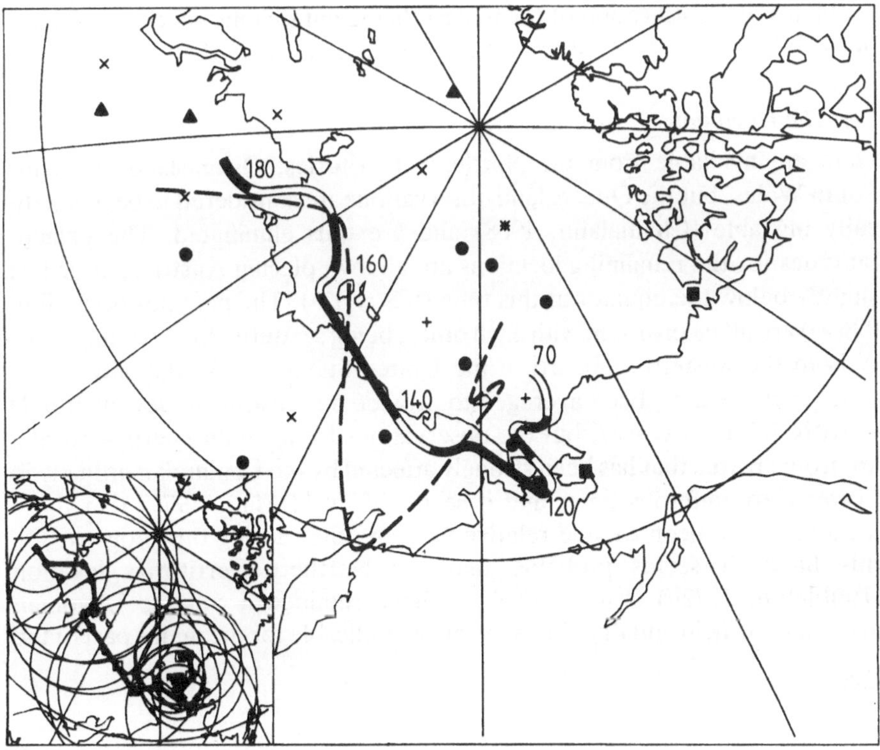

(b)

(c)

Palaeomagnetism

1. Preferably at least five separate sites have been sampled, thus tending to reduce the effect of secular variations and other geomagnetic fluctuations.
2. The samples have been subjected to demagnetization procedures to establish their magnetic stability and if single component or not.
3. The age of the magnetization is thought to correspond with the age of the rocks themselves or to be known from other dating methods.
4. The sites are all located in the same tectonic block and are unlikely to have moved relatively to each other since they were originally magnetized.

Such criteria are inadequate as inhomogeneous and anisotropic data may have been included, but such properties are frequently either not measured or not reported. The minimum number of observations is much too small to ensure adequate averaging of secular variations, and it has generally been necessary to assume that the observed magnetizations are primary when no fold or tilt tests are available. Despite such inadequacy, it has still been impossible to apply these criteria rigidly. Much of the data from the USSR, for example, are not readily available in sufficient detail to make full evaluation. In these, and many other cases, the construction of the polar wandering path has necessarily included a large element of subjective assessment. In order to illustrate some of these problems, and the way in which these types of problem have been tackled, three examples are given prior to a brief outline of the polar wandering paths for the major continental blocks – the Devonian of Australia, as an example of problems in defining a tectonic unit; the Devonian of Europe as an example of overprint magnetizations; and the Jurassic–Cretaceous of North America as an example of problems of identification of the more reliable data. Some specific problems in establishing the Pre-Cambrian paths are discussed later (Section 9.2.3(e)).

(a) *The Devonian of Australia*

Data are available from the Northern Territories, Queensland and New South Wales, but the Queensland observations are considered to be magnetically unstable (Chamalaun, 1968) and are thus eliminated. The palaeolatitudes for the remaining locations are all low, placing Australia at or just slightly below the equator at this time ($8°S \pm 11°$). The pole positions (Fig. 9.8(a)) are all inconsistent with each other, being scattered from southeastern Asia to the western seaboard of the United States. Conventionally, these pole positions have been averaged to provide an Australian Devonian pole of 64°N 180°E, $\alpha_{95} = 27°$ but the observations from southeastern Australia are from an area that has been strongly affected by the Kanimblan orogeny in Lower Carboniferous times and it seems likely (Tarling, 1979) that these locations have been rotated relative to each other during this orogeny. On this basis, it seems probable that the Northern Territories location (Embleton, 1972b) is more likely to have retained a genuine Devonian geomagnetic field and that this is the most applicable for the major part of the

220

Australian tectonic plate. The age of this tectonically stable observation, the Mereenie Sandstones, is not firm and it could be Silurian in age. However, the fact that the palaeolatitudes for all observations are similar would suggest that it is of probable Lower Devonian age. In this particular instance, therefore, a pole position can be postulated but it must still be regarded as being poorly defined.

(b) *The Devonian of Europe*

A large number of observations are available from rocks of this age in Europe, but are widely scattered, even after excluding areas that may have been affected by the Alpine orogeny (Tarling, 1969). The poles, however, still form groups corresponding to different regions (Fig. 9.8(b)) which could be interpreted as indicating a series of different tectonic plates for this time. However, most poles from the Moscow, Urals and southern USSR (west of the Urals) areas lie close to the mean European Permian positions, while those from northern European USSR (Timan) lie near the equator. Pole positions for the rest of western Europe show a large scatter, broadly spreading from the east of Japan, southwards towards the equator, but mostly more easterly than the Timan grouping. It is possible that the Timan group does, in fact, indicate slight anticlockwise rotation of this region, relative to the rest of Europe, in post-Devonian times, but the remaining data seem more likely to represent a varying degree of overprinting (Creer, 1968; Tarling *et al.*, 1976; Storetvedt *et al.*, 1978; Storetvedt and Carmichael, 1979; Tarling, 1980a), with the poles furthest away from the Carboniferous–Permian pole positions possibly being least remagnetized, and those near the Late Palaeozoic grouping having been remagnetized by various processes during Carboniferous and Permian times. On this basis, the simple mean pole position for the western European data would be misleading and the true pole is assessed as lying near the equator, somewhere between New Guinea and the Timan group of poles. Clearly such an assessment is highly subjective and requires to be substantiated by detailed vector analyses of all the available data, combined with fold and tilt tests to assess the age of the remanence.

(c) *The Jurassic–Cretaceous of North America*

All data from the Western Cordillera need to be excluded from an assessment of the polar wandering path for North America as this area has been subjected to extensive rotation and dislocation during Tertiary times (Section 9.6) but most other parts of the continent can be assumed to have been part of the stable North American plate during Mesozoic and later times. Even with such exclusions, only six of the ten Cretaceous poles are closely grouped near western Alaska (Fig. 9.8(c)), of which most of these are from igneous rocks and most are of Lower Cretaceous age; two of the remaining four are not very distant from the main grouping, but the other two must be considered to be separate from the Cretaceous grouping. For Jurassic times, the situation is even worse. Equal numbers of sedimentary and igneous formations have been sampled, but there is no clear grouping, even though most poles lie near to northern Siberia. Vugteveen *et al.* (1981) considered only eight Jurassic–Cretaceous pole determinations to be relevant for cratonic North America (Fig. 9.8(c)). Steiner (1978) has proposed that major movements of the North American craton occurred in the Jurassic, associated with the development of the Central Atlantic. However, the main episode affecting the polar path is likely to be the Laramide orogeny some 80 million years ago associated with the stagnation of ocean-floor spreading in the Caribbean and the 'capture' of North Atlantic spreading by the northwards extension of the South Atlantic spreading system (Tarling, 1980b). This would suggest a major motion of North America relative to the pole would be expected in Late Cretaceous times rather than within Jurassic times.

The present Jurassic–Cretaceous palaeomagnetic data from stable North America must therefore be considered only poorly defined. Unfortunately no real checks, using other criteria, can be made on the available assessment, as the palaeolatitude differences are likely to be obscured by any changes caused by changing land–sea and oceanic circulation patterns. Although reconstructions of Laurentia (Section 9.6) can be undertaken independently of the terrestrial palaeomagnetic data, the European data for this period are, if anything, even less reliable than the North American determinations. Van der Voo and French (1974) evaluated a common polar wandering path for the reconstructed continents that allowed the two data sets to be combined, possibly reducing random sources of error. However, it is evident that the precise pole positions for these two critical periods cannot be regarded as established and it seems premature, therefore, to attempt to define the polar wandering path for even shorter periods of time (Fig. 9.8(c)).

(d)   *Phanerozoic (last 590 million years) polar curves*

The main features of the curves for each of the main continental blocks are only briefly outlined – mainly to indicate which periods are the most well established for each continental unit. In the maps of the individual curves (Fig. 9.9), the circles of 95% confidence have been calculated on those data points that are considered to be the most likely representatives of the pole position, relative to that continent, for that time. The pole position and confidence limits (Table 9.3) are calculated on the total data whenever possible, i.e. using all pole determinations for sites located on that cratonic block of the continent and reputed to be of that specific age. This is generally practicable for poles corresponding to ages of less than 300 million years or so. However, if the poles for any particular time have an oval distribution, this could represent either a progressive movement of the continent relative to the pole during that time interval or the presence of more than one magnetic component. If the oval distribution tends to 'string' towards a younger part of the polar wandering curve, it is generally interpreted as indicating the presence of later components of magnetization and the mean pole position has then been determined (or estimated) to lie within the oval, but located furthest from the younger part of the curve. When the poles are merely widely dispersed, the mean direction has sometimes been used, or the pole has been assessed from a consideration of the location of the younger parts of the polar wandering path. A further subjective element has been the assumption that the motion of a continent will tend to show little or no change unless it has collided with another continental block, i.e. sudden changes in the direction or rate of motion along polar wandering curve will normally be considered to be associated with a time of an orogenic event in that continental block. For example,

Table 9.3 Phanerozoic mean pole positions. The poles used in Fig. 9.9 are given in the left-hand column, while alternate poles are given in the right-hand column; the latter are mostly based on straight averages of the data which fulfil the criteria listed in the text. It is emphasized that even these criteria are inadequate as no allowance is made for anisotropy and much data have not been adequately analysed for the presence of multicomponents. Additionally, thermal demagnetization has not always been used or, when used, the effects of chemical changes were not always recognized in the original study. However, proper criteria would eliminate most of the available data. (W = centre of Lower and Upper Cambrian group – Watts *et al.*, 1980a,b; CP = estimated Carboniferous–Permian overprinted group; OS = estimated Ordovician–Silurian overprinted group; URA = Urals; UK = Britain).

Age	N	Lati-tude	Longi-tude	α_{95}	N	Lati-tude	Longi-tude	α_{95}
N. America								
Cretaceous	5	68.4	185.1	5.8				
Jurassic	13	73.4	98.6	9.6				
Triassic	40	60.8	96.1	3.1				
Permian	43	44.5	116.3	2.7				
Carboniferous	33	37.6	125.1	3.4				
Devonian	4	−18.8	133.9	36.4	27	39.4	128.0	13.9
Silurian	E	0.0	120.0	–	13	13.6	127.8	16.0
Ordovician	4	−13.9	114.4	16.8	15	20.9	136.0	22.3
Cambrian	3	−34.4	187.6	9.5	E	10.0	160.0	– (W)
					20	38.9	128.1	7.7 (CP)
					15	1.7	161.4	9.4 (OS)
Europe								
Cretaceous	E	72.0	173.0	–	–	72.0	173.3	5.0
Jurassic	E	67.0	155.0	–				
Triassic	69	54.7	159.8	4.2				
Permian								
(Upper)	73	44.4	166.3	1.3				
(Middle)	21	39.1	163.3	3.9				
(Lower)	62	40.8	165.6	2.0				

Palaeomagnetism

Table 9.3 – *continued*

Age	N	Latitude	Longitude	α_{95}	N	Latitude	Longitude	α_{95}
Carboniferous								
(Upper)	37	39.3	162.7	2.8				
(Middle)	24	28.8	174.4	4.3				
(Lower)	17	12.9	137.8	16.1				
Devonian	E	10.0	145.0	–				
Silurian	E	5.0	150.0	–	–	−0.5	135.0	7.9
Ordovician	E	0.0	150.0	–	–	15.6	148.7	8.1 URA
Cambrian	E	−10.0	150.0	–	–	17.4	180.3	12.4 UK
Angara (Siberia)								
Cretaceous	33	76.1	175.7	4.8				
Jurassic	16	76.7	143.2	13.3				
Triassic	67	49.3	147.4	2.6				
Permian	36	45.4	146.5	6.4				
Carboniferous	E	30.0	150.0	–				
Devonian	E	25.0	155.0	–				
Silurian	8	−4.3	121.2	19.0				
Ordovician								
(Middle)	11	−21.8	129.7	4.0				
(Lower)	12	−40.2	132.3	6.6				
Cambrian								
(Aldan)	13	−46.4	162.5	4.8				
(Siberia)	11	−36.8	130.8	2.8				
South America								
Cretaceous	12	85.6	203.0	4.6				
Jurassic	8	86.6	41.0	10.5				
Triassic	9	84.1	69.1	5.9	19	85.5	46.1	7.6
Permian	6	60.0	173.5	6.4	17	77.2	165.8	6.9
Carboniferous	7	60.7	169.2	11.5	8	58.3	175.8	30.6
Devonian	6	−4.5	135.5	27.6				
Silurian	–	–	–	–				
Devonian	7	−16.6	159.4	27.4	7	−9.7	159.0	26.7
Cambrian A	2	−3.5	213.4	25.0	5	−38.2	201.7	40.8
B	2	−50.5	205.4	7.7				
Africa								
Cretaceous	13	68.3	237.3	6.5				
Jurassic	17	67.9	251.8	5.8				
Triassic	10	66.0	246.9	9.0				
Permian	3	37.6	229.4	29.1	5	44.4	237.4	17.6
Carboniferous	E	20.0	220.0	–				
Devonian	1	0.5	205.0	–				
Silurian	–	–	–	–				
Ordovician	1	−14.0	156.0	–	3	60.5	159.8	>90
Cambrian	5	−18.4	156.2	12.2				

Table 9.3 – *continued*

Age	N	Lati-tude	Longi-tude	α_{95}	N	Lati-tude	Longi-tude	α_{95}
India								
Lower Tertiary	12	29.3	275.4	13.9				
Cretaceous	26	20.6	282.3	11.2				
Jurassic	–	–	–	–				
Triassic	4	13.8	304.4	19.3				
Permian	7	−6.5	309.3	89.2				
Carboniferous	–	–	–	–				
Devonian	2	2.4	327.5	>90				
Silurian	1	−30.0	348.0	–				
Cambrian	2	27.5	213.0	–	4	49.8	211.9	72.3
Australia								
Cretaceous	4	53.7	336.8	10.1				
Jurassic	9	48.0	345.0	10.1				
Triassic	3	46.5	339.7	24.3				
Permian	3	41.7	307.4	10.5				
Carboniferous	2	33.7	315.0	47.3	6	69.4	337.0	27.9
Devonian	1	41.5	228.5	–	6	68.5	197.3	27.2
Silurian	2	36.0	219.1	40.7				
Ordovician	3	27.0	207.2	20.1	4	24.4	213.4	18.7
Cambrian	10	29.6	199.5	7.0	17	28.8	189.2	12.1
Antarctica								
Cretaceous	5	83.7	130.3	8.4	6	85.1	93.8	10.1
Jurassic	7	51.6	39.1	7.5				
Ordovician	3	(13.8	202.4	25.7)				

in North America a change in the curve could be expected to be related to the Laramide orogeny at the end of the Cretaceous. (It is unfortunately possible to find an appropriate orogenic event to coincide with most conceivable change in any polar wandering curve so that such a criterion cannot be used without consideration of the nature of the orogeny; most of the world's continents were drastically affected by the Variscan (Hercynian) orogeny some 300 million years ago in Late Carboniferous–Early Permian times (Zwart and Dornsiepen, 1978) and North America and Europe were strongly affected by the Taconic and Acadian orogenies in the Lower Palaeozoic.) The curves presented here are therefore subjective evaluations, especially for periods older than some 300 million years, but it is emphasized that, with the exception of Antarctica, each curve has been assessed individually, i.e. without consideration of the implications for continental reconstructions (Section 9.6). Other recent reviews of the available data have been given by McElhinny (1973a), Anderson and Schwyzer (1977) and Irving (1977, 1981).

Palaeomagnetism

(i) *North American Phanerozoic polar wandering path*

Most of the early palaeomagnetic data were derived from the mid-western states, Colorado–Utah–Arizona–Wyoming, and most of the well-dated, reliably established palaeomagnetic data still come from this region (Fig. 9.9(a)). The Cenozoic observations are mostly from the Western Cordillera and hence potentially rotated during the Late Tertiary motions in this area (Section 9.6.3), but increasing observations are becoming available from intracratonic volcanism. The Jurassic–Cretaceous uncertainties have been discussed earlier, and are clearly only poorly to moderately well defined for the cratonic region. These Cretaceous poles are loosely grouped immediately north of Alaska, while those of Jurassic age are even more loosely grouped between Alaska and northern Siberia. Triassic data are abundant, mostly reflecting the presence of stably magnetized red sandstones in many areas, particularly the southwestern United States. These give a group of poles extending from the Urals to southern China, with a well-defined mean, and Permian poles are similarly well defined, but centred on northern China. The Carboniferous poles are almost entirely of Upper Carboniferous (Pennsylvanian) age and are quite well grouped near Japan. While there are sometimes indications of magnetic overprinting in some of the pole determinations for rocks less than 300 million years old, the pole positions for Lower Carboniferous and older rocks indicate the presence of complex magnetizations, partially reflecting that many of the observations have been obtained in the eastern United States and have been affected by later orogenic episodes. The Cambrian poles, for example, are scattered over about half a hemisphere. Watts *et al.* (1980a,b) have argued, cogently, that the Upper Cambrian poles lie in a region near 10°N 160°E, as do Lower Cambrian poles, but Mid-Cambrian poles are thought to lie near 50°N 110°E, implying a clockwise rotation of North America by some 60°, followed by an exactly equal and opposite rotation. While such motions cannot be excluded, it seems intuitively unlikely that continental movements should occur in such a manner and that the Mid-Cambrian position should also happen to coincide with the Carbo–Permian section of the earlier part of the curve. Nonetheless, such discrepancies only indicate the level of reliability and the complexity of the magnetization for such old rocks. (Other reviews of North American Phanerozoic data have been given by Van der Voo and French (1974), Vandenberg (1979), and Vugteveen *et al.* (1981).)

(ii) *European Phanerozoic polar wandering path*

The European data are the most extensive, per unit area, of all the continents (Fig. 9.9(b)) and the region has been subjected to long and intensive study. It is somewhat surprising, therefore, that the actual definition of the path is still so poor. The Cenozoic data are mostly derived from southwestern Europe, often within the Alpine tectonic region, but the main difficulty has been that

226

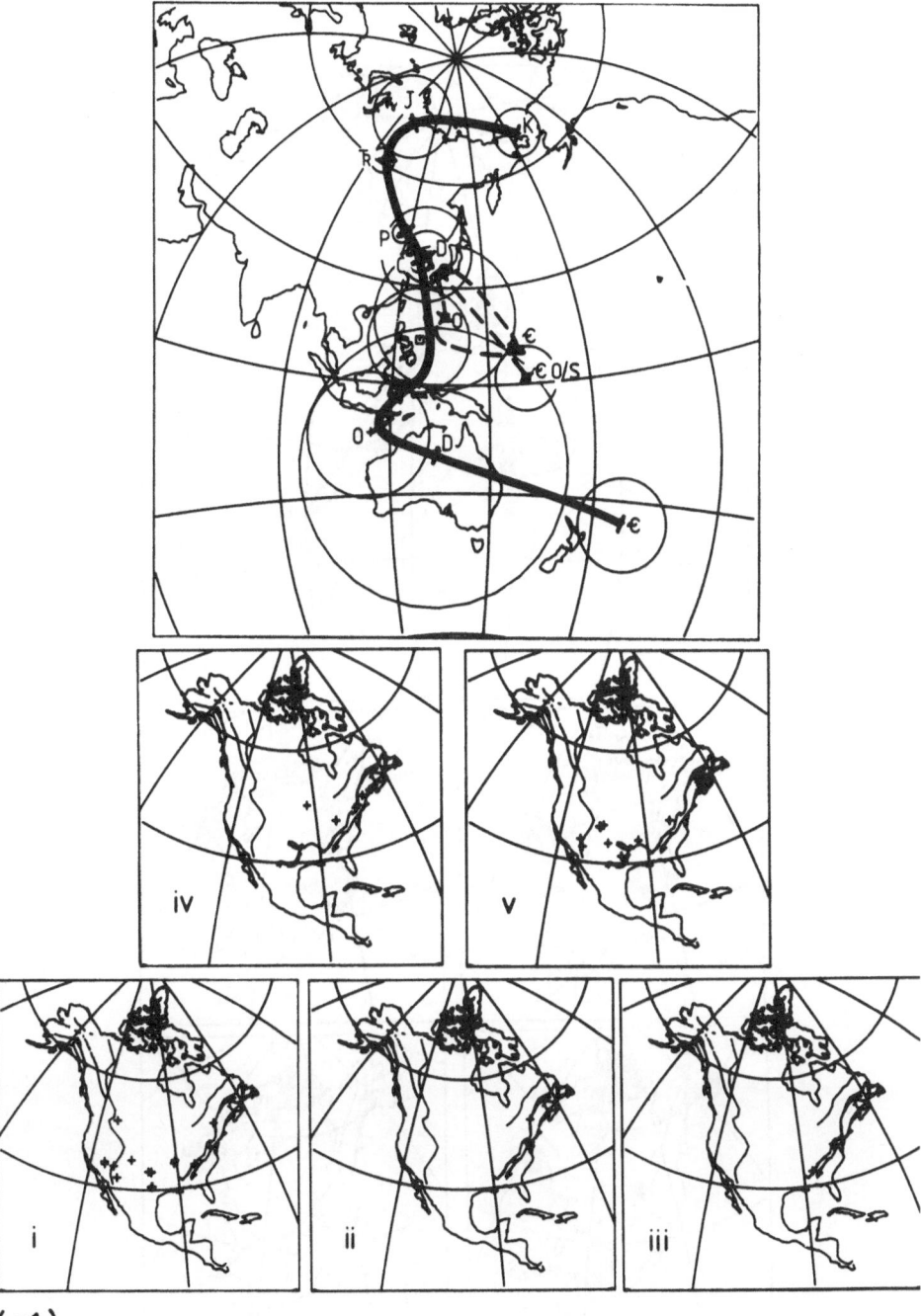

(a1)

Figure 9.9 Phanerozoic polar wandering curves for the major tectonic blocks. The thickest lines are the estimated position of the pole for different periods based on a combined objective and subjective evaluation of the available palaeomagnetic data. Where dashed, this curve represents a somewhat less subjective judgment of the probable position. (See further discussion in text.) The ages are Cretaceous (K), Jurassic (J), Triassic (T), Permian (P), Carboniferous (C), Devonian (D), Silurian (S), Ordovician (O) and Cambrian (€), with upper (u), middle (m) and lower (l) where appropriate. (a1) North America, together with the Watts *et al*. (1980a, b) estimation for the old section of the curve.

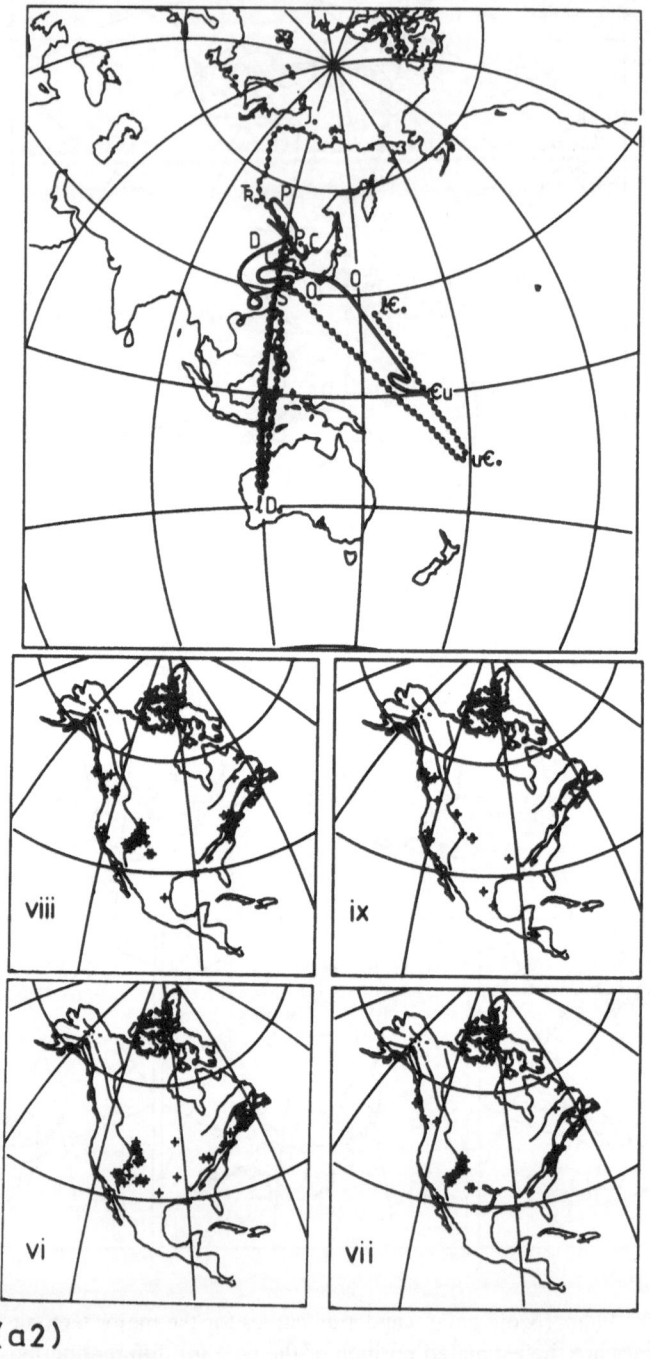

(a2)

Figure 9.9(a2)   Alternative North American curves according to Irving (1979) and Van der Voo *et al.* (1980).

228

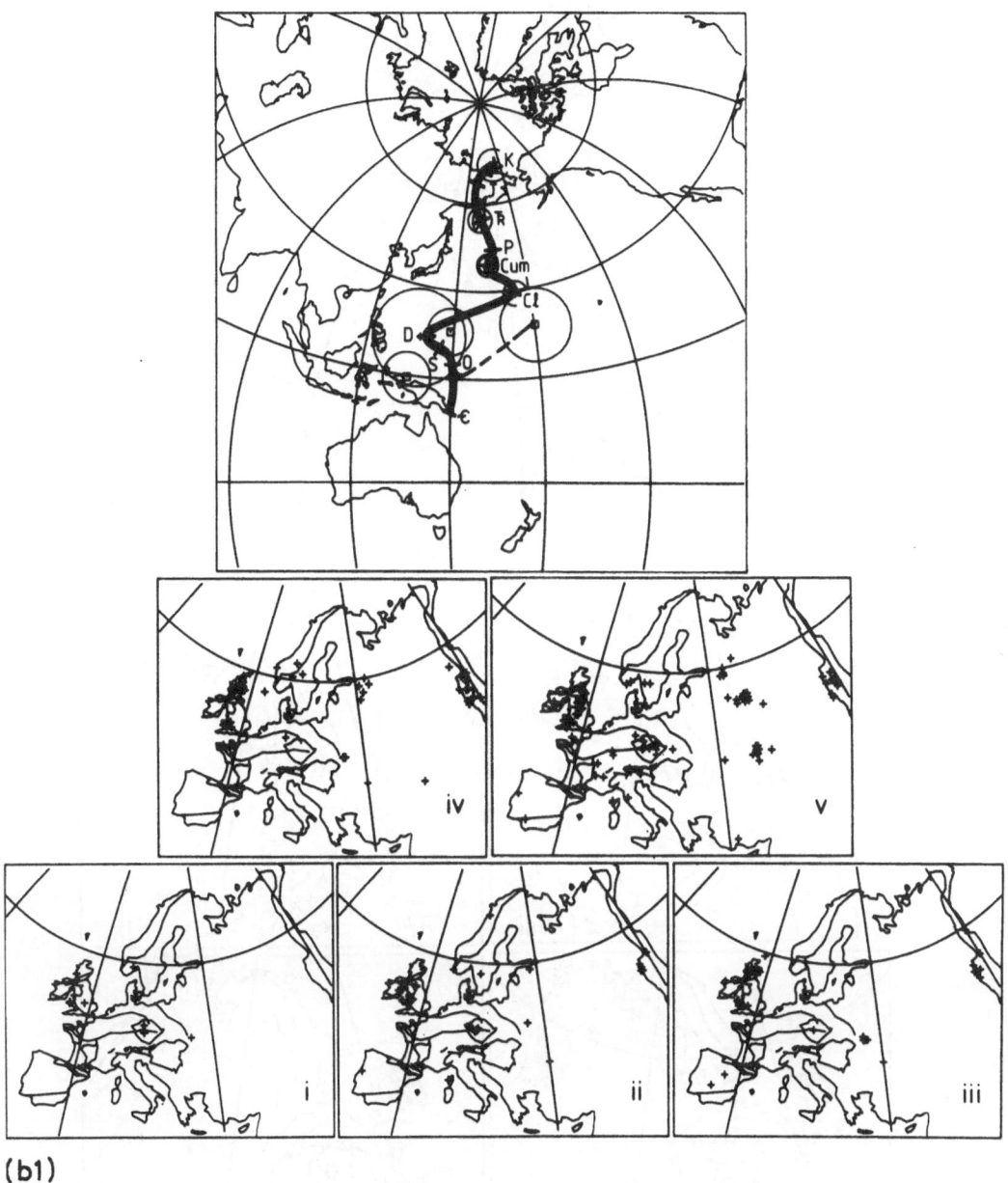

Figure 9.9 (b1) Europe, with (b2) (overleaf) estimations by Irving (1977) and Vandenberg (1979).

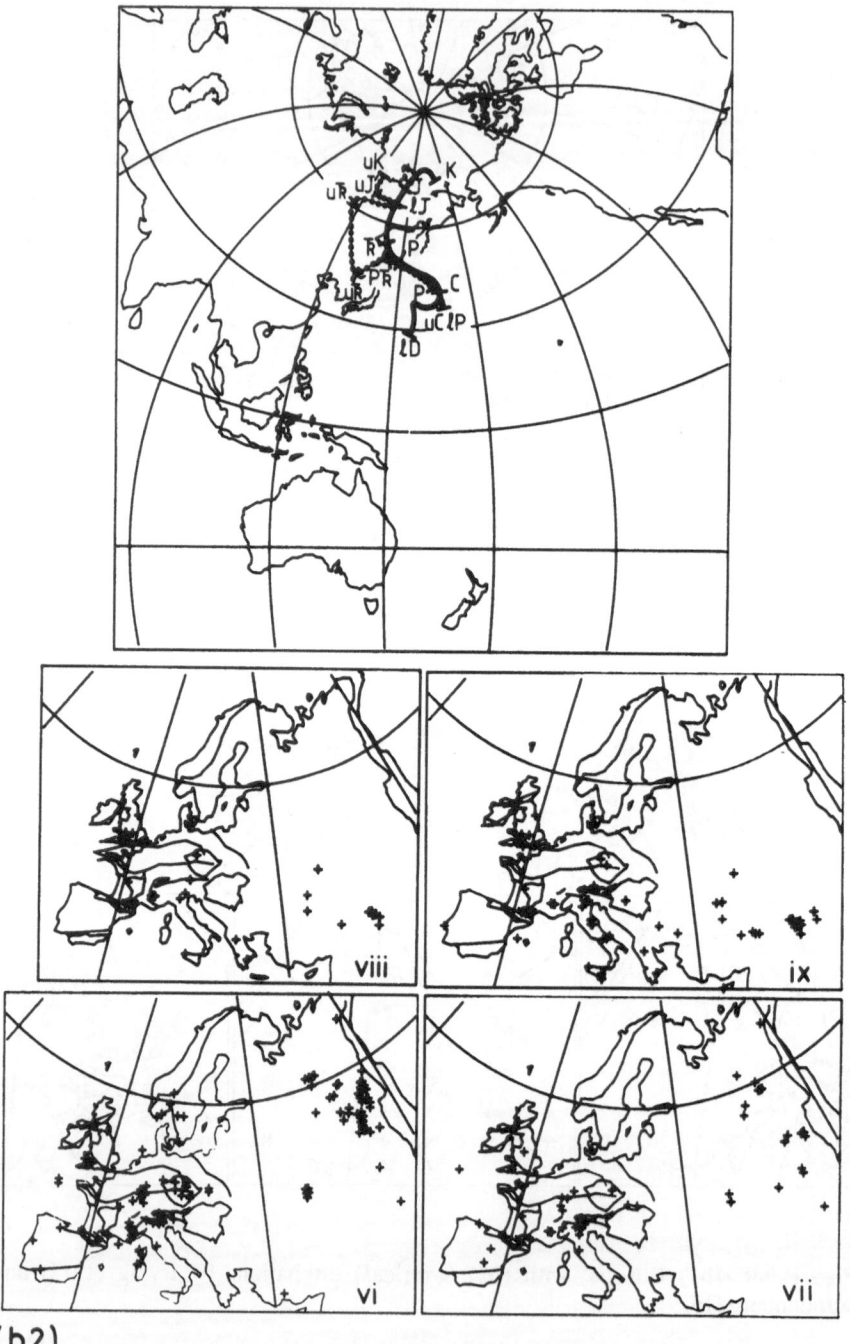

(b2)

the observations are almost entirely based on volcanic rocks, the radiometric age and stratigraphic age of which are only poorly established. The most reliable observations, excluding the Recent period, are for the Tertiary igneous province in Britain, but these represent a duration of only some 3–5 million years, and the actual radiometric ages are only generalized at some 60 million years in view of extensive metasomatic activity. The Jurassic and Cretaceous data have been mentioned in relationship with those of North America (Section 9.2.3(c)), and the pole positions, for tectonically stable Europe, must be considered inadequately defined, with the most apparently reliable observations coming from areas, such as the Crimea, that are likely to have been affected by the Alpine orogeny. In contrast, the Permo–Triassic period has been intensively and extensively studied, with many of the earliest studies being based on red sandstones of this age. However, the fact that many of these determinations were made prior to analyses of partial demagnetization behaviour raises the question of their reliability. Nonetheless, later studies have generally confirmed the earlier, untreated observations and the data for this period can be considered to be reliable and the average pole positions, for at least the individual periods, are well established. Similarly the uppermost Carboniferous period seems to be well established, but older rocks are of less certain reliability, often having been affected to uncertain extents by the Hercynian orogeny in Upper Carboniferous–Lower Permian times. This major orogenic episode, which also occurred in Late Devonian times in some areas, affected most of southern and western Europe and there are many disputes concerning the degree of magnetic overprinting of Palaeozoic rocks, with some authors considering that such overprinting completely obscures any of the original remanence, and others claiming evidence for original magnetizations having been isolated within extremely tectonized regions. The truth probably lies somewhere between these two extremes, as discussed in the example of the European Devonian data (Section 9.6). In the path presented here (Fig. 9.9(b)), the arguments outlined in discussion of the Devonian data are applied, i.e. unless there are known geological factors against, it is generally considered that the palaeomagnetic poles furthest from the Late Palaeozoic (Permo–Triassic) or later pole groupings are likely to be the most realistic representatives of the older geomagnetic field. Clearly such evaluations are necessarily subjective, involving an assessment of the palaeomagnetic data and the tests carried out on the rocks, combined with an assessment of the region from which they were derived. Ultimately, the palaeomagnetic data for any one stable block must be mutually consistent and also agree with any geological evidence for the palaeolatitude and orientation of the block. However, it is not considered here that the Cambrian pole position is actually defined at all for the main tectonic unit – the Baltic Shield – and it is only possible to estimate the likely pole positions for some of the other Palaeozoic periods. (Other reviews include Van der Voo and French, 1974; Irving, 1979; Vandenberg, 1979.)

(iii) *Siberia (Angaran) Phanerozoic polar wandering path*

The USSR has a large number of palaeomagnetic laboratories and a very large amount of data must exist (Fig. 9.9(c)). Unfortunately most of this only appears in the form of the final tabulated results, which makes any independent assessment somewhat difficult, particularly as the Russians have only recently begun to use partial thermal and alternating magnetic field demagnetization procedures on a large scale, and other methods, such as direct current demagnetization, are of limited value. The major feature of the

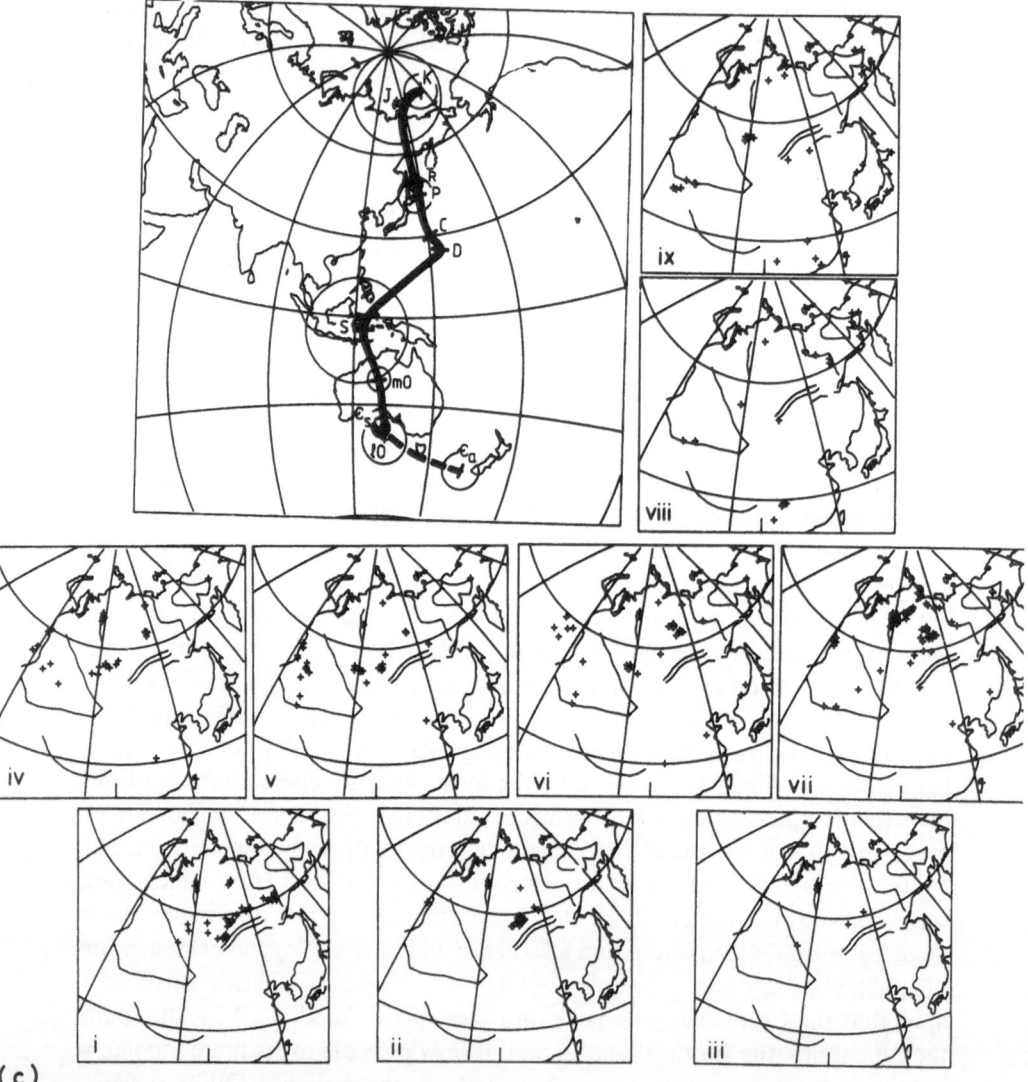

(c)

Figure 9.9(c) Siberia (Angara).

Siberia data is, however, the remarkable degree of internal consistency that has been obtained, although there are also regions for which such consistency seems to indicate extensive overprinting. In general, however, the pole groupings move systematically away from the preceding grouping and appear to be consistent with being genuine determinations of the older field directions, with secular variations and the like strongly suppressed, either by selection of the least scattered data, or by individual pole positions being based on a large number of observations and hence reducing the effects of secular variation and polarity transition effects.

(iv) *South American Phanerozoic polar wandering path*
The bulk of the palaeomagnetic data from this continent (Fig. 9.9(d)) come from Argentina, mainly from the foothills of the Andes, although data are, in fact, available from most countries. The Cretaceous poles form an oval distribution centred close to the present geographic pole and this distribution is closely mirrored by the Jurassic poles. For Triassic times, the average pole position remains close to the present rotational pole, but the distribution is more circular although rather loosely scattered. The Permian poles form an oval pattern extending from Spitsbergen (Svarlsbad) to Kamchatka and the Late Carboniferous poles also have an oval distribution, but this is further south, extending from 20°S of Kamchatka towards the present geographic pole. Older poles are more poorly defined and the ages have often been revised subsequent to their original study. The Siluro–Devonian poles lie between Japan and southern Australia, with the Australian location being considered to be the most likely location for the true pole, partially because this would then lead towards the mean Ordovician pole, probably located near northeastern Australia. The Cambrian poles appear to form two groups, some 50° apart, either or both of which may be genuine pole positions for this period. More recent work has mainly been concerned with establishing the tectonic evolution of the area, rather than determining the pole path for the cratonic areas.

(v) *African Phanerozoic polar wandering path*
Most of the initial data for this continent (Fig. 9.9(e)) were from southern Africa, while the most extensive recent studies have been in the north (Michard *et al.*, 1975; Martin *et al.*, 1978). Some of the northern data are compatible with those from the south, but others are not, reflecting the local effect of tectonic rotations in the northwestern part of the continent. When the data from the south are fairly scattered, observations from the north have been included in this analysis, unless they are clearly rotated by tectonic disturbances. The Cretaceous poles are, for example, widely scattered, but most lie in northwestern North America. The Jurassic and Triassic poles similarly lie in northern North America, but east of the Cretaceous grouping. The analysis for the Jurassic–Triassic periods includes many duplicated

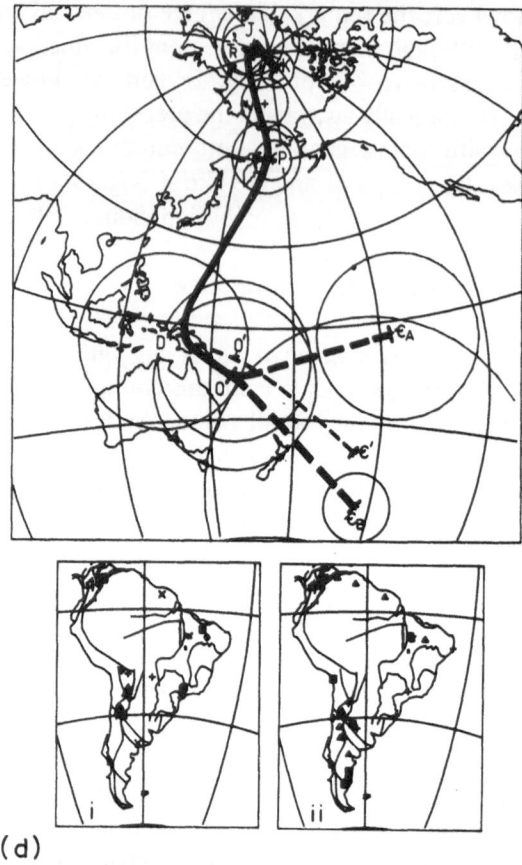

(d)

Figure 9.9(d) South America – site locations are for (i) Cambrian, Ordovician, Devonian, Carboniferous and Permian as triangle, upright cross, diamond, square and diagonal crosses respectively, and (ii) Triassic, Jurassic and Cretaceous as triangles, squares and crosses.

results, indicating the large quantity of observations using the Karroo Basalts erupted close to the Triassic–Jurassic boundary. Many of these are probably of Lower Jurassic age, *c.* 213 million years, as their eruption appears to have been linked with the initial formation of the Indian Ocean. Permian observations are now dominated by information from northwestern Africa, with poles strung between eastern Canada and Hawaii – some of which scatter may be due to tectonic effects – and the mean pole has been based on only three poles that appear to be most consistent with the older and younger observations, and also gives an anti-pole (south magnetic pole) location that would be consistent with the Lower Permian (Sakmarian) ice-sheet glaciation of the continent (McLachlan and Anderson, 1975). There are no definite Carboniferous poles for this continent and only one Siluro–Devonian pole position. The Cambro–Ordovician poles, however, mostly lie between

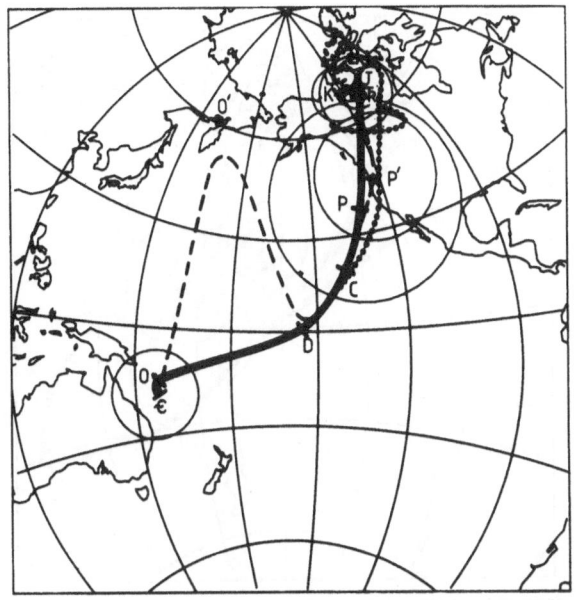

Figure 9.9(e) Africa, together with an estimation by Vandenberg (1979) as dotted. Site locations are (i) Cambrian, Ordovician and Devonian as square, triangle and filled diagonal cross, (ii) Carboniferous and Triassic as upright crosses and triangles and (iii) Triassic, Jurassic and Cretaceous as squares, triangles and crosses.

eastern Australia and southeast of New Zealand, although the actual location is poorly defined.

(vi) *Indian Phanerozoic polar wandering path*

Most recent studies have concentrated in the Himalayan region, mainly to determine the tectonic behaviour of this region during the collision between India and Asia. The pole path (Fig. 9.9(f)) is, however, particularly well defined for the time of eruption of the Decann Traps (Cretaceous), and some of the somewhat earlier Trap rocks, although quite good results have also been obtained from the sediments of Carbo–Permian age. Nonetheless, major gaps still exist for which there are no data and much of the observations

235

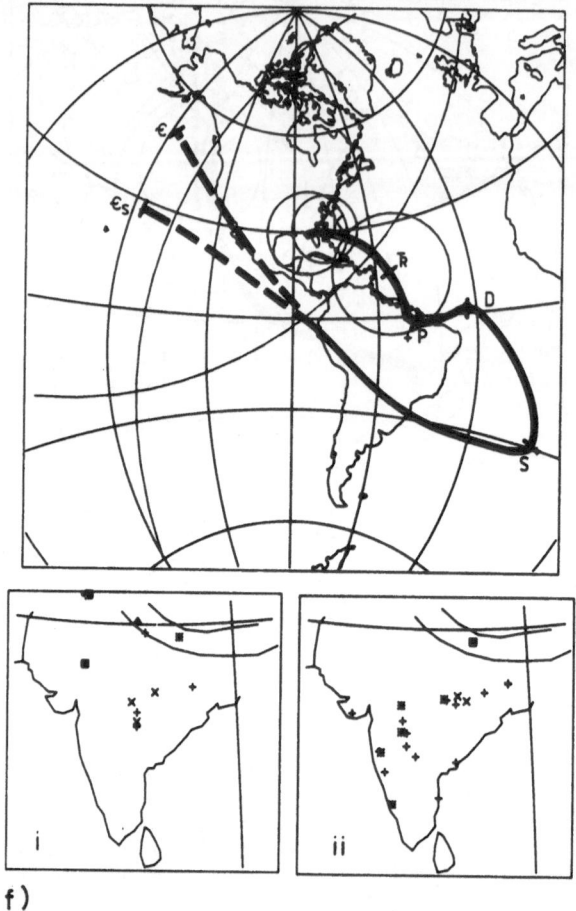

(f)

Figure 9.9(f) India, together with an estimate by Klootwijk and Peirce (1979) – site locations are (i) Cambrian, Devonian, Permian and Triassic as squares, triangles, upright and diagonal crosses, and (ii) Jurassic, Cretaceous and Lower Tertiary as squares, upright and diagonal crosses.

from the north of the continent are likely to have been affected, to varying extents, by the Alpine orogenic episode. (Other reviews include Klootwijk and Peirce, 1979; Athavale *et al.*, 1980; Klootwijk *et al.*, 1981.)

(vii) *Australian Phanerozoic polar wandering path*
The polar curve for this continent (Fig. 9.9(g)) is the most well defined of those for the southern continents, largely reflecting the specific planning of integrated studies in order to define it. The Mesozoic data are mostly from southeastern Australia; this region was tectonically stable during this period, but the definition of the pole position during the Late Palaeozoic, particularly in the Carboniferous, is still poor, as also for parts of the Lower Palaeozoic. On the present evidence, there appears to have been a major movement of

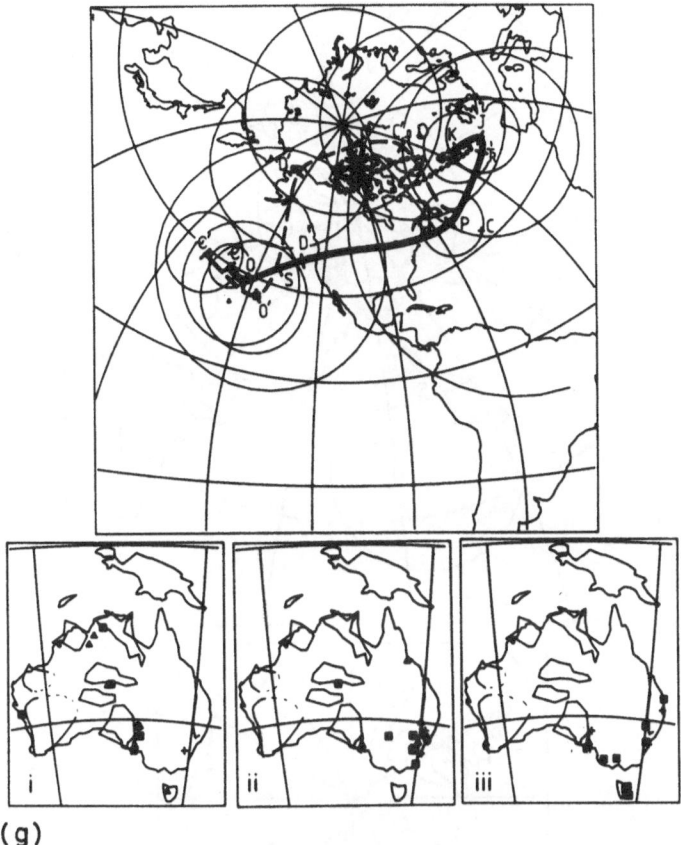

(g)

Figure 9.9(g)   Australia, with an estimate by Klootwijk and Peirce (1979). Site locations are (i) Cambrian, Ordovician and Silurian as squares, triangles and crosses, (ii) Devonian, Carboniferous and Permian as squares, triangles and crosses and (iii) Triassic, Jurassic and Cretaceous as crosses, squares and triangles.

the continent relative to the pole during Carboniferous times, as is also reflected in the palaeoclimatic evidence (Brown *et al.*, 1968). The problem of the Devonian data has been discussed previously (Section 9.2.3(a)). The Cambrian position is apparently well established. The quality of the palaeomagnetic and dating studies means that it is possible to use this continent as a key area in reassemblies of the continents (Section 9.6) and other geological contexts. (Other reviews include Klootwijk *et al.*, 1979; Klootwijk and Peirce, 1979.)

(viii) *Antarctic Phanerozoic polar wandering path*
Few data are available from this continent (Fig. 9.9(h)), with most of the reliable observations coming from Jurassic rocks, the Ferrar dolerites, in Eastern Antarctica (together with some older granitic intrusives, but with no tectonic central) and Cenozoic rocks from the Antarctic Peninsula. The

237

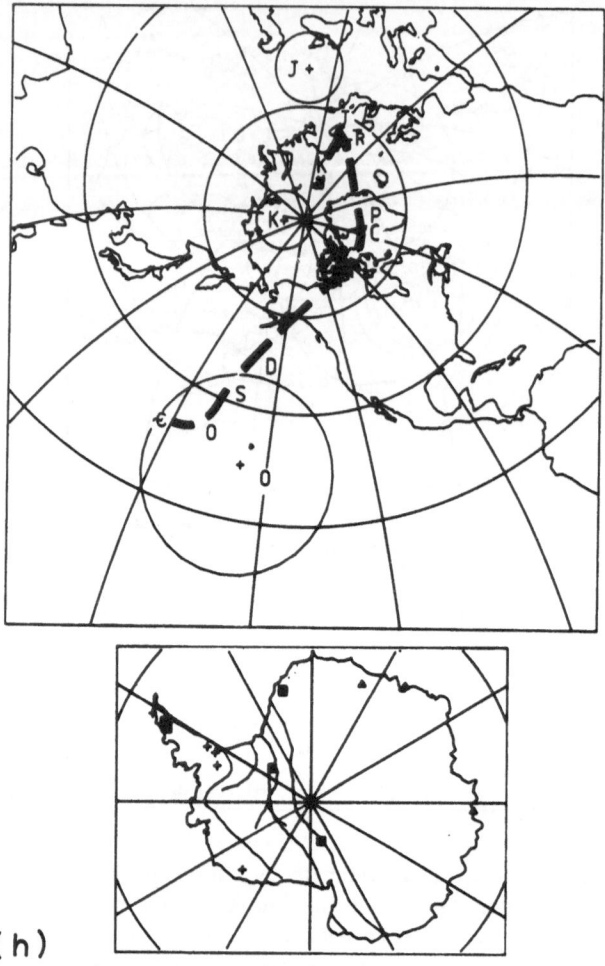

(h)

Fig. 9.9(h) Antarctica – no curve is available for this continent, but the Australian curve is shown after rotation to Antarctica per Norton and Molnar (1977), and the Antarctic data are shown relative to this curve. The site locations show Palaeozoic (mostly Ordovician), Jurassic and Cretaceous as triangles, squares and crosses.

latter, although critical for any understanding of the tectonic development of Western Antarctica and the Weddell Sea, will only be applicable to the major cratonic areas, Eastern Antarctica, when this tectonic evolution is known. At the moment, it seems best to estimate the Antarctic polar wandering path by fitting Antarctica to Australia and thus to make use of the much better established Australian path. This method obviously assumes that the fit of the two continents is known (Section 9.6), but there is little disagreement on this fit, although it is conceivable that the Australian Bight formed on the site of a previously transcurrent fault system that could give a systematic displacement for all pre-Middle Tertiary pole positions.

(e) *Pre-Cambrian polar curves*

It is evident that if there is major uncertainty about polar wandering paths in the Palaeozoic, such difficulties are likely to be even more emphasized in the Pre-Cambrian (Fig. 9.10) when the age of the rocks is much more difficult to determine because of the lack of adequate fossil control; the rocks are also much older and their original remanence is thus more likely to have decayed, in addition to the greater probability of burial with associated thermal and chemical changes. Indeed, many of the rocks must, themselves, have formed at depth and are only now exposed following extensive erosion. Additionally, there is rarely any tectonic control and it must often be assumed that the rocks are in their original position, for example, that igneous dykes were either intruded in their present orientation or the remanence is corrected assuming vertical intrusion. The range of time is also some seven times longer than that of the Phanerozoic rocks. The paucity of data thus raises problems of actually determining a polar wandering curve when dated observations are separated by intervals of a few hundred million years, during which time the continent could have moved in such a manner that the newer reversed pole became nearer to the older normal pole – a situation that is aggravated when the actual age determinations have associated errors of a few hundred million years – and it is conventional to assume the minimum amount of continental motion, i.e. where no data are available the poles are joined along the shortest path between them.

Despite such considerations, the Pre-Cambrian polar paths are known, broadly, for several continents, mainly North America (combined with Greenland and northwestern Britain after reconstruction to their Permian configurations), northern Europe (mainly around the Baltic), Africa and Australia. Data are also available for the USSR, India and South America, but are too few for adequate definition of their own individual polar wandering curves – although these can be incorporated with other continents, assuming, for example, that South America was always part of Africa during at least most of the Pre-Cambrian times.

The North American data are by far the most extensive and coherent, reflecting, in part, the freshness of the rocks as exposed following the Quaternary glaciation – although the same glaciation also covered most of the exposures with glacial clays. On the basis of extremely extensive and intensive studies, Irving and McGlynn (1976, 1981) and Irving (1979) have been able to define polar paths for Laurentia (N. America, Greenland and extreme N.W. Europe) within which the mean poles lie within bands that are some 10–20° wide (although the error circles are obviously very much greater). In such paths (Fig. 9.10(a), (b)), not all individual data are well dated, but their approximate ages are known, thereby defining the probable location of the pole between dated observations, and thus reducing the problems of polarity inversions causing uncertainty about which direction the path should take. This work also showed that the path was characterized by long periods of

239

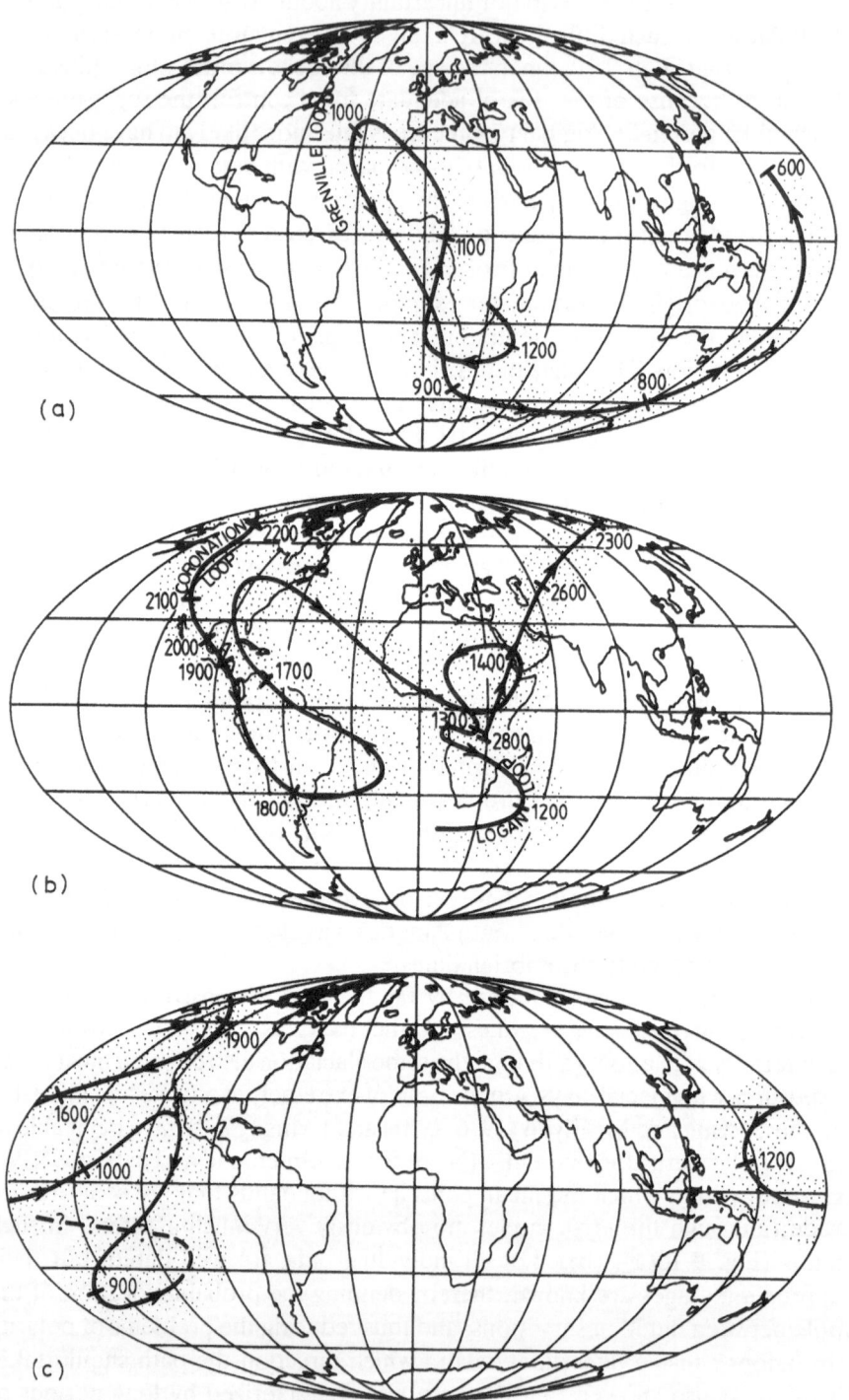

(a)

(b)

(c)

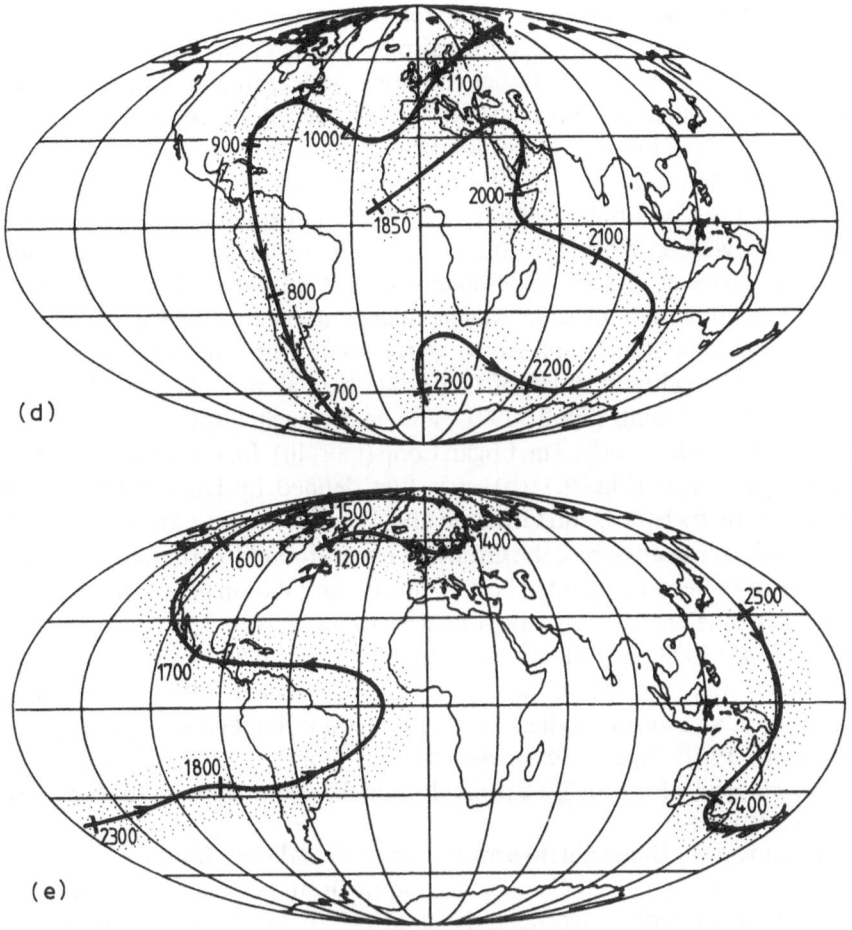

Figure 9.10 Pre-Cambrian polar wandering curves. (a, b) North America (after Irving and McGlynn, 1976, 1981; Irving, 1979). (c) Europe – the Baltic Shield (after Poorter, 1981). (d) Africa (mainly after McWilliams and Kröner, 1981). (e) Australia (after McWilliams and McElhinny, 1980). The estimated errors in the possible pole positions are shown as shaded, but the uncertainty mainly arises from either difficulty in age determination or the presence of multiple components of remanence, and individual 'errors' may well exceed those indicated.

slow, gradual motion of Laurentia relative to the pole, and then drastic changes in motion, giving rise to 'hairpins' in the polar curves (Irving and Park, 1972). The fact that such motions were genuine motions of the continent through different latitudes is evidenced by the palaeoclimatic evidence and the hairpins corresponding with major orogenic activity on the Laurentian continent. (Such observations are obviously critical in understanding Pre-Cambrian tectonics as they immediately show that the continents were in motion at such times and the changes in direction imply that the orogenic events were mainly associated with collisions between continental blocks in the same way as changes in the polar wandering curves for Phanerozoic times appear to correspond with continental collision events.) The reality of much of these data has, at various times, been the subject of argument (Spall, 1971) and many such disputes can only be resolved by further study. The Logan Loop (hairpin), for example, some 1200 million years ago (Fig. 9.11(b)) was first defined by DuBois (1962) and subsequently by Elston and Grommé (1974) and Irving and McGlynn (1976) and could have been part of the Grenville Loop around 1000 million years ago that has itself been interpreted both as due to the collision between Laurentia and 'Grenvillia', and also as due to local rotations and remagnetization effects (Irving *et al.*, 1972; McWilliams and Dunlop, 1975, 1978; Berger *et al.*, 1979; Roy and Robertson, 1979; Piper, 1980b; Dallmeyer and Sutter, 1980). Older loops have also been studied for their tectonic implications (Irving and McGlynn, 1979; Piper, 1980a; Schutts and Dunlop, 1981) although such drastic changes also provide extremely valuable time-markers for magnetic dating purposes.

The European Baltic Shield provides similar conditions for the exposure of Pre-Cambrian rocks as in Canada but, although studied early, they have not been so intensively investigated (Poorter, 1975, 1981; Pesonen and Neuvonen, 1981). Nonetheless, the curve is becoming established (Fig. 9.10(c)). The data from Africa are quite extensive (McElhinny and McWilliams, 1977; Embleton and Schmidt, 1979; Kröner *et al.*, 1980; Morel, 1981; McWilliams and Kröner, 1981; Morgan and Briden, 1981) but of variable quality in both palaeomagnetism and dating, partially reflecting the great depths of weathering that occur in this continent (Fig. 9.10(d)). The Australian curve (Fig. 9.10(e)) has been studied more recently and has a somewhat higher quality of both magnetic and radiometric dating (McElhinny and McWilliams, 1977; Schmidt and McDougall, 1977; McWilliams and McElhinny, 1980).

9.3 MAGNETIC FABRIC

The measurement of strain in rocks, using conventional petrofabric techniques, is a skilled and slow process, often restricted to relatively few observations by the time involved and the lack of adequate strain markers, such as deformed fossils, reduction spots, pebbles, etc. (Siddans, 1976; Donath and Wood, 1976; Owens and Bamford, 1976; Durrance *et al.*, 1978; Oertel, 1978), or crystalline alignments of, for example, the c-axes of quartz or the cleavages of micas (Lipshie *et al.*, 1976; Oertel and Ernst, 1978; Etheridge and Oertel, 1979). In contrast, magnetic fabric (Graham, 1954c; Hrouda, 1982) determinations can be made in a few minutes as opposed to hours and a very wide variety of rock types are suitable for analysis (Section 5.3.3). The main disadvantage is that the strain, as indicated by the magnetic susceptibility, can only be expressed as an ellipsoid and it is not possible to determine different components of the fabric in any one specimen (Section 6.6 and Fig. 6.13). It is convenient, however, that many petrofabric strain markers are themselves ellipsoidal and the strain is conventionally expressed as a strain ellipsoid (Siddans, 1980; Milton, 1980; Holm and Wood, 1980; Treagus and Treagus, 1981). There has also been wide agreement between the orientation of the ellipsoid, whether determined petrofabrically or magnetically (Von Rad, 1970; Hrouda, 1973, 1976a, 1978; Wood, Oertel, Singh and Bennett, 1976; Rathore, 1980a), although the correlation between the magnitudes of the ellipsoidal axes is not yet established, possibly reflecting the fact that the strain ellipsoid can be formed by a variety of mechanisms – crystal plane gliding, ionic diffusion, partial melting and recrystallization, actual physical deformation, twinning, etc. (Langdon, 1976; Rutter, 1976; Ramsay, 1976; Paterson, 1976; Schmid *et al.*, 1977; Lister *et al.*, 1978; Borradaile, 1981) – as well as being dependent on the physical differences between the grains and their matrix (De Paor, 1980).

The magnetic fabric method (Hrouda, 1982) is thus not yet as well established as many other petrofabric techniques and is somewhat less informative for any one specimen, but its speed and sensitivity, combined with the fact that it works on a wide variety of rock types, means that it has a very high potential in petrofabric studies, although it is only recently that such speed and precision have become available and many of the applications in sedimentary, igneous and metamorphic rocks have not yet been tested or evaluated. In the following sections, therefore, most of the comments are speculative, rather than based on direct observations.

9.3.1 Sediments and sedimentary rocks

Petrofabric studies of recently deposited sediments are few, largely reflecting the difficulty in collecting undisturbed samples and the need to impregnate the samples prior to most petrofabric analyses. Magnetic fabric studies of laboratory-deposited materials have been largely confined to sands and finer grained materials that have been deposited grain by grain on both horizontal and sloping surfaces using flumes (Section 4.3). Some studies of aeolian sands have also been made (Ellwood and Howard, 1981) and differences in fabric alone have proved diagnostic in glacial tills (Stupavsky *et al.*, 1974). The

243

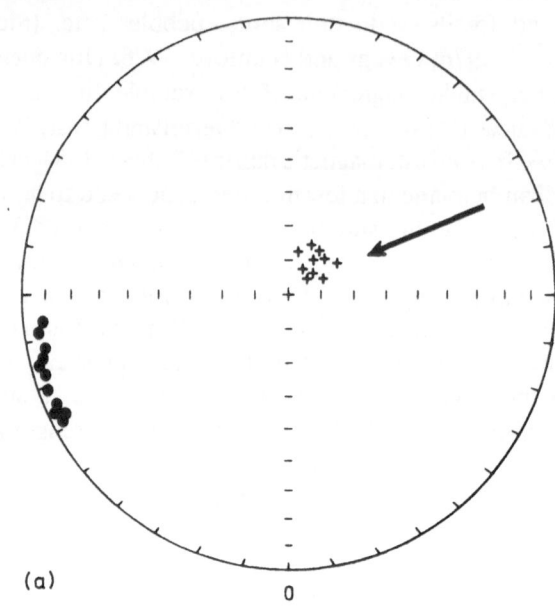

(a)

0

Figure 9.11 Some applications of magnetic fabric. (a) Flow directions can be determined from the magnetic fabric because, although the maximum axes (dots) tend to be aligned parallel to the flow direction, the fabric has an oblate form so that the maximum and intermediate susceptibility axes tend to form a girdle. However, the imbrication of the fabric means that the ellipse is tilted so that the minimum axes (crosses) are tilted from vertical in the upstream direction. Changes in the structure can then be examined as consolidation takes place, eventually leading towards a metamorphic fabric. In general the properties of sediments cannot be used as diagnostic of a primary deposition fabric, although there are specific limits which enable the unambiguous recognition of metamorphic fabric. (b) The orientation of the foliation plane associated with granites, their aureoles and the country rock can be rapidly determined from their magnetic susceptibility parameters, thus distinguishing between pre-orogenic strains and post- or syn-orogenic strains. (Bodmin Moor, Devon; after Kafafy, 1981.) (c) Dynamic deformation of the grains can also be used to determine the direction of motion on fault systems. Examination of the fabric associated with rocks of different competence also enables the history of the deformation to be determined. (The Periadriatic Line; after Rathore and Heinz, 1979.)

measurements have commonly been made using low-field torque methods (Section 5.3.3) and show that the magnetic grains are mostly aligned within or close to the bedding plane, with their longer axes in the direction of flow, with some degree of imbrication (Fig. 9.11(a)). More specifically, the magnetic fabric parameters (Section 6.6) fall within specific ranges, such as $0.06 \le q \le 0.67$ and the imbrication angle, the angle between horizontal and the plane of maximum–intermediate susceptibility, is less than $20°$ (Rees, 1966; Crimes and Oldershaw, 1967; Hamilton and Rees, 1970; Rees, 1979;

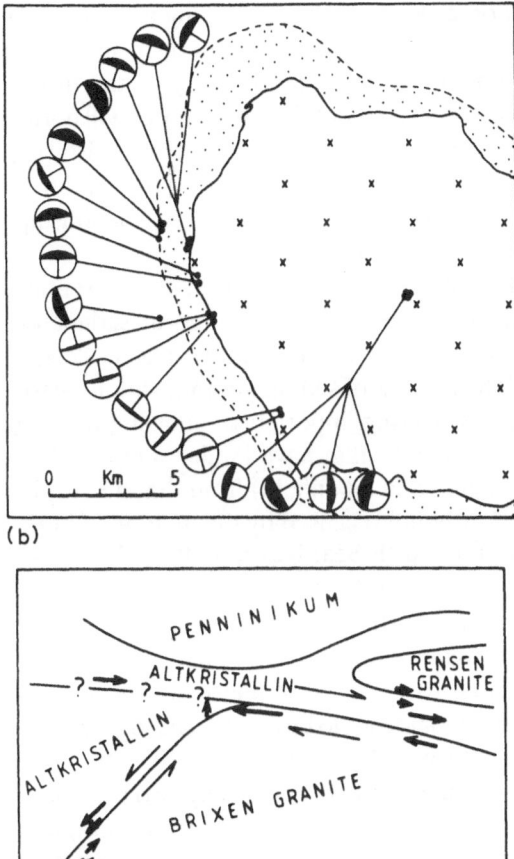

(b)

(c)

Ellwood, 1979a,b; Taira and Scholle, 1979; Reed, 1980; Owens, 1980), corresponding to the angle V being generally greater than 70° (Graham, 1966). These ranges have been widely adopted as being diagnostic of 'primary' sedimentary fabrics when found in natural sediments, while values outside these ranges can generally be attributed to coring disturbances, bioturbation and the like. In reality, these relationships are best expressed in stating that parameters outside these ranges correspond to 'secondary' fabrics while values within the range are consistent with being 'primary' but could include 'secondary' fabrics. Further distinction of the nature of the deposition conditions is also provided by laboratory experiments which indicate that low q values, less than 0.42, may be produced by strong, persistent bottom currents (Hamilton, 1967) or by shearing produced by deposition on a slope (Rees, 1968) but the former are characterized by an imbrication in which the long axes are inclined upwards by a few degrees in the downstream direction, while deposition on a slope results in the long axes being imbricated downwards in the downstream direction. On their own, such magnetic parameters could be ambiguous in terms of the polarity of the direction of flow, but regional study and other sedimentological features will normally allow further assessment of the actual sense of flow. There may also be problems in that different grain sizes tend to show different imbrication angles (Rees and Woodall, 1975) although further study of these relationships may lead to an unambiguous assessment of flow directions.

To evaluate and utilize magnetic fabrics, it is essential to examine the magnetic fabrics of sediments from a range of sedimentological environments, distinguishing between, for example, glacial deposits from stagnant and moving ice. The process of sedimentation, with its associated fabric, can also be of direct engineering importance, allowing the distinction between rapidly deposited sediments that have infilled a crater (left on the sea-floor after the escape of gas from an oilfield) from that due to slow sedimentation over a sink-hole (kettle-hole) caused by the melting of buried ice (the 'pockmark' problem in the North Sea, Newfoundland Banks and so forth). The movement of sand waves, a danger to drilling-platforms, could similarly be better evaluated from a study of the processes of their growth and movement. Even the simple recognition of primary sedimentary fabrics in consolidated or partially consolidated rock can be of major significance. Clearly the determination of flow directions leads to basin and palaeogeographic analyses for the time of deposition (Rusnak, 1957; Rees, 1965, 1979; Hrouda and Janák, 1971; Henry, 1973; Channell et al., 1979b) and may also provide a rapid method of determining the actual depositional conditions from rapid magnetic analysis of bore-core samples. At the simplest level, the existence of a lineated primary magnetic fabric indicates the existence of water currents at the time of deposition, and this observation is sufficient to distinguish, for example, between different modes of formation of organic shales. Hailwood and Sayre (1979) showed that Cretaceous (Albian–Aptian) black shales have a distinct primary fabric and therefore did not form in stagnant conditions but are more likely to have originated from a high organic influx as the sea transgressed over the vegetated continental shelves at this time. Clearly such a model does not exclude stagnation conditions for other organic rich shales, but it is a critical observation as such shales are unlikely to be major oil source rocks if formed in this way, although they provide potential hydrocarbon gas sources, as oils are formed almost entirely from algal materials rather than the higher organic life-forms that have different isotopic compositions (Section 9.7).

Unfortunately insufficient studies have yet been made on the effects of compaction on the primary fabric, although the recognition of such fabrics in Pre-Cambrian rocks (Rahman et al., 1975) and in Carboniferous calcareous nodules (Oertel and Curtis, 1972) indicates that such fabrics can be preserved. However, progressive compaction appears to have caused a shallowing of the magnetic remanence in the Mahogany Oil Shales (Richardson and Noltimier,

1980). Such shallowing is likely to be demonstrated even more effectively by the magnetic fabric that is based on the total magnetic mineral content, rather than the remanence that is carried by only a small fraction of the magnetic minerals.

The effect of diagenetic changes on the magnetic fabric is also unclear because no studies have yet been undertaken. Clearly the growth of new magnetic minerals (Henshaw and Merrill, 1980) may replace already oriented minerals, particularly the clays and micas, and hence pseudomorph the original fabric, but most growth is likely to be under simple hydrostatic loading and to show little obvious fabric. Of particular importance in terms of hydrocarbon migrations is the occurrence and formation of dolostones (Section 9.4) as they have a much greater permeability than limestones, even at pressures of 2.5 kb and ambient temperatures of 200°C (Friedman *et al.*, 1981). Clearly secondary dolostones will have a radically different fabric from that of primary, depositional dolostones and will therefore be readily distinguishable on its magnetic fabric. Additionally, such dolostone formation may be datable by the remanence acquired as the chemical changes took place (Sections 9.2 and 9.4).

9.3.2 Igneous rocks

When molten rocks flow, already crystallized grains will be subjected to similar stresses as in sedimentary conditions. Studies of the magnetic fabric should therefore reveal similar features to those observed in sediments, allowing flow directions and the magnitude of the stresses to be determined. As in sediments, such alignment can be of the grain shapes or of the crystalline axes. Flow fabrics have been recognized in many granites from alignments of crystalline axes and shapes, such as those of the feldspars, and identical patterns have been found in magnetic fabric studies (King, 1966; Ellwood and Whitney, 1980; Birch, 1979) although most studies have not established whether the fabric is carried by magnetite or haematite, and the reported susceptibility values are more consistent with magnetite and hence shape rather than crystalline alignment (Section 4.7). Studies of many British granites (Rathore, 1979a, 1980b; Kafafy, 1981, 1982) indicate that flow fabrics can be determined in both magnetite and haematite-bearing granites. It is, however, often difficult to distinguish such fabrics from later, low-grade metamorphic effects or from alignments caused by emplacement stresses. In general, flow fabrics in granite are of lower magnitudes than those caused by other stresses (Hrouda and Chlupačová, 1980). Many granites show strong to intense lineation near their margins (Rathore, 1979a, 1980b; Birch, 1979) and such studies can clearly be of direct relevance to the mode of emplacement (Figure 9.11(b)), thereby also placing constraints on the temperature and probable depth of the original intrusion. The original fabric in such granites may also be strongly modified by late-stage metasomatic activity. In some cases this appears to have occurred in such a way as to intensify the original fabric (Hrouda, *et al.*, 1971a), probably by the mimetic growth of magnetite in previously oriented mica minerals. However, it is also probable that such later activities may obscure the original fabric. Any full interpretation of the fabric therefore needs to be co-ordinated with petrological studies of the ferromagnetic minerals. In general, however, the high degree of incompetence of granites means that they will be largely unaffected by subsequent stress fields and are likely to preserve their original fabric.

While most magnetic studies have been undertaken on granites and their aureoles, some preliminary studies have also been made of dykes and lavas. Flow banding in intermediate to acidic lavas and dykes can often be readily identified in the field, but flow profiles in basaltic rocks are harder to define by standard techniques, such as vesicle deformation (Coward, 1980). The magnetic fabric can be quite readily defined in basalts, being usually carried by magnetite, and can often lie between 5 and 10% (Abouzakhm, 1974). Studies of columnar basalts (Brown *et al.*, 1964; Symons, 1967c; Ellwood and Fisk, 1977; Ellwood, 1979c) appear to be conflicting, with some evidence for flow and some for strains resulting from contractional cooling. It may well be, however, that these studies have been insufficiently precise to make such evaluations, particularly when there are difficulties in quantitative comparisons of the various parameters available (Section 6.5). Ellwood (1978) and Ellwood and Watkins (1976) have used the anisotropic properties to distinguish between extrusive and intrusive basalts, while Halvorsen (1974) used them to differentiate between sills and ring dykes.

9.3.3 Metamorphic rocks

As rocks are stressed, both physical and chemical changes occur (Section 4.4) although the mechanisms involved are poorly understood and vary from one rock type to another as well as with the nature and rate of imposition of the stress. The magnetic fabric method is, however, most effective in rocks subjected to small strains as magnetite, being a cubic mineral, is easily deformed and the differences in susceptibility between the axes are greatest over the small strain ranges than over the large strains; a 0.1% change from a pure sphere (1:1 to 1.001:1) being very readily detected from susceptibility changes, but a 10% increase in strain of an already severely strained ellipsoid (10:1 to 11:1) may be barely detectable from the resultant susceptibility changes. This property means that the technique is most sensitive over ranges at which most other petrofabric techniques are insensitive. It is thus hard to establish correlation factors for the axial differences between the magnetic anisotropy and the strain ellipsoids, although the directions of the different axes are usually in excellent agreement. Rathore (1975, 1979b, 1980a), Kneen (1976) and Wood *et al.* (1976) have determined that the relationship between the axial ratios of the anisotropy ellipsoid (k_i, k_j) and the strain ellipsoid (l_i, l_j)

has a power relationship of the form $(k_i/k_j) = (l_i/l_j)^a$ (for $i = 1, 2, 3; j = 1, 2, 3$ and $i = j$) and found similar exponents in both haematite and magnetite bearing slates ($a = 0.142$ to 0.145), but it is not considered that this is a universal relationship.

In view of the sensitivity of magnetic fabric technique to low strain the optimum use of the technique is in low-grade metamorphic rocks, especially when no other strain markers are present. It should be possible, for example, to determine whether the initial fabric is random, the March (1932) model, or has a structure on which a later stress field operates. Intuitively, it seems likely that most rocks, prior to the imposition of a tectonic strain, will normally have some degree of fabric. Sedimentary rocks will, for example, have primary sedimentary fabrics unless overprinted by diagenetic processes. Preliminary studies, although not conclusive, appear to show that sedimentary fabric parameters can persist in weakly deformed rocks (Borradaile and Tarling, 1981; Kligfield, Owens and Lowrie, 1981), but this illustrates the fact that such parameters are only consistent with being sedimentary and that high values of such parameters can only be used to exclude a sedimentary origin for the fabric (Abouzakhm and Tarling, 1975; Urrutia-Fucugauchi and Tarling, 1982). Clearly it is vital to establish the nature of the processes causing such fabric to form. Hrouda (1976a) and Hrouda and Janák (1976) have proposed that magnetic grains physically rotate within a ductile matrix as slates form, but it is probable that different mechanisms apply to magnetite than to haematite since they have different crystalline structures, and that the prevailing conditions, such as the fluid content, the composition of both the large grains and the matrix and so forth, are also critical. Nonetheless in all low-grade metamorphic rocks there has been found an identity between the orientations of the magnetic foliation and the slaty cleavage or the schistosity, and similar agreement between the magnetic fabric orientation and the strain fabrics is found in high-grade rocks (Khan, 1962; Heller, 1973; Rathore, 1975; Hrouda, 1976; Henry, 1977a,b). However, the magnitudes of the magnetic fabric and susceptibilities in high-grade rocks often show remarkable variation (Hrouda et al., 1971a, 1978; Kafafy, 1982). Hrouda et al. (1978) suggest that this variation may be explainable in terms of the ineffectiveness of the orientation mechanism in relationship to the magnetic minerals, but it may well be that the increased chemical changes in higher grade rocks also drastically affect the total susceptibility, oxidizing magnetite to haematite in one area and reducing haematite to magnetite in others. In some cases such new minerals will reinforce the existing fabric, but in others may obscure it. A further difficulty is that many of the parameters defining the magnitude of the magnetic anisotropy are sensitive to the total susceptibility (Section 6.6).

Few studies have been made of the magnetic fabric of rocks in high-pressure, low-temperature regimes. Ozima and Kinoshita (1964) examined the magnetic fabric in an andesite at different distances from a fault zone, with an apparent increase in magnitude towards the fault. Most other studies (Birch, 1979) indicate little or no fabric within igneous rocks near fault zones, although mylonites clearly show major deformational fabrics (Goldstein, 1980). The lack of clear stress effects in igneous rocks bordering fault zones probably reflects the incompetence of igneous rocks under such conditions. (An igneous rock will not exhibit ductility, preventing grain rotation, unless it is also subjected to heating, such as during burial and orogenic deformation (Wagner, Hedley, Steen, Tinkler and Vaugnat, 1981).) The study of the magnetic fabric of rocks of different competence, associated with a fault zone, can be used to evaluate the history of many of its deformational phases. The argument, used by Rathore and Heinz (1979), Rathore and Becke (1980) and Rathore (1980b) in studies along the Periadriatic (Gailtal) Line in Austria, is that the most incompetent rocks, granites, will carry the oldest of the preserved fabrics (Fig. 9.11(c)). Progressively more competent rocks will have been successively reset by younger phases of lower strain. Comparison of the orientation and nature of the susceptibility ellipsoids in rocks of different competence can thus be used to evaluate a history of deformation along such persistent features.

The speed and simplicity of the magnetic fabric technique now means that it is practicable to undertake regional surveys and thus examine regional fabric patterns. Rathore (1975, 1980b) and Ellwood et al. (1980) examined the relationships between the fabric in granites, their aureoles, and the country rocks in the Hercynian orogenic belt of southwestern England and north-western France and the Appalachians. This allowed a distinction between pre-orogenic, syn-orogenic, and post-orogenic granites and aureoles on the basis that post-orogenic features would have no regional structures, pre-orogenic would have identical structure to the country rocks, and syn-orogenic granites would show partial relationships with the country rocks (Fig. 9.11(b)). Again, the examination of rocks with different degrees of competence could be used to evaluate a regional tectonic evolution in the same way as studies along fault zones.

Similar considerations also apply to studies of ore deposits (Hargraves, 1959). Those that have been subjected to metamorphism or have originated in association with metasomatic activity, will carry fabrics related to the meta-morphic stress fields, and also record subsequent tectonic activity. Those that have formed in a very low stress field may still show crystalline anisotropy (Porath and Chamalaun, 1966), but not stress deformational fabrics. Such fabric studies can thus be used to ascertain the origin and subsequent tectonic development of such ores.

9.4   SEDIMENTOLOGICAL APPLICATIONS

When studies originally commenced on geological sediments (Section 1.2), it was found that many sediments were characterized by directions that did not always correspond to the geomagnetic field at the time of their deposition. It was thus advisable to avoid these materials in studies leading to geomagnetic or tectonic evaluations. Such caution is, of course, still justified, but conversely, it is possible to investigate magnetically the processes that have caused such deviations in sediments. Magnetic studies, particularly when combined with fabric analyses (Section 9.3), can thus be used to study and possibly date (Section 9.2) the physical and chemical processes that affect sediments from the time of their deposition to the time that they became lithified and metamorphosed. The variation of magnetic minerals within soils can itself be of value in deciphering their history (Mullins, 1977).

Most common iron-bearing minerals, such as olivines, pyroxenes, amphiboles and micas, are relatively easily decomposed during erosion and transport (Section 4.3), generally forming magnetite, haematite, maghemite, goethite, limonite and iron-rich clays – mainly depending on the degree of oxidation and hydration. The compositions of the magnetic minerals deposited within a sediment in a lake (or other sedimentary environment) depend primarily on the compositions of the rocks in the headwaters of streams feeding into the basin, but their concentration is also strongly controlled by the hydrological regime of the river system. In a dry year, for example, water motions will be slow, allowing plenty of time for oxidation and hydration, and for the removal of the more soluble minerals. Conversely, in a wet year, the fast-flowing rivers will tend to carry relatively immature detritus into the sedimentary environment. It can be expected, therefore, that the composition and concentration of magnetic minerals in successive years still tend to fluctuate in response to changes in the hydrological regime that are predominantly controlled by climatic variability. Particular years may thus be characterized by sedimentary layers with significantly greater or smaller concentrations of magnetic minerals. Thus the magnetic properties of such sediments can be expected to vary and provide a means of correlating sediments deposited at the same time. Variations in the intensity of the horizontal component of natural remanence and of the low-field susceptibility can now be undertaken continuously on unopened cores of fresh water or marine sediments (Sections 4.3 and 5.3). Thompson (1977), Oldfield et al. (1978) and Walling et al. (1979) have found that good correlations can be obtained using both the intensity and susceptibility, although the correlations based on susceptibilities appeared to be the most diagnostic. A feature of some studies has been the importance of forest fires in the vicinity (Bloemendal et al., 1979) as these result in changes in the magnetic mineralogy of the soil (Tite, 1972) that are reflected in major changes in the susceptibility of sediments deposited shortly afterwards. As discussed previously (Sections 7.2, 8.3 and 9.2), it is also possible to undertake correlations based upon changes in the direction of the remanence in such sediments.

The advantage of using such properties is, of course, the possibility of

251

providing an absolute date for the sediment based on a correlation with geomagnetic 'master curves' for that locality (Sections 7.2.3 and 9.2.3). There are thus a wide variety of ways in which the magnetic properties of lake and pond sediments can be used for attempting correlations. However, these all suffer, as do other methods, from the irregularity of deposition in such lakes or similar environments. Most sediment is deposited at the mouth of streams and rivers entering the lake, with only the more finely comminuted materials being dispersed widely. The movement of sediments from the delta into deeper areas often takes place by turbidity currents or by slumping which may occur several years after the original deposition. Similarly, bioturbation may mix sediments from different levels to varying degrees, making any identification of a precise layer more difficult. An additional source of difficulty using magnetic methods is the development of early diagenetic magnetic minerals (Section 4.3), particularly those associated with organo-chemical reactions (Giovanoli, 1979; Kirschvink and Lowestam, 1979). Conversely, of course, it is also possible to use such studies of the magnetic properties of sediments, combined with standard limnological techniques, to determine the extent and significance of such physical and chemical factors. In particular, magnetic identification of some of these minerals can be undertaken (Section 3.3) while the sediments are still in their sedimentary environment within the core, thus avoiding the major problems that can arise due to the very rapid oxidation of such minerals on exposure to air (Lyons et al., 1979). Similar studies of the mineralogy can be undertaken in sediments that have accumulated more slowly, such as those of the deep oceans (Emerson et al., 1980; Klinkhammer, 1980) where similar problems of oxidation can occur when attempting to examine minerals.

Variations in the intensity of remanence and in susceptibility will also tend to reflect climatic changes, but on a long time-scale as oceanic circulation has a time-scale of some 2000 years and each sample of deep ocean sediment is also likely to represent accumulation during several thousand years. It is unlikely, however, that the composition of the magnetic minerals being deposited was changed significantly, except by major meteoritic showers or extensive volcanic eruptions, and so most observed changes are more likely to reflect changes in the rate of deposition of non-detrital materials, such as carbonates, thus diluting the content of detrital grains during warmer periods when carbonate deposition was enhanced. Such a relationship may partially account for the observed correlation between intensity of remanence and oxygen isotopes (Section 9.7) but it provides a readily determined variable that can be utilized for correlation over wide areas, almost irrespective of the cause of their variation. In such sediments, the rate of accumulation is similar or greater than the time-scale of secular variations, and so correlations based on secular changes are not possible but changes in the polarity of the geomagnetic field provide precise time-markers that are invaluable in dating many sedimentary features, such as bioturbation, climatic changes, the growth of manganese nodules, fossil extinctions and so forth (Sections 9.2, 9.7 and 9.8).

As sediments accumulate, they are also subjected to increasing pressure from overlying detritus, with consequent compaction and dewatering (Section 4.3). The physical effect of these processes is indicated in the remanence of lake sediments, at least, by an approximately linear decrease in the inclination of remanence (Fig. 8.5). Clearly study of such compaction processes should be combined with magnetic fabric studies (Section 9.3), but it is already clear that similar effects can be observed in a range of sediments (Richardson and Noltimier, 1980). The processes involved in such compactions are still only poorly understood, yet the timing and nature of such processes are of importance in terms of the ability of fluids, such as hydrocarbons, to migrate from a compacting sedimentary sequence. This feature may well also be datable from the magnetic effects associated with such chemical changes (Section 9.2).

In addition to the physical effects associated with deposition, the record of secular changes will also be strongly affected by chemical changes (Section 4.3), such as those associated with organo-chemical effects (Henshaw and Merrill, 1980). Again, the fact that such changes affect the magnetization means that their presence can be readily determined magnetically and it is also possible that their environment can be evaluated (Mullins, 1977) as iron oxides and hydroxides are such sensitive indicators of E_h and P_h conditions (Section 4.3.2). It is also important that magnetic identification is possible while the sediment is still within the sampling tube, before the effects of drying out and atmospheric contamination have occurred. Continued diagenesis ultimately leads to metamorphic changes, but many diagenetic changes are associated with the destruction and creation of new magnetic minerals, so these processes can be studied magnetically, both in terms of the mineralogy and in terms of the possible age of such changes. In order to illustrate such applications, examples of different sediments will be outlined below. It is emphasized, however, that these are essentially illustrations of principles that are applicable to all sediments, not merely the specific examples examined.

9.4.1 Sandstones and siltstones

Red sandstones and siltstones were amongst the first consolidated rocks to be studied palaeomagnetically (Section 1.2) and still provide one of the main sources of palaeomagnetic data, yet the origin of their remanence, on which any interpretation depends, has been highly controversial (Helsley and Steiner, 1974; Baag and Helsley, 1974; Walker, 1976; Elston and Purucker, 1979; Walker et al., 1981). However, when palaeomagnetic studies are used in conjunction with sedimentological studies (Turner, 1980, 1981), it is possible to dissect the different components of remanence and often date them magnetically (Section 9.2). Sandstones (quartz arenites) are defined as consolidated rocks composed of particles between 0.006 and 2 mm in diameter, of which detrital quartz grains form at least 50%, and usually with a matrix of either carbonate or silica cement. Red sandstones are defined (Turner, 1980) as having a colour between 5R and 10R on the Rock Colour Chart (Goddard, 1951), although brown sandstones in a drab sequence are frequently mapped as red. Most red sandstones, in any case, include layers of other colours, white and green especially, that appear to reflect only slight differences in the oxidation conditions during either deposition or diagenesis that have prevented the formation or promoted the removal of the very fine-grained haematite that accounts for the pigmentation in most red beds.

Sandstones may clearly have a depositional remanence that originated as they were deposited subaqueously or subaerially as sands. Sands have a high porosity, 20–60%, which would also mean that post-depositional rotations would be facilitated (Section 4.3). These factors would suggest that any depositional remanences are unlikely to be preserved, yet it has been claimed that certain red beds, such as the Triassic Moenkopi Formation of the western United States, have patterns of changes in direction of remanence that simulate those expected from secular variations (Baag and Helsley, 1974) and therefore indicate the possible preservation of a depositional, or very rapid post-depositional, remanence. Diagenetic changes will eventually take place in any sandstone accumulation and these chemical changes will generally affect the detrital grains as well as causing both the removal and introduction of iron-bearing fluids. Detrital magnetite, for example, is often oxidized to form well-crystallized (specularite) or polycrystalline (martite) haematite (Collinson, 1965b, 1974; Deer et al., 1966; Van Houten, 1968; Turner and Archer, 1975; Turner and Ixer, 1976; Claesson and Turner, 1980). Specularite and martite can also form from the breakdown of other iron-bearing minerals – olivines, pyroxenes, amphiboles and micas, particularly biotite. Microcrystalline haematite, 0.3 to 2 μm, is also commonly most closely associated with heavy mineral bands, further indicating that much of the authigenic haematite in red sandstones has been derived from detrital grains which they also tend to pseudomorph (Ixer, Turner and Waugh, 1979; Turner, 1981; Walker et al., 1981). Pseudomorphs also occur after pyrite, suggesting that the growth of authigenic haematites often takes place after the pyritization stage of diagenesis (Section 4.3). The growth of authigenic haematite will therefore result in the acquisition of a chemical remanence, the age of which can sometimes be isolated (Turner, 1981) and dated magnetically (Section 9.2). Associated with the diagenesis will normally be the migration of fluids, most of which have been derived intrastratally and frequently contain iron-bearing solutions which may then be deposited as cements (Collinson, 1980). Such pigmentations are usually associated with very fine-grained haematite, usually below 0.03 μm in diameter, and hence below their superparamagnetic threshold (Banerjee, 1971). Such fine-grained cement is thus impossible to date by palaeomagnetic methods.

It is clear from this brief review, that changes in the iron oxides, and hence the magnetic properties, occur at many different times during their evolution from sands (Pye, 1981; Tauxe *et al.*, 1980) to consolidated sandstones (Turner, 1980). It is thus possible to investigate these processes from an identification of the magnetic minerals and, if the different components of remanence can be isolated, to attempt dating of the different processes (Schwarz, 1969; Van der Voo *et al.*, 1978; Turner, 1980, 1981) from their associated directions of remanence (Section 9.2).

Turner (1980) recognizes three major types of sandstone on the basis of their palaeomagnetic characteristics (Fig. 9.12). Type A show uniform, stable results, consistent with acquisition of their remanence very shortly after deposition; type B show multicomponents of remanence, some of which can be attributed to the time of formation of the red beds; and type C have well-grouped directions, but are unrelated to their original formation. However, any effective study requires close co-ordination of sedimentological and palaeomagnetic techniques. For example, it seems possible that a

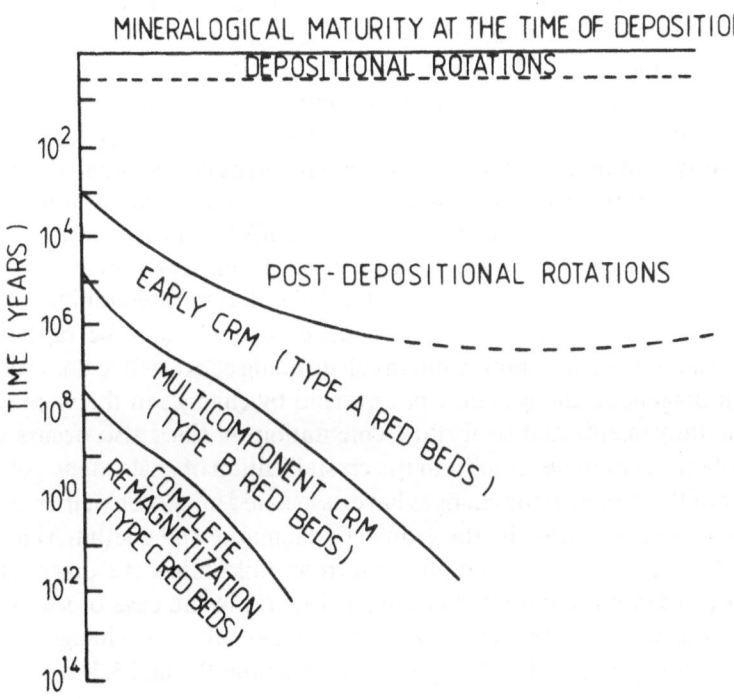

Figure 9.12 The magnetic characteristics of red sandstones. Although complex, it is possible to classify red sandstones on the basis of their magnetic features and hence to assess their probable diagenetic history (after Turner, 1981). Similar considerations can also be applied to other forms of sediment, such as limestones.

high initial organic content may well inhibit oxidation of the surrounding grains until the organic matter has decomposed and the hydrocarbons have been released during katagenesis. Under such circumstances, the formation of authigenic haematite may well relate to the time of cessation of katagenesis, possibly some 100 million years later than the original deposition of the sediment (Tarling *et al.*, 1976).

It seems probable, therefore, that the remanence in most, but certainly not all, sandstones will be related to oxidation processes, some of which may be penecontemporaneous with deposition and others with diagenetic processes that may take place at any time, possibly several million to hundreds of millions of years after their original deposition. Combined sedimentological and palaeomagnetic studies do, however, provide an elegant and potentially precise way of differentiating which processes have been of the most importance in different sediments and also offer the potential possibility of their magnetic dating.

9.4.2 Carbonates

Limestones ($CaCO_3$) and dolostones ($Ca, Mg(CO_3)_2$) generally have only a weak remanence, less than 10^{-4} A m^{-1}, and the remanence of the purer varieties has only been adequately measurable during the last few years. They are usually formed by organic processes, but contain a wide range of detrital grains (Bathurst, 1975). On deposition, the detrital grains, usually predominantly carbonates, are generally surrounded by comminuted shell debris that forms a carbonate mud, micrite. In most modern environments, the micrite is predominantly aragonitic (organic $CaCO_3$) which is unstable at low pressures and generally converts to calcite during early diagenesis. The initially high, but variable, porosity (40–70%) is thus reduced rapidly so that it is possible that any depositional remanence features may be rapidly preserved. Later diagenesis may result in calcite being converted to dolomite, but all such diagenetic changes may be reversed by changes in the water table, tectonic movements and so forth. Cementation by silica also occurs and is generally thought to develop from the crystallization of chalcedony gels.

Most of these diagenetic changes have associated migration of iron-bearing compounds as evidenced by the growth of haematite and goethite rims, such as those around dolomite rhombs as ferroan dolostones are diagenetically altered (Al-Hashimi and Hemingway, 1973). As in the case of sandstones, this means that the processes and mineralogy of such changes can be examined and possibly dated magnetically (Sections 9.2 and 3.3).

9.4.3 Coal

Peat deposits usually contain detrital grains, but are clearly subjected to considerably more compaction effects as the peat converts to coal, than in most other sedimentary environments. Sulphur is present in association with the organic matter and is combined by bacterial activity during diagenesis to form pyrite which often replaces fossil remains (Casagrande *et al.*, 1979). Many of the original iron compounds are therefore converted to non-magnetic forms, although changes in acidity and oxidation states may lead to the formation of magnetite or haematite during later stages of diagenesis. There is, therefore, some potential for sedimentological study of such processes, but few studies have yet been attempted (Noltimier and Ellwood, 1977).

9.4.4 Evaporites

The formation of evaporites in basins is generally accompanied by some detrital grains, if only blown in by winds from the surrounding deserts. Few studies have yet been made of these sediments as it is expected that the original remanence will be strongly distorted by the extensive recrystal-lization and mobility of these materials. It does seem possible, however, to examine their deformation patterns by means of their magnetic fabric (Section 9.3).

9.4.5 'Sedimentary' ores

The origin of many of these deposits is problematical (James, 1969; Tarling, 1973b, 1974a, 1981; Badham, 1981). The Pre-Cambrian Banded Iron Ores are generally thought to have been deposited as a result of micro-organism activity and would thus be likely to have a chemical remanence very closely related to their time of formation, but secular variations have not yet been demonstrated between the different bands and it seems probable that extensive recrystallization of the magnetic minerals also takes place during diagenesis. Many of the richer ore deposits, in any case, are characterized by enrichment by supergene weathering that will clearly have obliterated most, if not all, of the primary magnetization. Clearly, these rocks form ideal materials for magnetic study and dating of such chemical changes. More problematic is the origin of the lead and zinc sulphides of the Mississippi type of ore deposit. Few studies have yet been undertaken as most of these deposits are only weakly magnetic and often affected by later weathering. Nonetheless, changes in the magnetic mineralogy are associated with their formation and it is often thought that they have a moderately high tempera-ture origin. While there are arguments both for and against such origins (Tarling, 1973b), these deposits are a major metal resource that would amply justify further examination using palaeomagnetic techniques. More recent deposits, such as the sedimentary ironstones, bauxites and so forth, all have a complex mineralogy that can be partially evaluated from a study of the magnetic minerals alone, but the complexities of these deposits have, so far, inhibited much active research into their magnetic behaviour (Tarling, 1974a).

9.5 IGNEOUS AND METAMORPHIC ROCKS

Ferromagnetic minerals are present in almost all igneous and metamorphic rocks. Their sensitivity to changes in the chemistry, particularly the redox conditions, and the cooling rate make them ideal materials to investigate the original compositions and their subsequent changes. Conveniently such minerals are also sensitive to stresses imposed, providing information on the mode of emplacement (Hargraves, 1980) and subsequent deformation (Sections 9.3.1 and 9.3.2). Additionally, the magnetic properties can often be studied remotely using field magnetometers or susceptibility meters, making such investigations cheaper than many conventional techniques. Nonetheless remarkably few studies have been undertaken, even on a preliminary scale, into the ways in which palaeomagnetic techniques can be used to supplement standard analyses, some of which have direct economic importance (Brimhall, Geissman and Kelly, 1979). In the following sections, only a few of the potential applications of these methods are outlined and even these can only be regarded as indications of the ways in which palaeomagnetic methods can be applied to a range of geological problems in igneous and metamorphic rocks.

9.5.1 Composition, redox conditions and oceanic rocks

Iron oxides are particularly sensitive to the chemistry, cooling rate and especially the redox conditions during crystallization and subsequent magmatic evolution. Despite their ubiquitousness and sensitivity, the ferro-magnetic minerals have been surprisingly little used to evaluate the geo-chemical and physical development of magmatic or metamorphic products, with most studies being predominantly concerned with oceanic rocks, particularly the 'weathering' of basalts on mid-oceanic ridges. Nonetheless, the magnetic properties of all igneous and metamorphic rocks can be used to ascertain the geochemical conditions under which they formed. For example, spatial variations in the concentration of magnetite and haematite can be determined in order to establish the homogeneity of granitic masses; some are characterized by magnetite alone, others by haematite and some show mixtures of magnetite and haematite (Kafafy, 1982). Each characteristic indicates different redox and chemical compositions, the lateral variations showing possible relationships with the metamorphic aureole, flow structures (Section 9.3.2) and the like. Igneous rocks formed at higher levels in the crust similarly contain magnetic mineral reflecting the chemical conditions, but also their structure may be informative, rapidly cooled magmas being character-ized by fine grains with solid-solution compositions. Such solid solutions may eventually spontaneously exsolve as a function of time, or exsolution can take

258

place as a result of late-stage metasomatic activity and, eventually, weathering. Ellwood (1981) found that the Milledgeville granite (Georgia) showed unstable magnetization when weathered, resulting from the reduction of primary magnetite to sub-domain sizes and the formation of maghemite, and the presence of magnetic properties associated with these minerals could be used to assess, possibly in the field, the potential strength of rocks for dam foundations.

Studies of oceanic basalts are complicated by the common occurrence of low-temperature changes and serpentinization. In particular, the origin (Duncan and Green, 1980), composition and structure of unaltered oceanic basalts are not clearly established. Apart from an intrinsic petrological interest, recent studies of the magnetic properties of samples of oceanic basalts and of ophiolites (thought to be obducted remnants of previous oceanic floors) have been mainly aimed at

1. Finding whether ophiolites from mid-oceanic ridges can be distinguished magnetically from those that have formed in other circumstances, such as in the marginal basins, and
2. Attempting to determine the magnetic source for the oceanic magnetic anomalies.

The present oceanic crust (Cann, 1974; Kuznir and Bott, 1976) beneath the water and sediments is formed of some 0.5–1 km of lavas, often with pillow structures underlain by a sheeted dyke complex of thickness 0.5–1 km. These all overlie the main layer, 5 km of gabbro, and a basal layer, possibly less than 0.5 km thick, of layered olivines and pyroxenes. Banerjee (1980) suggested that the upper pillow lavas, which are in zeolite grade of metamorphism, have a characteristic natural intensity of remanence of about 1 A m^{-1}, with a lower intensity in the lower lavas which are in greenschist grade of metamorphism. The intensity of remanence of the dykes is some 0.1 A m^{-1}, on average, while the low-field susceptibility is between 1 and 0.1 SI in both the lavas and dykes. While these results are preliminary, it appears that these properties do differ somewhat from those in ophiolites thought to have formed in marginal basins. However, the ranges of both intensity and susceptibility are not yet well established and are strongly influenced by submarine weathering (halmyrolysis). These low-temperature processes are also thought largely to account for the systematic decrease in the amplitude of magnetic anomalies away from the mid-oceanic ridge crests. The 'weathering' process can be seen in dredged samples from changes in colour, a decrease in the abundance of fresh olivines and the growth of alteration rims, and the effects are shown clearly in rare earth element analyses (Frey et al., 1968; Ludden and Thompson, 1978). Although the $FeO:Fe_2O_3$ ratio tends to remain constant, the main effect on the ferromagnetic minerals appears to be an oxidation of titanomagnetites to titanomaghemites (Ozima et al., 1974; Tucker and O'Reilly, 1980a; Prévot et al., 1981; Moskowitz and Banerjee, 1981; Özdemir and Banerjee, 1981;

Palaeomagnetism

Böhlke *et al.*, 1981). Such low-temperature changes are mostly constrained to the upper lavas, although the increasing evidence for deep penetration of hydrothermal circulation (Lister, 1974, 1977) raises the probability of extensive alteration at depth. However, the onset of greenschist grade metamorphism in the lower lavas causes the breakdown of titanomaghemites to form titanium poor titanomagnetites with paramagnetic titanium rich ilmenohaematites (Beske-Diehl and Banerjee, 1979; Banerjee, 1980). (While still largely unstudied, similar changes in the ferromagnetic mineralogy can be expected in non-oceanic metamorphic situations.)

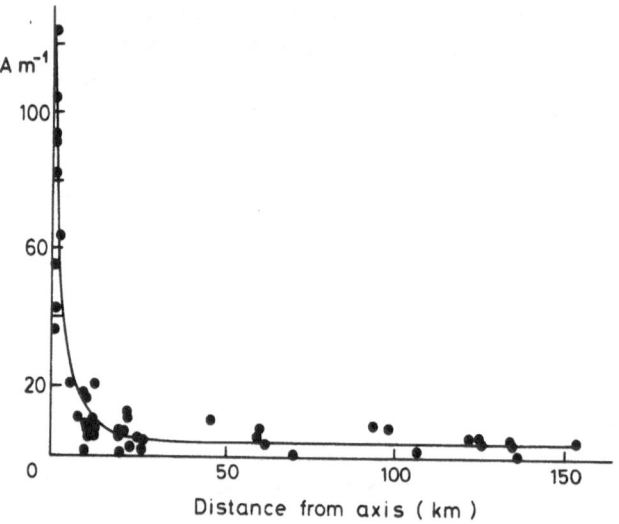

Figure 9.13 The decrease in the intensity of remanence away from a spreading ridge. (After Irving, 1970.)

The very strong magnetic anomaly on the crest of the oceanic ridges corresponds with very high intensity of magnetization in basaltic samples (Fig. 9.13) dredged from them, generally 20–120 A m^{-1} in the North Atlantic (Irving, 1970; Carmichael, 1970). Although some of this magnetization spontaneously decays as the rocks age (Prévot, 1981) this effect is much smaller than the observed decrease in amplitude of the oceanic magnetic anomalies (Marshall and Cox, 1972) which appears to be largely controlled by the submarine weathering processes in the basaltic lavas. However, this implies that the source of the anomalies lies within the uppermost basaltic layer (Anderson *et al.*, 1980). While this part of the oceanic crust is clearly a major contributor to the anomaly, the overlapping of lavas of different polarity would decrease the sharpness in the change from one polarity to the next to a much greater extent than is observed. The dykes, which are intruded in a remarkably regular manner, always within a few metres of the crest of the oceanic ridge (Cann, 1974), provide a better alternative magnetic source, but appear to be too weakly magnetized in relation to their susceptibility. While such a relationship does not prevent the detection of magnetic anomalies, the signal would be degraded to a greater extent than is actually observed (Harrison, 1976). The gabbros of the main layer have, so far, mainly been studied from dredged samples, mostly obtained near fracture zones. These areas are thought to be much more highly serpentinized than other parts of the oceanic crust (Tarling, 1978a) and the low intensity of remanence of these samples (Banerjee, 1980) may not be typical. The problem of the source of magnetic anomalies in the oceans thus remains largely unsolved, requiring samples from undoubted dykes and gabbros far from transform fault systems. Clearly further studies of submarine weathering (especially maghematization – Moskowitz, 1980; Prévot et al., 1981) and serpentinization (Lienert and Wasilewski, 1979) are also required, using a combination of geochemical, petrological and palaeomagnetic techniques.

9.5.2 Emplacement temperatures

As most lavas are erupted at temperatures in excess of 1000°C and solidify between about 950 and 850°C, they will not acquire their remanence until after they have been emplaced. However, tephra (Walker et al., 1980; Wilson et al., 1980; Fisher et al., 1980; Wilson and Self, 1980) contains clasts and particles that may be emplaced at much lower temperatures, possibly even at ambient temperatures, such as in the form of mud flows (Fink et al., 1981; Steele, 1981). The distinction between hot and cold emplacement can therefore be critical (Aramaki and Akimoto, 1957). Any one clast or particle may thus have been emplaced at or above its Curie temperature(s), or below its highest Curie temperature. In the former case, it can only be shown that the minimum possible emplacement temperature was that of the highest Curie temperature. In the latter case, the particle or clast will have acquired two components of remanence: one acquired at high temperatures prior to emplacement, and a lower temperature component acquired after emplacement. The lower temperature component will be parallel to the geomagnetic field in the locality at the time of emplacement (Geissman, 1980). The separation of these two components (Section 6.4) can thus be used to determine the emplacement temperature. The properties of the higher temperature component can also be of interest as it may be consistent, indicating that it cooled in situ within the volcanic cone prior to eruption, or its components may be scattered reflecting cooling while turning into different position during its passage through the air. Kent et al. (1981) determined emplacement temperatures for the 79 AD Vesuvius eruption as between 350 and 400°C. while Hoblitt and Kellogg (1979) used a modification of this method to obtain emplacement temperatures for St Helen's tephra to within 25°C. Hoblitt and Kellogg, however, based their method (Section 6.4) on clasts with an original high temperature of magnetization that was emplaced at angles greater than 90° from the ambient geomagnetic field direction in which acquired as the clast cooled subsequent to emplacement. The summation of the vectorial intensities of natural remanence, during step thermal demagnetization, therefore peaked at the temperature at which the antiparallel component was removed, i.e. at the emplacement temperature. This method can, of course, only be used where the high-temperature component has a consistent magnetization and by determining its orientation prior to determination of the assessment of emplacement temperatures. Analyses of vectors have shown moderately consistent emplacement temperatures in the Phlegrean Field tephras near Naples (Incoronato *et al.*, 1981), and systematic decreases in temperatures of emplacement, away from the vent, have been observed in the Minoan tephra on Thira, Greece (Downey, 1982, 1983).

The main difficulties are usually in determining the precise temperature at which the low-temperature component has been removed, thus isolating the high-temperature component (Section 6.4), particularly if the higher temperature component does not have a consistent direction. This can be aggravated by individual clasts having a range of temperatures when emplaced, i.e. their surface temperature is likely to be lower than their internal temperature. Such variations can, of course, be investigated magnetically and may provide useful information on the method by which clasts cool during transport. A further complication is the passage of hot fluids through the tephra after deposition which can give rise to localized heating and chemical changes, such as accompany zeolitization.

9.5.3 Metamorphic aureoles, thermal contacts and depth of burial

The metamorphic aureole of a granite can usually be readily distinguished close to the granite, but the boundary with the country rocks is gradational depending on the temperature and depth of emplacement, and the composition and nature of the country rocks. The outer limits of such zones are, however, frequently characterized by iron migration, with associated growth of magnetic minerals. In many granites, the effective boundary of their metamorphism aureole can thus be determined by means of the magnetic susceptibilities. In some areas, the increase in magnetite content is sufficient to cause a magnetic anomaly (Fig. 9.14) that can be detected simply by standard surveying magnetometers (Speer, 1981), but more gradational contacts require more precise determination of the susceptibilities using either field susceptibility meters or by collection of samples and laboratory analysis. The distinctive style of the magnetic fabric in these aureoles (Section 9.3) is similarly quite distinct from that of their country rocks (Kafafy, 1981, 1982).

The width of such metamorphic zones can also be variable, even for a single dyke. The width of the aureole, defined by iron enrichment, of the Early Tertiary Cleveland–Armathwaite Dyke in northern Britain, varies from a few

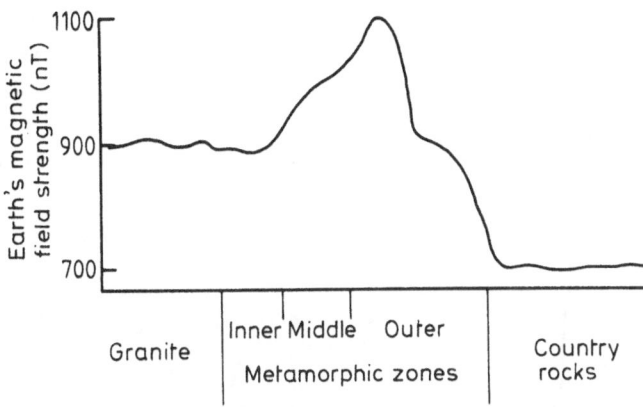

Figure 9.14 The magnetization in metamorphic aureoles. While the magnetic fabric can provide valuable information on the mode of intrusion of igneous bodies, the changes in the magnetic mineralogy associated with the heating and circulation of hot fluids mean that the outer limit of the metamorphic aureole are characterized by a higher susceptibility than other areas, thus delineating a boundary that can be detected by use of field magnetometers or by laboratory measurements of susceptibility. (After Speer, 1981; Kafafy, 1981.)

centimetres to 300 m (Thomas, 1969; Giddings, 1969) and appears to correlate with the permeability of the country rocks, being wider in sandstones than limestones. The determination of the blocking temperature spectrum away from a dyke (Fig. 9.15) can similarly indicate whether cooling was by purely thermal conduction (Jaeger, 1964; McClelland-Brown, 1981) or involved a strong convective transport, probably associated with metasomatic activity (Buchan *et al.*, 1980).

Regional metamorphism can also arise as rock sequences are carried to greater and greater depths in the Earth. This usually arises from the action of tectonic forces, probably related to a cooling of the Earth's crust (Beaumont, 1978) although the actual reasons are not yet clear. At greater depths, the rocks are subjected to increased pressure from the overlying rocks, although this is not thought to have much effect on their magnetization (Section 2.3). The main palaeomagnetic effect arises from

1. The chemical changes associated with the dehydration of some layers and hence the circulation of fluids throughout the entire sequence, and
2. The increase in temperature which is mainly controlled by the local geothermal gradient in the area.

The circulation of fluids can cause either oxidation or reduction, depending on the local E_h/P_h conditions, with associated changes in the magnetic mineralogy, thus resulting in an acquisition of a chemical magnetization,

263

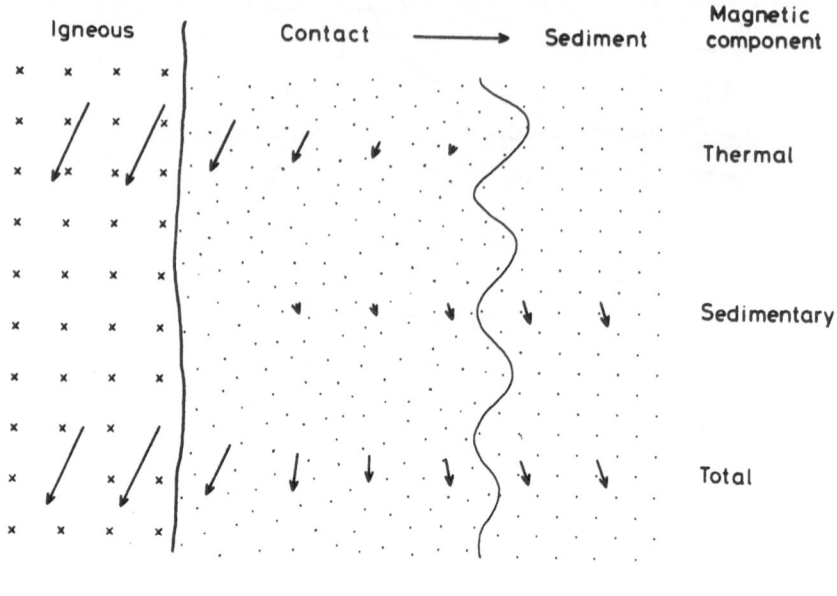

Figure 9.15 Components of remanence in a metamorphic zone. The intrusion of a dyke bakes the country rock out to distances that depend on the total heat available and the method of transport, i.e. conduction or convection. Determination of the changing ratio of these components enables an assessment to be made of the relative importance of these mechanisms and also of the possible depth at which the intrusion originally took place.

often over a prolonged interval as different interactions take place over the period of burial and cementation. If such secondary magnetizations do not completely overprint the original magnetization, then it is possible to compare and contrast the initial remanence with that acquired later, possibly identifying the secondary mineralogy involved. However, even if the initial remanence is completely destroyed, it should be possible to define the directions of the geomagnetic field during the time of acquisition of the chemical remanence and hence date it (Section 9.2) by reversal polarities or pole positions. In many cases, it is likely that such chemical remanences are dominant and obscure both the original and thermal remanence associated with burial. However, if such chemical changes are small, it is possible to isolate the secondary remanence acquired thermally as the sequence gradually cooled as it was being uplifted back to the surface. This can also be dated, but the examination of this remanence should also allow the estimation of the maximum temperature to which the rock has been subjected (Briden, 1965; Pulliah *et al.*, 1975; Buchan *et al.*, 1977; Van der Voo *et al.*, 1978; Schmidt and Embleton, 1981; Buchan and Schwarz, 1981) using identical considerations as in determining volcanic emplacement temperatures (Section 9.5.2). Such maximum temperatures will, in fact, be higher than the actual temperatures to which the rocks have been subjected as the remanence will have been acquired over a prolonged period, probably several million years, and will thus comprise a moderate temperature viscous remanence (Section 2.3) that can only be corrected to the true maximum temperature if the duration of burial at that temperature is known. In practice, such considerations are not very relevant as many other processes are similarly time-dependent. The release of hydrocarbons (katagenesis) or their destruction (Stoneley and Bailey, 1981), takes place at lower temperatures over periods very much more prolonged than occur during laboratory experimentation. The effective destruction of all hydrocarbons at some 170°C, would thus also have occurred if the maximum moderate temperature remanence was of a similar or greater value. Such considerations are, of course, of major interest in terms of the potential hydrocarbon content of rocks on continental edges in which tectonism, associated with the break-up of continents, has resulted in local uplifts and downwarps (Schmidt and Embleton, 1981). Furthermore, such palaeomagnetic studies offer the potential of evaluating the temperatures of the circulating fluids, in addition to estimating the maximum depth of subsidence. The situation is, however, complex and it is essential to distinguish between chemically and thermally imposed secondary magnetization in such conditions; their use would clearly be optimized in conjunction with geological (Fyfe, 1976) and radiometric methods (Gleadow and Lovering, 1978; Morley *et al.*, 1981).

9.6 STRUCTURAL APPLICATIONS

9.6.1 Introduction

There has been a tremendous expansion in the use of palaeomagnetic methods to detect and measure the relative rotation of different tectonic units on both a global and subcontinental scale. In virtually all cases the data requirements are similar. The individual samples must be stably magnetized and the stable component must reflect the direction of the local geomagnetic field at a known time. Sufficient samples must have been collected so that the effect of secular variations can be averaged to yield a palaeomagnetic pole that can be considered to represent the axis of the Earth's rotation at the time when the magnetization was acquired. The 'averaging' requirement is obviously greater in igneous rocks, which often preserve a 'spot reading' of secular variation, than in sedimentary rocks in which long-term diagenetic magnetizations often average out such variations (Section 4.3) even within a single specimen. Finally sufficient data must be available to assess the presence of directions acquired during anomalous geomagnetic field behaviour such as polarity transitions or excursions (Section 8.4). Such requirements cannot always be fully satisfied, in particular the assessment of the age of magnetization, and the fundamental test is then whether the tectonic interpretations make geological sense, which is often a subjective evaluation. Nonetheless it is frequently possible actually to measure the total degree of differential motion involved, and also to determine the magnitude and timing of individual phases of motion.

The basic argument in such interpretations is that the average geomagnetic field corresponds closely to that of an axial geocentric dipole. At any one time, therefore, rocks becoming magnetized will acquire directions that will conform to this field. If one tectonic unit is subsequently moved relative to the others, then all samples taken from it will have magnetic vectors pointing to the same pole position, but this will differ from the pole position determined for all other areas by the amount of tectonic movement that occurred. Comparison of the pole positions will thus enable a measure of the relative motions between the tectonic block and other areas of the world. In practice, this means that the movement of one cratonic block relative to another (continental drift) can be determined, or the relative motions of individual tectonic units within a continent. If rocks covering a span of ages can be examined in both tectonic units, then it becomes possible to determine the time and magnitude of the component motions. For example, comparison of Cretaceous pole positions can distinguish the total magnitude of post-Cretaceous relative motion, but this could involve periods in the Tertiary when there was no relative motion and also times when the motion was in the opposite direction to the overall pattern. By comparing pole positions throughout the Tertiary, each phase of motion could, in principle, eventually be defined.

Having determined the motion, there may still be a problem in defining the area affected. In areas of considerable crustal mobility, such as nappes within orogenic zones, small elements may be moving as discrete, separate units. In a faulted area individual fault blocks may each have their own unique history of rotation. In the Lebanon, for example, Freund and Tarling (1979) were able to define individual fault blocks on a scale of 50×200 km, each of which behaved separately, although partaking in an overall pattern to absorb the transcurrent motion as the Dead Sea fault gradually died away. The size of such units can, of course, be on all scales, with individual slivers of rock each behaving differently within a fault zone. It would, in such cases, be wrong to extrapolate the behaviour of any one block as showing the total regional pattern of fault motion. Similarly in folded terrain, the supracrustal layers may be highly mobile and can move separately from their basement (allochthonous), or they may be rigidly attached to their basement (autochthonous). In the latter case, samples from any one locality will provide information on the behaviour of the entire autochthonous block, but this will not be the case for allochthonous regions. In practice, it is often difficult to assess the degree of autochthonousness from standard geological studies. An important aspect of palaeomagnetic studies is thus to determine the extent to which any one region is autochthonous by determining the extent of the area over which rocks of identical age yield an identical palaeomagnetic pole, i.e. after corrections for all known tectonic movements since the magnetization was acquired. In comparing different continents, this will normally mean that they will both have an identical polar wandering path when they have been reassembled into their previous positions relative to each other and that their polar paths will separate for times before, or after they were in that assemblage. In this regard it is important that a section of their polar wandering paths be compared, rather than a single pole position that only defines the orientation and distance relative to the palaeo-axis of rotation. Blocks with similar palaeo-latitudes may thus be matched, but their total relationship can only be established by comparison of their actual polar wandering paths.

In the following sections, a few of many examples of structural applications of palaeomagnetism will be outlined to illustrate the types of problem that can and have been tackled palaeomagnetically. Comparisons of continental blocks and the existence of super-continents will be considered first (9.6.2), followed by (9.6.3) examples of structural studies on a large but sub-continental scale (Van der V· ·u and Channell, 1980). The same techniques and principles can, howeve· , ne applied on even smaller scales (9.6.4) or, at the other extreme (9.6.5), on a planetary or even cosmic scale.

9.6.2 Intracontinental movements

The concept of a unity of all the 'southern' continents – South America, Africa, India, Australia and Antarctica – is an ancient one. Similarities of the flora in India, South Africa and Australia led Blandford (1875) to suggest a need for geographic connections between these areas, although it was Suess (1885) who synthesized the ideas of Blandford (1875) and Medlicott and Blandford (1879) into the concept of a southern land mass, Gondwanaland (based on the Permian flora in the Indian Kingdom of the Gonds – Blandford *et al.*, 1859), that was, by 1901, extended to comprise all these southern continents and to exist from at least Permian times until, originally, Miocene times. As it was then considered that the continents could not move because of the rigidity of the Earth's mantle, the floral and faunal similarities of these areas were necessarily explained in terms of 'land-bridges', i.e. parts of the ocean floors were then above sea-level and provided land connections that have since been submerged (Schuchert, 1910, 1928). The acceptance of continental drift partially followed increased palaeontological evidence for very close similarities between these continents, although it mainly followed the dating of the intervening ocean floors by magnetic anomaly dating (Section 9.2.2), and the geometric fitting of the continental edges (Carey, 1958; Bullard *et al.*, 1965). Such fits were also found to be consistent with the matching of the polar wandering curves as then known (Creer, 1958; Creer *et al.*, 1958). Subsequent studies of both geophysical and geological features have confirmed the fit of South America and Africa up to some 120 million years ago. There has been similar agreement during the last few years on the fit of India, Australia, New Zealand and Antarctica, following initial uncertainties. The fit of western and eastern Gondwanaland has been more uncertain, with most workers accepting a fit similar to that proposed by Du Toit (1937) and quantified by Smith and Hallam (1970). Unfortunately the magnetic anomaly patterns in the western Indian Ocean are poorly defined, various geometric fits can all be made and the palaeontological evidence is too imprecise to define the actual fit. Most of the evidence, until palaeomagnetic studies of the Indian ocean floor can be made, must therefore rely on the matching of the polar wandering curves for the continents involved (Fig. 9.16).

There is less disagreement on the fit of Africa and South America than for any other continents. This mainly reflects the tight geometric controls, rather than better geophysical or geological data. Most discussion of the evolution of the South Atlantic (Tarling, 1980b, 1982b) has centred on the record for changes in the rate and direction of motion of these continents as the Atlantic Ocean opened, rather than the initial fit, so that the later reconstructions (Norton and Sclater, 1979) do not differ greatly from those of Bullard *et al.* (1965). Taking the postulated polar wandering paths for these two continents (Table 9.3 and Section 9.2), and rotating South America, with its polar wandering curve, to fit against Africa (Fig. 9.16(a)) shows only that the palaeomagnetic data are consistent with this reconstruction. For much of the Mesozoic, the South American curve lies to the east of the African curve, but their corresponding error circles (Table 9.3) are largely, but not entirely, overlapping. The agreement between the curves appears to be better for the Palaeozoic data, but the definition of the pole positions, for both continents, is poor. The fit of the curves does, however, indicate that the Ordovician pole position (based on all available data) for South America is spurious.

Figure 9.16 Gondwanaland reconstructions and polar wandering curves. In all these reconstructions, Africa has been left in its present position and its curve (Fig. 9.9(e)) is shown solid and the other continental polar wandering curves have been rotated to be coincident with particular geometric fits for Gondwanaland. (a) There is wide agreement on the fit of South America to Africa, in this case that of Norton and Sclater (1979), and its curve is shown dotted. This allows an evaluation of the data available and hence the reliability of reconstructions based on palaeomagnetic data alone. (b–e) Eastern Gondwanan fits are shown, with the Australian curve as dotted and the Indian curve as a narrow line. The reconstructions shown inset are based on (b) Smith and Hallam (1970), (c) Tarling in Soper *et al.* (1967a), (d) Norton and Sclater (1979) and (e) Veevers *et al.* (1980).

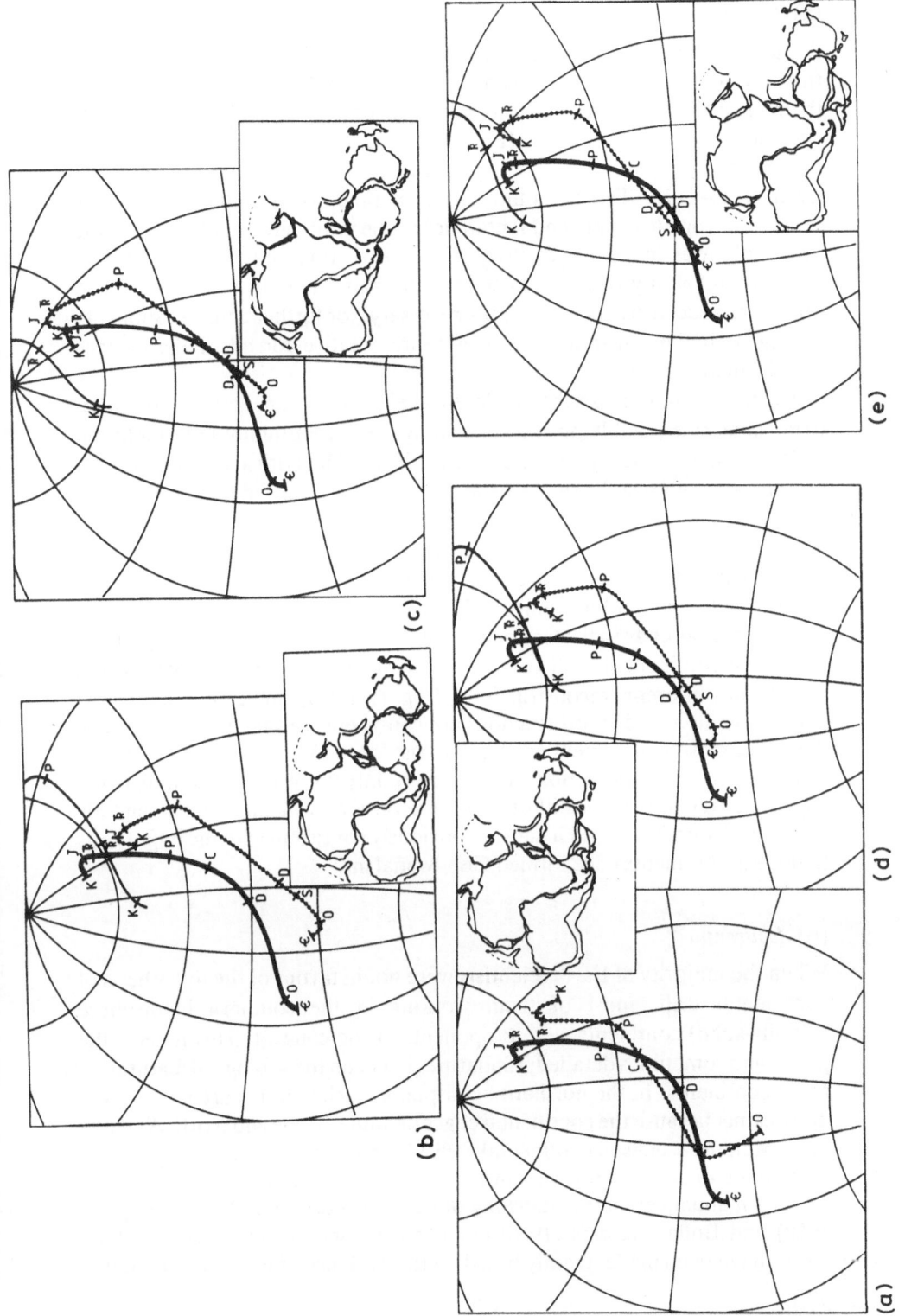

(a)

(b)

(c)

(d)

(e)

Palaeomagnetism

(a) *Gondwanaland*

Although the fit of India, Australia and Antarctica has been more uncertain than the fit of South America and Africa, there is general agreement about their general disposition and this seems to be supported by the available oceanic magnetic anomalies (Fig. 9.16). Unfortunately, the data from Antarctica are sparse and those from India are poor, except for the Late Cretaceous–Early Tertiary (and this is a time when India had already separated from Australia and Antarctica). The poor agreement between the three polar wandering curves, using all reconstructions, is thought to indicate the poor reliability of the data from India and Antarctica, rather than to obviate the reconstruction. It is thus necessary, for further considerations, to assume that the Australian curve is representative for all of the eastern Gondwanan continents.

The fit of eastern and western Gondwanaland has been, and still continues to be, in dispute (Embleton and McElhinny, 1975; McElhinny and Embleton, 1976; Tarling and Kent, 1976; Tarling, 1980c). The critical observations of oceanic anomalies still need to be made in the western Indian Ocean as the available anomaly data are sparse and the correlations that have been suggested are often poor and disputable. Similarly, analyses of the palaeomagnetic data have been used as being adequate to provide an unambiguous fit (Embleton and McElhinny, 1975; McElhinny and Embleton, 1976). However, as already indicated for each of these continents, the polar path for individual continents is still very poorly defined. Study of the polar curves based on different reconstructions (Fig. 9.16(c)) indicates that most commonly discussed reconstructions are consistent with the palaeomagnetic data. The closest agreement appears to exist for the Veevers *et al.* (1980) and Tarling (1981) reconstructions, but it is certainly not yet possible to exclude *any* of the reconstructions illustrated here on the available palaeomagnetic evidence. Nonetheless, such data are obviously consistent with the hypothesis that these continents once formed Gondwanaland!

(b) *Laurentia*

When the majority of Earth scientists were not in favour of the hypothesis of continental drift, most of them working in the southern hemisphere (Gondwanan) continents were proponents of the concept. This reflects the gross, and sometimes detailed, similarities between the geological features on these continents. In the northern hemisphere, such features are not so clear and distinct because the continental edges of both Europe and North America have been the scenes of continental collisions and interactions during at least two major orogenic episodes during the last 500 million years. Following Carey's manual geometric matching of North America and Europe (Carey, 1958) and Bullard *et al.*'s (1965) computed reconstruction, various refinements have been made, mainly based on the evidence of magnetic anomalies and the tracing of transcurrent faults. The details of such matching are, in fact, critical as, for example, the Bullard *et al.* (1965) reconstruction results in a closure of the Mediterranean region by some 2000 km, when compared with that proposed by Tarling (in Soper *et al.*, 1976b) which is similar to that of Le Pichon *et al.* (1977). It was also the palaeomagnetic data which were initially used to determine the actual time of separation of the continents (Section 1.2) by comparison of the polar wandering paths. However, the present data are perplexing (Fig. 9.17), as the curves are mutually inconsistent on the Bullard *et al.*, Tarling, and Le Pichon *et al.* reconstructions (and also with Sclater *et al.*, 1977), yet both polar curves are well defined for both continents, certainly for Upper Carboniferous to Triassic times (c. 300 to 200 million years ago). The polar paths, if taken literally, imply that the North Atlantic was approximately one-third open during this period, which is clearly unrealistic. The Bullard *et al.* (1965) model is certainly marginally better than that of other reconstructions, although the curves are still distinctly different. Such a situation clearly raises problems concerning the actual precision of much of the available data, although most of the older, undemagnetized Triassic observations have, for example, since been confirmed by more sophisticated analyses. A further possibility is inaccuracies in the basic axial geocentric dipole model, although the intracontinental evidence appears to confirm the general reliability of this model (Section 8.5). Yet another explanation could be that of a changing radius for the Earth (Section 9.8) – but this also seems to be excluded both by palaeomagnetic data as well as other evidence. As with the Gondwanan continents, therefore, it can only be stated that the data are consistent with the proposed reconstructions.

(c) *Pangaea*

It has been proposed that the continents of the world were once all part of the same continental block, Pangaea, surrounded by a world-ocean, Panthalassa. The geological evidence is certainly consistent with most of the world's continental blocks being in close proximity in Permian times, although the location of parts of eastern and southeastern Asia are problematical – China, for example, probably not then forming part of this block, although probably with a poor land connection with Australia (Tarling, 1971b, 1972; Ridd, 1971). Examination of the polar wandering paths can be used to demonstrate the aggregation of Asia (McElhinny, 1973b) but, in view of the previous considerations of Gondwanaland and Laurentia, such evaluations need to be treated cautiously, although there are no grounds for doubting their authenticity. Similarly, the main Asiatic block, Siberia (Angaraland) may not have finally joined 'Pangaea' until late Permian or Triassic times.

Examination of the available polar wandering curves (Morel and Irving, 1981; Piper, 1982) must clearly be treated with caution if well-established reconstructions are not well defined by the available palaeomagnetic evidence. Nonetheless, the general trends of the curves are very well established

271

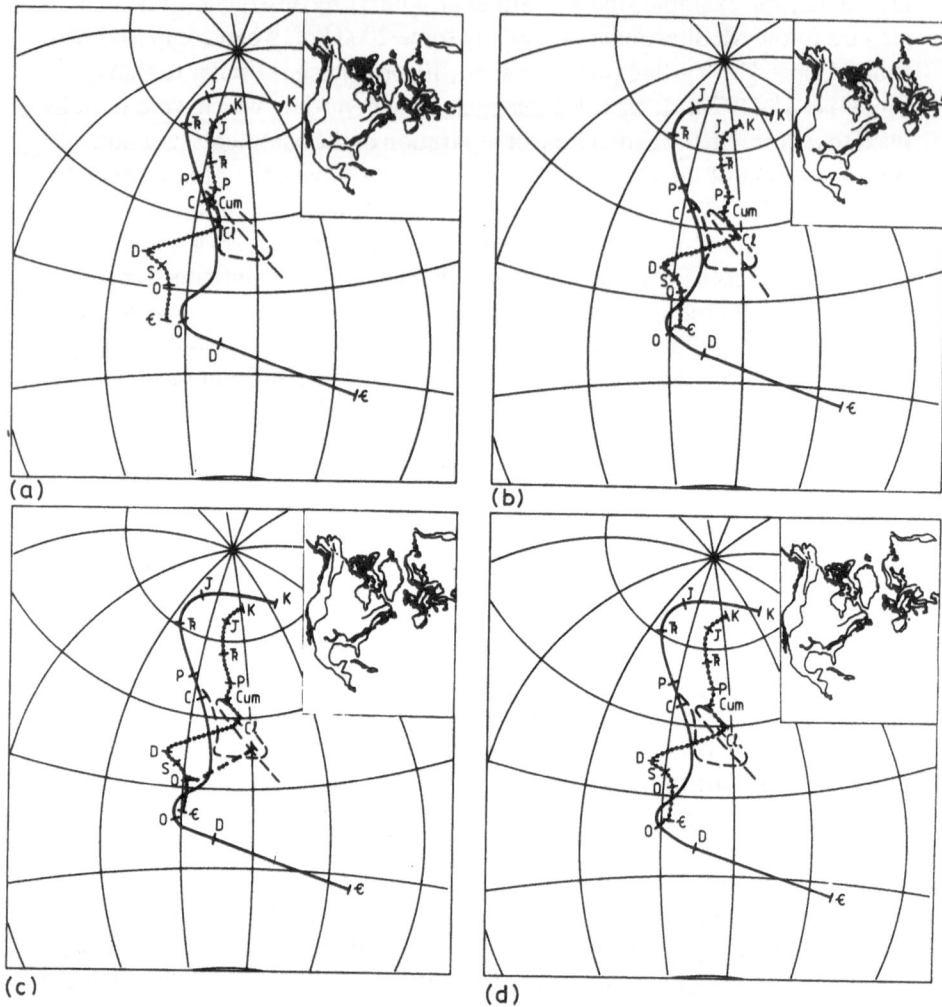

Figure 9.17 Laurentian reconstructions and polar wandering curves. In these reconstructions, the European polar wandering curve (dotted) is rotated to the North American curve (Fig. 9.9(a)). (a) Bullard *et al.*, 1965, (b) Tarling in Soper *et al.*, 1976a, (c) Sclater *et al.*, 1977, (d) Le Pichon *et al.*, 1977. The original Bullard *et al.* 1965 reconstruction is clearly more consistent with the available palaeomagnetic data, but it is important to stress that the match is remarkable in that there is only poor agreement even for times, such as Upper Carboniferous to Triassic, when a large quantity of good-quality palaeomagnetic data are available. This could suggest that the criteria used for selection of the data for analysis were not adequate, but certainly indicates the degree of caution that must be exercised in palaeomagnetically based reconstructions based on even fewer data.

and allow generalizations to be made about the relative motions of the two major continental blocks (Tarling, 1979) although, obviously, local areas caught between the two blocks may show radically different individual motions. The Taconic and Caledonian orogenies in the Ordovician–Silurian probably marked yet another recreation of Laurentia, and the Taconic orogeny, in the southeastern United States, may mark a collision with either Africa or a major island arc system that lay between them. Following this orogeny, the 'Central Atlantic' probably opened to only some 300–400 km by Lower Devonian times. During the Devonian, both supercontinents had a clockwise motion, with Gondwanaland moving somewhat faster northwards than Laurentia in the Western Mediterranean–Central Atlantic region. Laurentia probably collided with the Kolmya block in Devonian times, increasing the size of Laurentia in the Arctic region (this block eventually to leave in the Jurassic, forming the Canadian Arctic). The Acadian orogeny in the northwestern Appalachians may have been a collision of Laurentia with either Africa or another island arc system – possibly also accounting for some of the Maurentide orogenic developments in western Africa. The result of these collisions seems to have been to make western Gondwanaland have an essentially northward motion in Middle–Upper Devonian times, while Laurentia had begun to rotate anticlockwise and was moving northwards, further opening a 'Central Atlantic' (Tethyan) ocean between Laurentia and Gondwanaland during the Lower Carboniferous. However, the anticlockwise motion of Laurentia meant that southern North America was rapidly closing with the clockwise-rotating Gondwanaland – the two colliding in the region of the present Gulf of Mexico, and Iberia and Armorica (southern Europe) possibly acting as hinge areas to such motions. The southern North America collision with northern South America–West Africa caused a drastic change in the relative motions of the two super-continents, probably also affected by a locking of the promontories of southern Europe (Iberia, etc.) with those of Africa (the Adriatic Peninsula). This collision and locking resulted in a brittle response on these super-continents, causing brittle fracturing that blocked out what were to become our present continental outlines and causing zones of weakness that resulted in major basin formation. Both super-continents now began rotating clockwise and the area of the Mediterranean became one of major transcurrent motion, locally influenced by the presence of individual tectonic blocks, such as Bohemia, Iberia, etc.

Such a development of the two super-continents, although speculative, is consistent with the faunal and climatic evidence as well as the broad palaeomagnetic evidence and would strongly imply that Gondwanaland and Laurentia have always acted as two completely separate tectonic units (plates) that were occasionally in contact, enabling inter-continental dispersions of fauna and flora at specific times, mainly controlled by the specific palaeogeography of those times (Section 9.7).

273

(d) *Pre-Cambrian tectonics*

Comparison of the Pre-Cambrian curves for different tectonic units is particularly interesting. In Africa, most of the available evidence indicates that there was little or no displacement between the different cratonic nuclei during the Pan-African orogeny, 750–550 million years ago. This suggests that inter-plate collisions were not involved (Briden, 1976; Embleton and Schmidt, 1979; Daly *et al.*, 1980). Comparisons of the Pre-Cambrian curves for the Gondwanan continents only gives rise to a generalized consistency, mainly reflecting the problems of dating, rather than the palaeomagnetic data or uncertainty about the reconstructions (Piper *et al.*, 1973; Piper, 1976; McWilliams, 1977) – and comparisons between Laurentia and Gondwana (Piper, 1976; Embleton and Schmidt, 1979) have been made, with apparent evidence for similarities in their polar wandering curves indicating the existence of a single continent through most of the Proterozoic (Piper, 1982). Such assessments must be regarded as preliminary at this stage as the palaeomagnetic evidence, even for the unity of Gondwanaland or Laurentia in the Late Proterozoic, is not very convincing although obviously consistent with the geological evidence for such uniformity. At this stage, the main importance of the available data is probably the assessment of the nature of intercratonic movements, the determinations of the depths of emplacement and burial and so forth, associated with Pre-Cambrian orogenic events (Tarling, 1978d, 1982b). It seems probable that drastic revisions in the known polar wandering curves are likely to be achieved with the greater facility to isolate different components of magnetization within Pre-Cambrian rocks – although the problem of dating the time of acquisitions of each of the components will always remain.

9.6.3 Large-scale intracontinental tectonics

Two main areas will be considered as examples of applications of palaeomagnetic studies in unravelling tectonic patterns on scales of several hundred kilometres. In each case – the Western Cordillera of North America and the Adriatic Peninsula in the Mediterranean – active studies are still in progress, so that the reviews are not intended as definitive, but are cited as examples of how palaeomagnetism has been and continues to be applied on this type of scale.

(a) *The Western Cordillera of North America*

This is an extremely complex tectonic area, with extensive faulting, thrusting and batholithic intrusions having occurred during the last 200 million years. The region was originally considered to be a geosyncline (large-scale depression) within which the sedimentary detritus from cratonic North America accumulated and sporadically underwent uplift and igneous intrusion. Early palaeomagnetic studies in the area showed that stably magnetized rocks were characterized by radically different directions of remanence from those expected. Miocene age (10–20 million years) rocks would, for example, be expected to have directions similar to that for the axial geocentric dipole field as the amount of continental drift since that time is small, some 200 km (1–2°), yet deflections of 40–50° were found. These were initially attributed to anomalous geomagnetic field effects (Cox, 1957), although now interpreted, following Irving (1964) and Beck (1975, 1976), as being largely controlled by tectonic movements affecting the sampling areas. Such rotations are now recognized throughout the length of the Western Cordillera (Fig. 9.18(a)) from Alaska to Central America. The data are now extensive (Watkins, 1965a; Stone, 1967, 1980; Clark, 1969; Packer and Stone, 1972, 1974; Symons, 1973, 1977a,b; Beck, 1975, 1976, 1980; Stone and Packer, 1977, 1979; Simpson and Cox, 1977; Hillhouse, 1977; Irving, 1979; Greenhaus and Cox, 1979; Hannah and Verosub, 1979, 1980; Van der Voo and Channell, 1980; Van der Voo et al., 1980; Wilson and Cox, 1980; Beck and Plumley, 1980; Magill et al., 1981; Vugteveen et al., 1981) and their general consistency and age ranges clearly preclude a geomagnetic origin.

The scale of the individual tectonic units is not, as yet, well defined. Consistent directions of remanence occur in association with individual batholiths (Symons, 1977a), and there are general similarities about the directions, for the same time, in large areas, such as the Oregon Coast Range. Many observations could, however, be attributed only to small-scale units. Many of the data are also difficult to interpret tectonically because of the lack of external control. It is generally assumed, for example, that batholiths have not been tilted since emplacement. Deviations in declination are not therefore generally attributable to tilts but to rotations about a vertical axis (Section 6.3), often combined with a horizontal translation of the region. In southern Alaska, for example, sites in the Wrangell Mountains show evidence for a clockwise rotation and a translation from very much further south, such as from the Franciscan terrain of California (as proposed geologically by Jones et al., 1977), or even from South America, in Cretaceous times.

The timing of the rotations is also poorly defined, with Miocene age rocks showing indications for rotations in most areas, and major displacements affecting almost all Mesozoic age rocks. At the moment, however, the size of the individual blocks needs to be adequately defined before individual polar wandering paths can be established for each unit. Preliminary analyses indicate episodic rotations and translations throughout the Mesozoic and Cenozoic. The fact that, with the exception of southeastern Alaska, all the deviations indicate a clockwise rotation now raises the question whether small units behaved as 'roller bearings' as movement took place along the bounding fault line, and the extent to which basement rocks also moved in order to provide regional similarities. At the moment, it would appear that several units, such as the Washington and Oregon Coastal Ranges, have behaved as an island arc system (Dickinson, 1970) with each volcanic unit having a unique rotation while also participating in the general movement as the arc closed against cratonic North America during the subduction of the Farallon and Juan da Fuca plates (Atwater, 1970). Conceivably individual units sometimes behaved separately from their neighbours and sometimes acted together (Heptonstall, 1977), but this will require even more intensive study, particularly covering a wide age span for each putative tectonic unit. Considerably more attention will also have to be paid to the probability of later remagnetization (Hannah and Verosub, 1980), but such studies can readily be extended southwards into Mexico (Urrutia-Fucugauchi, 1981a), the rest of Central America and South America (Skerlec and Hargraves, 1980; Palmer, Hayatsu and MacDonald, 1980a,b), and Antarctica (Valencio, 1979; Kellogg, 1980; Burns et al., 1980; Watts and Bramall, 1981). As well as applying to all modern subduction zones, similar considerations obviously apply to the different tectonic units of the Appalachians, the Caledonian and Variscan orogenic belts in Europe and so forth. The older orogenic systems are, of course, even more likely to be characterized by later remagnetizations, but the broad indications are for similar tectonic behaviour (Tarling, 1979), implying that similar broad tectonic controls were operating in earlier times.

(b) *The Adriatic Peninsula*

Fundamental to any tectonic interpretation of the Mediterranean region during the Mesozoic and Cenozoic is an assessment of the Adriatic promontory (Argand, 1924). Geologically, the Italian area is complex, with nappes, carbonate platforms, and compressed and distorted sedimentary troughs, exposed rocks being mostly of Cretaceous age, but frequently resting on Triassic evaporites that form mobile layers overlying more rigid basement rocks of probably Hercynian age. Many of the upper layers may thus be mobile (allochthonous), even if the basement remains fixed, while it is also possible that basement movements could occur with little apparent surficial response. Several units are generally, but not universally, regarded as resting directly on their basement or as actually comprising the basement i.e. they are autochthonous. Most of these autochthonous blocks (Fig. 9.18(b)) have been sampled palaeomagnetically, albeit on a preliminary scale, providing information on mainly Cretaceous rocks. In all these areas, the measured Cretaceous directions are in agreement with each other and with those expected for Africa (Fig. 9.18(b)), and this agreement is also indicated in the Permo–Triassic data, although these are mostly from the north (Vandenberg and Wonders, 1980). The internal agreement persists to the present, but the block clearly became independent of Africa in Tertiary times. The available evidence thus confirms that these three areas are autochthonous and implies that their basements form part of a single tectonic plate from Permian times to the present. This was probably largely created in the Hercynian orogeny (Permian) but its northern extent is not yet defined adequately as similar old, rigid blocks within the Alps have been extensively remagnetized during Alpine deformations (Flugel et al., 1978). Tectonic maps frequently indicate that the northern boundary of the Adriatic promontory lies along the Insubric Line (part of the Periadriatic lineament), but palaeomagnetic and magnetic fabric studies of these rocks bordering this major fault system (Rathore and Heinz, 1979; Ellis, 1979) indicate that it is an ancient feature and that rocks to the north also belong to the Adria tectonic unit during the Mesozoic and Cenozoic. This means that the northern boundary must lie beneath the Northern Calcareous Alps. These rocks have been derived from the south, and their obvious nappe nature is further confirmed by palaeomagnetic studies (Mauritsch and Frisch, 1978), and they now overlie the 'African–European' boundary.

Between the now-confirmed autochthonous units of Italy, the palaeomagnetic vectors are widely scattered (Fig. 9.18(b)), although often consistent with each other over several square kilometres. Many of these areas are known to have been drastically affected by surficial movements and palaeomagnetism is currently defining the actual motions involved in such allochthonous behaviour (Channell and D'Argenio, 1980). It is important to note that, in many cases, the palaeolatitudes appear inconsistent with the present relationship with the autochthonous blocks. It is unlikely that these indicate differences in the original palaeogeographic distribution of these units as they are often variable, even from the same allochthonous unit. It seems more likely that such effects represent intrastratal deformation during nappe formation and also that the simple tectonic correction normally applied in palaeomagnetic studies (Section 5.5) is inadequate so that some of the displacement of declinations may mirror intrastratal dipping fold axes, in addition to regional tectonic rotation.

In a wider view, the Italian data provide a major factor in interpreting the evolution of the entire Mediterranean region (Channell et al., 1979b; D'Argenio et al., 1980). Palaeomagnetic data from Hungary (Márton and Márton, 1981) indicate that the E. Pannonian Basin also has had largely 'African' affinities during most of its Mesozoic development. Corsica, Sardinia and the Iberian Peninsula are similarly showing major motions, relative to both Africa and Europe (Nairn and Westphal, 1968; De Jong et al., 1969; Van der Voo, 1969; Zijderveld et al., 1970; Klootwijk and Vandenberg, 1975; Westphal et al., 1976; Storetvedt and Petersen, 1976; Storetvedt and Markhus, 1978; Vandenberg, 1980), although magnetic overprinting obscures much of the evidence in Corsica and Sardinia. Fuller understanding obviously requires more detailed studies, even within those areas already investigated, but any understanding of the evolution of the Mediterranean region as a whole requires an extension of studies into northwestern Africa in particular as few extant models take sufficient cognizance of the known geological evolutions of the Atlas Mountains, and the palaeomagnetic control in this area is sparse for such a complex area (Nairn et al., 1981). Similarly the spatial consequences of opening up the allochthonous areas into their previous geography are only just being examined (D'Argenio *et al.*, 1980) and have major implications for palaeogeographic recombinations for the entire region (Section 9.7).

Similar studies can obviously be undertaken in many other areas of the world, and recent studies in the Himalayan region (McElhinny *et al.*, 1978a; Klootwijk *et al.*, 1979; Klootwijk and Conaghan, 1979; Klootwijk and Peirce, 1979; Klootwijk, 1979a,b, 1980; Bingham and Klootwijk, 1980; Klootwijk and Bingham, 1980; Klootwijk *et al.*, 1981) are leading to a much greater understanding of the mechanism and behaviour of tectonic units during continental collisions. The problem with many such preliminary studies is that the extent of the tectonic unit to which the data apply is not always clear, particularly if there is a risk that the sampled area may be allochthonous. Nonetheless, even preliminary studies can place major constraints on models for the evolution of any particular area and also allow a better assessment of the key areas for future studies.

Older orogenic belts can be studied similarly, but are obviously more difficult to understand and interpret as the geological picture is often less clear due to erosion of many of the structures and the exposed portions may well have been subjected to deep burial before being raised to their present level. A wide variety of studies have now been made relating to the Taconic and Acadian orogenies in eastern North America (Kent and Opdyke, 1978, 1980; Kent, 1979; Brown, 1979; Brown and Kelly, 1980; Watts *et al.*, 1980a; Rao and Van der Voo, 1980) and to the Variscan–Hercynian and Caledonian orogenic systems in Europe (Morris, 1976; Morris and Tanner, 1977; Tarling, 1979, 1980a; Ries, 1979; Ries *et al.*, 1980; Auvray *et al.*, 1980). At this stage, it is clear that major movements can be detected in association with each of these mountain episodes, but many of the data are in dispute – mostly

---

Figure 9.18 Intracontinental deformation. Only two of many examples of displacements within the present continents are shown here. (a) The Western Cordillera of North America. The locations of sites in the Western Cordillera are shown as crosses and their pole positions are shown as dots. These are virtually all rotated clockwise from the Jurassic (J) and Cretaceous (K) pole positions for cratonic North America. (b) The 'Adria Promontory' in the Mediterranean. The Adria Promontory has been rotated into its palaeolatitude and orientation as if it were part of Africa in Cretaceous times. The individual location mean declinations are shown as arrows, with those from known autochthonous units (shaded) as large, compared with those from known or suspected allochthonous areas. The palaeolatitude, determined from the inclination, is shown as a number near the arrow. The data are extensive but of variable quality and quantity (de Boer 1965; Baumann and Krs, 1967; Schult, 1973; Lowrie and Alvarez, 1974; Barberi *et al.*, 1974; Manzoni, 1975; Gregor *et al.*, 1975; Channell and Tarling, 1975; Förster *et al.*, 1975; Klootwijk and Vandenberg, 1975; Channell, 1977; Krs *et al.*, 1977; Krs, 1978; Krs and Smid, 1979; Lowrie *et al.*, 1980a; Márton and Márton, 1981) – some arrows thus represent a single site, others represent the mean of a group of sites from the same or very close localities.

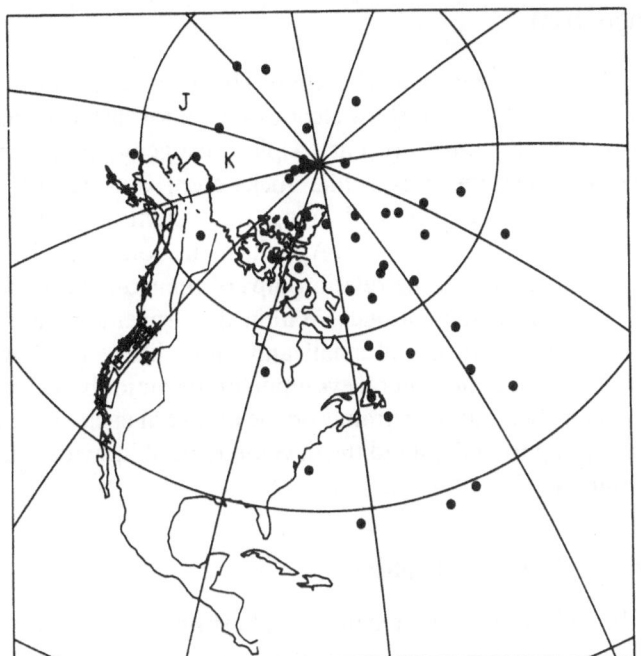

(a)

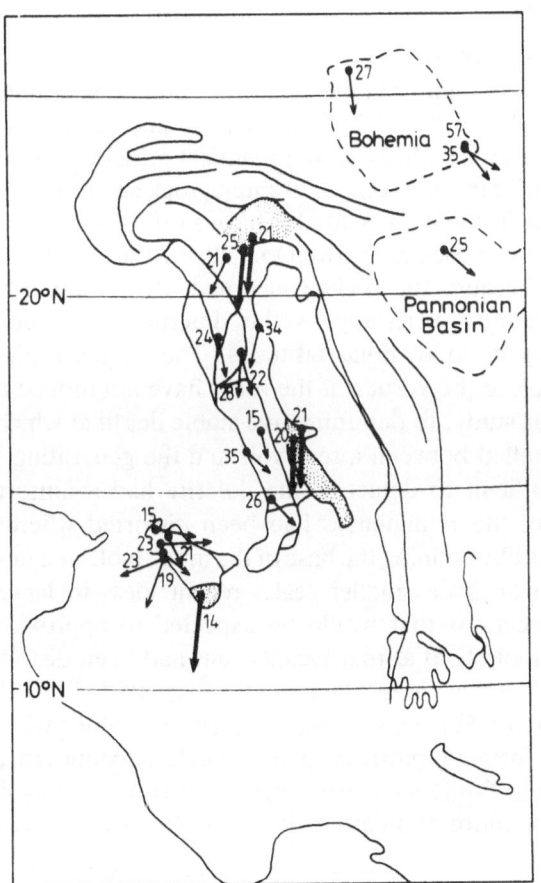

(b)

concerning the age of their magnetization. However, even rocks remagnetized during the orogeny itself show clear evidence for dislocations similar to those observed in the more recent 'Alpine' orogeny, indicating broadly similar tectonic characteristics, while rocks from outside the immediate orogenic zone are increasingly defining the extent and importance of 'Arcadia' in the northern and central Appalachian evolution. The priority, at the moment, is to establish the relationships of the main continental blocks involved in these orogenic episodes. Within the orogenic belts themselves, the tectonic units are probably small and more likely to show complex magnetic histories, and the general evolution of the major blocks is thus more important in placing major constraints on the nature of each orogenic episode prior to attempting to understand the development of individual units within the larger framework.

9.6.4 Small-scale tectonic applications

Although the palaeomagnetic technique is best known for continental-scale tectonic studies, such as continental drift, the principles can be applied throughout all levels for which measurements can be made. The distinction between autochthonous and allochthonous units, nappe structures, etc., has been mentioned previously (Section 9.6.3). Individual fault blocks of the order of 50×200 km can be identified and their rotations determined (Sections 9.6.1 and 9.6.3) and the nature and age of folding determined (Morris and Tanner, 1977). Using identical principles, it is also possible to establish relative motions on, for example, rock and soil creep, land slips, and the like. In each case it is essential to have a reference datum for comparison. In one example of rock creep (McDougall and Green, 1958), the creep could be detected because the rocks concerned, the Tasmanian Dolerite, were magnetized almost vertically as the dolerite had been extruded while Australia was in south polar latitudes. The magnetization of bore cores should, therefore, be vertical if the rocks have not moved (Fig. 9.19(a)). In this particular study, to determine the stable depth at which hydraulic pipes could be installed between a reservoir and the generating station, the rock creep was difficult to detect as the dolerite had a uniform homogeneous structure, but the remanences had been distorted where rock creep had occurred, thereby defining the base of the metastable and unstable areas (Fig. 9.19(a)). On an even smaller scale, recent loess in Israel showed stable remanent directions that would be expected to approximate to the axial geocentric dipole field at that locality, but had been distorted by soil creep between successive loess layers (Bruins, 1976). Similarly land slips (Scheidegger, 1975) have associated slip surfaces along which rotation of the rocks occurs forming, particularly in permafrost conditions, very large-scale land slippages. While such areas have normally been avoided when being examined for more standard palaeomagnetic studies, comparison of the

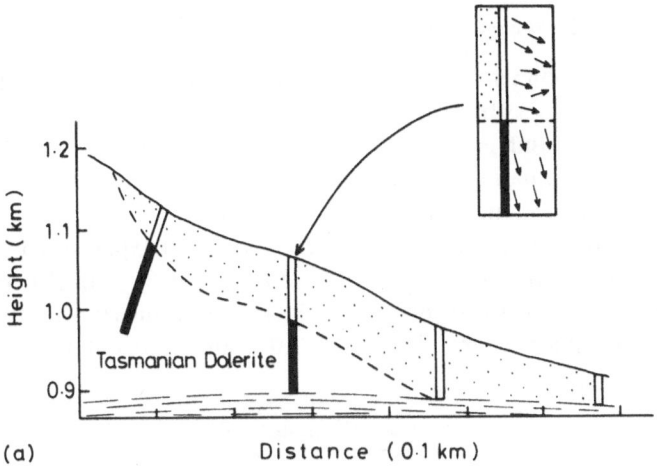

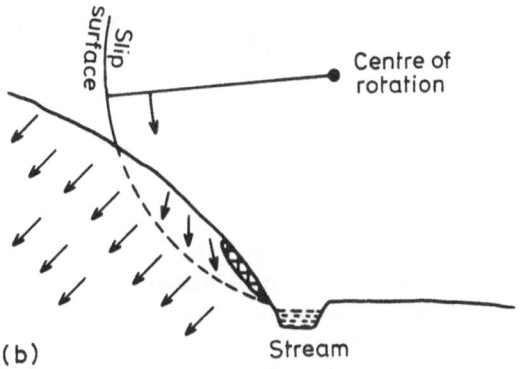

Figure 9.19 The stability of slopes. (a) Drill cores through the Tasmanian Dolerite, Australia, showed few diagnostic properties for determining the mechanical stability of the rocks for engineering structures. Conveniently the magnetization of the dolerite is almost vertical as it is of Jurassic age and was magnetized when Australia was close to the magnetic pole. Studies of the changing directions of remanence therefore provided a method of distinguishing mechanically stable and unstable regions. (After McDougall and Green, 1958.) (b) Similar situations are also likely to be involved in land slip situations, and it has been possible to measure soil-creep in loess deposits (Bruins, 1976).

directions within the partially slumped regions and the undisturbed rocks should allow definition of the centre of rotation of the instability of the slope (Fig. 9.19(b)) and hence be applied within geological engineering conditions. In a different context, Verosub and Moores (1981) consider that anomalously shallow inclinations in oceanic basalts (Natland, 1978) may well arise from similar shallow angle listric faulting.

9.6.5 Very large-scale applications

The largest scale on which palaeomagnetic techniques can be applied is at a cosmic level! The possibility that the universal gravitational constant, G, may not be constant, but could vary on long time-scales, could be of major importance in evaluating the significance of the red shift of the light from distant galaxies. On a more parochial scale, it has been suggested that many terrestrial features may have resulted from an Earth expansion as a result of changes in the 'constant' (Carey, 1958; Owen, 1976) – most ocean basins postulated as forming by an increase in the Earth's radius by one-third during the last 200 million years. There are, of course, cogent reasons for not

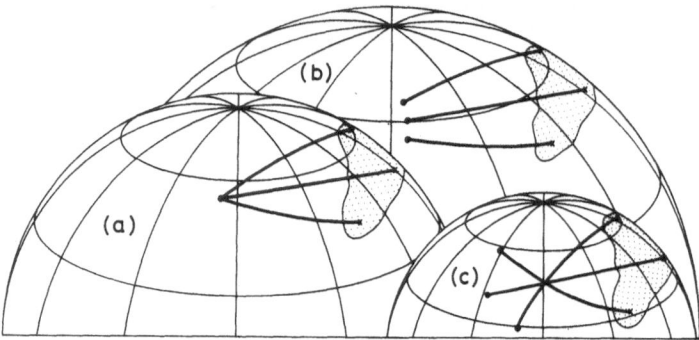

Figure 9.20 Pole positions for a changing earth radius. If the pole positions for three locations of the same age on a single tectonic block are determined, they should, within experimental error, coincide when calculated on an Earth of the present radius (a), but if the magnetization was acquired when the Earth was (b) larger, or (c) smaller the former causes the calculated poles to 'undershoot', the latter causes them to 'overshoot'. The degree of undershooting or overshooting increases the scatter and gives a measure of the extent to which the Earth's radius has changed with time.

considering that G has, in fact, changed by more than $\pm 1\%$ during the last 4.5 billion years as other planetary surfaces of this age do not show such expansion features (McElhinny *et al.*, 1978b). The hypothesis of a changing Earth's radius can, however, be tested palaeomagnetically (Egyed, 1959; Ward, 1963, 1966; Schmidt and Clark, 1980) by examining the scatter of palaeomagnetic poles of similar age for single tectonic units of the Earth's surface. As the Earth expands, the surface rocks also expand, but the change in linear distance between points on the same tectonic plate will remain essentially constant as the Earth, as a whole, expands. The tectonic plate is thus displaced radially towards or away from the Earth's centre (Fig. 9.20). The increase in curvature will, of course, result in some flattening as it adjusts to the new curvature, but this effect is small, less than 2° for a 25% change in radius (Schmidt and Clark, 1980). However, the locus of the individual poles,

282

calculated for different Earth radii, will vary considerably for such radii change. The pole positions, if of the same age, should be identical for the correct radius at the time that the sites were magnetized. Comparison of the degree of polar scatter should therefore be a measure of past Earth radii (Ward, 1963). Comparison of the precision for different radii indicates no significant change in the radius over the last 400 million years or so (Ward, 1963; Van Andel and Hospers, 1968a,b; McElhinny *et al.*, 1978b), but cannot rule out changes of 5–10%, reflecting the paucity of statistical tests for comparing *k* values for low numbers and the problem of defining areas that have since behaved as a single tectonic unit and yet are large enough to provide a reasonable definition of the actual mean pole position.

On a smaller scale, the fact that polar wandering paths (Fig. 9.10) can be established, albeit with uncertainty, throughout most of Pre-Cambrian time, the last 3000 million years (Section 9.2.3(e)), is of major geological and geophysical importance in itself. As polar paths arise, not because of the movement of the Earth's average geomagnetic pole, but because of the movement of the continents relative to the rotational and geomagnetic pole, it means that continental motions have taken place throughout the recorded history of the Earth. Such a conclusion is not particularly surprising if it is accepted that the motion arises from radiogenically driven convective motions in the Earth's mantle (Holmes, 1927; Tarling, 1978b), but is of both academic and economic importance in understanding the origin and development of ore deposits throughout geological time (Tarling, 1981). The fact that such curves are genuine reflections of such motions is thus confirmed by the formation of specific ore deposits and sediments at specific palaeolatitudes as defined by the palaeomagnetic data (Section 9.7). The apparent correlation between sharp changes in the direction of the polar wandering path and orogenic episodes in North America (Irving and Park, 1972) is not difficult to verify as most palaeomagnetic data are from dykes intruded during tensional phases that may be related to relaxation phenomena following collision between different continental blocks. If verified by subsequent work, then this would suggest that plate tectonic processes were operating throughout geological time, although their surface expression may well differ in detail.

The radiogenic heat production in the early Pre-Cambrian was probably some four or five times greater than today (Lambert, 1976), with the consequence that mantle convection would be approximately twice as rapid as today since the rate of convection, to a first approximation, depends on the square root of the heat production (Tozer, 1977). This could be expressed in terms of a greater number of mid-oceanic ridges in order to release the heat at the surface and also in the probability of somewhat greater motions of the continents. Unfortunately, the palaeomagnetic data provide no control on the absolute movements of any tectonic block in terms of its ancient longitudes, although relative longitude movements can be defined between different blocks. This means that the rate of plate motion can only be determined for latitudinal changes and any estimates for the Pre-Cambrian, in particular, are limited by the accuracy of radiometric dating as errors of several hundred million years can be present during which movements of some 90° could be attained even on present rates of motion. Ullrich and Van der Voo (1981) using such analyses determined that there was evidence for periods of both faster and slower movements of the plates in the Pre-Cambrian when compared with the last 200 million years, but the motions were not consistently higher. As they pointed out, the latitudinal motion can only provide an absolute minimum for such plate velocities. In a similar way, Gordon *et al*. (1979) showed that there appeared to be no relationship between continental sizes and their rate of motion – a conclusion that would be consistent with the concept of mantle-wide convective currents in which the continents themselves would form relatively insignificant parts of the total system.

In this context, it is important to emphasize that it is impossible to distinguish between 'true' polar wandering and continental (plate) motions unless it can be shown that there are systematic differences between the palaeomagnetically determined pole and palaeoclimatic indicators. Despite this the available polar wandering curves are sometimes examined to determine common components, which are then attributable to actual polar movements (Jurdy, 1981). At the present day, however, there must be a net motion of all the continents towards the Pacific as it is in this area that virtually all the world's crustal subduction is taking place. Similarly, the Pacific plate itself must have a predominantly northwesterly motion as the northern and western Pacific appear to have more active subduction zones than in the south and east. The summation of all present plate motions must, therefore, be towards a locus in the northwestern Pacific area. It is arguable whether such a motion can be termed 'true' polar motion in the sense normally applied in considering actual movements, in space, of the Earth's axis of rotation (Jardetzky, 1958; Weyer, 1978).

9.7 BIOLOGICAL, WEATHER, CLIMATIC, PALAEONTOLOGICAL AND PALAEOGEOGRAPHIC ASPECTS

There are strong connections between climatic zones and the distribution of different life forms on the present Earth, and such relationships can be shown to have existed in the past (Section 9.7.3). There are also some indications that changes in the geomagnetic properties can also have direct effects on both life forms (Section 9.7.1) and the weather (Section 9.7.2). Many of such indications are, at best, speculative, but may provide important insights into otherwise puzzling phenomena, such as the ability of many animals to navigate even in the absence of visible markers. However, any attempt to evaluate past climatic patterns must be based, fundamentally, on an assessment of past palaeolatitudes and palaeogeographies. As such, the main contribution of palaeomagnetic studies is likely to be in making such reconstructions and determining, in particular, the distribution of life forms under such conditions. Such evaluations are particularly important in palaeontological evaluation as the classification of fossils, although ideally objective, is necessarily subjective and can only be undertaken with an evaluation of all the available factors, in particular the possibility of interchanges between specific areas.

9.7.1 Biological aspects

There is increasing evidence for subtle influences of geomagnetic changes on animal and plant behaviour, although most of these observations are still poorly controlled and the mechanisms accounting for such interactions are not yet understood (Sheppard and Eisenbad, 1977). It is commonly stated, for example, that plants, such as corn, grow faster when in a magnetic field stronger than that of the Earth, but this does not appear to be caused by any direct magnetization of seeds because magnetized, as opposed to non-magnetized, grains do not seem to show any differences in their growth rates. Neither do the grains themselves appear to be capable of being magnetized, but it is now clear that many animals and insects contain magnetic particles in their brains or necks that appear to have an influence on their behaviour. In 40 different species of birds, the magnetized zones in the brains or necks of migratory birds were much more strongly magnetized than in non-migratory species (Presti and Pettigrew, 1980), suggesting that these regions may be fundamental to migratory behaviour. Bees also contain minute, strongly magnetic cells (Gould *et al.*, 1978) as do various invertebrates (Arendse, 1978) and even bacteria (Blakemore, 1975). It seems probable that such stably magnetic zones provide a reference for magnetic orientation for migratory sparrows or homing pigeons (Moore, 1980), possibly acting mainly as a background mechanism when visual systems cannot be operated. It is conceivable that such things as 'dowsing' for minerals and so forth may, in fact, be based on such facilities in human beings, and may account for the greater 'sense of direction' in some people than in others.

Similar direct effects of the geomagnetic field on animals have been proposed (Sheppard and Eisenbad, 1977), such as the higher incidence of heart attacks during magnetic storms (Malin and Srivistava, 1979), but such correlations are not universal (Knox *et al.*, 1979) and, even if real, may well be indirect. Similar statistical correlations have been proposed between geomagnetic activity, particularly polarity changes, and changes in the rate of speciation in the geological record. Studies of polarity changes and the fossil abundances of marine micro-organisms, particularly radiolaria (Fig. 9.21), indicate that new species of these organisms often appear close to the time of a polarity transition and other species become extinct at such times (Harrison

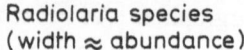

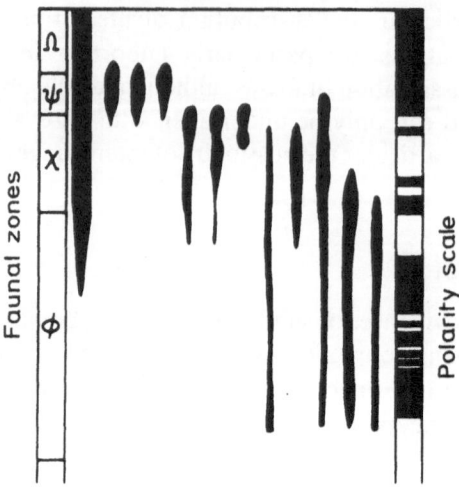

Figure 9.21 Radiolarian extinctions. The abundance of different species appears to change drastically at times of polarity reversal, but it is puzzling why some species can live, apparently unchanged, through several transitions, prior to extinction. (After Opdyke *et al.*, 1966.)

and Funnell, 1964; Opdyke *et al.*, 1966; Hays and Opdyke, 1967; Watkins and Goodell, 1967; Hays *et al.*, 1969). Different species of radiolaria, and other micro-organisms, commonly appear or disappear throughout the geological record and it is thus difficult to establish the exactness of such correlations, particularly as the same species can persist through numerous polarity transitions before becoming extinct. Attempts at statistical analyses of the probability of such extinctions, or appearances, are variable, even on the same data. Hays (1971) assessed the probability of six out of eight radiolarian species becoming extinct by chance at one polarity transition as $1:10^3$, while Mann (1972), on the same data, evaluated the same probability as $1:10^6$. Plotnick (1980) re-analysed the data and found that there was an almost 50:50 probability of such extinctions occurring at that time. Such disagreements illustrate the difficulty in establishing the exact time of such appearances or extinctions in oceanic sediments and also in evaluating the time range that should be considered. Obviously this problem requires urgent assessment and incorporation of a much wider data base but it may well be that, if there is a genuine geomagnetic correlation, then it is likely to be related to changes in the total intensity of the geomagnetic field (see below) and hence speciation rates are also likely to have been enhanced during geomagnetic excursions in which intensity reductions may also take place (Section 8.4.2). The times of such excursions are still very badly defined, even for the last 10 000 years, so any statistical correlations must, therefore, be treated cautiously.

If established, such correlations are unlikely to be direct as the rates of change of the strength and direction of the geomagnetic field are still long compared with the rates of adaptation of most organisms to environmental change. It seems more likely that such a correlation would reflect an indirect control, particularly related to the decrease in strength of the field. This would cause a reduction in the height of the ionosphere and hence allow a greater penetration of harmful cosmic and solar radiation (Uffen, 1963, 1965). These could cause much higher mortality and also an increase in natural mutation rates. The total increase in such radiation is, however, calculated to be small, less than 10%, even for a total loss of the geomagnetic field (Black, 1967; Waddington, 1967; Harrison, 1968) because of the shielding effect of the atmosphere. Marine organisms, for which there are, at the moment, the most data, would also be very effectively protected by the filtering action of sea-water. Another possibility is that ionospheric changes would also have climatic effects (Section 9.7.2) that could have major environmental effects, resulting in changing speciation rates. Such changes would, again, be on a scale of a few thousand years, but do not seem to be detectable by other climatic indicators. It is also particularly puzzling, if the correlation is established, why certain species should be able to persist through several transitions prior to their eventual extinction.

Direct relationships between geomagnetic changes and rates of evolution on a geological time-scale are thus not clearly established either on a statistical or causative basis. Nonetheless, the fact that palaeomagnetic techniques allow past geographies to be reconstructed (Section 9.6) means that these techniques are fundamental in establishing the significance of climatic control on now fossilized organisms. In particular it seems likely that the changes in land–sea distributions will, on occasion, give rise to sudden large-scale climatic changes. For example, the entrance of Arctic water into the Atlantic during Tertiary times, or the development of the circum-Antarctic oceanic circulation system following the final separation of South America from Antarctica (Section 9.6), may be responsible for drastic environmental changes occurring over a period of only a few years, thus accounting for some mass extinctions observed in the fossil record.

9.7.2 Weather and climatic aspects

Over the last few decades, several correlations have been proposed between changes in the geomagnetic field strength and various aspects of climatic change, Wollin *et al.* (1971, 1973), in particular, claiming that a higher magnetic intensity either produces or is associated with a colder climate. As most geomagnetic observatories are usually near to meteorological stations, it is possible to examine records from both types of observatory for such trends.

Palaeomagnetism

Wollin *et al.* (1973) used records from a variety of stations to demonstrate increases in magnetic field associated with decreases in temperature and vice versa, and extended this correlation into more remote times by means of oceanic sediment records in which the intensity of natural remanence fluctuations correlated with temperature changes recorded by oxygen isotopes. Such correlations are, in fact, weak (Williams and Garety, 1978; Sternberg and Damon, 1979) and not, of course, necessarily causative as changes in the solar radiation can alter both the magnitude of ionospheric currents (and hence the observed magnetic field) and the surface temperature, as can variations in the eccentricity of the Earth's orbit (Wollin *et al.*, 1978). In sediments, the change in intensity of magnetization will depend, not merely on the geomagnetic field strength, but also on the rates of sedimentation – higher sedimentation of calcium carbonates taking place during warmer periods and hence diluting the strength of the natural remanence (Amerigian, 1974). Similarly erosion rates, and hence grain sizes and the total quantity of sediment, will also vary in response to local climatic changes. At the moment, therefore, there appears to be little strong evidence for a direct control of weather patterns by geomagnetic influences (Chave and Denham, 1979), although it is likely that there is some interaction associated with changes in the ionosphere being reflected in circulation patterns of the atmosphere. Roberts and Olson (1973), for example, suggested that the onset of troughs in the westerly cyclonic waves in the North Pacific was triggered by magnetic storms in the preceding 3 or 4 days, and that their subsequent development was influenced by other geomagnetic disturbances.

An important consequence of changes in the properties in the ionosphere is the influence this may have on the generation of ^{14}C as this is fundamental to dating methods based on this isotope (De Vries, 1958; Willis *et al.*, 1960; Damon, 1970; Bucha, 1970a,b). Bucha (1970a) concluded that ^{14}C production was increased when the geomagnetic moment was decreased. The causative relationship probably is that there is a greater penetration of solar and cosmic radiation when the Earth's magnetic field is weaker and hence a greater interaction between the solar wind and the atmosphere, resulting in greater ^{14}C production. Analyses of carbon isotopes in tree rings (Stuiver, 1965, 1971; Suess, 1965, 1971; Damon *et al.*, 1966; Ferguson, 1970) show very short-term 'wiggles' which initially were thought to be instrumental or due to

---

Figure 9.22 Palaeoclimatic indicators. A wide range of geological materials can be used to infer the ancient palaeoclimate and there is a strong correlation between their palaeolatitude, determined from their palaeomagnetic properties, and their likely mode of formation or origin. Unfortunately, no one palaeoclimatic indicator has sufficiently high precision to determine the true axiality of the axial geocentric dipolar mode for the geomagnetic field, but the fact that all such palaeoclimatic indicators are consistent with this model means that palaeomagnetic data can be used to infer the palaeoclimatic conditions for specific rock types of fossil flora and fauna.

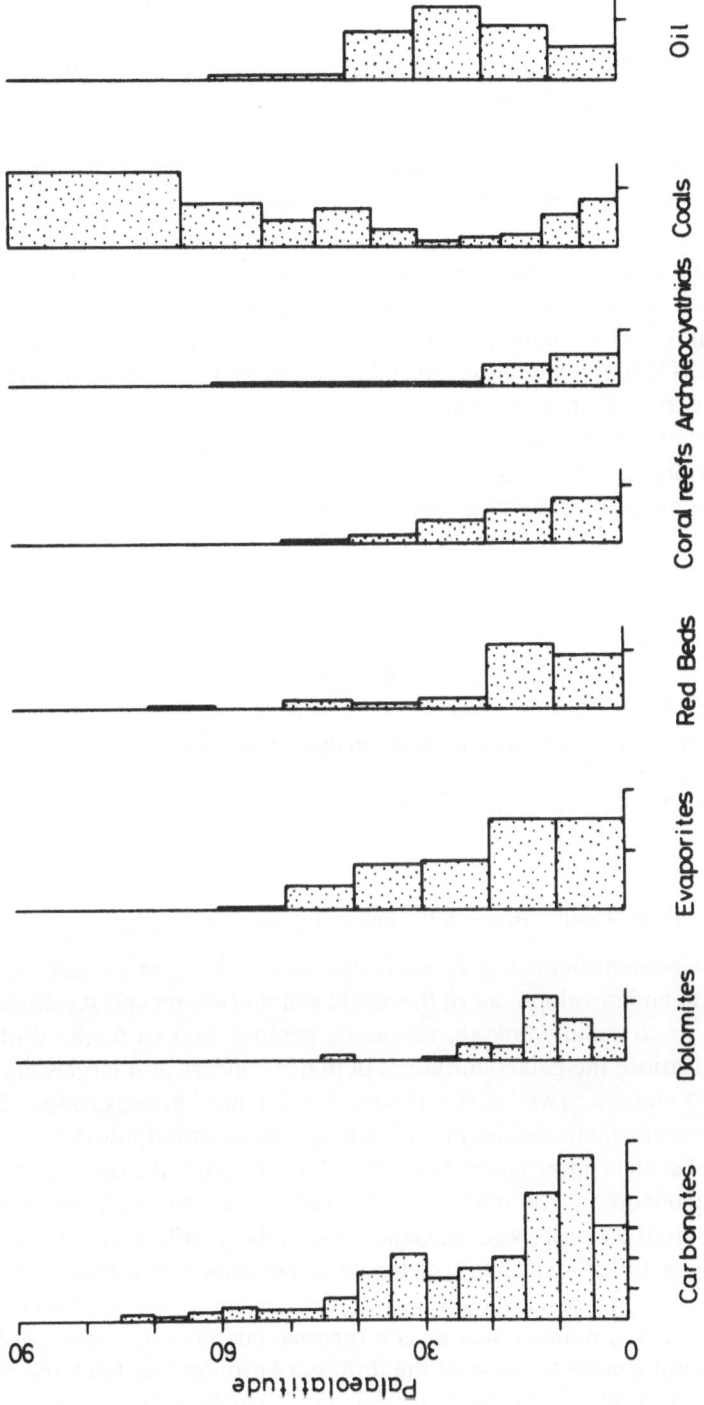

diffusion problems, except by Suess (1970a,b), but their reality is now clearly established (De Jong *et al.*, 1979) although their cause is uncertain. An obvious possibility for such fluctuations could be the incidence of magnetic storms, but there is no clear evidence for a sun-spot cycle in the records (Damon *et al.*, 1973) and the replenishment of the ionosphere apparently takes place rapidly, within 14 days, of a magnetic storm (Kersley *et al.*, 1978). It is, nonetheless, generally considered that such 'wiggles' are caused by fluctuations in the atmospheric reservoir (Tans *et al.*, 1979) and may thus be related to influences on the ionosphere, but variations in the properties of the internally generated geomagnetic field appear to be on much too long a time-scale to be a cause for such brief fluctuations. However, the long-term ^{14}C trends may be influenced by geomagnetic changes (Fig. 8.6), although Bucha's (1970a) evaluation must be considered only a very preliminary attempt to explain such trends. The decrease in the total moment in the geomagnetic field is only poorly known during the last 400 years, and mostly only for the last 100 years (Section 8.3) and there is no clear evidence that this is part of a sinusoidal variation with a periodicity of 8000 years – other than the fact that it would appear to fit with the ^{14}C variations. Unfortunately, palaeointensity determinations are still few for archaeological times and also concentrated in particular locations (Section 7.2), making any evaluation of the long-term behaviour of the total geomagnetic field strength distinctly uncertain, despite Bucha's (1967, 1970a,b) attempts. It may, nonetheless, be no coincidence that the long-term periodicity in ^{14}C, some 8000 years, is comparable to the periodicity in geomagnetic secular variation, and it could also be that the ^{14}C variations are the most informative evidence for changes in the total magnetic moment of the geomagnetic field.

9.7.3 Palaeontology, palaeolatitudes and palaeogeography

The palaeoenvironment is fundamental to any attempt to understand the evolution and development of the world's biota (Raven and Axelrod, 1974), as well as to sedimentology, economic geology and so forth. Within this generalization, the palaeolatitude is of major concern as it largely dominates the likely climate as well as the nature of sedimentation and erosion. The fact that palaeomagnetic studies provide a basis for palaeolatitude determinations is thus of critical importance (Fig. 9.22). Furthermore, the combined analysis of the palaeomagnetic data with all available geological data enables an evaluation of the past size of the land masses, the distribution of land and sea, and thus a fuller assessment of the total environment within which other factors can then be examined (Fig. 9.23). For example, the oceanic circulation patterns have a major effect on the regional climate and these can be estimated from a consideration of the different latitudes and land–sea distributions (Robinson, 1973). Such models can then be tested in terms of the

290

predictions which can then be made for the nature of sedimentation and the flora and fauna in different regions.

Palaeontological studies of the distribution of various fossil species demonstrated, even in the 19th century, the need for most of the major continents to have been contiguous at various times during the last 300 million years – long before the palaeomagnetic dating of the ocean floors and the re-positioning of continents into the previous relationships. However, such evaluations are necessarily highly subjective as the distribution and recognition of species of flora and fauna is difficult and often influenced by consideration of which characteristics can be used to distinguish one species from another. Terrestrial fauna are more difficult to study than marine because of the poor preservation conditions on land, and thus a paucity in the available number for study and many areas of the world for which there is no record as the appropriate preservation conditions did not exist. To a large extent, therefore, palaeontological and palaeomagnetic studies are complementary to each other, with palaeomagnetic methods giving a greater precision in the location of different areas, and the fossil distributions indicating the relationships between them. Two land areas can, for example, be close to each other, but their fauna could be very different because of the existence of a sea barrier between them – as in the case of Wallace's Line in the present western Pacific (Whitmore, 1982). Despite such possibilities, the fossil data usually provide a vital check on the palaeomagnetic positioning and can, for example, indicate that the areas were not contiguous and therefore the palaeomagnetic data from one of the areas concerned may be suspect. There are numerous examples of such combined studies, usually by palaeontologists plotting fossil distributions on palaeomagnetic reconstructions, for example Middlemiss *et al.* (1971); Tarling and Runcorn (1973); Ross (1974); Hallam (1973); Tarling (1979, 1980a); Panchen (1980) and Turner and Tarling (1982). Unfortunately, in some of these cases, the authors have tended to suspect the palaeontological data rather than the palaeomagnetic data, although the converse

---

Figure 9.23 (Overleaf)   Phanerozoic reconstructions. For the last 200 million years, reconstructions of the past relationships of the continents can be determined from the palaeomagnetism of the ocean floor. These allow the positions of the continents to be determined relative to each other by comparing magnetic anomalies of the same age, and these reconstructions can then be adjusted to their palaeolatitude positions using continental palaeomagnetic data. For older times, the reconstructions can only be based on the continental palaeomagnetic data and then tested against the geological evidence. (a) Eocene, c. 50 million years ago; (b) Cretaceous, c. 100 million years; (c) Jurassic, c. 150 million years; (d) Upper Carboniferous, c. 300 million years; (e) Lower Devonian, c. 400 million years, and (f) Cambrian, c. 550 million years. The dotted outlines on the older reconstructions correspond to the estimated locations of Manchuria, China, Sundaland (southeast Asia) and Central Asian blocks now just north of India.

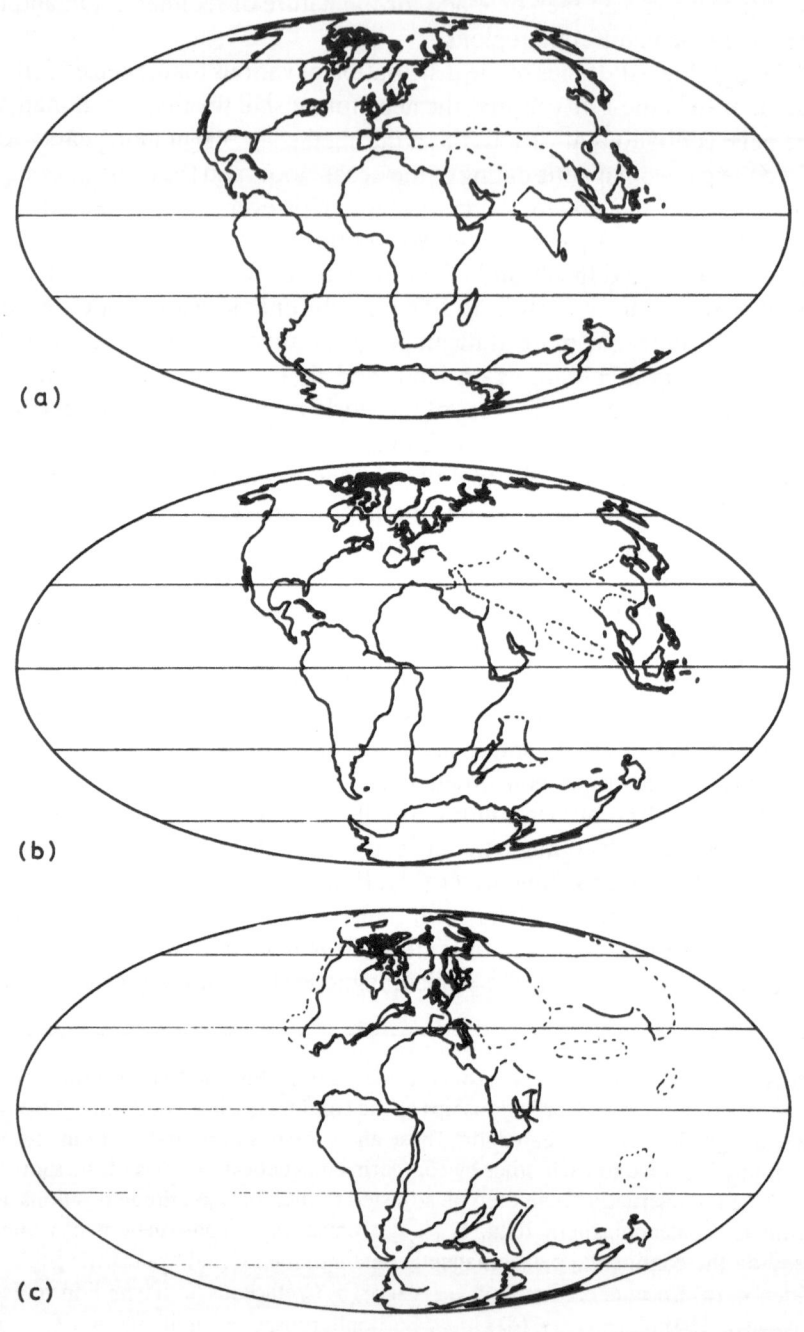

(a)

(b)

(c)

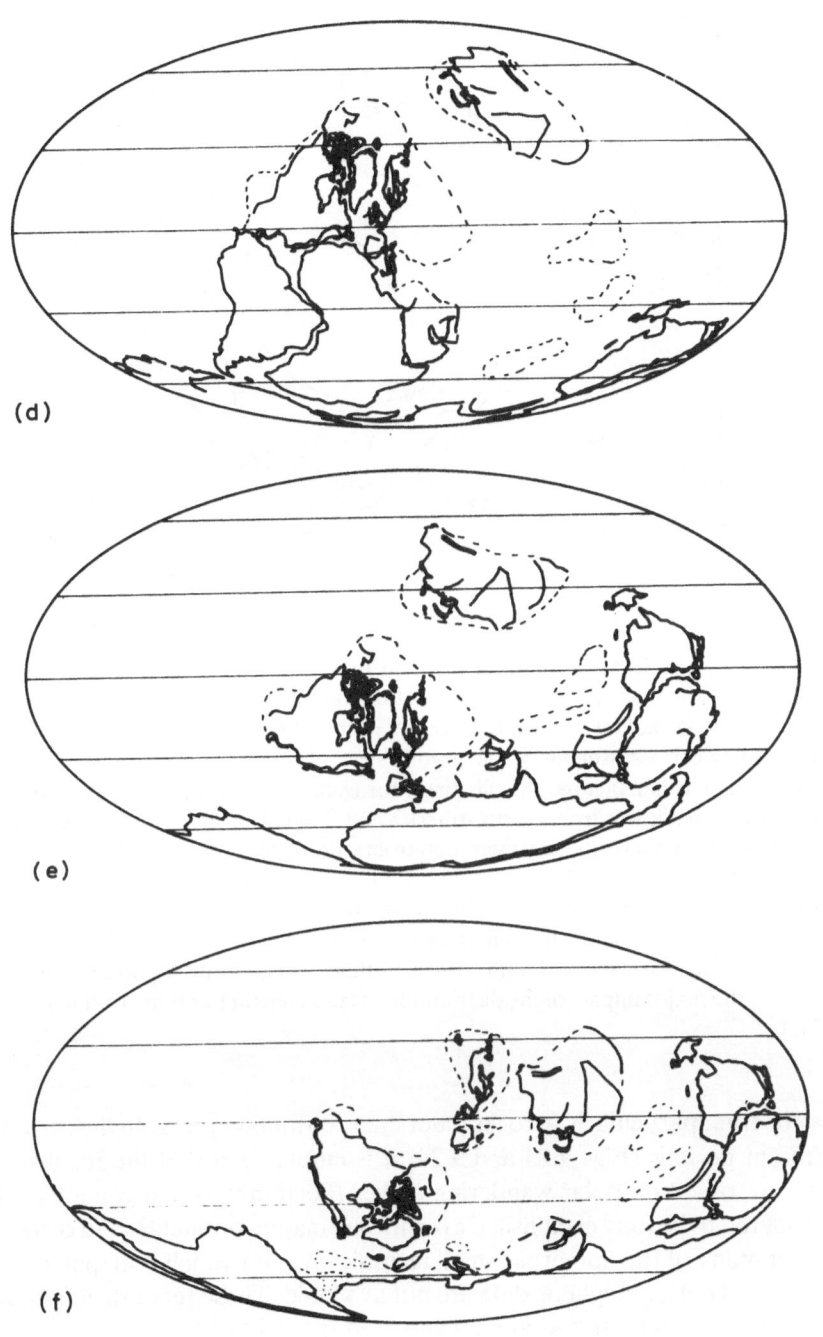

(d)

(e)

(f)

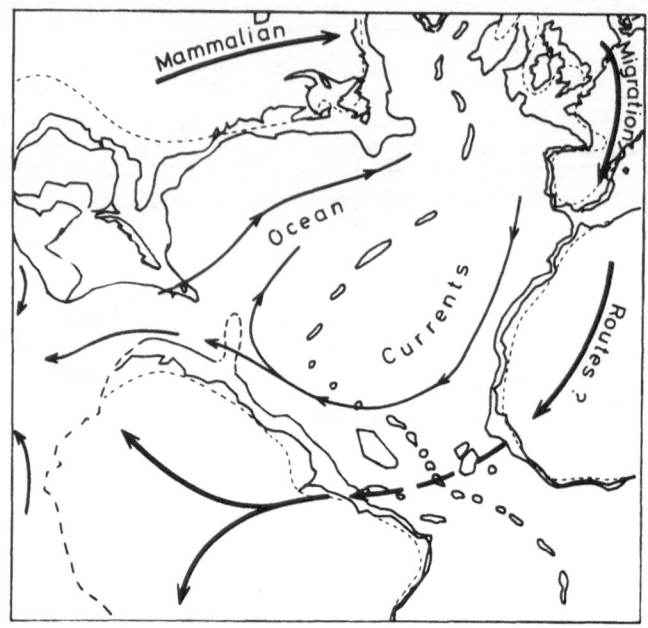

Figure 9.24 The Central–South Atlantic in Lower Oligocene Times, c. 35 million years ago. While palaeomagnetic studies of the oceans and continents can locate the position of continental blocks relative to each other for different periods, any palaeontological assessment of the significance of such positions must take into account many other factors. At this time, for example, South America was further from Africa than it was from North America, yet it seems probable that islands in the Atlantic allowed relatively easy migration from Europe to Africa and South America. An assessment of the probable oceanic circulation system for this time would also indicate that floating materials from North America would be less likely to reach South America than would debris floating from Africa. Terrestrial migration was thus more likely between Africa and South America than from North to South America, reflecting the major impact of the distribution of land areas at this time. (After Tarling, 1981.)

may be true, particularly in rocks older than 300 million years. In the maps for different periods (Figs 9.23 and 9.24) it is emphasized that the reconstructions are based on polar wandering curves (Section 9.2) that cannot yet be considered to be fully established even for the major continents, and certainly not for many of the numerous small tectonic units for which both palaeontological and palaeomagnetic data are not available. These reconstructions are thus intended to be tested and amended in the light of both newer palaeomagnetic data and palaeontological–palaeoclimatic evidence.

Such correlations clearly indicate that the Earth's climatic zonation has tended to be broadly similar through most of the last 2 billion years at least (Nairn, 1964, 1972; Furon, 1972; Frakes, 1979). This is to be expected, of

Geological applications

course, because of the dominance of solar insolation. However, the situation is extremely complex when examined further. The influence of the distribution of land and sea can drastically alter the climatic regime in different areas. The very existence of ice sheets at the poles may also be a major constraint on the climatic pattern (Tarling, 1978c). Comparison with other planetary atmospheres suggests, for example, that the removal of the present ice sheets could lead to the development of a low-pressure polar region instead of our present high-pressure region. At times when there were no ice sheets, such as during the Jurassic–Cretaceous periods, it seems probable that the world's atmospheric circulation pattern could have been significantly different from that today, although still preserving the basic latitudinal variations. Clearly many such considerations are, at best, speculative, but greater control on the distribution of the major continental land masses, supplemented by mapping of the land–sea patterns upon them, seems likely to lead to a much greater understanding of both past and present climatic controls.

The advance and retreat of ice sheets during the last 3–5 million years are clearly unrelated directly to plate tectonic movements as these are extremely small and consistent during this time, as indicated by the ocean-floor magnetic anomalies. There seems little doubt, therefore, that the fundamental control on such advances and retreats is related to regular cycles in the obliquity of the Earth's axis to the solar plane (Milankovitch, 1938) and it is possible to construct a time-scale on this basis (Evans, 1971) that is apparently consistent with the geological, palaeoclimatic and palaeomagnetic (Idnurm and Cook, 1980) observations. The conditions within which such Milankovitch cycles become important are less clear and it seems probable that specific conditions of land–sea distributions must exist for such ice sheets to be initiated. Once initiated, their short term (10^3–10^4 years) evolution is controlled by astronomical factors. One criterion appears to be for a major land mass to be in high latitudes, as in Antarctica today, or for an oceanic area to be virtually landlocked within high latitudes, as in the present Arctic. However, neither situation is the sole prerequisite as Antarctica has lain in high polar latitudes for most of the last 300 million years, according to the palaeomagnetic evidence, and the onset of ice sheet glaciation (as distinct from local mountain glaciation) appears to have only followed the development of circum-Antarctic oceanic circulation pattern, i.e. following the separation of Australia from Antarctica some 48 million years ago, and then the final separation between South America and the Antarctic Peninsula during Oligocene times that eventually opened sufficiently to allow continuous circulation within the Miocene (c. 15 million years ago). The later initiation of ice sheet glaciation in the northern hemisphere appears to depend on the entry of Atlantic waters into the Arctic region as a result of the gradual opening of the North Atlantic and the sinking of the Iceland–Faeroes Ridge (Tarling, 1978c). Similar considerations indicate that earlier ice sheet glaciations need extensive land area (or enclosed ocean basins) in specific palaeogeographic

295

conditions whereby moist air is available to initiate the ice sheet, after which it becomes self-governing, until the constraining conditions are relaxed, after which the ice sheet stagnates and disappears rapidly. The main 'problematic' ice sheet evidence is that for the Eocambrian, some 600 million years ago (Harland, 1964; McElhinny *et al.*, 1974; Harland and Herod,1975). Various explanations have been suggested for the apparent synchroneity – changes in the obliquity of the Earth's axis of rotation (Williams, 1974a,b, 1975, 1981), rapid polar wandering so that most areas were in high polar latitudes during a fairly short time span (Piper, 1973), or world-wide glaciation (Spencer, 1971; Tarling, 1974b) – or, of course, that they were not synchronous anyway (Crawford and Daily, 1971). The latter is particularly important as such levels provide useful inter-continental chronostratigraphic markers if they are synchronous (Morris, 1977).

More economically interesting is the evidence for palaeolatitude control over certain economically important reserves, such as evaporites and lead–zinc stratiform ores (Section 9.4), and within this same context, hydrocarbons (Bailey and Stoneley, 1981; Stoneley, 1981; Stoneley and Bailey, 1981). Unfortunately the source rocks of hydrocarbons are only rarely known for certain, such as the Kimmeridge Shales in the northern North Sea or the algal limestones in the Michigan Belle River Mills field (Gill, 1973) and so most studies have necessarily concentrated on the palaeolatitude of the reservoir rocks within which the hydrocarbons are now found (Irving and Gaskell, 1962; Deutsch, 1965; Irving *et al.*, 1974; Moody, 1975). On the assumption that hydrocarbons become mobilized shortly after their formation, they will tend to be trapped within formation of broadly similar age to their source rocks (with many notable exceptions, such as in western Venezuela, parts of Australia, and Libya) and the palaeolatitude distribution thus strongly favours the formation of hydrocarbons in low latitude palaeogeographic environments (Tarling, 1973a), almost certainly closely similar to the sabkha (salty marshes bordering tropical latitude seas) conditions for major evaporite formation (Irving and Briden, 1962; Kinsman, 1969).

9.8 EXTRATERRESTRIAL STUDIES

The study of the magnetic fields of the other planets is fundamental to any understanding of their origin, and important in evaluating the possible processes operating within the Earth which are causing the observed geomagnetic behaviour (Chapter 8). Most of this information has been obtained by satellite probes carrying magnetometers, although studies of meteorites and lunar samples can be carried out directly. Such direct studies are particularly critical as they indicate the nature of past magnetic fields and are thus essential to any studies of the interactions with the solar wind and hence atmospheric and climatic evolutionary processes on other planets (Cordell, 1980) as well as on the Earth (Section 9.7). Studies of lunar samples, in particular, have demonstrated the existence of a previous intrinsic magnetic field that no longer exists on the Moon.

Lunar magnetic studies are unique in combining both remote sensing and hand samples, although it seems probable that the Russians may soon remotely acquire and return samples from Mars and so allow their magnetic properties to be measured directly. The early lunar fly-by and orbiter missions carried magnetometers which indicated that the Moon had virtually no dipolar magnetic field (Russell *et al.*, 1974). Studies of the magnetization induced by the solar wind and in the magnetic 'tail' of the Earth also indicated very low permeability, consistent with predominantly paramagnetic rocks, with only an extremely weak ferromagnetism. In contrast, lunar samples show distinct ferromagnetic properties, mainly associated with multidomain iron in the basalts, with a very small contribution from nickel and cobalt, and intensities of remanence of the order of 10^{-6} A m^2 kg^{-1} (Strangway *et al.*, 1970; Pearce *et al.*, 1971; Collinson *et al.*, 1973). The breccias, with a somewhat higher iron content, are more strongly magnetized, c. 10^{-4} A m^2 kg^{-1}, and the ferromagnetism is predominantly associated with single and multidomain iron. Although the lunar samples were unoriented their magnetization indicated uniformity over at least a few cubic centimetres. Magnetometers also indicated that this uniformity of magnetization must extend to at least some 1 km depth (Dyal *et al.*, 1970; Fuller, 1974) while magnetometers carried by *Apollo 15* and 17 sub-satellites showed areas of consistent magnetization, but also rapid changes of magnetization over distances of c. 1 km. These magcons (magnetic concentrations) are particularly associated with certain far-side craters (Anderson and Wilhelms, 1979; Hood *et al.*, 1981), such as Van de Graaff (Fig. 9.25) and Aitken, and can only be interpreted in terms of lunar surface rocks possessing uniform magnetizations extending to depths greater than 1 km. As the Moon no longer possesses an intrinsic dipolar magnetic field, various models have been proposed to account for such uniform magnetizations during the first billion years or so of the Moon's development. These can be broadly classified into three main groups: those external to the Moon, those of local lunar origin, and those of a global scale and internal to the Moon, although none of these explanations can be considered fully satisfactory (Daily and Dyal, 1979).

1. External models involve the lunar surface cooling and acquiring a thermal remanence in the magnetic field of either the Sun (Nagata *et al.*, 1972) or the Earth (Alfvén and Lindberg, 1974). In the case of the Sun, it is widely considered that it went through a T-tauri stage (Tarling, 1978b), characterized by rapid mass loss and intense solar magnetic fields. However, this stage is relatively short and would not persist throughout the period of mare basalt formation and, in any case, is generally postulated to account for the distribution of the angular momentum in the solar system and differentiation in the gas/dust from which the solar system formed. On this basis, the T-tauri stage would be completed long before

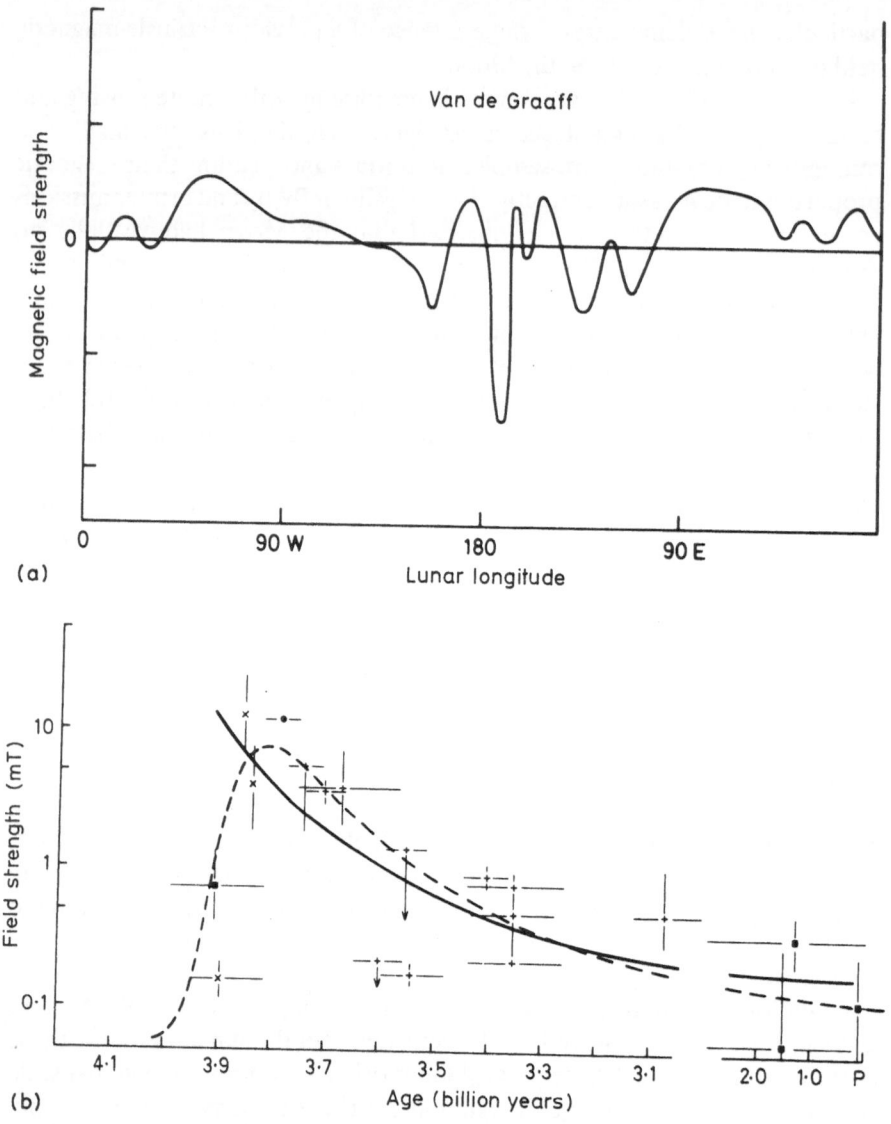

Figure 9.25 Extra-terrestrial applications. (a) The magnetization of the lunar crust is clearly demonstrated by the anomalies associated with particular craters, such as Van de Graaff, which are termed magcons (magnetic concentrations). (b) Samples of lunar rocks of varying age indicate a consistent decrease in the intensity of the lunar field, possibly following an initial rise (after S. M. Cisowski, personal communication). (c) The extra-terrestrial magnetic field could be quite strong if the crust were (ii) uniformly magnetized in, for example, a very strong solar field. The decrease in the lunar field suggests that a more likely explanation is (i) that an initial dynamo in the core gave rise to a magnetization of the crust. After the dynamo ceased, the crust remained magnetized and, where cratered, gives rise to magcons, but only a very weak external field would be observed.

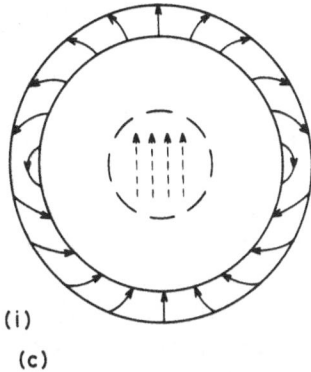

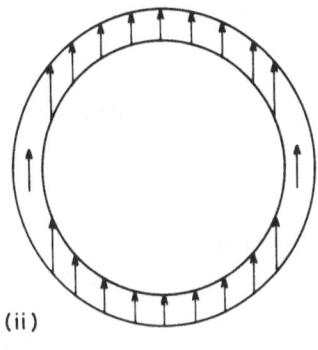

(c)

(i)

(ii)

the eruption of the first mare basalts, 3900 million years ago. The Earth's magnetic field, at the time of the eruption of the lunar mare basalts (pre- and early Archaean times) is thought to have been of similar magnitude as today (Section 8.5) while the Moon cannot have been nearer than some 2.5–3.0 Earth radii, the Roche limit, as otherwise both the Earth and Moon would have been largely disrupted by mutual tidal interactions. This means that the terrestrial field in the nearest possible orbit of the Moon must have been in much too weak a field to account for the observed intensities of magnetization.

2. Local pockets of molten iron may have existed at the time of formation of the lunar mare basalts and could have formed sources of local magnetic fields by means of

    (i) Dynamo action (Pearce *et al.*, 1972; Murthy and Banerjee, 1973),
    (ii) Induction effects related to much stronger solar winds (Sonett and Colburn, 1967; Schwartz *et al.*, 1969), or
    (iii) Thermoelectric currents (Dyal *et al.*, 1977).

All these processes are poorly understood, but would appear to be insufficient to account for the observed intensity of magnetization. In addition, the lunar crust appears to be much more uniformly magnetized and relatively few craters actually show associated magnetic anomalies – including most of the mascons (mass concentrations – Solomon and Head, 1980) in which most of these processes would, presumably, be enhanced. On an even more local scale, the impact of meteorites could give rise to shock-induced piezoremanence (Section 2.2), particularly in the lunar soils and breccias (Nagata *et al.*, 1970, 1971; Cisowski *et al.*, 1973, 1977). While such effects may be extremely important on a local scale (Lin, 1979), it seems unlikely that they can account for the scale of uniformity of magnetization that has been indicated by surface and satellite observations.

3. Postulations for the existence of an early lunar magnetic field also require the Moon to have a fundamentally iron core, albeit small. Such a core could have become magnetized in the very early history of the Moon during the T-tauri stage of solar evolution by either 'deposition' of iron or isothermally (Runcorn and Urey, 1973; Strangway *et al.*, 1973). While this original core magnetization must have decayed, the Moon's crustal shell could have acquired a remanent magnetization while cooling in this primordial field. However, for most plausible circumstances, the shell would still possess a net residual dipolar moment although too weak to be detected in the currently available data (Stephenson, 1976a). Alternatively, if the core was originally molten, it could have generated a lunar dipolar field by dynamo action in a similar manner to that of the Earth. This would require most of the Moon to become at least close to its melting point to form a liquid iron core, as is also suggested by the extent of differentiation required to create the lunar highlands. As the Moon evolved, the dynamo could undergo reversals but would eventually cease as the core solidified. The lunar seismic data are generally interpreted to indicate the absence of a core (Toksoz, 1979) although it is possible that the epicentres have been mislocated and revised positions for meteorite impacts could be consistent with the existence of a core (Sellers, 1979). A systematic decay in the intensity of the lunar field is, in fact, tentatively indicated (Fig. 9.25) by palaeointensity determinations of lunar samples of different age (Stephenson *et al.*, 1975). However, the weakness of the remanence in most lunar samples and the problems of preventing chemical changes during palaeointensity determinations (Section 5.6) mean that there are often large apparent differences in palaeointensity determinations even from rocks of similar age and type. Such a model would, however, account for most of the observed features. The magcons would then be explained in terms of a magnetized shell (Fig. 9.25), punctured by craters (Runcorn, 1975, 1978), although it is still not clear why more craters are not characterized by magnetic anomalies. Nonetheless, such a model places very severe restraints on all postulates for the early development of the Moon and hence all the other planets.

The magnetic fields of the other planets have not yet been fully established (Russell, 1980). Neither Mars nor Venus has a clearly defined magnetic field, although they both have a magnetosphere in which the interaction of the solar wind with their atmospheres results in the generation of external currents which dominate any intrinsic field. The total magnetic moment of *Venus* must be less than 10^{19} A m^2 (Russell, 1980) while that of *Mars* cannot exceed 10^{18} A m^2 (Dolginov *et al.*, 1976; Russell, 1979). Samples of surface materials are required to establish whether these terrestrial planets previously had intrinsic magnetic fields in the same way as the Moon. *Mercury*, however, still has a clear internal field, with dipole and quadrupole components, and a moment between $2\text{--}3 \times 10^{19}$ A m^2 (Whang, 1977; Ng and Beard, 1979) and 6×10^{19} A m^2 (Slavin and Holzer, 1979) that could either be generated by internal processes within its core or a residual field left in the crustal shell (Stephenson, 1976b) created by a now defunct core dynamo. The surface field would therefore be only about 1% of that of the Earth (Ness, 1979). It has been known for many decades that *Jupiter* has a magnetic field as its radio emissions are polarized. Its magnetic moment is 1.55×10^{27} A m^2, with strong quadrupole and octopole components (Acuña and Ness, 1975; Smith *et al.*, 1976). There is also some evidence for the occurrence of secular variations (Hide and Malin, 1979) as Jupiter's field changes between different fly-by missions. *Saturn* is also now known to have similar radio emissions to Jupiter and a highly dipolar magnetic moment of some 4.7×10^{25} A m^2 (Smith *et al.*, 1976).

Meteorites have been available for magnetic study long before space travel commenced (Stacey *et al.*, 1961; Weaving, 1962; Gus'kova, 1963) yet surprisingly little is still known about the processes by which they have acquired their remanence. Such studies are critical to many studies of the physical conditions during the early formation of the solar system as isotopic studies indicate that these mostly formed during the very early stages of condensation of the solar system and the carbonaceous chondritic meteorites are generally thought to be the most accessible samples of the most primitive solar system (Larimer, 1978; Marvin and Wasson, 1979; Zaikowski, 1980). About 90% of observed meteorite falls are of stony meteorites, aerolites, with iron meteorites (almost entirely iron and nickel) forming the remaining 10%. Some 90% of the aerolites contain small rounded grains, chondrules, whose composition and isotope ratios show that these are the most primitive forms of meteorite. The origin of meteorites is, however, uncertain. They may have been derived from proto-planetary bodies or even disintegrated planets within the asteroid belt or from the vicinity of Earth–Mars orbits (Anders, 1964; Wetherill, 1976). The solution of their origin could well be provided from an understanding of the way in which they have acquired their remanences and thus establish the nature of magnetic fields during this critical period of planetary formation. All meteorites may have a strongly magnetized outer skin which acquired a thermal remanence while entering the Earth's atmosphere. In the case of the Allende meteorite, which fell in 1969, the crust is thin, c. 1 mm, but is an order of magnitude more strongly magnetized than the interior (Butler, 1972).

The chondritic meteorites can be divided into five main chemical categories although the total iron content remains approximately constant. These chemical classifications can be readily defined by their saturation moments (Nagata, 1979) which decrease systematically from E to H, L, LL and C types, reflecting changing nickel:iron ratios, and in the case of the C type (carbonaceous) the occurrence of magnetite. It is, in fact, the carbonaceous chondritic meteorites that show the highest stability of remanence to alternating magnetic fields, followed by most achondrites. In general, the remaining non-carbonaceous chondrites tend to be magnetically unstable. Palaeo-intensity studies necessarily assume that the remanence was acquired thermally and mostly yield palaeofields of the order of 10^{-4} T for the chondrites and 10^{-5} T for the achondrites (Nagata, 1979). There also appear to be distinct differences between the magnetization carried by the chondrules and that of the surrounding meteoritic material as Banerjee and Hargraves (1972) found consistent components in standard-sized samples only at temperatures below 130–150°C, while studies of the chondrules themselves indicate blocking temperatures in excess of 500°C and these have apparently cooled in a field between 2 and 7×10^{-4} T (Lanoix et al., 1978), but there is only consistency between the magnetization of individual chondrules at low temperatures (Sugiura et al., 1979).

Many more studies are still required before a proper assessment of the available meteoritic data can be made. It certainly appears that the primitive chondrules cooled in a strong stellar field prior to their incorporation into meteorites, and that these acquired a remanence in a lower ambient field, while the achondritic meteorites acquired a magnetization in even weaker fields. The fields indicated at the time of formation of the chondrules seem to be extremely high for feasible solar magnetic fields in the orbits of the terrestrial planets, yet the components appear to be too uniform to have been acquired during lightning discharges during their cooling – various other alternatives have been suggested (Sugiura et al., 1979). Shock reheating often appears to be of major importance (Brecher and Fuhrman, 1979) but it is premature to consider any one explanation as being satisfactory at this stage of investigation.

References

Abouzakhm, A. G. (1974). *Magnetic anisotropy studies in the Whin Sill.* M.Sc. dissertation, University of Newcastle upon Tyne.

Abouzakhm, A. G. and Tarling, D. H. (1975). Magnetic anisotropy and susceptibility from northwestern Scotland. *J. Geol. Soc., Lond.*, **131**, 647–652.

Abrahamsen, N. (1967). Some paleomagnetic investigations in the Faroe Islands. *Medd. Dansk, Geol. For.*, **17**, 371–384.

Abrahamsen, N. (1973). Archaeomagnetic tilt correction on bricks. *Archaeometry*, **15**, 267–274.

Abrahamsen, N. and Knudsen, K. L. (1979). Indication of a geomagnetic low-inclination excursion in supposed Middle Weichselian interstadial marine clay at Tubjerg, Denmark. *Phys. Earth Planet. Ints*, **18**, 238–246.

Abrahamsen, N. and Readman, P. W. (1980). Geomagnetic variations recorded in older (≥23 000 BP) and younger *Yoldia* Clay (~14 000 BP) at Nørre Lyngby, Denmark. *Geophys. J. R. astr. Soc.*, **62**, 329–344.

Acuña, M. H. and Ness, N. F. (1975). Jupiter's main magnetic field measured by Pioneer 11. *Nature*, **253**, 327–328.

Addison, F. T. (1982a). *A magnetic study of diagenesis in carbonate sediments.* Ph.D. thesis, University of Newcastle upon Tyne.

Addison, F. T. (1982b). A magnetic study of diagenesis in the Pendleside Limestone Group. (Abs.). *Geophys. J. R. astr. Soc.*, **69**, 291.

Ade-Hall, J. M. (1964). A correlation between remanent magnetism and petrological and chemical properties of Tertiary basalt lavas from Mull, Scotland. *Geophys. J. R. astr. Soc.*, **8**, 403–423.

Ade-Hall, J. and Wilson, R. L. (1963). Petrology and natural remanence of the Mull lavas. *Nature*, **198**, 659–660.

Ade-Hall, J. and Wilson, R. L. (1969). Opaque petrology and natural polarity in Mull (Scotland) Dykes. *Geophys. J. R. astr. Soc.*, **18**, 333–352.

Ahoroni, A., Frei, E. H. and Schiber, M. (1962). Curie point and origin of weak ferromagnetism in haematite. *Phys. Rev.*, **127**, 439–441.

Aitken, M. J. (1958). Magnetic dating. *Archaeometry*, **1**, 16–20.

Aitken, M. J. (1974). *Physics and Archaeology*, 2nd Edn, Clarendon Press, Oxford, p. 291.

Aitken, M. J. and Hawley, H. N. (1967). Archaeomagnetic measurements in Britain – IV. *Archaeometry*, **10**, 129–134.

Aitken, M. J., Alcock, P. A., Bussell, G. D. and Shaw, C. J. (1981). Archaeomagnetic determination of the past geomagnetic intensity using ancient ceramics: Allowance for anisotropy. *Archaeometry*, **23**, 53–64.

Alfvén, H. (1950). *Cosmical Electrodynamics*, Clarendon Press, Oxford.

Alfvén, H. and Lindberg, L. (1974). Magnetization of celestial bodies with special application to the primeval Earth and Moon. *Moon*, **10**, 323–325.

Al-Hashimi, W. S. and Hemingway, J. E. (1973). Recent dolomitization and the origin of the rusty crusts of Northumberland. *J. Sedi. Petrol.*, **43**, 82–91.

303

Palaeomagnetism

Allen, J. R. L. (1969). Some recent advances in the physics of sedimentation, *Proc. Geol. Ass.*, **80**, 1–42.

Al'thausen, O. N. (1949). On certain regularities in production of thermoremanence magnetism in magnetites. *Izv. Akad. Nauk SSSR, Ser. Geogr. Geofiz.*, **13**, 359–362.

Alvarez, W., Arthur, M. A., Fischer, A. G., Lowrie, W., Napoleone, G., Silva, I. P. and Roggenthen, W. M. (1977). Upper Cretaceous–Palaeocene magnetic stratigraphy at Gubbio, Italy: V type section for the Late Cretaceous Palaeocene geomagnetic reversal time scale. *Geol. Soc. Amer. Bull.*, **88**, 383–389.

Amerigian, C. (1974). Sea-floor dynamic processes as the possible cause of correlations between paleoclimate and paleomagnetic indices in Deep-sea sedimentary cores. *Earth Planet. Sci. Letters*, **21**, 321–326.

Amerigian, C., Watkins, N. D. and Ellwood, B. B. (1974). Brunhes Epoch geomagnetic secular variation on Marion Island: Contribution to evidence for a long-term regional geomagnetic secular variation maximum. *J. Geomagn. Geoelect.*, **26**, 429–441.

Anders, E. (1964). Origin, age and composition of meteorites. *Space Sci. Rev.*, **3**, 583–714.

Anderson, J. M. and Schwyzer, R. U. (1977). The biostratigraphy of the Permian and Triassic, Part 4 – Palaeomagnetic evidence for large-scale intra-Gondwanan plate movements during the Carboniferous to Jurassic. *Trans. Geol. Soc. S. Afr.*, **80**, 211–234.

Anderson, K. A. and Wilhelms, D. E. (1979). Correlation of lunar farside magnetized regions with ringed impact basins. *Earth Planet. Sci. Letters*, **46**, 107–112.

Anderson, R. N., Spariosu, D. J., Weissel, J. K. and Hayes, D. E. (1980). The interrelation between variations in magnetic anomaly amplitudes and basalt magnetization and chemistry along the Southeast Indian Ridge. *J. Geophys. Res.*, **85**, 3883–3898.

Aramaki, S. and Akimoto, S. I. (1957). Temperature estimation of pyroclastic deposits by natural remanent magnetism. *Amer. J. Sci.*, **255**, 619–627.

Arendse, M. C. (1978). Magnetic field detection is distinct from light detection in the invertebrates *Tenebrio* and *Talitrus*. *Nature*, **274**, 358–362.

Argand, E. (1924) La Tectonique de l'Asie. *Proc. Internat. Geol. Cong.*, **13**, p. 171.

As, J. A. (1960). Instruments and measuring methods in palaeomagnetic research. *Med. Verh., Kon. Ned, Meteorol. Instit.* No. 78, p. 56.

As, J. A. and Zijderveld, J. D. A. (1958). Magnetic cleaning of rocks in palaeomagnetic research. *Geophys. J. R. astr. Soc.*, **1**, 308–319.

Aspinall, A., Feather, S. W. and Renfrew, C. (1972). Neutron activation analysis of Aegean obsidians, *Nature*, **237**, 333–334.

Atwater, T. (1970). Implications of plate tectonics for the Cenozoic tectonic evolution of western North America. *Geol. Soc. Amer. Bull*, **81**, 1513–1536.

Auvray, B., Mace, J., Vidal, P. and Van der Voo, R. (1980). Rb-Sr dating of the Plouézic volcanics, N. Brittany: implications for the age of red beds (Series rouges) in the northern Armorican Massif. *J. Geol. Soc., Lond.*, **137**, 207–210.

Baag, C. G. and Helsley, C. E. (1974). Evidence for penecontemporaneous magnetization of the Moenkopi Formation. *J. Geophys. Res.*, **79**, 3308–3320.

Bacon, F. (1605). *The Advancement of Learning* (ed. G. W. Kitchin), Dent, London, p.246.

Bacon, R. (1267). *Opus Tertium*.

Badham, J. P. N. (1981). The origins of ore deposits in sedimentary rocks. In *Economic Geology and Geotectonics* (ed. D. H. Tarling), Blackwell, Oxford, pp. 149–191.

Baer, A. J. (1979). An unconventional Proterozoic palaeomagnetic polar wander path for North America, and its advantages. *Phys. Earth Planet. Ints*, **19**, 100–105.

Bagina, O. L. and Petrova, G. N. (1977). Determination of palaeomagnetic field intensity using anhysteretic magnetisation. *Phys. Earth Planet. Ints*, **13**, 360–362.

Bagnold, R. A. (1966) An approach to the sediment transport problem from general physics. *Prof. Papers, US Geol. Surv.*, **422–I**, p. 37.

Bailey, R. J. and Stoneley, R. (1981). Petroleum: Reservoirs and conclusions. In *Economic Geology and Geotectonics* (ed. D. H. Tarling), Blackwell, Oxford, pp. 73–98.

Balsley, J. R. and Buddington, A. F. (1958). Iron–titanium oxide minerals, rocks and aero-magnetic anomalies of the Adirondack area, New York. *Econ. Geol.*, **53**, 777–805.

Banerjee, S. K. (1971). Characteristic difference between TRM and CRM in ultrafine haematite, *Eos*, **52**, 191–192.

Banerjee, S. K. (1980). Magnetism of the oceanic crust: evidence from ophiolite complexes. *J. Geophys. Res.*, **85**, 3556–3566.

Banerjee, S. K. and Hargraves, R. B. (1972). Natural remanent magnetization of carbonaceous chondrites and the magnetic field in the early solar system. *Earth Planet. Sci. Letters*, **17**, 110–119.

Banerjee, S. K., Lund, S. P. and Levi, S. (1979). Geomagnetic record in Minnesota lake sediments – Absence of the Gothenburg and Erieau excursions. *Geology*, **7**, 588–591.

Banerjee, S. K. and Mellema, J. P. (1974). A new method for the determination of palaeointensity from the ARM properties of rocks. *Earth Planet. Sci. Letters*, **23**, 177–184.

Banerjee, S. I. and O'Reilly, W. (1967). The behaviour of ferrous ions in iron–titanium spinels. *J. Phys. Chem. Solids*, **28**, 1323–1335.

Barber, F., Civetta, L., Gasparini, P., Innocenti, F., Scandone, R. and Villari, L. (1974). Evolution of a section of the Africa–Europe Plate Boundary: Palaeomagnetic and volcanological evidence from Sicily. *Earth Planet. Sci. Letters*, **22**, 123–132.

Barbetti, M. and Flude, K. (1979a). Palaeomagnetic field strengths from sediments baked by lava flows of the Chaîne des Puys, France. *Nature*, **278**, 153–156.

Barbetti, M. and Flude, K. (1979b). Geomagnetic variation during the late Pleistocene period. *Nature*, **279**, 202–205.

Barbetti, M. and McElhinny, M. (1972). Evidence of a geomagnetic excursion 30,000 yr BP. *Nature*, **229**, 327–330.

Barbetti, M. F. and McElhinny, M. W. (1976). The Lake Mungo geomagnetic excursion, *Phil. Trans. R. Soc., Lond.*, **A281**, 515–542.

Barbetti, M. F., McElhinny, M. W., Edwards, D. J. and Schmidt, P. W. (1977). Weathering processes in baked sediments and their effects on archaeomagnetic field-intensity measurements. *Phys. Earth Planet. Ints*, **13**, 346–354.

Barghorn, S. (1981). Magnetic-polarity stratigraphy of the Miocene type Tesique Formation, Santa Fe Group, in the Espanola Valley, New Mexico. *Geol. Soc. Amer. Bull.*, **92**, 1027–1041.

Palaeomagnetism

Barraclough, D. R. (1974). Spherical harmonic analyses of the geomagnetic field for eight epochs between 1600 and 1910. *Geophys. J. R. astr. Soc.*, **36**, 497–513.

Barraclough, D. R. (1982). Historical observations of the geomagnetic field. *Phil. Trans. R. Soc., Lond.*, **A306**, 71–78.

Barraclough, D. R., Harwood, J. M., Leaton, B. R. and Malin, S. R. C. (1975). A model of the geomagnetic field at Epoch 1975. *Geophys. J. R. astr. Soc.*, **43**, 645–659.

Barton, C. E. and Burden, F. R. (1979). Modifications to the Mackereth Corer. *Limnol. Oceanogr.*, **24**, 977–983.

Barton, C. E. and McElhinny, M. W. (1979). Detrital remanent magnetisation in five slowly redeposited long cores of sediment. *Geophys. Res. Letters*, **6**, 229–232.

Barton, C. E. and McElhinny, M. W. (1981). A 10,000 yr geomagnetic secular variation record from three Australian maars. *Geophys. J. R. astr. Soc.*, **67**, 465–485.

Barton, C. E., McElhinny, M. W. and Edwards, D. J. (1980). Laboratory studies of depositional DRM. *Geophys. J. R. astr. Soc.*, **61**, 355–377.

Barton, C. E., Merrill, R. T. and Barbett, M. (1979). Intensity of the Earth's magnetic field over the last 10,000 years. *Earth Planet. Sci. Letters*, **20**, 96–110.

Bathal, R. S. (1971). Magnetic anisotropy in rocks. *Earth. Sci. Res.*, **7**, 227–253.

Bathurst, R. G. C. (1975). *Carbonate Sediments and their Diagenesis*, 2nd Edn, Elsevier, Amsterdam, p. 658.

Baumann, L. von and Krs, M. (1967). Palaeomagnetische Altersbestimmungen an einigen Mineral-paragenesen des Freiberger Lagerstattenbezirhes. *Geologie*, **16**, 765–780.

Beaumont, C. (1978). The evolution of sedimentary basins on a viscoelastic lithosphere: theory and examples. *Geophys. J. R. astr. Soc.*, **55**, 471–497.

Beck, Jr., M. E. (1975). Remanence magnetism of the Twin Sisters dunite intrusion and implications for the tectonics of the Western Cordillera. *Earth Planet. Sci. Letters*, **26**, 263–268.

Beck, M. (1976). Discordant paleomagnetic pole positions as evidence of regional shear in the Western Cordillera of North America. *Amer. J. Sci.*, **276**, 694–712.

Beck, M. E. (1980). Paleomagnetic record of plate-margin tectonic processes along the western edge of North America. *J. Geophys. Res.*, **85**, 7115–7131.

Beck, M. E. and Plumley, P. W. (1980). Paleomagnetism of intrusive rocks in the Coast Range of Oregon: Microplate rotations in middle Tertiary time. *Geology* **8**, 573–577.

Berger, G. W., York, D. and Dunlop, D. J. (1979). Calibration of Grenvillian palaeopoles by ^{40}Ar/^{39}Ar dating. *Nature*, **277**, 46–48.

Berggren, W. A. (1972a). Cenozoic biostratigraphy and paleobigeography of the North Atlantic. *Init. Reps Deep Sea Drilling Project*, **12**, 965–1001.

Berggren, W. A. (1972b). A Cenozoic time scale – some implications for regional geology and paleobigeography. *Lethaia*, **5**, 195–215.

Berggren, W. A. and Hollister, C. D. (1977). Plate tectonics and paleocirculation – Commotion in the ocean. *Tectonophys.*, **38**, 11–48.

Berggren, W. A. and Van Couvering, J. A. (1974). The Late Neogene: Biostratigraphy, geochronology and paleoclimatology of the last 15 million years in marine and continental sequences, *Palaeogeog., Palaeoclimat., Palaeoecol.*, **16**, 1–216.

Bernal, J. D. (1965). *Science in History*. Watts, London, p. 1039.

306

Beske-Diehl, S. and Banerjee, S. K. (1979). An example of magnetic properties as indicators of alteration in ancient oceanic lithosphere – the Othris Ophiolite. *Earth Planet. Sci. Letters*, **44**, 451–462.

Bingham, C. (1964). *Distributions on the sphere and the projective plane*. Ph.D. Thesis, Yale University.

Bingham, D. K. and Klootwijk, C. T. (1980). Palaeomagnetic constraints on Greater India's underthrusting of the Tibetan Plateau. *Nature*, **284**, 336–338.

Birch, F. S. (1979). Magnetic fabric of the Exeter Pluton, New Hampshire. *J. Geophys. Res.*, **84**, 1129–1137.

Black, D. I. (1967). Cosmic ray effects and faunal extinctions at geomagnetic reversals. *Earth Planet. Sci. Letters*, **3**, 225–236.

Blackett, P. M. S. (1947). The magnetic field of massive rotating bodies. *Nature*, **159**, 658–666.

Blackett, P. M. S. (1952). A negative experiment relating to magnetism and the earth's rotation. *Phil. Trans. R. Soc., Lond.*, **A245**, 309–370.

Blakely, R. J. and Christiansen, R. L. (1978). The magnetization of Mount Shasta and implications for virtual geomagnetic poles determined from seamounts. *J. Geophys. Res.*, **83**, 5971–5978.

Blakemore, R. (1975). Magnetotactic bacteria. *Science*, **190**, 377–379.

Blandford, H. F. (1875). On the age and correlations of the Plant-bearing series of India, and the former existence of an Indo-oceanic continent. *Quart. J. Geol. Soc., Lond.*, **31**, 519–541.

Blandford, W. T., Blandford, H. F. and Theobald, W. (1859). On the geological structure and relations of the Talcheer Coal Field, in the District of Cuttack. *India. Geol. Surv. Mem.*, **1**, 33–89.

Bloemendal, J., Oldfield, F. and Thompson, R. (1979). Magnetic measurements used to assess sediment influx at Llyn Goddionduon. *Nature*, **280**, 50–53.

Blong, R. J. and Gillespie, R. (1978). Fluvial transported charcoal gives erroneous ^{14}C ages for recent deposits. *Nature*, **271**, 739–741.

Bogue, S. W. and Coe, R. S. (1981). Paleomagnetic correlation of Columbia River basalt flows using secular variation. *J. Geophys. Res.*, **86**, 11883–11897.

Böhlke, J. K., Honnorez, J., Honorez-Guerstein, B.-M., Muehlenbachs, K. and Petersen, N. (1981). Heterogeneous alteration of the Upper Oceanic Crust: Correlation of rock chemistry, magnetic properties, and O isotope ratios with alternative patters in basalts from Site 396B DSDP. *J. Geophys. Res.*, **86**, 7935–7950.

Bonhommet, N. and Babkine, J. (1967). Sur la presence d'aimantations inversées dans la chaines des Puys, *C.R. Acad. Sci., Paris*, **264**, 92–94.

Bonhommet, N. and Zahringer, J. (1969). Palaeomagnetism and potassium argon age determinations of the Laschamp geomagnetic polarity event. *Earth Planet. Sci. Letters*, **6**, 43–46.

Borradaile, G. J. (1981). Particulate flow of rock and the formation of cleavage. *Tectonophys.*, **72**, 305–321.

Borradaile, G. J. and Tarling, D. H. (1981). The influence of deformation mechanisms on magnetic fabrics in weakly deformed rocks. *Tectonophys.*, **77**, 151–168.

Bouroz, A., Einor, O. L., Gordon, M., Meyen, S. V. and Wagner, R. H. (1975). Proposals for an International chronostratigraphic classification of the Carboniferous. *C.R. 8ᵉ Congres. Carbonifère, Moscow* **1**, 36–69.

Boyle, R. (1691). Chymico magnetical experiments and observations. In *Experimenta*

& *Observationes Physicae*, London, Chapter 1.

Braginskii, S. I. (1972). Spherical analyses of the main geomagnetic field, 1500–1800. *Geomagn. Aeron.*, *USSR*, **12**, 524–529.

Brecher, A. and Fuhrman, M. (1979). Magnetism, shock and metamorphism in chondritic meteorites. *Phys. Earth Planet. Ints*, **20**, 350–360.

Bressler, S. L. and Elston, D. P. (1980). Declination and inclination errors in experimentally deposited specularite-bearing sand, *Earth Planet. Sci. Letters*, **48**, 227–232.

Briden, J. C. (1965). Ancient secondary magnetizations in rocks. *J. Geophys. Res.*, **70**, 5205–5221.

Briden, J. C. (1972). A stability index of remanent magnetism. *J. Geophys. Res.*, **77**, 1401–1405.

Briden, J. C. (1976). Applications of palaeomagnetism to Proterozoic tectonics. *Phil. Trans. R. Soc., Lond.*, **A280**, 405–416.

Briden, J. C. and Arthur, G. R. (1981). Precision of measurement of remanent magnetization. *Canad. J. Earth Sci.*, **18**, 527–538.

Briden, J. C. and Ward, M. A. (1966). Analyses of magnetic inclination in borecores. *Pure Appl. Geophys.*, **63**, 133–152.

Brill, R. H. (1970). Lead and oxygen isotopes in ancient objects. *Phil. Trans. R. Soc., Lond.*, **A279**, 143–164.

Brimhall, G. H., Geissman, J. W. and Kelly, W. C. (1979). Paleomagnetic documentation of the early, high-temperature mineralization at Butle, Montana. *Eos*, **60**, 241.

Brown, D. A., Campbell, K. S. W. and Crook, K. A. W. (1968). *The geological evolution of Australia and New Zealand*, Pergamon, Oxford, p. 409.

Brown, H. C., Khan, M. A. and Stacey, F. D. (1964). A search for flow structure in columnar basalt using magnetic anisotropy measurements. *Geophys. J. R. astr. Soc.*, **57**, 61–65.

Brown, L. (1979). Paleomagnetic results from northern Maine and the western limit cf 'Avalon' in the Mid-Paleozoic. *Geophys. Res. Letters*, **6**, 821–824.

Brown, L. L. and Kelly, W. M. (1980). Paleomagnetic results from northern Maine – reinterpretations. *Geophys. Res. Letters*, **7**, 1109–1111.

Bruins, H. J. (1976). *The origin, nature and stratigraphy of paleosols in the loessial deposits of the N.W. Negev (Netwot, Israel)*. M.Sc. thesis, Hebrew University, Jerusalem.

Brunhes, B. (1906). Récherches sur le directions d'aimantation des roches volcaniques. *J. Phys.*, **5**, 705–724.

Bucha, V. (1967). Archaeomagnetic and palaeomagnetic study of the magnetic field of the Earth in the past 600,000 years. *Nature*, **213**, 1005–1007.

Bucha, V. (1970a). Influence of the Earth's magnetic field on radiocarbon dating. In *Radiocarbon Variations and Absolute Chronology* (ed. Olsen), Wiley–Interscience, New York, 501–511.

Bucha, V. (1970b). Evidence for changes in the Earth's magnetic field intensity. *Phil. Trans. R. Soc., Lond.*, **A269**, 47–55.

Bucha, V. (1973). The continuous patter of variation of the geomagnetic field in the Quaternary and their causes. *Stud. Geophys. Geol.*, **17**, 218–231.

Bucha, V., Taylor, R. W., Berger, R. and Haury, E. Q. (1970). Geomagnetic intensity: changes during the past 3000 years in the Western Hemisphere. *Science*, **168**, 111–114.

Buchan, K. L., Berger, G. W., McWilliams, M. O., York, D. and Dunlop, D. J. (1977). Thermal overprinting of natural remanent magnetization and K/Ar ages in metamorphic rocks. *J. Geomagn. Geoelect.*, **29**, 401–410.

Buchan, K. L. and Schwarz, E. J. (1981). Uplift estimated from remanent magnetization: Munro area of Superior Province since 2150 Ma ago. *Canad. J. Earth Sci.*, **18**, 1164–1173.

Buchan, K. L., Schwarz, E. J., Symons, D. T. A. and Stupavsky, M. (1980). Remanent magnetization in the contact zone between Columbia Plateau flows and feeder dikes: Evidence for groundwater layer at time of intrusion. *J. Geophys. Res.*, **85**, 1888–1898.

Bullard, E. C. (1949a). The magnetic field within the Earth. *Proc. R. Soc., Lond.*, **A197**, 433–453.

Bullard, E. C. (1949b). Electromagnetic induction in a rotating sphere. *Proc. R. Soc., Lond.*, **A199**, 413–443.

Bullard, E. C. (1955). The stability of a homopolar dynamo. *Proc. Cambridge Phil. Soc.*, **51**, 744–760.

Bullard, E. C., Everett, J. E. and Smith, A. G. (1965). The fit of the continents around the Atlantic. *Proc. Roy. Soc., Lond.*, **A258**, 41–51.

Bullard, E. C., Freedman, C., Gellman, H. and Nixon, J. (1950). The westward drift of the Earth's magnetic field. *Phil. Trans. R. Soc., Lond.*, **A243**, 67–92.

Burakov, K. S., Gurary, G. Z., Khramov, A. N., Petrova, G. N., Rassanova, G. V. and Rodinov, V. P. (1976). Some peculiarities of the virtual pole position during reversals. *J. Geomagn. Geoelect.*, **28**, 295–307.

Burek, P. J. (1964). Korrelation revers magnatisierter Gesteinfolgen in Ober Bundsandstein, S.W. Deutschland. *Geol. Jahrb.*, **84**, 591–616.

Burlatskaya, S. and Nachasova, I. (1977). Archaeomagnetic determinations of geomagnetic field elements. *Meter. Mirov. Tsentra Dannykh, B, Sov. Geophys. Comm. Acad. Sci.* USSR, Moscow.

Burmester, R. F. and Bates, R. G. (1979). A spurious remanence in non-tumbling demagnetization. *Eos*, **60**, 815.

Burnham, R. J. P. and Tarling, D. H. (1975). Magnetization of shards as an assistance to the reconstruction of pottery vessels. *Studies in Conservation*, **20**, 152–158.

Burns, K. L., Rickard, M. J., Belbin, L. and Chamalaun, F. (1980). Further Palaeomagnetic confirmation of the Magellanes Orocline. *Tectonophys.*, **63**, 75–90.

Busse, F. H. (1979). Some new results on spherical dynamos, *Phys. Earth Planet. Ints*, **20**, 152–157.

Butler, R. F. (1972). Natural remanent magnetization and thermomagnetic properties of Allende meteorite. *Earth Planet. Sci. Letters*, **17**, 23–70.

Cain, B., Payne, M. A., Shulik, S., Donahue, J., Rollins, H. B. and Schmidt, V. A. (1979). The recovery of paleomagnetic polarities from cyclothemic sediments in the Carboniferous Appalachian Basin, USA. *Geophys. Res. Letters*, **6**, 261–264.

Campos, H. S. and Hallam, A. (1979). Diagenesis of English Lower Jurassic limestones as inferred from oxygen and carbon isotope analysis. *Earth Planet. Sci. Letters*, **45**, 23–31.

Cande, S. C. (1976). A palaeomagnetic pole from Late Cretaceous marine magnetic anomalies in the Pacific. *Geophys. J. R. astr. Soc.*, **44**, 547–566.

Cande, S. C., Larson, R. L. and LaBrecque, J. L. (1978). Magnetic lineations in the Pacific Jurassic Quiet Zone, *Earth Planet. Sci. Letters*, **41**, 434–440.

Cann, J. R. (1974). A model for oceanic crustal structure developed. *Geophys. J. R. astr. Soc.*, **39**, 169–187.

Cann, J. R., Dixon, J. E. and Renfrew, C. (1969). Obsidian analysis and the obsidian trade. In *Science in Archaeology* (eds D. Brothwell and E. S. Higgs), Thames and Hudson, London, pp. 578–591.

Cann, J. R. and Renfrew, C. (1964). The characterization of obsidian and its application to the Mediterranean region. *Proc. Prehist. Soc.*, **30**, 111–132.

Carey, S. W. (1958). A tectonic approach to continental drift, in *Continental Drift Symposium* (ed. S. W. Carey), Univ. Hobart, Tasmania, 177–355.

Carmichael, C. M. (1970). The mid-Atlantic ridge near 45°N, VII. Magnetic properties and opaque mineralogy of dredged samples. *Canad. J. Earth Sci.*, **7**, 239–256.

Casagrande, D. J., Idowu, G., Griedman, A., Rickert, P., Siefert, K. and Schlenz, D. (1979). H_2S incorporation in coal precursors: origins of organic sulphur in coal. *Nature*, **282**, 599–600.

Chamalaun, F. H. (1968). The magnetization of the Dotswood Red Beds (Queensland). *Earth Planet. Sci. Letters*, **3**, 439–443.

Champion, D. E., Dalrymple, G. B. and Kuntz, M. A. (1981). Radiometric and paleomagnetic evidence for the Emperor reversed event at 0.46 ± 0.05 m.y. in basalt lava flows from the eastern Snake River Plain, Idaho. *Geophys. Res. Letters*, **8**, 1055–1058.

Channell, J. E. T. (1977). Palaeomagnetism of limestones from the Gargano Peninsula (Italy), and the implication of these data. *Geophys. J. R. astr. Soc.*, **51**, 605–16.

Channell, J. E. T. (1978). Dual magnetic polarity measured in a single bed of Cretaceous pelagic limestone from Sicily. *J. Geophys.*, **44**, 613–622.

Channell, J. E. T. and D'Argenio, B. (1980). The evolution of the Calabrian Arc. *Instit. Geograf. Nacional, Madrid, Spec. Publ.*, **201**, 129–147.

Channell, J. E. T., D'Argenio, B. and Horvath, F. (1979a). Adria, the African promontory in Mesozoic Mediterranean palaeogeography. *Earth Sci. Rev.*, **15**, 213–292.

Channell, J. E. T., Heller, F. and van Stuijvenberg, J. (1979b). Magnetic susceptibility anisotropy as an indicator of sedimentary fabric in the Garnigel Flysch. *Eclog. geol. Helv.*, **72**, 781–787.

Channell, J. E. T., Lowrie, W. and Medizza, F. (1979c). Middle and Early Cretaceous magnetic stratigraphy from the Cismon Section, Northern Italy. *Earth Planet. Sci. Letters*, **42**, 153–166.

Channell, J. E. T. and Medizza, F. (1981). Upper Cretaceous and Palaeogene magnetic stratigraphy and biostratigraphy from the Venetian (Southern) Alps. *Earth Planet. Sci. Letters*, **55**, 419–432.

Channell, J. E. T. and Tarling, D. H. (1975). Palaeomagnetism and the rotation of Italy. *Earth Planet. Sci. Letters*, **25**, 177–188.

Chapman, S. (1967). Perspective. In *Physics of Geomagnetic Phenomena* (eds S. Matsushita and W. H. Campbell), Academic Press, New York, **1**, 3–28.

Chapman, S. and Bartels, J. (1940). *Geomagnetism*, Vols 1 and 2, Oxford University Press.

Chapman, S. and Bartels, J. (1951). *Geomagnetism*, Vols 1 and 2, 2nd ed, Clarendon Press, Oxford, pp. 1049.

Chave, A. D. and Denham, C. R. (1979). Climatic changes, magnetic intensity variations and fluctuations of the eccentricity of the Earth's orbit during the past

2,000,000 years and a mechanism which may be responsible for the relationship – a discussion. *Earth Planet. Sci. Letters*, **44**, 150–152.

Chevallier, R. (1925). L'Aimantation des laves de l'Etna et l'orientation du champ terrestre en Sicile du XIIe au XVIIe siècle. *Ann. Phys.*, **4**, 5–162.

Cisowski, S. M. (1980). The relationship between the magnetic properties of terrestrial igneous rocks and the composition and internal structure of their component Fe-oxide grains. *Geophys. J. R. astr. Soc.*, **60**, 107–122.

Cisowski, S. M. and Fuller, M. (1978). The effect of shock on the magnetism of terrestrial rocks. *J. Geophys. Res.*, **83**, 3441–3458.

Cisowski, S., Fuller, M. D., Rose, M. E. and Wasilewski, P. J. (1973). Magnetic effects of experimental shocking of lunar soil. Proc. 3rd Lunar Sci. Conf., *Geochim. Cosmochim. Acta*, **3**, 3003–3017.

Cisowski, S. M., Hale, C. and Fuller, M. D. (1977). On the intensity of ancient lunar fields. Proc. 8th Lunar Sci. Conf., *Geochim. Cosmochim. Acta*, **1**, 725–750.

Claesson, K. C. and Turner, P. (1980). Diagenetic magnetization of the Caerfai Bay Shales (Cambrian), South Wales. *Geophys. J. R. astr. Soc.*, **60**, 95–106.

Clark, A. J., Hammo-Yassi, N. and Tarling, D. H. (1983). British archaeomagnetic data: A review. (In preparation.)

Clark, H. C. (1969). Remanent magnetization cooling history, and paleomagnetic record of the Marys Peak Sill, Oregon. *J. Geophys. Res.*, **74**, 3143–3160.

Clark, H. C. and Kennett, J. P. (1973). Paleomagnetic excursion recorded in latest Pleistocene deep-sea sediments, Gulf of Mexico. *Earth Planet. Sci. Letters*, **19**, 267–274.

Clark, R. M. and Thompson, R. (1980). Author's reply to a comment on 'An objective method for smoothing palaeomagnetic data'. *Geophys. J. R. astr. Soc.*, **60**, 315–317.

Clegg, J. A., Almond, A. and Stubbs, P. H. S. (1954a). Remanent magnetism of some sedimentary rocks in Great Britain. *Phil. Mag.*, **45**, 583–598.

Clegg, J. A., Almond, A. and Stubbs, P. H. S. (1954b). Some recent studies of the pre-history of the Earth's magnetic field. *J. Geomagn. Geoelect.*, **4**, 194–199.

Clegg, J. A., Deutsch, E. R., Everitt, C. W. F., and Stubbs, P. H. S. (1957). Some recent palaeomagnetic measurements made at Imperial College, London. *Academic Physics*, **6**, 219–231.

Clegg, J. A., Deutsch, E. R. and Griffiths, D. H. (1956). Rock magnetism in India. *Phil. Mag.*, Ser. 8, **1**, 419–531.

Clegg, J. A., Radakrishnamurty, C. and Sahasrabudhe, P. W. (1958). Remanent magnetism of the Rajmahal traps of northeastern India. *Nature*, **181**, 830–831.

Clement, B. M., Kent, D. V. and Opdyke, N. D. (1982). Brunhes-Matuyana polarity transition in three deep-sea sediment cores. *Phil. Trans. R. Soc., Lond.*, **A306**, 113–119.

Coe, R. S. (1974). The effect of magnetic interactions on palaeointensity determinations by the Thelliers' Method. *J. Geomagn. Geoelect.*, **26**, 311–317.

Coe, R. S. (1979). The effect of shape anisotropy on TRM direction. *Geophys. J. R. astr. Soc.*, **56**, 369–383.

Coe, R. S. and Grommé, C. S. (1973). A comparison of three methods of determining geomagnetic paleointensities. *J. Geomag. Geoelectr.*, **25**, 415–435.

Coey, J. M. D., Morrish, A. H. and Sawatsky, G. A. (1971). A Mössbauer study of conduction in magnetite. *J. Phys.*, **C1-32**, 271–273.

Collinson, D. W. (1965a). Depositional remanent magnetization in sediments. *J.*

Geophys. Res., **70**, 4663–4668.

Collinson, D. W. (1965b). Origin of remanent magnetization and initial susceptibility of certain Red Sandstones. *Geophys. J. R. astr. Soc.*, **9**, 203–217.

Collinson, D. W. (1967). Chemical demagnetization, in *Methods in Palaeomagnetism* (ed. D. W. Collinson, K. M. Creer and S. K. Runcorn), Elsevier, Amsterdam, pp. 306–310.

Collinson, D. W. (1974). The role of pigment and specularite in the remanent magnetism of Red Sandstones. *Geophys. J. R. astr. Soc.*, **38**, 253–264.

Collinson, D. W. (1979). On the possibility of using lunar fines to determine the intensity of the ancient lunar magnetic field. *Phys. Earth Planet. Ints*, **20**, 312–316.

Collinson, D. W. (1980). An investigation of the scattered remanent magnetization of the Dunnet Head sandstone. *Geophys. J. R. astr. Soc.*, **62**, 393–402.

Collinson, D. W. (1983). *Methods in Palaeomagnetism and Rock Magnetism*, Chapman and Hall, London, p. 500.

Collinson, D. W., Creer, K. M., Irving, E. and Runcorn, S. K. (1957). Palaeomagnetic measurements in Great Britain – I. Measurement of the permanent magnetization of rocks. *Phil. Trans. R. Soc., Lond.*, **A250**, 73–82.

Collinson, D. W., Creer, K. M. and Runcorn, S. K. (eds) (1967). *Methods in Palaeomagnetism*. Elsevier, Amsterdam.

Collinson, D. W. and Molyneux, L. (1967). An instrument for the measurement of isotropic initial susceptibility of rock samples. In *Methods in Palaeomagnetism* (eds D. W. Collinson, K. M. Creer and S. K. Runcorn), Elsevier, Amsterdam, pp. 368–371.

Collinson, D. W. and Runcorn, S. K. (1960). Polar wandering and continental drift evidence from Paleomagnetic observations in the United States. *Bull. Geol. Soc. Amer.*, **71**, 915–958.

Collinson, D. W., Stephenson, A. and Runcorn, S. K. (1973). Magnetic properties of Apollo 15 and 16 rocks. Proc. 4th Lunar Sci. Conf., *Geochim. Cosmochim. Acta*, **3**, 2963–2976.

Colombo, U., Fugherazzi, G., Gazzarrini, F., Lanzavechia, G. and Sirowi, G. (1964). Studio nill'ossidazione delle magnetite. *La Chimica e l'Industria*, **46**, 357–362.

Condomines, M. (1978). Age of the Olby–Laschamp geomagnetic polarity event. *Nature*, **276**, 257–258.

Cordell, B. M. (1980). Martian climatic change: a magnetic trigger. *Geophys. Res. Letters*, **7**, 1065–1068.

Coupland, D. H. and Van der Voo, R. (1980). Long-term nondipole components in the geomagnetic field during the last 130 m.y. *J. Geophys. Res.*, **85**, 3529–3548.

Coward, M. P. (1980). The analysis of flow profiles in a basaltic dyke using strained vesicles. *J. Geol. Soc., Lond.*, **137**, 605–615.

Cowling, T. G. (1934). The magnetic field of sunspots. *Mon. Not. R. astr. Soc.*, **94**, 39–48.

Cox, A. V. (1957). Remanent magnetism of lower to middle Eocene basalt flows from Oregon. *Nature*, **179**, 685–686.

Cox, A. (1961). Anomalous remanent magnetization of basalt. *US Geol. Surv. Bull.*, **1083E**, 131–160.

Cox, A. (1962). Analysis of present geomagnetic field for comparison with Paleomagnetic results. *J. Geomagn. Geoelect., Kyoto*, **13**, 101–112.

Cox, A. (1969). Geomagnetic reversals. *Science*, **163**, 237–245.

Cox, A. (ed.) (1973). *Plate Tectonics and Geomagnetic Reversals*, W. H. Freeman, San Francisco, p. 702.

Cox, A. (1975). The frequency of geomagnetic reversals and the symmetry of the nondipole field. *Revs Geophys. Space Phys.*, **13**, 35–52.

Cox, A. (1981). A stochastic approach towards understanding the frequency and polarity bias of geomagnetic reversals. *Phys. Earth Planet. Ints*, **24**, 178–190.

Cox, A. and Dalrymple, G. B. (1967a). Statistical analysis of geomagnetic reversal data and the precision of potassium-argon dating. *J. Geophys. Res.*, **72**, 2603–2614.

Cox, A., and Dalrymple, G. B. (1967b). Geomagnetic polarity epochs: Nunivak Island, Alaska. *Earth Planet Sci. Letters*, **3**, 173–177.

Cox, A. and Doell, R. R. (1960). Review of paleomagnetism. *Geol. Soc. Amer. Bull.*, **71**, 645–768.

Cox, A., Doell, R. R. and Dalrymple, G. B. (1963a). Geomagnetic polarity epochs and Pleistocene geochronology. *Nature*, **198**, 1049–1051.

Cox, A., Doell, R. R. and Dalrymple, G. B. (1963b). Geomagnetic polarity epochs: Sierra Nevada II. *Science*, **142**, 382–385.

Cox, A., Doell, R. R. and Dalrymple, G. B. (1964a). Geomagnetic polarity epochs. *Science*, **143**, 351–352.

Cox, A., Doell, R. R. and Dalrymple, G. B. (1964b). Reversal of the Earth's magnetic field. *Science*, **144**, 1537–1543.

Cox, A., Doell, R. R. and Dalrymple, G. B. (1965). Quaternary paleomagnetic stratigraphy. *The Quaternary of the United States*, 817–830.

Cox, A., Doell, R. R. and Dalrymple, G. B. (1968). Radiometric time-scale for geomagnetic reversals. *Quart. J. Geol. Soc., Lond.*, **124**, 53–67.

Cox, K. C., Bell, J. D. and Pankhurst, R. J. (1979). *The Interpretation of Igneous Rocks*. Allen and Unwin, London, p. 450.

Crawford, A. R. and Daily, B. (1971). Probable non-synchroneity of Late Precambrian Glaciations. *Nature*, **230**, 111–112.

Crecelius, E. A., Carpenter, R. and Merrill, R. T. (1973). Magnetism and magnetic reversals in ferromanganese nodules. *Earth Planet. Sci. Letters*, **17**, 391–396.

Creer, K. M. (1957a). Palaeomagnetic investigations in Great Britain – IV. The natural remanence magnetization for certain stable rocks from Great Britain. *Phil. Trans. R. Soc., Lond.*, **A250**, 111–129.

Creer, K. M. (1957b). Palaeomagnetic investigations in Great Britain – V. The remanent magnetization of unstable Keuper Marls. *Phil. Trans. R. Soc., Lond.*, **A250**, 130–143.

Creer, K. M. (1958). Preliminary palaeomagnetic measurements from South America. *Ann. Geophys.*, **15**, 373–390.

Creer, K. M. (1959). A.C. demagnetization of unstable Triassic Keuper Marls from S.W. England. *Geophys. J. R. astr. Soc.*, **2**, 261–275.

Creer, K. M. (1961). Superparamagnetism in Red Sandstones. *Geophys. J. R. astr. Soc.*, **5**, 16–28.

Creer, K. M. (1962). The dispersion of the geomagnetic field due to secular variation and its determination for remote times from palaeomagnetic data. *J. Geophys. Res.*, **67**, 3461–3476.

Creer, K. M. (1967). Rock magnetic investigations at low temperatures. In *Methods in Palaeomagnetism* (eds D. W. Collinson, K. M. Creer and S. K. Runcorn), Elsevier, Amsterdam, pp. 514–528.

Palaeomagnetism

Creer, K. M. (1968). Palaeozoic palaeomagnetism. *Nature*, **219**, 246–250.

Creer, K. M. (1977). Geomagnetic secular variations during the last 25 000 years; an interpretation of data obtained from rapidly deposited sediments. *Geophys. J. R. astr. Soc.*, **48**, 91–110.

Creer, K. M. (1981). Long-period geomagnetic secular variations since 12,000 yr BP. *Nature*, **292**, 208–212.

Creer, K. M., Embleton, B. J. and Valencio, D. A. (1969). Comparison between the Upper Palaeozoic and Mesozoic palaeomagnetic poles for South America, Africa and Australia. *Earth Planet. Sci. Letters*, **7**, 228–292.

Creer, K. M., Irving, E., Nairn, A. E. M. and Runcorn, S. K. (1958). Palaeomagnetic results from different continents and their relation to the problem of continental drift. *Ann. Geophys.* **15**, 492–501.

Creer, K. M., Irving, E. and Runcorn, S. K. (1954). The direction of the geomagnetic field in remote epochs in Great Britain. *J. Geomagn. Geoelect.*, **6**, 163–168.

Creer, K. M., Irving, E. and Runcorn, S. K. (1957). Palaeomagnetic investigations in Great Britain. *Phil. Trans. R. Soc., Lond.*, **A250**, 144–156.

Creer, K. M. and Kopper, J. S. (1976). Secular oscillations of the geomagnetic field recorded by sediments deposited in caves in the Mediterranean region. *Geophys. J. R. astr. Soc.*, **45**, 35–58.

Creer, K. M., Molyneux, L., Vernet, J. P. and Wagner, J. J. (1975). Palaeomagnetic dating of 1-metre cores of sediment from Lake Geneva. *Earth Planet. Sci. Letters*, **28**, 127–132.

Creer, K. M., Readman, P. W. and Jacobs, A. M. (1980). Palaeomagnetic and palaeontological dating of a section of Gioia Tauro, Italy: identification of the Blake Event. *Earth Planet. Sci. Letters*, **50**, 289–300.

Creer, K. M., Readman, P. W. and Papamarinopoulos, S. (1981). Geomagnetic secular variations in Greece through the last 6000 years obtained from lake sediment studies. *Geophys. J. R. astr. Soc.*, **66**, 193–219.

Creer, K. M. and Sanver, M. (1967). The use of the sun compass. In *Methods in Palaeomagnetism* (eds D. W. Collinson, K. M. Creer and S. K. Runcorn), Elsevier, Amsterdam, pp. 11–15.

Creer, K. M. and Tucholka, P. (1982). Secular variation as recorded in lake sediments: a discussion of North American and European results. *Phil. Trans. R. Soc., Lond.*, **A306**, 87–102.

Creer, K. M. and Valencio, D. A. (1969). Palaeomagnetic and rock magnetic studies on the Cenozoic Basalts from Western Argentina. *Geophys. J. R. astr. Soc.*, **19**, 113–146.

Crimes, T. P. and Oldershaw, M. A. (1967). Palaeocurrent determinations by magnetic fabric measurements on the Cambrian rocks of St Tudwal's peninsula, North Wales. *J. Geol.* **75**, 217–232.

Crossling, B. F. (1970). Seamount magnetism. In *The Sea*, Vol. 4, Pt I (ed. A. E. Maxwell), Wiley–Interscience, New York, pp. 129–156.

Curtis, C. D. (1977). Sedimentary geochemistry: environments and processes dominated by involvement of an aqueous phase. *Phil. Trans. R. Soc.*, **A286**, 353–372.

Curtis, C. D. (1980). Diagenetic alteration in black shales. *J. Geol. Soc. Lond.*, **137**, 189–194.

Dagley, P. and Lawley, E. (1974). Palaeomagnetic evidence for the transitional behaviour of the geomagnetic field. *Geophys. J. R. astr. Soc.*, **36**, 577–598.

Daily, W. D. and Dyal, P. (1979). Theories for the origin of lunar magnetism. *Phys. Earth Planet. Ints*, **20**, 255–270.

Dallmeyer, R. D. and Sutter, J. F. (1980). Acquisition chronology of remanent magnetization along the 'Grenville Polar Paths': Evidence from $^{40}Ar/^{39}Ar$ ages of hornblende and biotite from the Whitestone Diorite, Ontario. *J. Geophys. Res.*, **85**, 3177–3186.

Dalrymple, G. B. (1972). Potassium–argon dating of geomagnetic reversals and North American glaciations. In *Calibration of Hominoid Evolution* (eds W. W. Bishope and J. A. Miller), Scottish Academic Press, Edinburgh, pp. 107–134.

Daly, L. (1981). Des aimantations partielles aux aimantations superposées: Éspoires et difficultes. *Phys. Earth Planet. Ints*, **24**, 218–227.

Daly, L., Morel, P., Irving, E. and Black, R. (1980). Situation paléomagnetique des cratons nord-africains à la fin de l'orogènese panafricaine. *C. R. Acad. Sci., Paris*, **290**, 407–410.

Damon, P. E. (1970). Climatic versus magnetic perturbation of the atmospheric ^{14}C reservoir. In *Radiocarbon Variations and Absolute Chronology* (ed. J. V. Olsson), Almqvist Wiksells, Uppsala, 571–593.

Damon, P. E., Long, A. and Grey, D. C. (1966). Fluctuations of atmospheric C^{14} during the last six millennia. *J. Geophys. Res.*, **71**, 1055–1063.

Damon, P. E., Long, A. and Grey, D. C. (1970). Arizona radio-carbon dates for dendrochronologically dated samples. In *XII Nobel Symp.* (ed. I. U. Ollsson), Wiley, New York, pp. 615–618.

Damon, P. E., Long, A. and Wallick, E. I. (1973). On the magnitude of the 11-year radiocarbon cycle. *Earth Planet. Sci. Letters*, **20**, 300–306.

Dankers, P. (1981). Relationship between median destructive field and remanent coercive forces for dispersed natural magnetite, titanomagnetite and haematite. *Geophys. J. R. astr. Soc.*, **64**, 447–461.

Dankers, P. H. M. and Zijderveld, J. D. A. (1981). Alternating field demagnetization of rocks, and the problem of gyromagnetic remanence. *Earth Planet. Sci. Letters*, **53**, 89–92.

D'Argenio, B., Horvath, F. and Channell, J. E. T. (1980). Palaeotectonic history of Adria, the African promontory. In 'Geologie des chaînes alpines. Issues de la Tethys', 26th Cong. Géol. Int., *Mem. B.R.G.M.*, **115**, 331–351.

David, P. (1904). Sur la stabilité de la direction d'aimantation dans quelques roches volcaniques. *C. R. Acad. Sci., Paris*, **138**, 41–42.

Davis, J. C. (1973). *Statistics and Data Analysis in Geology*, Wiley, New York, p. 550.

Davis, K. E. (1981). Magnetite rods in plagioclase as the primary carrier of stable NRM in ocean floor gabbros. *Earth Planet. Sci. Letters*, **55**, 190–198.

Day, R. (1977). TRM and its variation with grain size: A review. *J. Geomagn. Geoelect.*, **29**, 223–266.

De Boer, J. (1965). Palaeomagnetic indications of megatectonic movements in the Tethys. *J. Geophys. Res.*, **70**, 931–944.

Deer, W. A., Howie, R. A. and Zussman, J. (1966). *An Introduction to the Rock-Forming Minerals*, Wiley, New York, p. 528.

De Geer, G. (1940). Geochronologia Suecica: Principles. *K. Sven. Vetenskap, Handl.*, **18**, 1–367.

De Jong, A. F. M., Mook, W. G. and Becker, B. (1979). Confirmation of the Suess wiggles: 3200–3700 BC. *Nature*, **280**, 48–49.

De Jong, K. A., Manzoni, M., Stavenga, T., Van Dijr, F., Van der Voo, R. and Zijderveld, J. D. A. (1973). Palaeomagnetic evidence for rotation of Sardinia during the Early Miocene. *Nature*, **243**, 281–283.

De Jong, K. A., Manzoni, M. and Zijderveld, J. D. A. (1969). Palaeomagnetism of the Algharo Trachyandesites. *Nature*, **224**, 67–69.

Delesse, A. (1849). Sur le magnétisme polaire dans les mineraux et dans les roches. *Ann. Chim. Physique*, **25**, 194–209.

Denham, C. R. (1974). Counter-clockwise motion of paleomagnetic directions 24,000 years ago at Mono Lake, California. *J. Geomagn. Geoelect.*, **26**, 487–498.

Denham, C. R. (1975). Blake polarity episode in two cores from the Greater Antilles Outer Ridge, *Earth Planet. Sci. Letters*, **29**, 422–434.

Denham, C. R. and Cox, A. (1971). Evidence that the Laschamp polarity event did not occur 13,300–30,400 years ago. *Earth Planet. Sci. Letters*, **13**, 181–190.

De Paor, D. G. (1980). Some limitations of the R_f/ϕ technique of strain analysis. *Tectonophys.*, **64**, T29–31.

Deutsch, E. R. (1965). The paleolatitude of Tertiary Oil Fields. *J. Geophys. Res.*, **70**, 5193–5203.

Deutsch, E. R., Rao, K. V., Laurent, R. and Seguin, M. K. (1977). New evidence and possible origin of native iron in ophiolites of eastern Canada. *Nature*, **269**, 684–685.

De Vries, H. (1958). Variation in concentration of radiocarbon with time and location in earth. *Kon. Med. Akad. Wet., Proc.*, **B61**, 94.

Dickinson, W. R. (1970). Relations of andesites, granites, and derivative sandstones to arc-trench tectonics. *Revs Geophys. Space Phys.*, **8**, 813–860.

Dodson, M. A. (1973). Closure temperature in cooling geochronological and petrological systems. *Contr. Mineral. Petrol.*, **40**, 259–274.

Dodson, M. A. and McClelland-Brown, E. A. (1980). Magnetic blocking temperatures of single-domain grains during slow cooling. *J. Geophys. Res.*, **85**, 2625–2637.

Dodson, R. E. (1979). Counterclockwise precession of the geomagnetic field vector and westward drift of the non-dipole field. *J. Geophys. Res.*, **84**, 637–644.

Dodson, R. E. (1980). Late Tertiary secular variation of the geomagnetic field in the North Atlantic. *J. Geophys. Res.*, **85**, 3606–3622.

Doell, R. R. (1955). Paleomagnetic study of rocks from the Grand Canyon of the Colorado River. *Nature*, **176**, 1167.

Doell, R. R. (1956). Remanent magnetisation of the Upper-Miocene 'Blue' Sandstones of California. *Trans. Amer. Geophys. Union.*, **37**, 156–167.

Doell, R. R. (1957). Crystallization magnetization. *Adv. Phys.*, **6**, 327–332.

Doell, R. R. and Cox, A. (1967a). Measurement of natural remanent magnetization at the outcrop. In *Methods in Palaeomagnetism* (eds D. W. Collinson, K. M. Creer and S. K. Runcorn), Elsevier, Amsterdam, pp. 159–162.

Doell, R. R. and Cox, A. (1967b). Analysis of alternating field demagnetization equipment. In *Methods in Palaeomagnetism* (eds D. W. Collinson, K. M. Creer and S. K. Runcorn), Elsevier, Amsterdam, pp. 241–253.

Doell, R. R. and Cox, A. (1971). Pacific geomagnetic secular variation. *Science*, **171**, 248–254.

Doell, R. R., Dalrymple, G. B. and Cox, A. (1966). Geomagnetic polarity epochs: Sierra Nevada data, 3. *J. Geophys. Res.*, **71**, 531–541.

Dolginov, Sh. Sh., Yeroshenko, Ye. G., and Zhuzgov, L. N. (1976). The magnetic field of Mars according to the data from the Mars 3 and Mars 5. *J. Geophys. Res.*, **81**, 3353–3362.

Donath, F. A. and Wood, D. S. (1976). Experimental evaluation of the deformation path concept. *Phil. Trans. R. Soc., Lond.*, **A283**, 187–201.

Donovan, D. T. and Jones, E. J. W. (1979). Causes of world-wide changes in sea level. *J. Geol. Soc., Lond.*, **136**, 187–192.

Downey, W. S. (1981). Preliminary palaeomagnetic results relating to the Santorini eruption and the destruction of Minoan Crete. (Abs.) *Geophys. J. R. astr. Soc.*, **65**, 269.

Downey, W. S. (1982). Palaeointensity results from burnt mud-brick samples from Cretan Late Minoan Palaces. (Abs.) *Geophys. J. R. astr. Soc.*, **69**, 292.

Downey, W. S. (1983). *Magnetic studies on Santorini tephra and Minoan Cretan archaeological materials.* Ph.D. thesis, University of Newcastle upon Tyne.

Drake, R. E., Curtis, G. H., Cerling, T. E., Cerling, B. W. and Hampel, J. (1980). KBS tuff dating and geochronology of tuffaceous sediments in the Koobi Fora and Shungura Formations, East Africa. *Nature*, **283**, 368–372.

DuBois, P. M. (1962). Paleomagnetism and correlation of Keweenawan rocks. *Geol. Surv. Canada Bull.*, **71**, 1–75.

DuBois, R. L. (1975). Secular variation in southwestern United States as suggested by archeomagnetic studies. In *Magnetic Fields: Past and Present* (ed. R. M. Fisher *et al.*), Goddard Space Flight Center, New York, pp. 133–144.

Ducruix, J., Courtillot, V. and Le Mouël, J. L. (1980). The late 1960s secular variation impulse, the eleven year magnetic variation and the electrical conductivity of the deep mantle. *Geophys. J. R. astr. Soc.*, **61**, 73–94.

Duff, B. A. (1979). Peaked curves for haematite-bearing rocks and concentrates. *Phys. Earth Planet. Ints*, **19**, 1–4.

Duff, B. A. (1980). The palaeomagnetism of Jersey volcanics and dykes, and the Lower Palaeozoic apparent polar wander path for Europe. *Geophys. J. R. astr. Soc.*, **60**, 355–375.

Duncan, R. A. and Green, D. H. (1980). Role of multistage melting in the formation of oceanic crust. *Geology*, **8**, 22–26.

Dunlop, D. J. (1969). Interactions in rocks and the reliability of paleointensity data. *Earth Planet Sci. Letters*, **7**, 178–182.

Dunlop, D. J. (1973a). Superparamagnetic and single-domain threshold sizes in magnetite. *J. Geophys. Res.*, **78**, 1780–1793.

Dunlop, D. J. (1973b). Theory of the magnetic viscosity of lunar and terrestrial rocks. *Revs Geophys. Space Phys.*, **11**, 855–901.

Dunlop, D. J. (1979). On the use of Zijderveld vector diagrams in multicomponent paleomagnetic studies. *Phys. Earth Planet. Ints*, **20**, 12–24.

Dunlop, D. J. (1980). Paleomagnetism of the Archean Poohbah Lake Alkaline Complex, Northwestern Ontario, *Eos*, **61**, 214.

Dunlop, D. J., Stacey, F. D. and Gillingham, D. E. W. (1974). The origin of thermoremanent magnetization: Contribution of pseudo-single-domain magnetic moments. *Earth Planet. Sci. Letters.*, **21**, 288–294.

Dunlop, D. J. and Stirling, J. M. (1977). 'Hard' viscous remanent magnetization (VRM) in fine-grained hematite. *Geophys. Res. Letters*, **4**, 163–166.

Dunlop, D. J. and West, G. F. (1969). An experimental evaluation of single domain theories. *Revs Geophys.*, **7**, 709–757.

Dunn, D. A. and Moore, T. C. (1981). Late Miocene–Pliocene (magnetic epoch 9 – Gilbert magnetic epoch) calcium carbonate stratigraphy of the equatorial Pacific: Summary. *Geol. Soc. Amer. Bull. I.*, **92**, 104–107.

Durrance, E. M., Meads, R. E., Ballard, R. R. B. and Walsh, J. N. (1978). Oxidation state of iron in the Littleham Mudstone Formation of the New Red Sandstone Series (Permian–Triassic) of southeast Devon, England, *Geol. Soc. Amer. Bull.*, **89**, 1231–1240.

Du Toit, A. L. (1937). *Our Wandering Continents*, Oliver and Boyd, London, p. 366.

Dyal, P., Parkin, C. W. and Daily, W. D. (1977). Global lunar crust: Electrical conductivity and thermoelectric origin of remanent magnetism. Proc. 8th Lunar Sci. Conf., *Geochim. Cosmochim. Acta*, **1**, 767–783.

Dyal, P., Parkin, C. W. and Sonett, C. P. (1970). Apollo 12 magnetometer: Measurement of a steady field on the surface of the Moon. *Science*, **169**, 762–764.

Dzyaloshinski, I. (1958). A thermodynamic theory of 'weak' ferromagnetism of anti-ferromagnetics. *J. Phys. Chem. Solids*, **4**, 241–255.

Edwards, J. (1980a). Comparisons between the generation and properties of rotational remanent magnetization and anhysteretic remanent magnetization. *Geophys. J. R. astr. Soc.*, **62**, 379–392.

Edwards, J. (1980b). An experiment relating to rotational remanent magnetization and frequency of demagnetising field. *Geophys. J. R. astr. Soc.*, **60**, 283–288.

Egyed, L. (1959). The expansion of the Earth in connection with its origin and evolution. *Geophysica*, **7**, 13–22.

Einarsson, T. (1957a). Uber den Wert alter Sedimente für paläomagnetische zweche. *Neues Jahrb. Geologie u. Palaontologie Monatsh. Geophys. Abs.*, **172–172**, 51.

Einarsson, T. (1957b). Magneto-geological mapping in Iceland with the use of a compass. *Adv. Phys.*, **8**, 232–239.

Einarsson, T. and Sigurgeirsson, T. (1955). Rock magnetism in Iceland. *Nature*, **175**, 892.

Ellis, P. F. (1979). *Palaeomagnetic studies on the Gailtal Line Area (Austria/Italy)*. Ph.D. thesis, University of Newcastle upon Tyne, p. 559.

Ellwood, B. B. (1978a). Measurement of anisotropy of magnetic susceptibility: A comparison of the precision of torque and spinner magnetometer systems for basaltic specimens. *J. Phys. E Sci. Inst.*, **11**, 71–75.

Ellwood, B. B. (1978b). Flow and emplacement direction determined for selected basaltic bodies using magnetic anisotropy measurements. *Earth Planet. Sci. Letters*, **41**, 254–264.

Ellwood, B. B. (1979a). Particle flocculation: one possible control on the magnetization of deep-sea sediments. *Geophys. Res. Letters*, **6**, 237–240.

Ellwood, B. B. (1979b). Sample shape and magnetic grain sizes: two possible controls on the anisotropy of magnetic susceptibility variability in deep-sea sediments. *Earth Planet. Sci. Letters*, **43**, 309–314.

Ellwood, B. B. (1979c). Anisotropy of magnetic susceptibility variations in Icelandic columnar basalts. *Earth Planet. Sci. Letters*, **42**, 209–212.

Ellwood, B. B. (1981). Weathering effects on the magnetic properties of the Milledgeville granite, Georgia. *Earth Planet. Sci. Letters*, **55**, 311–316.

Ellwood, B. B., Amerigian, C.,Watkins, B. D. and Self, S. (1973). Brunhes Epoch geomagnetic secular variation in the Azores. *Eos*, **54**, 254.

Ellwood, B. B. and Fisk, M. R. (1977). Anisotropy of magnetic susceptibility variations in a single Icelandic columnar basalt. *Earth Planet. Sci. Letters*, **35**, 116–122.

Ellwood, B. B. and Howard, J. H. (1981). Magnetic fabric development in an experimentally produced barchan dune. *J. Sedi. Petrol.*, **51**, 97–100.

Ellwood, B. B. and Leddbetter, M. T. (1977). Antarctic bottom water fluctuations in the Vema Channel: Effects of velocity changes on particle alignment and size. *Earth Planet. Sci. Letters*, **35**, 189–198.

Ellwood, B. B. and Watkins, N. D. (1976). Comparison of observed intrusive to extrusive ratios in Iceland and the Troodos Massif with results of experimental emplacement mode analysis of DSDP igneous rocks. *J. Geophys. Res.*, **81**, 4152–4156.

Ellwood, B. B. and Whitney, J. A. (1980). Magnetic fabric of the Elberton Granite, Northeast Georgia. *J. Geophys. Res.*, **85**, 1481–1486.

Ellwood, B. B., Whitney, J. A., Wenner, D. B., Mose, D. and Amerigian, C. (1980). Age, paleomagnetism and tectonic significance of the Elberton Granite Northeast Georgia Piedmont. *J. Geophys. Res.*, **85**, 6521–6533.

Elsasser, W. M. (1939). On the origin of the Earth's magnetic field. *Phys. Rev.*, **55**, 489–498.

Elsasser, W. M. (1950). Causes of motions in the Earth's core. *Trans. Amer. Geophys. Union*, **31**, 454–462.

Elsasser, W. M. (1955). Hydromagnetism. I. A review. *Amer. J. Phys.*, **23**, 590–609.

Elsasser, W. M. (1956a). Hydromagnetism. II. A review. *Amer. J. Phys.*, **24**, 85–110.

Elsasser, W. M. (1956b). Hydromagnetic dynamo theory. *Revs Mod. Phys.* **28**, 135–163.

Elston, D. P. and Bressler, S. L. (1980). Paleomagnetic poles and polarity zonation from the Middle Proterozoic Belt Supergroup, Montana and Idaho. *J. Geophys. Res.*, **85**, 339–355.

Elston, D. P. and Grommé, C. S. (1974). Precambrian polar wandering from Unkas Group and Nankoweep Formation, Eastern Grand Canyon, Arizona. *Abs. with Prog., Geol. Soc. Amer.*, **6**, 440–441.

Elston, D. P. and Purucker, M. (1979). Detrital magnetization in red beds of the Moenkopi formation. *J. Geophys. Res.*, **84**, 1653–1665.

Embleton, B. J. J. (1972a). The palaeomagnetism of some Proterozoic–Cambrian sediments from the Amadeus Basin, Central Australia. *Earth Planet. Sci. Letters*, **17**, 217–226.

Embleton, B. J. J. (1972b). The palaeomagnetism of some Palaeozoic sediments from Central Australia. *J. Proc. R. Soc. New South Wales*, **105**, 86–93.

Embleton, B. J. J. and McElhinny, M. W. (1975). The palaeoposition of Madagascar: Palaeomagnetic evidence from the Isalo Group. *Earth Planet. Sci. Letters*, **27**, 329–341.

Embleton, B. J. J., McElhinny, M. W., Crawford, A. R. and Luck, G. R. (1974). Palaeomagnetism and the tectonic evolution of the Tasman Orogenic Zone. *J. Geol. Soc. Australia*, **21**, 187–194.

Embleton, B. J. J. and Schmidt, P. W. (1979). Recognition of common Precambrian polar wandering reveals a conflict with plate tectonics. *Nature*, **282**, 705–707.

Emerson, S., Jahnke, R., Bender, M., Froelich, P., Klinkhammer, G., Bowser, C. and Setlock, G. (1980). Early diagenesis in sediments from the Eastern Equatorial Pacific, I. Pore water nutrient and carbonate results. *Earth Planet. Sci. Letters*, **49**, 57–80.

Erlenkeuser, H. (1980). ^{14}C age and vertical mixing of deep-sea sediments. *Earth Planet. Sci. Letters*, **47**, 319–326.

Etheridge, M. A. and Oertel, G. (1979). Strain measurements from phyllosilicate preferred orientation – a precautionary note. *Tectonophys.*, **60**, 107–120.

Evans, M. E. and McElhinny, M. W. (1966). The paleomagnetism of the Modipe Gabbro. *J. Geophys. Res.*, **71**, 6053–6063.

Evans, M. E. and McElhinny, M. W. (1969). An investigation of the origin of stable remanence in magnetite-bearing igneous rocks. *J. Geomagn. Geoelect.*, **21**, 757–773.

Evans, M. E., McElhinny, M. W. and Gifford, A. C. (1968). Single domain magnetite and high coercivities in a gabbroic intrusion. *Earth Planet. Sci. Letters*, **4**, 142–146.

Evans, P. (1971). Towards a Pleistocene time-scale. In *The Phanerozoic Time-Scale – a supplement*. Geol. Soc., London, Spec. Publ. **5**, 123–356.

Everitt, C. W. F. and Clegg, J. A. (1962). A field test of palaeomagnetic stability. *Geophys. J. R. astr. Soc.*, **6**, 312–319.

Faller, A. M. (1975). Palaeomagnetism of the oldest Tertiary basalts in the Kangerdlugssuag area of East Greenland. *Medd. Dansk. geol. Foren.*, **24**, 173–178.

Faller, A. M. and Soper, N. J. (1979). Palaeomagnetic evidence for the origin of the coastal flexure and dyke swarm in central E. Greenland. *J. Geol. Soc., Lond.*, **136**, 737–744.

Ferguson, C. W. (1970). Concept and techniques of dendrochronology. *Sci. Methods Medieval Archaeol.*, 183–200.

Fink, J. H., Malin, M. C., D'Alli, R. E. and Greeley, R. (1981). Rheological properties of mudflows associated with the Spring 1980 eruptions of Mount St. Helens Volcano, Washington. *Geophys. Res. Letters*, **8**, 43–46.

Fisher, R. A. (1953). Dispersion on a sphere. *Proc. R. Soc.*, **A217**, 295–305.

Fisher, R. V., Smith, A. L. and Roobol, M. J. (1980). Destruction of St Pierre, Martinique, by ash-cloud surges, May 8 and 20, 1902. *Geology*, **8**, 472–476.

Fitch, F. J., Hooker, P. J., Miller, J. A. and Brereton, N. R. (1978). Glauconite dating of Palaeocene-Eocene rocks from East Kent and the time-scale of Palaeogene volcanism in the North Atlantic region. *J. Geol. Soc., Lond.*, **135**, 499–512.

Flinn, D. (1962). On folding during three-dimensional progressive deformation. *Quart. J. Geol. Soc., Lond.*, **118**, 385–433.

Flugel, H. W., Mauritsch, H. J., Heinz, F. and Frank, W. (1980). Palaeomagnetische und radiometrische Dater aus dem Grazer Palaozoikum. *Mitt. österr. geol. Geo.*, **71/72**, 201–211.

Folgerhaiter, G. (1894–1895). Origine del magnetismo nelle roccie vulcaniche del Lazio et al. *Att. d. Reala Accad. Lincei*, **3**, 53, 117, 165; **4**, 78, 203.

Folgerhaiter, G. (1899a). Sur les variations seculaires de l'inclinason magnetique dans antiquité. *J. Phys.*, **8**, 5–16.

Folgerhaiter, G. (1899b). Sur les variations séculaires de l'inclinason magnétique dans l'antiquité. *Archives Sci. phys. naturelles*, **8**, 660.

Förster, H., Soffel, H. and Zinsser, H. (1975). Palaeomagnetism of rocks from the Eastern Alps from North and South of the Insubrian Line. *N. Jb. Geol. Paläont.*, **149**, 112–127.

Forsyth, J. B., Hedley, I. G. and Johnson, C. E. (1968). The magnetic structure and hyperfine field of goethite. *J. Phys.*, **1**, 179–188.

Fox, J. M. W. and Aitken, M. J. (1980). Cooling-rate dependence of thermoremanent magnetization. *Nature*, **283**, 462–463.

Frakes, L. A. (1979). *Climates throughout Geologic Time*. Elsevier, Amsterdam, p. 310.

Freed, W. K. (1977). The virtual geomagnetic field path during the Brunhes/Matuyama polarity change when viewed from equatorial latitudes. *Eos*, **58**, 380.

Freed, W. K. and Healy, N. (1974). Excursions of the Pleistocene geomagnetic field recorded in Gulf of Mexico sediment. *Earth Planet. Sci. Letters*, **24**, 99–104.

Freer, R. and O'Reilly, W. (1980). The diffusion of Fe^{2+} ions in spinels with relevance to the process of maghemitization. *Mineral. Mag.*, **43**, 889–899.

Freund, R. and Tarling, D. H. (1979). Preliminary Mesozoic palaeomagnetic results from Israel and inferences for a microplate structure in the Lebanon. *Tectonophys.*, **60**, 189–205.

Frey, F. A., Haskin, M. A., Poetz, S. A. and Haskin, L. A. (1968). Rare earth abundances in some basic rocks. *J. Geophys. Res.*, **73**, 6085–6098.

Friedman, G. M., Reeckmann, S. A. and Borak, B. (1981). Carbonate deformation mechanisms in the world's deepest wells (~9 km). *Tectonophys.*, **74**, T15–19.

Fromm, E. (1970). An estimation of errors in the Swedish varve chronology. In *Radiocarbon Variations and Absolute Chronology* (ed. Olsson) Wiley–Interscience, New York, pp. 163–172.

Fucugauchi, J. U. (1979). Preliminary apparent polar wander path for Mexico. *Geophys. J. R. astr. Soc.*, **56**, 227–235.

Fucugauchi, J. U. (1980a). Further reliability tests for determinations of palaeo-intensities of the Earth's magnetic field. *Geophys. J. R. astr. Soc.*, **61**, 243–251.

Fucugauchi, J. U. (1980b). Palaeointensity determination and K-Ar dating of the Tertiary north-east Jalisco volcanics (Mexico). *Geophys. J. R. astr. Soc.*, **63**, 601–618.

Fuller, M. D. (1970). Geophysical aspects of paleomagnetism. *Crit. Rev. Solid State Phys.*, **1**, 137.

Fuller, M. (1974). Lunar magnetism. *Revs Geophys. Space Phys.*, **12**, 23–70.

Fuller, M., Williams, I. and Hoffman, K. A. (1979). Paleomagnetic records of geo-magnetic field reversals and the morphology of the transitional fields. *Revs Geophys. Space Phys.*, **17**, 179–203.

Funnell, B. M. (1964). The Tertiary period. In *The Phanerozoic Time-Scale* (ed. Harland *et al.*) Geological Society, London, pp. 179–191.

Furon, R. (1972). *Éléments de Paléoclimatologie*, Librairie Vuibert, Paris, p. 216.

Fyfe, W. S. (1976). Chemical aspects of rock deformation. *Phil. Trans. R. Soc., Lond.*, **A283**, 221–228.

Gale, N. H. (1978). *Lead isotopes and Aegean metallurgy*. Thera and Aegean World, London, **1**, 529–545.

Gallagher, K. J., Feitnecht, W. and Mannwiller, U. (1968). Mechanism of oxidation of magnetite to γ-Fe_2O_3. *Nature*, **217**, 1118–1121.

Games, K. P. (1977). The magnitude of the palaeomagnetic field: a new non-thermal, non-detrital method using sun-dried bricks. *Geophys. J. R. astr. Soc.*, **48**, 315–330.

Games, K. P. (1980). The magnitude of the archaeomagnetic field in Egypt between 3000 and 0 BC. *Geophys. J. R. astr. Soc.*, **63**, 45–56.

Gauss, C. F. (1833). *Intensitas vis magneticae terrestris ad mensuram absolutam revocata*. Gottingen.

Gauss, C. F. (1839). *Algemaine Theorie des Erdmagnetismus.*

Geissman, J. W. (1980). Paleomagnetism of ash-flow tuffs: microanalytical recognition of TRM components. *J. Geophys. Res.*, **85**, 1487–1499.

Gellibrand, H. (1839). A discourse mathematical on the variation of the magneticall needle. Together with its diminution lately discovered. London.

Gheradi, S. (1862). *Nuovo Cimento*, **16**, 384.

Giddings, J. (1969). *A palaeomagnetic investigation of selected British geological units.* M.Sc. dissertation, University of Newcastle upon Tyne.

Gilbert, W. (1600). *De Magnete.*

Gill, D. (1973). *Stratigraphy facies, evolution and diagenesis of productive Niagaran Guelph Reefs and Cayugan Sabkha deposits, the Belle River Mills Field, Michigan Basin.* Ph.D. thesis, University of Michigan, Ann Arbor, p. 282.

Gillot, P. Y., Labeyrie, J., Laj, C., Valladas, G., Guerin, G., Poupeau, G. and Delibrias, G. (1980). Age of the Laschamp paleomagnetic excursion revisited. *Earth Planet. Sci. Letters*, **42**, 444–450.

Giovanoli, F. (1979). A comparison of the magnetization of detrital and chemical sediments from Lake Zurich. *Geophys. Res. Letters*, **6**, 233–235.

Gleadow, A. J. W. and Lovering, J. F. (1978). Geochronology of King Island, Bass Strait, Australia: relationship to continental rifting. *Earth Planet. Sci. Letters*, **37**, 429–437.

Goddard, E. N. (1951). *Rock Color Chart*, Geological Society of America, New York.

Goldstein, A. G. (1980). Magnetic susceptibility anisotropy of mylonites from the Lake Char Mylonite Zone, southeastern New England. *Tectonophys.*, **66**, 197–211.

Goldstein, M. A., Strangway, D. W. and Larson, E. E. (1969). Paleomagnetism of a Miocene transition zone in southwestern Oregon. *Earth Planet. Sci. Letters*, **7**, 231–239.

Goodwin, R. G. (1980). Magnetic correlation of Holocene glaciolacustrine sediments, Glacier Bay, Alaska. *Eos*, **61**, 1194.

Gordon, R. G. and Cox, A. (1980a). Paleomagnetic test of the Early Tertiary plate circuit between the Pacific Basin plates and the Indian plate. *J. Geophys. Res.*, **85**, 6534–6546.

Gordon, R. G. and Cox, A. (1980b). Calculating palaeomagnetic poles for oceanic plates. *Geophys. J. R. astr. Soc.*, **63**, 619–640.

Gordon, R. G., McWilliams, M. O. and Cox, A. (1979). Pre-Tertiary velocities of the continents: A lower bound from paleomagnetic data. *J. Geophys. Res.*, **84**, 5480–5486.

Gordus, A. A., Wright, G. A. and Griffin, J. B. (1968). Obsidian sources characterised by neutron activation analysis. *Nature*, **233**, 242–245.

Goree, W. S. and Fuller, M. D. (1976). Magnetometers using R.F.-driven squids and their application in rock magnetism and paleomagnetism. *Revs Geophys. Space Phys.*, **14**, 591–608.

Gose, W. A., Pearce, G. W., Strangway, D. W. and Larson, E. E. (1972). Magnetic properties of Apollo 14 breccias and their correlation with metamorphism. *Geochim. Cosmochim. Acta*, **36**, Suppl. 3, 2387–2395.

Gould, J. L., Kirschvink, J. L. and Deffeyes, K. S. (1978). Bees have magnetic remanence. *Science*, **201**, 1026–1028.

Grabovsky, M. A. (1952). Magnetic properties of ferrites and their significance in geophysics. *Bull. Acad. Sci. URSS*, No. 5, 41–46.

Grabovsky, M. A. (1953). Thermoremanent magnetism of rocks. *Izv. Akad. Nauk SSSR, Ser. Geofiz.*, No. 3, 214–224.

Grabovsky, M. A. and Pushkov, A. H. (1954). On the origin of inverse remanent magnetization in rocks. *Izv. Akad. Nauk SSSR, Ser. Geofiz.*, No. 4, 320–329.

Graham, J. W. (1949). The stability and significance of magnetism in sedimentary rocks. *J. Geophys. Res.*, **54**, 131–167.

Graham, J. W. (1954a). Rock magnetism and the Earth's magnetic field during Paleozoic time. *J. Geophys. Res.*, **59**, 215–222.

Graham, J. W. (1954b). Tracing the Earth's magnetic field in geologic time. *J. Geomag. Geoelect.*, **6**, 215.

Graham, J. W. (1954c). Magnetic susceptibility anisotropy an unexploited petrofabric element. *Geol. Soc. Amer. Bull.*, **75**, 1257–1258.

Graham, J. W. (1955). Evidence of polar shift since Triassic time. *J. Geophys. Res.*, **60**, 329–347.

Graham, J. W. (1966). Significance of magnetic anisotropy in Appalachian sedimentary rocks. In *The Earth beneath the Continents* (eds J. S. Steinhart and T. J. Smith), *Geophys. Monog. Amer. Geophys. Union*, **10**, 627–648.

Graham, J. W. and Torreson, O. W. (1951). Contrasting magnetizations of flat-lying and folded Palaeozoic sediments. *Trans. Amer. Geophys. Union*, **32**, 336.

Graham, K. W. T. and Hales, A. L. (1957). Palaeomagnetic measurements on Karoo Dolerites. *Adv. Phys.* **6**, 149–161.

Graham, K. W. and Hales, A. L. (1961). Preliminary palaeomagnetic measurements on Silurian sediments from South Africa. *Geophys. J. R. astr. Soc.*, **5**, 318–325.

Granar, L. (1958). Magnetic measurements on Swedish varved sediments. *Arkiv. f. Geofysik*, **3**, 1–40.

Gravenor, C. P., Stupavsky, M. and Symonds, D. T. A. (1973). Paleomagnetism and its relationship to till deposition. *Canad. J. Earth Sci.*, **10**, 1068–1078.

Greenhaus, M. R. and Cox, A. (1979). Paleomagnetism of the Morro Rock–Islay Hill Complex as evidence for crustal block rotations in Central Coastal California. *J. Geophys. Res.*, **84**, 2392–2400.

Greenslate, J. (1978). Marine manganese concentration growth rates: non-radiometric considerations. *Geophys. Res. Letters*, **5**, 237–239.

Gregor, C. B., Nairn, A. E. M. and Negendank, J. F. W. (1975). Palaeomagnetic investigations of the Tertiary and Quaternary Rocks: IX The Pliocene of Southeast Sicily and some Cretaceous Rocks from Capo Passero. *Geol. Rundsch.*, **64**, 948–958.

Griffiths, D. H. (1953). Remanent magnetism of varved clays from Sweden. *Nature*, **172**, 539.

Griffiths, D. H. (1954). The remanent magnetism of varved clays from Sweden. *J. Geomagn. Geoelect.* **6**, 217–220.

Griffiths, D. H. (1955). The remanent magnetism of varved clays from Sweden. *Mon. Not. R. astr. Soc., Geophys. Suppl.*, **7**, 103–114.

Griffiths, D. H. and King, R. F. (1954). Natural magnetization of igneous and sedimentary rocks. *Nature*, **173**, 1114–1117.

Griffiths, D. H., King, R. F. and Rees, A. I. (1962). The relevance of magnetic measurements on some time grained silts to the study of their depositional process. *Sedimentology*, **1**, 134–144.

Griffiths, D. H., King, R. F., Rees, A. I. and Wright, A. E. (1960). The remanent magnetism of some recent varved sediments. *Proc. R. Soc., Lond.*, **A256**, 359–383.

Griffiths, D. H., King, R. F. and Wright, A. E. (1957). Some field and laboratory studies of the depositional remanence of Recent sediments. *Adv. Phys.*, **6**, 306–316.

Griffiths, D. H., King, R. F. and Wright, A. E. (1958). An assessment of the difficulties involved in using Quaternary varved sediments for palaeomagnetic studies of the secular variation. *Ann. Geophys.*, **14**, 515–518.

Grommé, C. S. and Hay, R. L. (1963). Magnetization of basalt of Bed 1, Olduvai Gorge, Tanganyika. *Nature*, **200**, 560–561.

Grommé, S., Mankinen, E. A., Marshall, M. and Coe, R. S. (1979). Geomagnetic paleointensities by the Thelliers' method from submarine pillow basalts: effects of seafloor weathering. *J. Geophys. Res.*, **84**, 3553–3575.

Grossling, B. F. (1970). Seamount Magnetism. In *The Sea*, Vol. 4, part I (ed. A. E. Maxwell) Wiley-Interscience, New York, pp. 129–156.

Gubbins, D. (1982). Finding core motions from magnetic observations. *Phil. Trans. R. Soc., Lond.*, **A303**, 247–254.

Gunn, N. M. and Murray, A. S. (1980). Geomagnetic field magnitude variations in Peru derived from archaeological ceramics dated by thermoluminescence. *Geophys. J. R. astr. Soc.*, **62**, 345–366.

Gus'kova, Ye. G. (1963). Investigation of natural remanence magnetization of stony meteorites. *Geomagn. Aeron. USSR.*, **3**, 308–312.

Haggerty, S. E. (1976a). Oxidation of opaque mineral oxides in basalts. In *Oxide Minerals*, **3**, Mineralogical Society of America, Hg 1–100.

Haggerty, S. E. (1976b). Opaque mineral oxides in terrestrial igneous rocks. In *Oxide Minerals*, **3**, Mineralogical Society of America, Hg 101–300.

Hagstrum, J. T., Van der Voo, R., Auvray, B. and Bonhommet, N. (1980). Eocambrian-Cambrian palaeomagnetism of the Armorican Massif, France. *Geophys. J. R. astr. Soc.*, **61**, 489–517.

Hailwood, E. A., Bock, W., Costa, L., Dupeuble, P. A., Muller, C. and Schnitker, D. (1979). Chronology and biostratigraphy of northeast Atlantic sediments, DSDP Leg 48. *Init. Reps, Deep Sea Drilling Project*, **48**, 1119–1141.

Hailwood, E. A., Hamilton, N. and Morgan, G. E. (1980). Magnetic polarity dating of tectonic events at passive continental margins. *Phil. Trans. R. Soc., Lond.*, **A294**, 189–208.

Hailwood, E. A., Løvlie, R. and Tarling, D. H. (in prep.). Correlation of Tertiary Basalts in East Greenland using palaeomagnetic directions.

Hailwood, E. A. and Mitchell, J. G. (1971). Palaeomagnetic and radiometric dating results from Jurassic intrusions in South Morocco. *Geophys. J. R. astr. Soc.*, **24**, 351–364.

Hailwood, E. A. and Sayre, W. O. (1979). Magnetic anisotropy and sediment transport directions in North Atlantic Early Cretaceous black shales and Eocene mudstones cored on DSDP leg 48. *Init. Reps Deep Sea Drilling Project*, **48**, 909–918.

Hailwood, E. A., Tarling, D. H., Mitchell, J. G. and Løvlie, R. (1973). Preliminary observations on the palaeomagnetism and radiometric ages of the Tertiary basalt sequence of Scoresby Sund, East Greenland. *Grøn. Geol. Unders. Rep.*, **58**, 43–47.

Halgedahl, S. L., Day, R. and Fuller, M. D. (1980). The effect of cooling rate on the intensity of weak-field TRM in single-domain magnetite. *J. Geophys. Res.*, **85**, 3690–3698.

Halgedahl, S. and Fuller, M. (1980). Magnetic domain observations of nucleation processes in fine particles of intermediate titanomagnetite. *Nature*, **288**, 70–72.

324

Hall, C. M. and York, D. (1978). K-Ar and ^{40}Ar/^{39}Ar age of the Laschamp geomagnetic polarity reversal. *Nature*, **274**, 462–464.

Hallam, A. (ed.) (1973). *Atlas of Palaeobiogeography*. Elsevier, Amsterdam, pp. 531.

Hallam, A. (1977). Secular changes in marine inundation of USSR and North America through the Phanerozoic. *Nature*, **269**, 769–772.

Hallam, A. (1978). Eustatic cycles in the Jurassic. *Palaeogeog., Palaeoclim., Palaeoecol.*, **23**, 1–32.

Hallam, A. and Bradshaw, M. J. (1979). Bituminous shales and oolitic ironstones as indicators of transgressions and regressions. *J. Geol. Soc., Lond.*, **136**, 157–164.

Hallam, B. R., Warren, S. E. and Renfrew, C. (1976). Obsidian in the West Mediterranean: Characterisation by neutron activation analysis and optical emission spectroscopy. *Proc. Prehist. Soc.*, **42**, 85–110.

Halley, E. (1692). On the cause of the change in the variation of the magnetic needle; with an hypothesis of the structure of the internal parts of the Earth. *Phil. Trans. R. Soc., Lond.*, **17**, 470–478.

Halls, H. C. (1979). Separation of multicomponent NRM: Combined use of difference and resultant magnetization vectors. *Earth Planet. Sci. Letters*, **43**, 303–308.

Halvorsen, E. (1974). The magnetic fabric of some dolerite intrusions, Northeast Spitsbergen; implications for their mode of emplacement. *Earth Planet. Sci. Letters*, **21**, 127–133.

Hamano, Y. (1980). An experiment on the post-depositional remanent magnetization in artificial and natural sediments. *Earth Planet. Sci. Letters*, **51**, 221–232.

Hamilton, N. (1963). Susceptibility anisotropy measurements on some Silurian siltstones. *Nature*, **197**, 170–171.

Hamilton, N. (1967). Laboratory redeposition studies – an appraisal of apparatus and technique. In *Methods in Palaeomagnetism* (eds D. W. Collinson, K. M. Creer and S. K. Runcorn). Elsevier, Amsterdam, pp. 597–603.

Hamilton, N. and King, R. F. (1964). Comparison of the bedding errors of artificially and naturally deposited sediments with those predicted from a simple model. *Geophys. J. R. astr. Soc.*, **8**, 370–374.

Hamilton, N. and Rees, A. I. (1970). The use of magnetic fabric in palaeocurrent estimation. In *Palaeogeophysics* (ed. S. K. Runcorn). Academic Press, London, New York, pp. 445–464.

Hammo-Yassi, N. B. and Andrews, P. (1981). Archaeomagnetic dating of the Earliest Man in Britain. (Abs.) *Geophys. J. R. astr. Soc.*, **65**, 268.

Hammo-Yassi, N. B. and Clark, A. J. (1982). A new archaeomagnetic curve for Britain. (Abs.) *Geophys. J. R. astr. Soc.*, **69**, 292.

Hancock, J. M. and Kauffman, E. G. (1979). The great transgression of the Late Cretaceous. *J. Geol. Soc., Lond.*, **136**, 175–186.

Hannah, J. L. and Verosub, K. L. (1980). Tectonic implications of remagnetized upper Paleozoic strata of the northern Sierra Nevada. *Geology*, **8**, 520–524.

Hardee, H. C. (1981). Thermal property measurements in a fresh pumice flow at Mt St Helens. *Geophys. Res. Letters*, **8**, 210–212.

Hargraves, R. B. (1959). Magnetic anisotropy and remanent magnetism in hemo-ilmenite from ore deposits at Allard Lake, Quebec. *J. Geophys. Res.*, **64**, 1565–1578.

Hargraves, R. B. (ed.) (1980). *Physics of Magmatic Processes*. Princeton University Press, N.J., p. 585.

Hargraves, R. B. and Onstott, T. C. (1980). Palaeomagnetic results from some Southern African Kimberlites, and their tectonic significance. *J. Geophys. Res.*, **85**, 3587–3596.

Hargraves, R. B. and Perkins, W. E. (1969). Investigations of the effect of shock on natural remanent magnetism. *J. Geophys. Res.*, **74**, 3576–3589.

Hargraves, R. B. and Roy, D. W. (1974). Paleomagnetism of anorthosite in and around the Charlevoix cryptoexplosion structure, Quebec. *Canad. J. Earth. Sci.*, **11**, 854–859.

Hargraves, R. B. and Young, W. M. (1969). Source of stable remanent magnetism in Lambertsville diabase. *Amer. J. Sci.*, **267**, 1161–1177.

Harland, W. B. (1964). Critical evidence for a great Infra-Cambrian Glaciation. *Geol. Rundsch.*, **54**, 45–61.

Harland, W. B., Cox, A., Llewellyn, P. G., Pickton, C. A. G., Smith, A. G. and Walters, R. (1982). *A Geological Time Scale*, Academic Press, London.

Harland, W. B. and Herod, K. N. (1975). Glaciations through time. In *Ice Ages: Ancient and Modern* (eds A. E. Wright and F. Moseley). Seel House Press, Liverpool, pp. 189–216.

Harold, M. R. (1960). Magnetic dating: kiln-wall fall out. *Archaeometry*, **3**, 45–47.

Harrison, C. G. A. (1968). Evolutionary processes and reversals of the Earth's magnetic field. *Nature*, **217**, 46–47.

Harrison, C. G. A. (1971). A seamount with a nonmagnetic top. *Geophys.*, **36**, 349–357.

Harrison, C. G. A. (1976). Magnetization of the oceanic crust. *Geophys. J. R. astr. Soc.*, **47**, 257–283.

Harrison, C. G. A. (1980a). Secular variation and excursions of the Earth's magnetic field. *J. Geophys. Res.*, **85**, 3511–3522.

Harrison, C. G. A. (1980b). Spreading rates and heat flow. *Geophys. Res. Letters*, **7**, 1041–1044.

Harrison, C. G. A. (1980c). Analysis of the magnetic vector in a single rock specimen. *Geophys. J. R. astr. Soc.*, **60**, 489–492.

Harrison, C. G. A. and Funnell, B. M. (1964). Relationship of paleomagnetic reversals and micropaleontology in two late Cainozoic cores from the Pacific Ocean. *Nature*, **204**, 566.

Harrison, C. G. A., Jarrard, R. D., Vacquier, V. and Larson, R. L. (1975). Paleomagnetism of Cretaceous Pacific seamounts. *Geophys. J. R. astr. Soc.*, **42**, 859–882.

Harrison, C. G. A., McDougall, I. and Watkins, N. D. (1979). A geomagnetic field reversal time scale back to 13.0 million years before present. *Earth Plant. Sci. Letters*, **42**, 143–152.

Harrison, C. G. A. and Somayajulu, B. L. K. (1966). Behaviour of the earth's magnetic field during a reversal. *Nature*, **212**, 1193–1195.

Harrison, C. G. A. and Watkins, N. D. (1979). Comparison of the offset dipole and zonal non-dipole geomagnetic field models using Icelandic paleomagnetic data. *J. Geophys. Res.*, **84**, 627–635.

Hart, M. B. and Tarling, D. H. (1974). Cenomanian palaeogeography of the North Atlantic and possible Mid-Cenomanian eustatic movements and their implications. *Palaeogeog., Palaeoclim., Palaeoecol.*, **15**, 95–108.

Hartmann, G. (1544). *Letter*. Published in *Sz. Ges. Endkunde*, **32**, 112–136 (1897).

Hassan, F. A. (1978). Sediments in archaeology: methods and implications for paleoenvironmental and cultural analysis. *J. Field Archaeol.*, **5**, 197–213.

Hays, J. D. (1971). Faunal extinctions and reversal of the Earth's magnetic field. *Geol. Soc. Amer. Bull.*, **82**, 2433–2447.

Hays, J. D. and Opdyke, N. D. (1967). Antarctic Radiolaria, magnetic reversals and climatic change. *Science*, **158**, 1001–1011.

Hays, J. D. and Pitman, W. C. (1973). Lithospheric plate motions, sea level changes and climatic and ecological consequences. *Nature*, **246**, 18–22.

Hays, J. D., Saito, T., Opdyke, N. D. and Burckle, L. H. (1969). Pliocene–Pleistocene sediments of the equatorial Pacific. *Geol. Soc. Amer. Bull.* **80**, 1481–1514.

Hedley, I. G. (1968). Chemical remanent magnetization of the FeOOH, Fe_2O_3 system. *Phys. Earth Planet. Ints*, **1**, 103–121.

Hedley, I. G. (1970). *The magnetic properties of fine grained ferric oxides and their hydrates*. Ph.D. thesis, University of Newcastle upon Tyne, p. 226.

Heirtzler, J. R., Dickson, G. O., Herron, E. M., Pitman, W. C. III and Le Pichon, X, (1968). Marine magnetic anomalies, geomagnetic field reversals, and motions of the ocean floor and continents. *J. Geophys. Res.*, **73**, 2119–2136.

Heller, F. (1973). Magnetic anisotropy of granitic rocks of the Bergell Massif (Switzerland). *Earth Planet. Sci. Letters*, **20**, 180–188.

Heller, F. (1980). Self-reversal of natural remanent magnetization in the Olby-Laschamp lavas. *Nature*, **284**, 334–335.

Heller, F. and Markert, H. (1973). The age of viscous remanent magnetization of Hadrian's Wall (Northern England). *Geophys. J. R. astr. Soc.*, **31**, 395–406.

Heller, F. and Petersen, N. (1982). The Laschamp excursion. *Phil. Trans. R. Soc., Lond.*, **A306**, 169–177.

Hellmann, G. (1896). *Neudrücke von Schriften and Karten über Meteorologie und Erdmagnetismus*. Asher, Berlin.

Helsley, C. E. (1969). Magnetic reversal stratigraphy of the Lower Triassic Moenkopi formation of western Colorado. *Geol. Soc. Amer., Bull.*, **80**, 2431–2450.

Helsley, C. E. and Steiner, M. (1969). Evidence for long intervals of normal polarity during the Cretaceous period. *Earth Planet. Sci. Letters*, **5**, 325–332.

Helsley, C. E. and Steiner, M. (1974). Paleomagnetism of the lower Triassic Moenkopi formation. *Geol. Soc. Amer. Bull.*, **85**, 457–464.

Henry, B. (1973). Studies of microtectonics, anisotropy of magnetic susceptibility and palaeomagnetism of the Permian Dôme de Barrot (France): Palaeotectonic and sedimentological implications. *Tectonophys.*, **17**, 61–72.

Henry, B. (1977a). Relations entre déformation et propriétés magnétiques dans les roches volcaniques des Alpes françaises. *Mem. B.R.G.M.*, **91**, 79–86.

Henry, B. (1977b). Structural implications of palaeomagnetic studies in the south-western border of the Pelvaux Massif (French Alps). *Trans. Amer. Geophys. Union*, **58**, 916.

Henry, B. (1980). *Contribution a l'étude des propriétés magnétiques de roches magnétiques des Alpes. Consequences structurales, regionales et génerales*. D.Sc. thesis, Université P. et M. Curie, Paris.

Henry, S. G. (1979). Chemical demagnetization methods, procedures, and applications through vector analysis. *Canad. J. Earth Sci.*, **16**, 1832–1841.

Henshaw, P. C. and Merrill, R. T. (1979). Characteristics of drying remanent magnetization in sediments. *Earth Planet. Sci. Letters*, **43**, 315–320.

Henshaw, P. C. and Merrill, R. T. (1980). Magnetic and chemical changes in marine sediments. *Revs Geophys. Space Phys.*, **18**, 483–504.

Heptonstall, W. B. (1977). Plate linkage mechanism to account for oroclinal deformation in the Western Cordillera of North America. *Nature*, **268**, 27–32.

Hess, H. H. (1960). Preprint of Hess (1962).

Hess, H. H. (1962). History of ocean basins. In *Petrologic Studies: A volume in honor of A. F. Buddington* (eds Engels *et al.*), Geological Society of America, New York, pp. 599–620.

Hesse, M. B. (1961). *Forces and Fields: The Concept of Action at a Distance in the History of Physics*, Melcon, London.

Hicken, A., Irving, E., Law, L. K. and Hastie, J. (1972). Catalogue of palaeomagnetic directions and poles. *Earth Sci. Branch, Dept Energy, Mines & Resources, Publ.* **45**, p. 135.

Hide, R. (1966). Free hydromagnetic oscillations of the Earth's core and the theory of geomagnetic secular variation. *Phil. Trans. R. Soc., Lond.*, **A259**, 615–648.

Hide, R. (1982). On the role of rotation in the generation of magnetic fields by fluid motions. *Phil. Trans. R. Soc., Lond.*, **A303**, 223–234.

Hide, R. and Horai, K. (1968). On the topography of the core–mantle interface. *Phys. Earth. Planet. Ints*, **1**, 305–308.

Hide, R. and Malin, S. R. C. (1970). Novel correlations between global features of the Earth's gravitational and magnetic fields. *Nature*, **225**, 605–609.

Hide, R. and Malin, S. R. C. (1979). The size of Jupiter's electrically conducting fluid core. *Nature*, **280**, 42–43.

Hide, R. and Roberts, P. H. (1979). How strong is the magnetic field in the Earth's liquid core? *Phys. Earth Planet. Ints*, **20**, 124–126.

Hillhouse, J. (1977). Paleomagnetism of the Triassic Nikolai greenstone, McCarthy Quadrangle, Alaska. *Canad. J. Earth. Sci.*, **14**, 2578–2592.

Hirooka, K. (1971). Archaeomagnetic study for the past 2,000 years in southwest Japan. *Mem. Fac. Sci., Kyoto Univ., Ser. Geol. Mineral.*, **38**, 167–207.

Hoblitt, R. P. and Kellogg, K. S. (1979). Emplacement temperatures of unsorted and unstratified deposits of volcanic rock debris as determined by paleomagnetic techniques. *Geol. Soc. Amer. Bull.*, **90**, 633–642.

Hoffman, K. A. (1979). Behaviour of the geodynamo during reversal: A phenomenological model. *Earth Planet. Sci. Letters*, **44**, 7–17.

Hoffman, K. A. (1982). The testing of geomagnetic reversal models: recent developments. *Phil. Trans. R. Soc., Lond.*, **A306**, 147–158.

Hoffman, K. A., Baker, J. R. and Banerjee, S. K. (1979). Combining palaeointensity methods: a dual-valued determination on lunar sample 10017, 135. *Phys. Earth Planet. Ints*, **20**, 317–323.

Hoffman, K. A. and Day, R. (1978). Separation of multi-component NRM: a general method. *Earth Planet. Sci. Letters*, **40**, 433–438.

Hoffman, K. A. and Fuller, M. (1978). Transitional field configurations and geomagnetic reversals. *Nature*, **273**, 715–718.

Holm, P. E. and Wood, D. S. (1980). The influence of strain heterogeneity on shape analytical methods. *Abs. with Prog., Geol. Soc. Amer.*, **12**, 449.

Holmes, A. (1927). Some problems of physical geology and the Earth's thermal history. *Geol. Mag.*, **64**, 263–278.

Hood, L. L., Russell, C. T. and Coleman, P. J. (1981). Contour maps of lunar remanent magnetic fields. *J. Geophys. Res.*, **86**, 1055–1069.

Horn, D. R., Horn, B. M. and Delach, M. N. (1973). Ocean manganese nodules –

metal values and mining sites. *N.S.F. Tech. Report, IDOE*, p. 57.

Hospers, J. (1951). Remanent magnetism of rocks and the history of the geomagnetic field. *Nature*, **168**, 1111–1112.

Hospers, J. (1953a). Reversals of the main geomagnetic field I. *Proc. Kon. Med. Akad. Wet., Ser. B*, **56**, 467–476.

Hospers, J. (1953b). Reversals of the main geomagnetic field II. *Proc. Kon. Med. Akad. Wet., Ser. B*, **56**, 477–491.

Hospers, J. (1954a). Reversals of the main geomagnetic field III. *Proc. Kon. Med. Akad. Wet., Ser. B*, **57**, 112–121.

Hospers, J. (1954b). Magnetic correlation in volcanic districts. *Geol. Mag.*, **91**, 352–360.

Hoye, G. S. (1981). Archaeomagnetic secular variation record of Mount Vesuvius. *Nature*, **291**, 216–218.

Hoye, G. S. (1982). A magnetic investigation of kiln wall distortion. *Archaeometry*, **24**, 80–84.

Hoye, G. S. and Evans, M. E. (1975). Remanent magnetizations in oxidized olivines. *Geophys. J. R. astr. Soc.*, **41**, 139–151.

Hrouda, F. (1973). A determination of the symmetry of the ferromagnetic mineral fabric in rocks on the basis of the magnetic susceptibility anisotropy measurements. *Geol. Beitr. Geophysik*, **82**, 390–396.

Hrouda, F. (1976a). The origin of cleavage in the light of magnetic anisotropy investigations. *Phys. Earth Planet. Ints*, **13**, 132–142.

Hrouda, F. (1976b). A model for the orientation process of ferromagnetic minerals in slates. *Earth Planet. Sci. Letters*, **33**, 101–110.

Hrouda, F. (1978). The magnetic fabric in some folds. *Phys. Earth Planet. Ints*, **17**, 89–97.

Hrouda, F. (1979). The strain interpretation of magnetic anisotropy in rock of the Nizky Jesenik Mountains. *Sborn. geol. Ved., Uzita Geofyz.*, **16**, 27–62.

Hrouda, F. (1982). Magnetic anisotropy of rocks and its application in geology and geophysics. *Geophys. Surv.*, **5**, 37–82.

Hrouda, F. and Chlupáčová, M. (1980). The magnetic fabric in the Nasavarky Massif. *Čas. Miner. Geol.*, **25**, 17–27.

Hrouda, F., Chlupáčová, M. and Rejl, L. (1971a). The mimetic fabric of magnetite in some foliated granodiorites, as indicated by magnetic anisotropy. *Earth Planet. Sci. Letters*, **11**, 381–384.

Hrouda, F. and Janák, F. (1971). A study of the haematite fabric of some red sediments on the basis of their magnetic susceptibility anisotropy. *Sedi. Geol.*, **6**, 187–199.

Hrouda, F. and Janák, F. (1976). The changes in shape of the magnetic susceptibility ellipsoid during progressive metamorphism and deformation. *Tectonophys.*, **34**, 135–148.

Hrouda, F., Janák, F. and Rejl, L. (1978). Magnetic anisotropy and ductile deformation of rocks in zones of progressive regional metamorphism. *Gerlaids Beitr. Geophys., Leipzig*, **87**, 126–134.

Hrouda, F., Janák, F., Rejl, L. and Weiss, J. (1971b). The use of magnetic susceptibility anisotropy for estimating the ferromagnetic mineral fabrics of metamorphic rocks. *Geol. Runds*, **60**, 1124–1142.

Hutchings, A. (1967). Computation of the behaviour of two- and three-axis rotating systems. In *Methods in Palaeomagnetism* (eds D. W. Collinson, K. M. Creer and

329

S. K. Runcorn), Elsevier, Amsterdam, pp. 222–236.

Hutson, W. H. (1980). Bioturbidation of deep-sea sediments: oxygen isotopes and stratigraphic uncertainty. *Geology*, **8**, 127–130.

Huxtable, J., Aitken, M. J. and Bonhommet, N. (1978). Thermoluminescence dating of sediment baked by lava flows of the Chaîne des Puys. *Nature*, **275**, 207–209.

Hyodo, M. and Yaskawa, K. (1980). Geomagnetic secular variation recorded in the stable magnetic remanence of the young sediments collected from Osaka Bay. *J. Geomagn. Geoelect.*, **32**, 581–584.

Idnurm, M. and Cook, P. J. (1980). Palaeomagnetism of beach ridges in South Australia and the Milankovitch theory of ice ages. *Nature*, **286**, 699–702.

Incoronato, A. (1981). A preliminary palaeomagnetic survey of the Phlegrean Fields, Italy. (Abs.) *Geophys. J. R. astr. Soc.*, **65**, 268.

Institute of Geological Sciences (1978). Annual mean values of geomagnetic elements 1978. *Geomagn. Bull. Inst. Geol. Sci.*, No. 10, p. 296.

Irving, E. (1956). The magnetisation of the Mesozoic dolerites of Tasmania. *Proc. R. Soc. Tasmania*, **90**, 157–168.

Irving, E. (1957a). The origin of the palaeomagnetism of the Torridonian Sandstones of north-west Scotland. *Phil. Trans. R. Soc., Lond.*, **A250**, 110.

Irving, E. (1957b). Directions of magnetization in the Carboniferous glacial varves of Australia. *Nature*, **180**, 280–281.

Irving, E. (1960a). Palaeomagnetic pole positions, Part I. *Geophys. J. R. astr. Soc.*, **3**, 96–111.

Irving, E. (1960b). Palaeomagnetic directions and pole positions, Part II. *Geophys. J. R. astr. Soc.*, **3**, 444–449.

Irving, E. (1961). Palaeomagnetic directions and pole positions, Part III. *Geophys. J. R. astr. Soc.*, **5**, 70–79.

Irving, E. (1962a). Palaeomagnetic directions and pole positions, Part IV. *Geophys. J. R. astr. Soc.*, **6**, 263–267.

Irving, E. (1962b). Palaeomagnetic directions and pole positions, Part V. *Geophys. J. R. astr. Soc.*, **7**, 263–274.

Irving, E. (1964). *Palaeomagnetism and its application to geological and geophysical problems*. Wiley, New York, p. 399.

Irving, E. (1965). Palaeomagnetic directions and polar positions, Part VII. *Geophys. J. R. astr. Soc.*, **9**, 185–194.

Irving, E. (1970). The Mid-Atlantic Ridge at 45°N. XVI. Oxidation and magnetic properties of basalt; review and discussion. *Canad. J. Earth Sci.*, **7**, 1528–1538.

Irving, E. (1977). Drift of the major continental blocks since the Devonian. *Nature*, **270**, 304–309.

Irving, E. (1979). Paleopoles and palaeolatitudes of North America and speculations about displaced terrains. *Canad. J. Earth. Sci.*, **16**, 669–694.

Irving, E. (1981). Phanerozoic continental drift. *Phys. Earth Planet. Ints*, **24**, 197–204.

Irving, E. and Briden, J. C. (1962). Paleolatitude of evaporite deposits. *Nature*, **196**, 425–428.

Irving, E. and Gaskell, T. F. (1962). The palaeogeographic latitude of oil fields. *Geog. J.*, **7**, 54–64.

Irving, E. and Green, R. (1958). Polar wandering relative to Africa. *Geophys. J. R. astr. Soc.*, **1**, 64–72.

Irving, E. and Hastie, J. (1975). Catalogue of palaeomagnetic directions and poles.

Geomagn. Series, Geomagn. Surv. Canada, No. 3, p. 42.

Irving, E. and Major, A. (1964). Post-depositional detrital remanent magnetization in a synthetic sediment. *Sedimentology*, **3**, 135–143.

Irving, E. and McGlynn, J. C. (1976). Proterozoic magnetostratigraphy and the tectonic evolution of Laurentia. *Phil. Trans. R. Soc., Lond.*, **A280**, 243–265.

Irving, E. and McGlynn, J. C. (1979). Paleomagnetism in the Coronation Geosyncline and arrangements of continents in the Middle Proterozoic. *Geophys. J. R. astr. Soc.*, **58**, 309–336.

Irving, E. and McGlynn, J. C. (1981). On the coherence, rotation and paleolatitude of Laurentia in the Proterozoic. In *Precambrian Plate Tectonics* (ed. A. Kröner), Elsevier, Amsterdam, pp. 561–598.

Irving, E., Molyneux, L. and Runcorn, S. K. (1966). The analysis of remanent intensities and susceptibilities of rocks. *Geophys. J. R. astr. Soc.*, **10**, 451–464.

Irving, E., North, F. K. and Couillard, R. (1974). Oil, climate, and tectonics. *Canad. J. Earth Sci.*, **11**, 1–17.

Irving, E. and Park, J. K. (1972). Hairpins and superintervals. *Canad. J. Earth Sci.*, **9**, 1318–1324.

Irving, E., Park, J. K. and Roy, J. L. (1972). Palaeomagnetism and the origin of the Grenville Front. *Nature*, **236**, 344–346.

Irving, E. and Parry, L. G. (1963). The magnetism of some Permian rocks from New South Wales. *Geophys. J. R. astr. Soc.*, **7**, 395–411.

Irving, E. and Pulliah, G. (1976). Reversals of the geomagnetic field, magnetostratigraphy, and relative magnitude of paleosecular variation in the Phanerozoic. *Earth-Sci. Revs*, **12**, 35–64.

Irving, E., Robertson, W. A., Stott, P. M., Tarling, D. H. and Ward, M. A. (1961b). Treatment of partially stable sedimentary rocks showing planar distribution of direction of magnetization. *J. Geophys. Res.*, **66**, 1927–1933.

Irving, E. and Stott, P. M. (1963). Palaeomagnetic directions and pole positions, Part VI. *Geophys. J. R. astr. Soc.*, **8**, 249–257.

Irving, E., Stott, P. M. and Ward, M. A. (1961a). Demagnetization of igneous rocks by alternating magnetic fields. *Phil. Mag.*, **6**, 225–241.

Ishikawa, Y. (1967). Magnetic properties of a single crystal of Fe_2TiO_4. *Phys. Letters*, **24A**, 725–727.

Ishikawa, Y. and Syono, Y. (1963). Order–disorder transformation and reverse thermoremanent magnetism in the $FeTiO_3$–Te_2O_3 system. *J. Phys. Chem. Solids*, **24**, 517–528.

Ising, G. (1943). On the magnetic properties of varved clay. *Arkiv. f. Matematik, Ast. Fysik*, **29A**, 1–37.

Ito, H. and Fuller, M. (1967). A paleomagnetic field study of the reversal process of the geomagnetic field. In: *Palaeogeophysics* (ed. S. K. Runcorn), Academic Press, London, 133–138.

Ixer, R. A., Turner, P. and Waugh, B. (1979). Authigenic iron and titanium oxides in Triassic red beds: (St Bees Sandstone), Cumbria, Northern England. *Geol. J.*, **14**, 179–192.

Jacobs, J. A. (1963). *The Earth's Core and Geomagnetism*, Pergamon, Oxford, p. 137.

Jacobs, J. A. and Alldredge, K. D. (1975). The Earth's interior and the Earth's rotation. In *Growth Rhythms and the History of the Earth's Rotation* (eds G. A. Rosenberg and S. K. Runcorn), Wiley, London, 337–351.

Jaeger, J. C. (1964). Thermal effects of intrusions. *Revs Geophys.*, **2**, 443–446.

James, C. H. (1969). *Sedimentary Ores: Ancient and Modern*, Spec. Publ., Geology, University of Leicester, p. 305.

Jardetzky, W. S. (1958). Polar wandering, shifting of the Earth's axis and dipole variations. *Ann. Geophys.*, **14**, 502–505.

Jenkyns, H. C. (1980). Cretaceous anoxic events: from continents to oceans. *J. Geol. Soc., Lond.*, **137**, 171–188.

Johnson, E. A., Murphy, T. and Torreson, O. W. (1948). The prehistory of the Earth's magnetic field. *Terr. Magn. Atmos. Elect.*, **53**, 349–372.

Johnson, H. P., Kinoshita, H. and Merrill, R. T. (1975). Rock magnetism and paleomagnetism of some North Pacific deep-sea sediments. *Geol. Soc. Amer. Bull.*, **86**, 412–420.

Johnson, N. M., Opdyke, N. D. and Lindsay, E. H. (1975). Magnetic polarity stratigraphy of Pliocene-Pleistocene terrestrial deposits and vertebrate faunas, San Pedro Valley, Arizona. *Geol. Soc. Amer. Bull.*, **86**, 5–12.

Johnston, M. J. S., Smith, B. E. and Burford, R. O. (1980). Local magnetic field measurements and fault creep observations on the San Andreas Fault. *Tectonophys.* **64**, 47–57.

Johnston, M. J. S. and Stacey, F. D. (1969a). Volcanomagnetic effect observed on Mt Ruapehu, New Zealand. *J. Geophys. Res.*, **74**, 6541–6544.

Johnston, M. J. S. and Stacey, F. D. (1969b). Transient magnetic anomalies accompanying volcanic eruptions in New Zealand. *Nature*, **224**, 1289–1290.

Jones, D. L., Robertson, I. D. M. and McFadden, P. L. (1975). A palaeomagnetic study of Precambrian dyke swarms associated with the Grate Dyke of Rhodesia. *Trans. Geol. Soc. S. Afr.*, **78**, 57–65.

Jones, D. L., Siberling, N. J., and Hillhouse, J. (1977). Wrangellia – A displaced terrane in northwestern North America. *Canad. J. Earth Sci.*, **14**, 2565–2577.

Jurdy, D. M. (1981). True polar wander. *Tectonophys.*, **74**, 1–16.

Kafafy, A. M. (1981). The magnetic susceptibility anisotropy of rocks from Shap region, Lake District, England. (Abs.) *Geophys. J. R. astr. Soc.*, **65**, 272.

Kafafy, A. M. (1982). *Magnetic studies of metamorphic aureoles of some British granites*. Ph.D. thesis, University of Newcastle upon Tyne.

Kato, Y. and Nagata, T. (1949). On the secular variation in geomagnetic declination in the Historic time of Japan. *Proc. 7th Pacific Sci. Cong.*, **2**, 562–564.

Kawai, N., Yaskawa, K., Nakajima, T., Torii, M. and Horie, S. (1972). Oscillating geomagnetic field with a recurring reversal discovered from Lake Biwa. *Proc. Japan Acad.*, **48**, 186–190.

Keating, B. H. and Helsley, C. E. (1978a). Magnetostratigraphy of Cretaceous age sediments from sites 361, 363 and 364. *Init. Reps Deep Sea Drilling Project*, **40**, 459–467.

Keating, B. H. and Helsley, C. E. (1978b). Magnetostratigraphic studies of Cretaceous sediments from DSDP site 369. *Init. Reps Deep Sea Drilling Project*, **41**, 983–986.

Keating, B. H. and Helsley, C. E. (1978c). Paleomagnetic results from DSDP hole 391C and the magnetostratigraphy of Cretaceous sediments from the Atlantic ocean floor. *Init. Reps Deep Sea Drilling Project*, **44**, 523–528.

Keating, B. and Sager, W. (1980). Watkins seamount: Preliminary paleomagnetic results. *J. Geophys. Res.*, **85**, 3567–3571.

References

Keefer, C. M. and Shive, P. N. (1981). Curie temperature and lattice constant reference contours for synthetic titanomaghemites. *J. Geophys. Res.*, **86**, 987–998.

Kellogg, K. S. (1980). Paleomagnetic evidence for oroclinal bending of the southern Antarctic Peninsula. *Geol. Soc. Amer. Bull.*, **9**, 414–420.

Kent, D. V. (1979). Paleomagnetism of the Devonian Onondaga Limestone Revisited. *J. Geophys. Res.*, **84**, 3576–3588.

Kent, D. V., Ninkovich, D., Pescatore, T. and Sparks, S. R. J. (1981). Palaeomagnetic determination of emplacement temperature of Vesuvius AD 79 pyroclastic deposits. *Nature*, **290**, 393–396.

Kent, D. V. and Opdyke, N. D. (1978). Paleomagnetism of the Devonian Catskill red beds: Evidence for motion of the coastal New England – Canadian maritime region relative to cratonic North America. *J. Geophys. Res.*, **83**, 4441–4450.

Kent, D. V. and Opdyke, N. D. (1980). Paleomagnetism of Siluro–Devonian rocks from eastern Maine. *Canad. J. Earth Sci.*, **17**, 1653–1665.

Kern, J. W. (1961a). Effects of moderate stresses on directions of thermoremanent magnetization. *J. Geophys. Res.*, **66**, 3801–3805.

Kern, J. W. (1961b). The effect of stress on the susceptibility and magnetization of a partially magnetized multidomain system. *J. Geophys. Res.*, **66**, 3807–3816.

Kern, J. W. (1961c). Stress stability of remanent magnetization. *J. Geophys. Res.*, **66**, 3817–3820.

Kersley, L., Hajeb-Hosseinier, H. and Edwards, K. J. (1978). Post-geomagnetic storm protonospheric replenishment. *Nature*, **271**, 429–430.

Khan, M. A. (1962). The anisotropy of magnetic susceptibility of some igneous and metamorphic rocks. *J. Geophys. Res.*, **67**, 2873–2885.

Khramov, A. N. (1958). *Palaeomagnetism and Stratigraphic Correlation*, Gostoptechizdat, Leningrad (Trans. p. 204, A.N.U. Australia). Trans. A. J. Lojkine, ed. E. Irving.

Khramov, A. N. (1967). The Earth's magnetic field in the late Palaeozoic. *Izv. Earth Phys.*, **1967**, 86–108.

Khramov, A. N. and Andreyeva, O. L. (1964). Application of the determination of a compensating field for finding the direction of the Primary magnetization of rocks. *Izv. Geophys. Ser.*, 552–555, 329–331.

Khramov, A. N. and Sholpo, L. Y. (1967). *Synoptic tables of USSR palaeomagnetic data*. (Trans. E. R. Hope from *Palaeomagnetism*.)

Kim, J. Y. and Carmichael, R. S. (1977). Hydrostatic pressure effects on saturation remanent magnetization, susceptibility and magnetic hardness of magnetite. *Eos*, **58**, 382.

King, R. F. (1955). The remanent magnetism of artificially deposited sediments. *Mon. Not. R. astr. Soc.*, **7**, 115–134.

King, R. F. (1966). The magnetic fabric of some Irish granites. *Geol. J.*, **5**, 43–66.

King, R. F. and Rees, A. I. (1962). The measurement of the anisotropy of magnetic susceptibility of rocks by the torque method. *J. Geophys. Res.*, **67**, 1565–1572.

King, R. F. and Rees, A. I. (1966). Detrital magnetism in sediments: An examination of some theoretical models. *J. Geophys. Res.*, **71**, 561–571.

Kinoshita, H. (1970). Lists of archeomagnetic and paleomagnetic results 1950–1970, in Japan, *J. Geomagn. Geoelect.*, **22**, 507–550.

Kinsman, D. J. J. (1969). Modes of formation, sedimentary associations, and diagnostic features of shallow-water and supratidal evaporites. *Amer. Assoc. Petrol.*

Geol. Bull., **53**, 830–840.

Kirschvink, J. L. (1980). The least-squares line and plane and the analysis of palaeo-magnetic data. *Geophys. J. R. astr. Soc.*, **62**, 699–718.

Kirschvink, J. L. and Lowestam, H. A. (1979). The role of biogenic minerals in the magnetization of sediments. *Eos*, **60**, 247.

Kligfield, R., Owens, W. H. and Lowrie, W. (1981). Magnetic susceptibility aniso-tropy, strain and progressive deformation in Permian sediments from the Maritime Alps (France). *Earth Planet. Sci. Letters*, **55**, 181–189.

Klinkhammer, G. P. (1980). Early diagenesis in sediments from the Eastern Equa-torial Pacific, II. Pore water metal results. *Earth Planet. Sci. Letters*, **49**, 81–101.

Klitgord, K. D., Huestis, S. P., Mudi, J. D. and Parker, R. L. (1975). An analysis of near-bottom magnetic anomalies: Sea-floor spreading and the magnetized layer. *Geophys. J. R. astr. Soc.*, **43**, 387–424.

Klootwijk, C. T. (1979a). A review of palaeomagnetic data from the Indo-Pakistani fragment of Gondwanaland. In *Geodynamics of Pakistan* (eds A. Farah and K. A. De Jong), Geological Survey of Pakistan, Quetta, pp. 41–80.

Klootwijk, C. T. (1979b). A summary of palaeomagnetic data from extra-peninsula Indo-Pakistan and south central Asia: implications for collision tectonics. In: *Structural Geology of the Himalaya* (ed. P. S. Saklaui) pp. 307–360.

Klootwijk, C. T. (1980). Early Palaeozoic palaeomagnetism in Australia. *Tectono-phys.*, **64**, 249–332.

Klootwijk, C. T. and Bingham, D. K. (1980). The extent of Greater India, III. Palaeomagnetic data from the Tibetan Sedimentary series, Thakkhola region, Nepal. Himalaya. *Earth Planet. Sci. Letters*, **51**, 381–405.

Klootwijk, C. T. and Conaghan, P. J. (1979). The extent of Greater India, I. Pre-liminary palaeomagnetic data from the Upper Devonian of the eastern Hindukush, Chitral (Pakistan). *Earth Planet. Sci. Letters*, **42**, 167–182.

Klootwijk, C. T., Nazirullah, R., De Jong, K. A. and Ahmed, H. (1981). A Palaeo-magnetic reconnaissance of Northeastern Baluchistan, Pakistan. *J. Geophys. Res.*, **86**, 289–306.

Klootwijk, C. T. and Peirce, J. W. (1979). India's and Australia's pole path since the late Mesozoic and the India–Asia collision. *Nature*, **282**, 605–607.

Klootwijk, C. T., Sharma, M. L., Gergan, J., Tirkey, B., Shah, S. K. and Agarwal, V. (1979). The extent of Greater India, II. Palaeomagnetic data from the Ladakh Intrusives at Kargil, Northwestern Himalayas. *Earth Planet. Sci. Letters*, **44**, 47–64.

Klootwijk, C. T. and Vandenberg, J. (1975). The rotation of Italy: Preliminary palaeomagnetic data from the Umbrian sequence, Northern Apennines. *Earth Planet. Sci. Letters*, **25**, 263–273.

Kneen, S. J. (1976). The relationship between the magnetic and strain fabrics of some haematite-bearing Welsh slates. *Earth Planet. Sci. Letters*, **31**, 413–416.

Knox, E. G., Armstrong, E., Lancashire, R., Wall, M. and Haynes, R. (1979). Heart attacks and geomagnetic activity. *Nature*, **281**, 564–565.

Kodama, K., Uyeda, S. and Isezaki, N. (1978). Paleomagnetism of Suiko seamount Emperor Seamount Chain. *Geophys. Res. Letters*, **5**, 165–168.

Koenigsberger, J. G. (1938). Natural residual magnetism of eruptive rocks. *Terr. Magn. Atmos. Elect.*, **43**, 119–127; 299–320.

Kono, M. (1974). Intensities of the Earth's magnetic field about 60 m.y. ago deter-mined from the Deccan Trap Basalts, India. *J. Geophys. Res.*, **79**, 1135–1151.

Kono, M. (1980). Statistics of Paleomagnetic inclination data. *J. Geophys. Res.*, **85**, 3878–3882.

Kono, M. and Morgan, W. J. (1978). Paleolatitude of Emperor seamounts from DSDP leg 55. *Eos*, **59**, 297.

Kono, M. and Tanaka, H. (1977). Influence of partial pressure of oxygen on thermoremanent magnetization of basalts. *Phys. Earth Planet. Ints*, **13**, 276–288.

Kono, M. and Ueno, N. (1977). Paleointensity determination by a modified Thellier method. *Phys. Earth Planet. Ints*, **13**, 305–314.

Kovacheva, M. (1980). Summarised results of the archaeomagnetic investigation of the geomagnetic field variation for the last 8000 yr in south-eastern Europe. *Geophys. J. R. astr. Soc.*, **61**, 57–64.

Kovacheva, M. (1982a). Geomagnetic secular variations according to archaeomagnetic data. *Phil. Trans. R. Soc., Lond.*, **A303**, 79–86.

Kovacheva, M. (1982b). Archaeomagnetic investigations of geomagnetic secular variations. *Phil. Trans. R. Soc., Lond.*, **A306**, 79–85.

Krause, F. (1982). Progress in dynamo theory: kinematic and dynamic models. *Phil. Trans. R. Soc., Lond.*, **A303**, 255–260.

Krause, F. and Radler, K. H. (1979). On the theory of the geomagnetic dynamo based on mean field electrodynamics. *Phys. Earth Planet. Ints*, **20**, 158–171.

Kristjansson, L. and Gudmundsson, A. (1980). Geomagnetic excursion in late-glacial basalt outcrops in south-western Iceland. *Geophys. Res. Letters*, **7**, 337–340.

Kristjansson, L. and McDougall, I. (1982). Some aspects of the late Tertiary geomagnetic field in Iceland. *Geophys. J. R. astr. Soc.*, **68**, 273–294.

Kröner, A., McWilliams, M. O., Germs, G. J. B., Reid, A. B. and Schalk, K. E. L. (1980). Palaeomagnetism of Late Precambrian to Early Proterozoic mixtite-bearing formations in Namibia (South West Africa) – The Nama Group and Blaubeker Formation. *Amer. J. Sci.*, **280**, 942–968.

Krs, M. (1978). Palaeomagnetic evidence of tectonic deformation of blocks in the Bohemian Massif. *Sborn. geol. Rad.*, **31**, 141–154.

Krs, M., Krsova, M. and Roth, Z. (1977). A palaeomagnetic study of Cenomanian – Lower Turonian sediments in the Maravskoslegske Beskydy Mts. *Vect. Ust, ust. geol.*, **52**, 323–332.

Krs, M. and Smid, B. (1979). Palaeomagnetism of Cretaceous rocks of the teschemite association, Outer West Carpathians of Czechoslovakia. *Sborn. Geol. ved, Uz. Geofyz.*, **16**, 7–25.

Ku, T. L., Broecker, W. S. and Opdyke, N. (1968). Comparison of sedimentation rates measured by paleomagnetic and the ionium methods of age determination. *Earth Planet. Sci. Letters*, **4**, 1–16.

Kundig, W., Boemmel, H., Konstabaris, G. and Lindquist, R. H. (1966). Some properties of supported small α-Fe_2O_3 particles determined with Mössbauer effect. *Phys. Rev.*, **142**, 327–333.

Kuznir, N. J. and Bott, M. H. P. (1976). A thermal study of the formation of oceanic crust. *Geophys. J. R. astr. Soc.*, **47**, 83–95.

Kuznir, N. J. and Tarling, D. H. (in press). *An Introduction to Applied Geophysics*, Academic Press, London.

Labeyrie, J. and Gillot, P. Y. (1977). Age of the Laschamp event re-evaluated. (Abs.) *5th Europ. Coll. Geochron, Cosmochron, Isotope Geol.*, Pisa.

Lagus, P. L. and Anderson, D. L. (1968). Tidal dissipation in the Earth and Planets.

Phys. Earth. Planet. Ints, **1**, 505–510.

Laj, C., Nordemann, D. and Pomeau, Y. (1979). Correlation function analyses of geomagnetic field reversals. *J. Geophys. Res.*, **84**, 4511–4515.

Lambeck, K. (1975). Effects of tidal dissipation in the oceans on the Moon's orbit and the Earth's rotation. *J. Geophys. Res.*, **80**, 2917–2925.

Lambeck, K. (1980). *The Earth's variable rotation: geophysical causes and consequences*. Cambridge University Press, p. 449.

Lambert, R. St J. (1976). Archean thermal regimes, crustal and upper mantle temperatures, and a progressive evolutionary model for the Earth. In *The Early History of the Earth* (ed. B. F. Windley), Wiley, London, 363–373.

Langdon, T. G. (1976). Grain boundary deformation processes. In *Deformation of Ceramic Materials* (eds R. C. Bradt and R. E. Tressler), Plenum, New York, 101–126.

Lanoix, M., Strangway, D. W. and Pearce, G. W. (1978). The primordial magnetic field preserved in chondrules of the Allende meteorite. *Geophys. Res. Letters*, **5**, 73–76.

Lapointe, P. L., Roy, J. L. and Morris, W. A. (1978). What happened to the high-latitude palaeomagnetic poles? *Nature*, **273**, 655–657.

Larimer, J. W. (1978). Meteorites: Relics from the Early Solar System. In *The origin of the Solar System* (ed. S. F. Dermott), Wiley, Chichester, pp. 346–393.

Larmor, J. (1919). How could a rotating body such as the sun become a magnet? *Rep. Brit. Assoc.*, 159–160.

Larochelle, A. (1965). The design of a spinner-type remanent magnetometer. *Geol. Surv. Canada*, Paper 64–43, p. 25.

Larson, E. E. and Strangway, D. W. (1966). Magnetic polarity and igneous petrology. *Nature*, **212**, 756–757.

Larson, E. E. and Strangway, D. W. (1968). Discussion of 'Correlation of petrology and natural magnetic polarity in Columbia plateau basalts' by R. L. Wilson and N. D. Watkins. *Geophys. J. R. astr. Soc.*, **15**, 437–441.

Larson, E. E., Watson, D. E. and Jennings, W. (1971). Regional comparison of a Miocene geomagnetic transition in Oregon and Nevada. *Earth Planet. Sci. Letters*, **11**, 391–400.

Larson, R. L. (1976). Late Jurassic and Early Cretaceous evolution of the Western Central Pacific Ocean. *J. Geomagn. Geoelect.*, **28**, 219–236.

Larson, R. L. and Helsley, C. E. (1975). Mesozoic Reversal sequence. *Revs Geophys. Space Phys.*, **13**, 174–176.

Larson, R. L. and Hilde, T. W. C. (1975). A revised time scale of magnetic reversals for the early Cretaceous and late Jurassic. *J. Geophys. Res.*, **80**, 2586–2594.

Larson, R. L. and Pitman, W. C. (1972). World-wide correlation of Mesozoic magnetic anomalies and its implications, *Geol. Soc. Amer. Bull.*, **83**, 3645–3662.

Larson, R. L. and Spiess, F. N. (1969). The East Pacific Rise crest: A near-bottom geophysical profile. *Science*, **163**, 66–67.

Latham, A. G., Schwarcz, H. P., Ford, D. C. and Pearce, G. W. (1979). Palaeomagnetism of stalagmite deposits. *Nature*, **280**, 383–385.

Learmouth, G. S. (1983). *The Sultangang Buddha*. Report of Birmingham City Museum (in press).

Leggett, J. K. (1978). Eustacy and pelagic regimes in the Iapetus Ocean during the Ordovician and Silurian. *Earth Planet. Sci. Letters*, **41**, 163–169.

Le Pichon, X., Sibuet, J.-C. and Francheteau, J. (1977). The fit of the continents around the North Atlantic. *Tectonophys.*, **38**, 169–209.

Lerberkmo, J. F., Evans, M. E. and Baadsgaard, H. (1979). Magnetostratigraphy, biostratigraphy and geochronology of Cretaceous–Tertiary boundary sediments, Red Deer valley. *Nature*, **279**, 26–30.

Levi, S. and Banerjee, S. K. (1976). On the possibility of obtaining relative paleo-intensites from lake sediments. *Earth Planet. Sci. Letters*, **29**, 219–226.

Liddicoat, J. C. (1982). Gauss-Matuyama polarity transition. *Phil. Trans. R. Soc., Lond.*, **A306**, 121–129.

Liddicoat, J. C. and Coe, R. S. (1979). Mono Lake geomagnetic excursion. *J. Geophys. Res.*, **84**, 261–271.

Liddicoat, J. C., Coe, R. S., Lambert, P. W. and Valastro, S. (1979). Paleomagnetic record in Late Pleistocene and Holocene dry lake deposits at Tlapacoya, Mexico. *Geophys. J. R. astr. Soc.*, **59**, 367–377.

Liddicoat, J. C., Opdyke, N. D. and Smith, G. I. (1980). Paleomagnetic polarity in a 930-m core from Searles Valley, California. *Nature*, **286**, 22–25.

Lienert, B. R. and Wasilewski, P. J. (1979). A magnetic study of the serpentinization process at Burro Mountain, California. *Earth Planet. Sci. Letters*, **43**, 406–416.

Lin, R. P. (1979). Constraints on the origins of lunar magnetism from electron reflection measurements of surface magnetic fields. *Phys. Earth Planet. Ints*, **20**, 271–280.

Lindsay, E. H., Jacobs, L. L. and Butler, R. F. (1978). Biostratigraphy and magneto-stratigraphy of Palaeocene terrestrial deposits, San Juan Basin, New Mexico. *Geology*, **6**, 425–429.

Lippmann, F. (1979). Stabilitatsbeziehunged der Tonminerale. *N. Jb. Miner. Abh.*, **136**, 287–309.

Lipshie, S. R., Oertel, G. and Christie, J. M. (1976). Measurement of preferred orientation of phyllosilicates in schists. *Tectonophys.*, **34**, 91–99.

Liritzis, Y. and Thomas, R. (1980). Palaeointensity and thermoluminescence measurements on Cretan kilns from 1300 to 2000 BC. *Nature*, **283**, 54–55.

Lister, C. R. B. (1974). On the penetration of water into Hot Rock. *Geophys. J. R. astr. Soc.*, **39**, 465–509.

Lister, C. R. B. (1977). Qualitative models of spreading-centre processes, including hydrothermal penetration. *Tectonophys.*, **37**, 203–218.

Lister, G. S., Paterson, M. S. and Hobbs, B. E. (1978). The simulation of fabric development in plastic deformation and its application to quartzite: The model. *Tectonophys.*, **45**, 107–158.

Loper, D. E. and Roberts, P. H. (1979). Are planetary dynamos driven by gravi-tational settling? *Phys. Earth Planet. Sci. Ints*, **20**, 192–193.

Løvlie, R., Krzywinski, K. and Kjøde, J. (1979). Impregnation of organic lake sedi-ments for palaeomagnetic measurements. *Phys. Earth Planet. Sci. Ints*, **20**, 22–24.

Lowes, F. J. and Runcorn, S. K. (1951). The analysis of the geomagnetic secular variation. *Phil. Trans. R. Soc., Lond.*, **A243**, 525–546.

Lowes, F. J. and Wilkinson, I. (1963). Geomagnetic dynamo: A laboratory model. *Nature*, **198**, 1158–1160.

Lowes, F. J. and Wilkinson, I. (1968). Geomagnetic dynamo: An improved laboratory model. *Nature*, **219**, 717–718.

Lowrie, W. and Alvarez, W. (1974). Rotation of the Italian Peninsula. *Nature*, **251**, 285–288.

Palaeomagnetism

Lowrie, W., Alvarez, W., Silva, I. P. and Monechi, S. (1980a). Lower Cretaceous magnetic stratigraphy in Umbrian pelagic carbonate rocks. *Geophys. J. R. astr. Soc.*, **60**, 263–281.

Lowrie, W., Channell, J. E. T. and Alvarez, W. (1980b). A review of magnetic stratigraphy investigations in Cretaceous pelagic rocks. *J. Geophys. Res.*, **85**, 3597–3605.

Lowrie, W., Channell, J. E. T. and Heller, F. (1980c). On the credibility of remanent magnetization measurements. *Geophys. J. R. astr. Soc.*, **60**, 493–496.

Lowrie, W. and Fuller, M. (1971). On the alternating field demagnetization characteristics of multidomain thermoremanent magnetization in magnetite. *J. Geophys. Res.*, **76**, 6339–6349.

Lowrie, W. and Kent, D. V. (1978). Viscous remanent magnetization in basaltic samples. *Init. Reps Deep Sea Drilling Project*, **34**, 479–484.

Ludden, J. N. and Thompson, G. (1978). Behaviour of rare earth elements during submarine weathering of theoleiitic basalt. *Nature*, **274**, 147–149.

Lyons, W. B., Gaudette, H. E. and Smith, G. M. (1979). Pore water sampling in anoxic carbonate sediments: oxidation artefacts. *Nature*, **277**, 48–49.

Macdonald, K. C., Miller, S. P., Huestis, S. P. and Spiess, F. N. (1980). Three-dimensional modelling of a magnetic reversal boundary from inversion of deep-tow measurements. *J. Geophys. Res.*, **85**, 3670–3680.

MacDonald, W. D. (1980). Net tectonic rotation, apparent tectonic rotation and the structural tilt correction in paleomagnetic studies. *J. Geophys. Res.*, **85**, 3659–3669.

MacFadden, B. J., Johnson, N. M. and Opdyke, N. D. (1979). Magnetic polarity stratigraphy of the Mio-Pliocene mammal-bearing Big Sandy Formation of Western Arizona. *Earth Planet. Sci. Letters*, **44**, 349–364.

Mackereth, F. J. H. (1958). A portable core sampler for lake deposits. *Limnol. Oceanogr.*, **3**, 181–191.

Mackereth, F. J. H. (1969). A short core sampler for subaqueous deposits. *Limnol. Oceanogr.*, **14**, 145–151.

Madden, T. and Le Mouel, J.-L. (1962). The recent secular variation and the motions at the core surface. *Phil. Trans. R. Soc., Lond.*, **A306**, 271–280.

Magill, J., Cox, A. and Duncan, R. (1981). Tillamook Volcanic Series: Further evidence for tectonic rotation of the Oregon Coast Range. *J. Geophys. Res.*, **86**, 2953–2970.

Malan, D. J. (1963). *Physics of Lightning*, English University Press, p. 176. London.

Malin, S. R. C. and Bullard, E. C. (1981). Direction of the Earth's magnetic field at London, 1570–1975. *Phil. Trans. R. Soc., Lond.*, **A299**, 357–423.

Malin, S. R. C. and Hide, R. (1982). Bumps on the core-mantle boundary: geomagnetic and gravitational evidence revisited. *Phil. Trans. R. Soc., Lond.*, **A306**, 281–287.

Malin, S. R. C. and Srivistava, B. J. (1979). Correlation between heart attacks and magnetic activity. *Nature*, **277**, 646–648.

Mankinen, E. A. and Dalrymple, G. B. (1979). Revised geomagnetic polarity time-scale for the interval 0–5 m.y. BP. *J. Geophys. Res.*, **84**, 615–626.

Mann, C. J. (1972). Faunal extinctions and reversals of the Earth's magnetic field: Discussion. *Geol. Soc. Amer. Bull.*, **83**, 330–335.

Manning, P. G., Williams, J. D. H., Charlton, M. N., Ash, L. A. and Birchall, T. (1979). Mössbauer spectral studies of the diagenesis of iron in a sulphide-rich

338

sediment core. *Nature*, **280**, 134–136.

Manzoni, M. (1975). Paleomagnetic evidence for rotation of northern Calabria. *Geophys. Res. Letters*, **2**, 427–429.

March, A. (1932). Mathematische Theorie de Regelung nach der Korngestalt bei affiner deformation. *Zs. Kristallogr.*, **81**, 285–297.

Mardia, K. V. (1972). *Statistics of Directional Data*. Academic Press, London, p. 357.

Marshall, M. and Cox, A. (1972). Magnetic changes in pillow basalt due to sea floor weathering. *J. Geophys. Res.*, **77**, 6459–6469.

Martin, D. L., Nairn, A. E. M., Noltimier, H. C., Petty, M. H. and Schmitt, T. J. (1978). Paleozoic and Mesozoic Paleomagnetic Results from Morocco. *Tectonophys.*, **44**, 91–114.

Martin, R. J. (1980). Is piezomagnetism influenced by microcracks during cyclic loading? *J. Geomagn. Geoelect.*, **32**, 741–755.

Márton, E. and Márton, P. (1981). Mesozoic palaeomagnetism of the Transdanubian Central Mountains and its tectonic implications. *Tectonophys.*, **72**, 129–140.

Marvin, U. B. and Wasson, J. T. (1979). Extraterrestrial samples: progress and prospects. *Geotimes*, **24**, 22–28.

Mason, R. G. (1958). A magnetic survey off the West Coast of the United States between latitudes 32° and 36°N and longitudes 121° and 128°W. *Geophys. J. R. astr. Soc.*, **1**, 320–329.

Mason, R. G. (1959). Geophysical investigations of the sea floor. *Liverpool & Manchester Geol. J.*, **2**, 389–409.

Mason, R. G. and Raff, A. D. (1961). Magnetic survey off the West Coast of North America, 32°N Latitude to 42°N Latitude. *Bull. Geol. Soc. Amer.*, **72**, 1259–1266.

Matuyama, M. (1929). On the direction of magnetization of basalt in Japan, Tyozen and Manchuria. *Proc. Imp. Acad. Japan*, **5**, 203–205.

Maurain, M. C. (1901). Propriétés des Dépots Électrolytiques de Fer obtenus dans un champ Magnetique. *J. Physique*, **10**, 123–135.

Mauritsch, H. J. and Frisch, W. (1978). Palaeomagnetic data from the central part of the Northern Calcareous Alps, Austria. *J. Geophys.*, **44**, 623–637.

Maxwell, A. E., von Herzen, R. P., Andrew, J. E., Boyle, R. E., Millow, E. D., Hsü, K. J., Percival, S. F. and Saito, T. (1970). *Init. Reps Deep Sea Drilling Project*, **3**, US Government, Washington, p. 806.

McClelland-Brown, E. A. (1981). Palaeomagnetic estimates of temperatures reached in contact metamorphism. *Geology*, **9**, 112–116.

McDougall, I. (1979). The present status of the geomagnetic polarity time scale. In *The Earth: Its Origin Structure and Evolution* (ed. M. W. McElhinny) Academic Press, London, pp. 543–566.

McDougall, I. and Green, R. (1958). The use of magnetic measurements for the study of the structure of Taluo slopes. *Geol. Mag.*, **95**, 252–260.

McDougall, I. and Tarling, D. H. (1963). Dating of polarity zones in the Hawaiian Islands. *Nature*, **200**, 54–56.

McDougall, J., Tarling, D. H. and Warren, S. E. (in press). The magnetic provenancing of some Mediterranean obsidians. *J. Archaeo. Sci.*

McElhinny, M. W. (1964). Statistical significance of the fold test in palaeomagnetism. *Geophys. J. R. astr. Soc.*, **8**, 338–340.

McElhinny, M. W. (1968a). Palaeomagnetic directions and pole positions, VIII. *Geophys. J. R. astr. Soc.*, **15**, 409–430.

Palaeomagnetism

McElhinny, M. W. (1968b). Palaeomagnetic directions and pole positions, IX. *Geophys. J. R. astr. Soc.*, **16**, 207–224.

McElhinny, M. W. (1969a). Notes on progress in geophysics. Palaeomagnetic directions and pole positions, X. *Geophys. J. R. astr. Soc.*, **18**, 305–327.

McElhinny, M. W. (1969b). The palaeomagnetism of the Permian of southeast Australia and its significance regarding the problem of intercontinental correlation. *Spec. Pubs Geol. Soc. Austral.*, **2**, 61–67.

McElhinny, M. W. (1970a). The palaeomagnetism of the Cambrian Purple sandstone from the Salt Range, West Pakistan. *Earth Planet. Sci. Letters*, **8**, 149–156.

McElhinny, M. W. (1970b). Palaeomagnetic directions and pole positions, Part XI. *Geophys. J. R. astr. Soc.*, **20**, 417–429.

McElhinny, M. W. (1971). Geomagnetic Reversals during the Phanerozoic. *Science*, **172**, 157–159.

McElhinny, M. W. (1972). Palaeomagnetic directions and pole positions, XIII. *Geophys. J. R. astr. Soc.*, **30**, 281–293.

McElhinny, M. W. (1973a). *Palaeomagnetism and Plate Tectonics*, Cambridge University Press, p. 358.

McElhinny, M. W. (1973b). Palaeomagnetism and plate tectonics of Eastern Asia. In *The Western Pacific: Island Arcs, Marginal Seas, Geochemistry* (ed. P. J. Coleman), University of Western Australia Press, 407–414.

McElhinny, M. W., Briden, J. C., Jones, D. L. and Brock, A. (1968). Geological and geophysical implications of palaeomagnetic results from Africa. *Revs Geophys.*, **6**, 201–238.

McElhinny, M. W. and Cowley, J. A. (1977). Palaeomagnetic directions and pole positions, XIV. *Geophys. J. R. astr. Soc.*, **49**, 313–356.

McElhinny, M. W. and Cowley, J. A. (1978). Palaeomagnetic directions and pole positions, XV. *Geophys. J. R. astr. Soc.*, **52**, 259–276.

McElhinny, M. W. and Cowley, J. A. (1980). Palaeomagnetic directions and pole positions, XVI. *Geophys. J. R. astr. Soc.*, **61**, 549–571.

McElhinny, M. W., Cowley, J. A. and Brown, D. A. (1979). Palaeomagnetic results from the USSR Supplement No. 1. *Research School Earth Sci., Austral. Nat. Univ., Publ.* 1377.

McElhinny, M. W., Cowley, J. A., Brown, D. A. and Wiburov, N. (1977). Palaeomagnetic Results from the USSR. *Research School Earth Sci., Austral. Nat. Univ., Publ.* 1268.

McElhinny, M. W., Cowley, J. A. and Edwards, D. J. (1978a). Palaeomagnetism of some rocks from Peninsular India and Kashmir. *Tectonophys.*, **50**, 41–54.

McElhinny, M. W. and Embleton, B. J. J. (1976). The palaeoposition of Madagascar: Remanence and magnetic properties of Late Palaeozoic sediments. *Earth Planet. Sci. Letters*, **31**, 101–12.

McElhinny, M. W., Giddings, J. W. and Embleton, B. J. J. (1974). Palaeomagnetic results and late Precambrian glaciations. *Nature*, **248**, 557–561.

McElhinny, M. W. and Gough, D. I. (1963). The palaeomagnetism of the Great Dyke of Southern Rhodesia. *Geophys. J. R. astr. Soc.*, **7**, 287–303.

McElhinny, M. W. and Luck, G. R. (1970). The palaeomagnetism of the Antrim Plateau Volcanics of Northern Australia. *Geophys. J. R. astr. Soc.*, **20**, 191–205.

McElhinny, M. W. and McWilliams, M. O. (1977). Precambrian geodynamics – A palaeomagnetic review. *Tectonophys.*, **40**, 137–159.

McElhinny, M. W. and Opdyke, N. D. (1964). The palaeomagnetism of the Precambrian Dolerite of Eastern Southern Rhodesia: An example of geologic correlation by rock magnetism. *J. Geophys. Res.*, **67**, 2465–2475.

McElhinny, M. W. and Senanayake, W. E. (1980). Palaeomagnetic evidence for existence of the geomagnetic field 3.5 Ga Ago. *J. Geophys. Res.*, **85**, 3523–3528.

McElhinny, M. W., Taylor, S. R. and Stevenson, D. J. (1978b). Limits to expansion of Earth, Moon, Mars and Mercury and to changes in the gravitational constant. *Nature*, **271**, 316–321.

McFadden, P. L. (1980a). The best estimate of Fisher's precision parameter κ, *Geophys. J. R. astr. Soc.*, **60**, 397–407.

McFadden, P. L. (1980b). Determination of the angle in a Fisher distribution which will be exceeded with a given probability. *Geophys. J. R. astr. Soc.*, **60**, 391–396.

McFadden, P. L. (1980c). Testing a palaeomagnetic study for the averaging of secular variation. *Geophys. J. R. astr. Soc.*, **61**, 183–192.

McFadden, P. L. and Jones, D. L. (1981). The fold test in palaeomagnetism. *Geophys. J. R. astr. Soc.*, **67**, 53–58.

McFadden, P. L. and Lowes, F. J. (1981). The discrimination of mean directions drawn from Fisher distributions. *Geophys. J. R. astr. Soc.*, **67**, 19–33.

McKerrell, H., Mejdahl, V., Francois, H. and Portal, G. (1974). Thermoluminescence and Glozel. *Antiquity*, **48**, 265–272.

McKerrow, W. S. (1979). Ordovician and Silurian changes in sea level. *J. Geol. Soc., Lond.*, **136**, 137–145.

McLachlan, I. R. and Anderson, A. M. (1975). The age and stratigraphic relationship of the glacial sediments in Southern Africa. In *Gondwana Geology* (ed. K. W. S. Campbell), Australian National University Press, Canberra, 415–422.

McNab, T. K., Fox, R. A. and Boyle, A. J. F. (1968). Some magnetic properties of magnetite (Fe_3O_4) microcrystals. *J. Appl. Phys.*, **39**, 5703–5711.

McNish, A. E. and Johnson, E. A. (1938). Magnetization of unmetamorphosed varves and marine sediments, *J. Terr. Mag.*, **43**, 401–407.

McWilliams, M. O. (1977). *Late Precambrian palaeomagnetism of Australia and Africa*. Ph.D. thesis, Australian National University, Canberra, 1977.

McWilliams, M. O. and Dunlop, D. J. (1975). Precambrian paleomagnetism: magnetizations reset by the Grenville orogeny. *Science*, **190**, 269–272.

McWilliams, M. O. and Dunlop, D. J. (1978). Grenville paleomagnetism and tectonics. *Canad. J. Earth Sci.*, **15**, 687–695.

McWilliams, M. O., Holcolm, R. T. and Champion, D. E. (1982). Geomagnetic secular variation from ^{14}C dated lava flows on Hawaii and the question of the Pacific non-dipole low. *Phil. Trans. R. Soc., Lond.*, **A306**, 211–221.

McWilliams, M. O. and Kröner, A. (1981). Paleomagnetism and tectonic evolution of the Pan-African Damara Belt, Southern Africa. *J. Geophys. Res.*, **86**, 5147–5162.

McWilliams, M. O. and McElhinny, M. W. (1980). Late Precambrian paleomagnetism of Australia: the Adelaide System. *J. Geol.*, **88**, 1–26.

Medlicott, H. B. and Blandford, W. T. (1879). *Manual of the Geology of India*, Part III, Government Printer, Calcutta.

Megumi, K. (1978). A problem in ^{210}Pb geochronologies of sediments. *Nature*, **274**, 885–887.

Melloni, M. (1853). Du magnètisme des roches. *C.R. Acad. Sci., Paris*, **37**, 966–968.

Menard, H. W. (1958). Geology of the Pacific Sea Floor. *Experimentia*, **15**, 205–213.

Palaeomagnetism

Menard, H. W. (1959). Geology of the Pacific Sea Floor. *Experimentia*, **15**, 205–213.

Menard, H. W. and Vacquier, V. (1958). Magnetic survey of the deep-sea floor off the coast of California. *Office Naval Research, Research Rev.*, 1–8.

Mercanton, P. L. (1918). Etat magnétique de quelques terres cuites pré-historiques. *C.R. Acad. Sci., Paris*, **166**, 681, 949.

Mercanton, P. L. (1926). Inversion de l'inclinaison magnétique terrestre aux âges géologiques. *Terr. Magn. Atmos. Elect.*, **31**, 187–190.

Mercanton, P. L. (1931). Inversion inclinaison magnétique aux âges géologique. *C.R. Acad. Sci., Paris*, **192**, 978.

Mercanton, P. L. (1932). Inversion inclinaison magnétique aux âges géologique. *C.R. Acad. Sci., Paris*, **194**, 1371.

Merrill, R. T. (1970). Low-temperature treatments of magnetite and magnetite-bearing rocks. *J. Geophys. Res.*, **75**, 3343–3349.

Merrill, R. T. (1981). Toward a better theory of thermal remanent magnetization. *J. Geophys. Res.*, **86**, 937–949.

Michard, A., Westphal, M., Bossert, A. and Hamzeh, R. (1975). Tectonique de blocs dans le socle Atlaso-Mesetien du Maroc; une nouvelle interpretation des données géoloques et paléomagnetiques. *Earth Planet. Sci. Letters*, **24**, 363–368.

Middlemiss, F. A., Rawson, P. F. and Newall, G. (eds) (1971). *Faunal Provinces in Space and Time* (Proc. 17th Inter-Univ. Geol. Cong.), Seel House Press, Liverpool, p. 236.

Milankovitch, M. (1938). Neue Engebrisse der astronischen Theorie der Klimaschwankungen. *Bull. Acad. Sci. math. nat. Belgr.*, **A4**, 1–41.

Milton, N. J. (1980). Determination of the strain ellipsoid from measurements on any three sections. *Tectonophys.*, **64**, T19–27.

Molostovsky, E. A., Pevzner, M. A., Petchersky, D. M., Rodionov, V. P. and Khramov, A. N. (1976). Phanerozoic magnetostratigraphic scale and geomagnetic field inversions regime. *Geomagnetic Researches 17, Nanka, Moscow* 45–52.

Molyneux, L. (1971). A complete result magnetometer for measuring the remanent magnetization of rocks. *Geophys. J. R. astr. Soc.*, **24**, 429–433.

Moody, J. D. (1975). Distribution and geological characteristics of giant oil fields. In *Petroleum and Global Tectonics* (eds A. G. Fischer and S. Judson), Princeton University Press, 307–320.

Moore, B. M. (1980). Is the homing pigeon's map geomagnetic? *Nature*, **285**, 69–70.

Morel, P. (1981). Palaeomagnetism of a Pan-African diorite: A Late Precambrian pole for western Africa. *Geophys. J. R. astr. Soc.*, **65**, 493–503.

Morel, P. and Irving, E. (1981). Palaeomagnetism and the evolution of Pangea. *J. Geophys. Res.*, **86**, 1858–1872.

Morgan, G. E. and Briden, J. C. (1981). Aspects of Precambrian palaeomagnetism with new data from the Lunpopo Mobile Belt and Kaapvaal craton of southern Africa. *Phys. Earth Planet. Ints*, **24**, 142–168.

Morin, J. (1950). Magnetic susceptibility of αFe_2O_3 and γFe_2O_3. *Phys. Rev.*, **78**, 819–820.

Morley, L. W. and Larochelle, A. (1964). Paleomagnetism as a means of dating geological events. *Roy. Soc. Canada, Spec. Publ.* **9**, 40–51.

Morley, M. E., Gleadow, A. J. W. and Lovering, A. J. F. (1981). Evolution of the Tasmanian Rift: Apatite fission track dating evidence from the southeastern Australian continental margin. In *Gondwana Five* (eds M. M. Cresswell and P.

Vella) Balkema, Rotterdam, pp. 289–293.

Mörner, N. A. and Lanser, J. (1975). Palaeomagnetism in deep-sea core A179-15. *Earth Planet. Sci. Letters*, **26**, 121–124.

Mörner, N. A., Lanser, J. P. and Hospers, J. (1971). Late Weichselian paleomagnetic reversal. *Nature, Phys. Sci.*, **234**, 173–174.

Morris, W. A. (1976). Transcurrent motion determined paleomagnetically in the northern Appalachians and Caledonides and the Acadian Orogeny. *Canad. J. Earth Sci.*, **13**, 1236–43.

Morris, W. A. (1977). Paleolatitude of glaciogenic upper Precambrian Rapitan group and the use of tillites as chronostratigraphic marker horizons. *Geology*, **5**, 83–104.

Morris, W. A. (1980). A palaeomagnetic study of Cambrian red beds from Cartaret, Normandy, France. *Geophys. J. R. astr. Soc.*, **62**, 577–590.

Morris, W. A., Schmidt, P. W. and Roy, J. L. (1979a). A graphical approach to polar paths; palaeomagnetic poles and global tectonics. *Phys. Earth Planet. Ints*, **19**, 85–99.

Morris, W. A., Schmidt, P. W. and Roy, J. L. (1979b). Reply to: An unconventional Proterozoic palaeomagnetic polar wander path for North America, and its advantages. *Phys. Earth Planet. Ints*, **20**, 71–73.

Morris, W. A. and Tanner, P. W. G. (1977). The use of paleomagnetic data to delineate the history of the development of the Connemara Antiform. *Canad. J. Earth Sci.*, **14**, 2601–2613.

Moskowitz, B. M. (1980). Theoretical grain size limits for single-domain pseudo-single-domain and multidomain behaviour in titanomagnetites ($x = 0.6$) as a function of low-temperature oxidation. *Earth Planet. Sci. Letters*, **47**, 285–293.

Moskowitz, B. M. (1981). Methods for estimating Curie temperatures of titanomaghemites from experimental J_s–T data. *Earth Planet. Sci. Letters*, **53**, 84–88.

Moskowitz, B. M. and Banerjee, S. K. (1981). A comparison of the magnetic properties of synthetic titanomaghemite and some oceanic basalts. *J. Geophys. Res.*, **86**, 11869–11882.

Mothersill, J. S. (1979). The paleomagnetic record of the Late Quaternary sediment of Thunder Bay. *Canad. J. Earth Sci.*, **16**, 1016–1023.

Mothersill, J. S. (1981). Late Quaternary paleomagnetic record of the Goderich Basin, Lake Huron. *Canad. J. Earth. Sci.*, **18**, 448–456.

Mullins, C. E. (1977). Magnetic susceptibility of the soil and its significance in soil science – a review. *J. Soil Sci.*, **28**, 223–246.

Munk, W. H. (1968). Once again – Tidal Friction. *Quart. J. R. astr. Soc.*, **9**, 352–375.

Murthy, V. R. and Banerjee, S. K. (1973). Lunar evolution: How well do we know it now? *Moon*, **7**, 149–171.

Mussett, A. E., Dagley, P. and Skelhorn, R. R. (1980). Magnetistratigraphy of the Tertiary igneous succession of Mull, Scotland. *J. Geol. Soc., Lond.*, **137**, 349–357.

Nagata, T. (1953). *Rock Magnetism*, Maruzan, Tokyo, p. 225.

Nagata, T. (1961). *Rock Magnetism*, 2nd Ed, Maruzan, Tokyo, p. 350.

Nagata, T. (1965). Main characteristics of Recent geomagnetic secular variation. *J. Geomagn. Geoelect.* **17**, 263–276.

Nagata, T. (1971). Introductory notes on shock remanent magnetization and shock demagnetization of igneous rocks. *Pure Appl. Geophys.* **89**, 159–177.

Nagata, T. (1979). Meteorite magnetism and the early solar system magnetic field. *Phys. Earth Planet. Ints*, **20**, 324–341.

Palaeomagnetism

Nagata, T., Akimoto, S. and Uyeda, S. (1951). Reverse thermo-remanent magnetism. *Proc. Jap. Acad.*, **27**, 643–645.

Nagata, T., Fisher, R. M., Schwerer, F. C., Fuller, M. and Dunn, J. R. (1971). Magnetic properties and remanent magnetization of Apollo 12 lunar materials and Apollo 11 lunar microbreccia. Proc. 2nd Lunar Sci. Conf., *Geochim. Cosmochim. Acta*, **3**, 2461–2476.

Nagata, T., Fisher, R. M., Schwerer, F. C., Fuller, M. D. and Dunn, J. R. (1972). Rock magnetism of Apollo 14 and 15 materials. Proc. 3rd Lunar Sci. Conf., *Geochim. Cosmochim. Acta*, **3**, 2423–2447.

Nagata, T., Ishikawa, Y., Kinoshita, H., Kono, M., Syono, Y. and Fisher, R. M. (1970). Magnetic properties and natural remanent magnetization of lunar materials. Proc. Apollo 11 Lunar Sci. Conf., *Geochim. Cosmochim Acta*, **3**, 2325–2340.

Nagata, T. and Kobayashi, K. (1958). Experimental studies on the generation of remanent magnetization of ferromagnetic minerals by chemical reactions. *Proc. Jap. Acad.*, **34**, 269–273.

Nagata, T., Rikitake, T. and Akasi, K. (1943). The natural remanent magnetism of sedimentary rocks. *Bull. Earthq. Res. Inst.*, **21**, 276–297.

Naidu, P. S. (1971). Statistical structure of geomagnetic field reversals, *J. Geophys. Res.*, **76**, 2649–2662.

Nairn, A. (1956). Relevance of palaeomagnetic studies of Jurassic rocks to continental drift. *Nature*, **178**, 935–936.

Nairn, A. E. M. (1957). A palaeomagnetic study of Jurassic and Cretaceous sediments. *Mon. Not. R. Astr. Soc., Geoph. Supp.*, **7**, 308–313.

Nairn, A. E. M. (1960a). Palaeomagnetic Results from Europe. *J. Geol.*, **68**, 295–306.

Nairn, A. E. M. (1960b). A palaeomagnetic survey of the Karroo System. *Overseas Geol. Min. Res.*, **7**, 398–410.

Nairn, A. E. M. (ed.) (1964). *Problems in Palaeoclimatology*, Interscience, London, p. 705.

Nairn, A. E. M. (1972). Palaeoclimatology: Present status. *Naturwissenschaften*, **59**, 388–393.

Nairn, A. E. M., Schmitt, T. J. and Smithwick, M. E. (1981). A palaeomagnetic study of the Upper Mesozoic succession in Northern Tunisia. *Geophys. J. R. astr. Soc.*, **65**, 1–18.

Nairn, A. E. M. and Westphal, M. (1968). Possible implications of the paleomagnetic study of Late Palaeozoic rocks of northwestern Corsica. *Palaeogeog., Palaeoclim., Palaeoecol.*, **5**, 179–204.

Nakajima, T. and Kawai, N. (1975). Secular geomagnetic variation in the Recent 60,000 years found from the Lake Biwa Sediments. *Rock Magn. Palaeogeophys.*, 34–38.

Nakajima, T., Yaskawa, K., Natsuhara, N., Kawai, N. and Horie, S. (1973). Very short geomagnetic excursion 18,000 yr BP. *Nature*, **244**, 8–10.

Nakamura, S. and Kikuchi, S. (1912). Remanent magnetism of volcanic rocks. *Proc. Tokyo Math. Phys. Soc.*, **6**, 268–275.

Natland, J. H. (1978). Comparison of chemical and magnetic stratigraphy of basement rocks at DSDP sites 332 and 395. *Init. Reps Deep Sea Drilling Project*, **45**, 657–677.

Neckham, A. (1187). *De Naturis Rerum*.

Needham, J. (1962). *Science and Civilization in China*, Vol. 4, Cambridge University Press, p. 434.

References

Néel, L. (1948). Propriétés Magnétiques de Ferrites; Ferrimagnétism et Antiferro-magnétisme. *Ann. Physique*, **3**, 137–198.

Néel, L. (1949). Théorie du trainage magnétique des ferromagnétiques en grains fins avec applications aux Terres Cuites. *Ann Geophys.*, **5**, 99–136.

Néel, L. (1951). L'inversion de l'aimantion permanente des roches. *Ann. Geophys.*, **7**, 90–102.

Néel, L. (1952a). Théorie du trainage magnétique de diffusion. *J. Phys. Radium*, **12**, 249–264.

Néel, L. (1952b). Antiferromagnetism and ferrimagnetism. *Proc. Phys. Soc.*, **A65**, 869–885.

Néel, L. (1955). Some theoretical aspects of rock magnetism. *Adv. Phys.*, **4**, 191–243.

Néel, L. (1961). Superantiferromagnétism dans les grains tres fins. *C. R. Acad. Sci., Paris*, **253**, 203–208.

Néel, L. (1962a). Propriétés Magnétiques des grains fins antiferromagnétiques: Super-paramagnétisme et superantiferromagnétisme. *J. Phys. Soc. Japan.* **17**, 676–685.

Néel, L. (1962b). Influence des couplages magnetocristallins sur le superantiferro-magnétisme des grains fins. *C. R. Acad. Sci., Paris*, **254**, 598–602.

Ness, G., Levi, S. and Couch, R. (1980). Marine magnetic anomaly timescales for the Cenozoic and Late Cretaceous: A precis, critique and synthesis. *Revs Geophys. Space Phys.*, **18**, 753–770.

Ness, N. F. (1979). The magnetic field of Mercury. *Phys. Earth Planet. Ints*, **20**, 209–217.

Ng, K. H. and Beard, D. B. (1979). Possible displacement of Mercury's dipole. *J. Geophys. Res.*, **84**, 2115–2117.

Niitsuma, N. (1971). Detailed study of the sediments recording the Matuyama – Brunhes geomagnetic reversal, *Tohoku Univ. Sci. Reps, Geol.*, **43**, 1–39.

Ninkovich, D. and Burckle, L. H. (1978). Absolute age of the base of the hominid-bearing beds in Eastern Java. *Nature*, **275**, 306–308.

Ninkovich, D., Opdyke, N., Heezen, B. C., and Foster, J. H. (1966). Paleomagnetic stratigraphy rates of deposition and tephrachronology in North Pacific Deep-Sea sediments. *Earth Planet. Sci. Letters*, **1**, 476–492.

Nöel, M. (1975). The palaeomagnetism of varved clays from Blekinge, southern Sweden. *Geol. Tor. Stockholm Forhandl.*, **97**, 357–367.

Nöel, M. (1976). *Palaeomagnetic studies of Swedish varved sediments*. Ph.D. thesis, University of Newcastle upon Tyne.

Nöel, M. (1980). Personal communication.

Nöel, M. (1980). Surface tension phenomena in the magnetization of sediments. *Geophys. J. R. astr. Soc.*, **62**, 15–25.

Nöel, M., Homonko, P. and Bull, P. (1979). The palaeomagnetism of sediments from Agen Allwedd, Powys. *Trans. British Cave Research Assoc.*, **6**, 85–92.

Nöel, M. and Tarling, D. H. (1975). The Laschamp geomagnetic event. *Nature*, **253**, 705–707.

Noltimier, H. C. (1971). Magnetic rock cylinders with negligible shape anisotropy. *J. Geophys. Res.*, **76**, 4035–4037.

Noltimier, H. C. and Colinvaux, P. A. (1976). Geomagnetic excursion from Imuruk Lake, Alaska. *Nature*, **259**, 197–200.

Noltimier, H. C. and Ellwood, B. B. (1977). The coal pole: Paleomagnetic results from Westphalian B, C and D Coals, Wales. *Trans. Amer. Geophys. Union*, **58**, 375.

Palaeomagnetism

Norman, R. (1581). *The New Attractive*. Published in *Rara Magnetism, 1269–1599*, Asher, Berlin 1897.

Norton, I. O. and Molnar, P. (1977). Implications of a revised fit between Australia and Antarctica for the evolution of the Eastern Indian Ocean. *Nature*, **267**, 338–340.

Norton, I. O. and Sclater, J. G. (1979). A model for the evolution of the Indian Ocean and the fragmentation of Gondwanaland. *J. Geophys. Res.*, **84**, 6803–6830.

Odin, G. S., Curry, D. and Hunziker, J. C. (1978). Radiometric dates from NW European glauconites and the Palaeogene time-scale. *J. Geol. Soc., Lond.*, **135**, 481–497.

O'Donovan, J. B. (1975). *Studies on synthetic analogues of some carriers of the palaeomagnetic record*. Ph.D. thesis, University of Newcastle upon Tyne.

Oertel, G. (1978). Strain determination from the measurement of pebble shapes. *Tectonophys.*, **50**, T1–7.

Oertel, G. and Curtis, C. D. (1972). Clay-ironstone concretion preserving fabrics due to progressive compaction. *Geol. Soc. Amer. Bull.*, **82**, 2597–2606.

Oertel, G. and Ernst, W. G. (1978). Strain and deformation in a multilayered fold. *Tectonophys.*, **48**, 77–106.

Oldfield, F., Appleby, P. G. and Battarbee, R. W. (1978). Alternative ^{210}Pb dating: results from the New Guinea Highlands and Lough Erne. *Nature*, **271**, 339–342.

Onstott, T. C. (1980). Application of the Bingham distribution function in paleomagnetic studies. *J. Geophys. Res.*, **85**, 1500–1510.

Opdyke, N. D. (1972). Paleomagnetism of deep-sea cores. *Revs Geophys. Space Phys.*, **10**, 213–249.

Opdyke, N. D., Burckle, L. H. and Todd, A. (1974). The extension of the magnetic time scale in sediments of the Central Pacific Ocean. *Earth Planet. Sci. Letters*, **22**, 300–306.

Opdyke, N. D., Glass, B., Hays, J. D. and Foster, J. (1966). Paleomagnetic study of Antarctic deep-sea cores. *Science*, **154**, 349–357.

Opdyke, N. D. and Henry, K. W. (1969). A test of the dipole hypothesis. *Earth Planet. Sci. Letters*, **6**, 139–151.

Opdyke, N. D., Kent, D. V. and Lowrie, W. (1973). Details of magnetic polarity transitions recorded in a high deposition rate deep-sea core. *Earth Planet. Sci. Letters*, **20**, 315–324.

Opdyke, N. D., Lindsay, E. H., Johnson, N. M. and Downs, T. (1977). The paleomagnetism and magnetic polarity stratigraphy of the mammal-bearing section of Anza Borrego State Park, California. *Quaternary Res.*, **7**, 316–329.

O'Reilly, W. (1969). Application of neutron diffraction and Mössbauer effect to rock magnetism. In *Application of Modern Physics to the Earth and Planetary Interiors* (ed. S. K. Runcorn), Wiley, London, pp. 479–484.

O'Reilly, W. (1976). Magnetic minerals in the crust of the Earth. *Rep. Prog., Phys.*, **39**, 857–908.

O'Reilly, W. and Banerjee, S. K. (1967). The mechanism of oxidation in titano-magnetites: A magnetic study. *Min. Mag.*, **36**, 29–37.

Othberg, K. L. (1973). Evidence for a geomagnetic excursion 20,000 years ago. *Eos*, **54**, 1075.

Otofuji, Y., Sasajima, S., Nishimura, S., Yokoyama, T., Hadiwisastra, S. and Hehunat, F. (1981). Palaeomagnetic evidence for the palaeoposition of Sumba Island, Indonesia. *Earth Planet. Sci. Letters*, **52**, 93–100.

Owen, H. G. (1976). Continental displacement and expansion of the Earth during the Mesozoic and Cenozoic. *Phil. Trans. R. Soc., Lond.*, **A281**, 223–291.

Owens, W. H. (1974). Mathematical model studies on factors affecting the magnetic anisotropy of deformed rocks. *Tectonophys.*, **24**, 115–131.

Owens, W. H. (1980). The orientation of grains in a sheared dispersion – reply. *Tectonophys.*, **66**, 274–377.

Owens, W. H. and Bamford, D. (1976). Magnetic, seismic and other anisotropic properties of rock fabrics. *Phil. Trans. R. Soc., Lond.*, **A283**, 55–68.

Özdemir, O. and Banerjee, S. K. (1981). An experimental study of magnetic viscosity in synthetic monodomain titanomaghemites: Implications for the magnetization of the ocean crust. *J. Geophys. Res.*, **86**, 11864–11868.

Özdemir, O. and O'Reilly, W. (1981). High-temperature hysteresis and other magnetic properties of synthetic monodomain titanomagnetites. *Phys. Earth Planet. Ints*, **25**, 406–418.

Özdemir, O. and O'Reilly, W. (1982). Magnetic hysteresis properties of synthetic monodomain titanomaghemites. *Earth Planet. Sci. Letters*, **57**, 437–447.

Ozima, M. and Kinoshita, H. (1964). Magnetic anisotropy of andesites in a fault zone. *J. Geomagn. Geoelect.*, **16**, 194–200.

Ozima, M. and Ozima, M. (1964). Acquisition mechanism of TRM – Low temperature characteristics of TRM and magnetization of some granites. *1964 Ann. Prog. Rep. Rock Magn. Res., Group Japan*, 21–29.

Ozima, M., Yoshima, M. and Kinoshita, H. (1974). Magnetic properties of submarine basalts and the implications on the structure of the oceanic crust. *J. Geomagn. Geoelect.*, **26**, 335–354.

Packer, D. R. and Stone, D. B. (1972). An Alaskan Jurassic paleomagnetic pole and the Alaskan Orocline. *Nature*, **237**, 25–36.

Packer, D. R. and Stone, D. B. (1974). Paleomagnetism of Jurassic rocks from southwestern Alaska and tectonic implications. *Canad. J. Earth Sci.*, **11**, 976–997.

Palmer, D. F., Henyey, T. L. and Dodson, R. E. (1979). Paleomagnetic and sedimentological studies at Lake Tahoe, California – Nevada. *Earth Planet. Sci. Letters*, **46**, 125–137.

Palmer, H. C., Hayatsu, A. and MacDonald, W. D. (1980a). The Middle Jurassic Camarac Formation, Arica, Chile: palaeomagnetism, K–Ar age dating and tectonic implications. *Geophys. J. R. astr. Soc.*, **62**, 155–172.

Palmer, H. C., Hayatsu, A. and MacDonald, W. D. (1980b). Palaeomagnetic and K–Ar age studies of a 6 km thick Cretaceous section from the Chilean Andes. *Geophys. J. R. astr. Soc.*, **62**, 133–153.

Panchen, A. L. (ed.) (1980). *The Terrestrial Environment and the Origin of Land Vertebrates*. Systematics Assoc. Spec. Vol. 15, Academic Press, London, p. 633.

Papamarinopoulos, S., Readman, P. W., Maniatis, Y. and Simopoulos, A. (1982). Magnetic characterisation and Mössbauer spectroscopy of magnetic concentrates from Greek lake sediments. *Earth Planet. Sci. Letters*, **57**, 173–181.

Parker, R. L. and Denham, C. R. (1979). Interpolation of unit vectors. *Geophys. J. R. astr. Soc.*, **58**, 685–687.

Parron, C. and Nahon, D. (1980). Red bed genesis by lateritic weathering of glauconitic sediments. *J. Geol. Soc., Lond.*, **137**, 689–693.

Paterson, M. S. (1976). Some current aspects of experimental rock deformation. *Phil. Trans. R. Soc., Lond.*, **A283**, 163–172.

Pearce, G. W. and Karson, J. A. (1981). On pressure demagnetization. *Geophys. Res. Letters*, **8**, 725–728.

Pearce, G. W., Strangway, D. W. and Gose, W. A. (1972). Remanence magnetization of the lunar surface. Proc. 3rd Lunar Sci. Conf., *Geochim. Cosmochim. Acta*, **3**, 2449–2464.

Pearce, G. W., Strangway, D. W. and Larson, E. E. (1971). Magnetism of two Apollo 12 igneous rocks. Proc. 2nd Lunar Sci. Conf., *Geochim. Cosmochim. Acta*, **3**, 2451–2460.

Peirce, J. W. and Clark, M. J. (1978). Evidence from Iceland on geomagnetic reversal during the Wisconsin Ice Age. *Nature*, **273**, 456–458.

Peregrinus, P. (1269). *Epistola de Magnete*.

Pesonen, L. J. and Halls, H. C. (1979). The paleomagnetism of Keweenawan dikes from Baraga and Marquette Counties, northern Michigan. *Canad. J. Earth Sci.*, **16**, 2136–2149.

Pesonen, L. J. and Neuvonen, K. J. (1981). Palaeomagnetism of the Baltic Shield – Implications for Precambrian tectonics. In *Precambrian Plate Tectonics* (ed. A. Kröner), Elsevier, Amsterdam, 623–648.

Pesonen, L. J. and Nevanlinna, H. (1981). Late Precambrian Keweenawan asymmetric reversals. *Nature*, **294**, 436–439.

Petersons, H. F. (1974). Changes in the geomagnetic dipole field for 1880–1970. *J. Geomagn. Geoelect.*, **26**, 363–364.

Petley, B. W. (1980). The ubiquitous SQUID. *Contemp. Phys.*, **21**, 407–630.

Pettijohn, F. J. (1957). *Sedimentary Rocks*, Harper, New York, p. 718.

Phillips, F. C. (1971). *The Use of Stereographic Projection in Structural Geology* (3rd Edn), E. Arnold, London, p. 90.

Phillips, J. D. (1977). Time variation and asymmetry in the statistics of geomagnetic reversal sequences. *J. Geophys. Res.*, **82**, 835–843.

Picard, M. D. (1964). Paleomagnetic correlation of units within Chugwater (Triassic) Formation, west-central Wyoming. *Bull. Amer. Ass. Petr. Geol.*, **48**, 269–291.

Piper, J. D. A. (1973). Latitudinal extent of Late Precambrian glaciations. *Nature*, **244**, 342–344.

Piper, J. D. A. (1976). Palaeomagnetic evidence for a Proterozoic super-continent. *Phil. Trans. R. Soc., Lond.*, **A280**, 469–490.

Piper, J. D. A. (1977). Magnetic stratigraphy and magnetic–petrologic properties of Precambrian Gardar lavas, South Greenland. *Earth Planet. Sci. Letters*, **34**, 247–263.

Piper, J. D. A. (1980a). Analogous Upper Proterozoic apparent polar wander loops. *Nature*, **283**, 845–847.

Piper, J. D. A. (1980b). Comments on 'Palaeomagnetism in the Coronation Geosyncline and arrangement of continents in the Middle Proterozoic' by E. Irving and J. C. McGlynn. *Geophys. J. R. astr. Soc.*, **62**, 473–477.

Piper, J. D. A. (1982). The Precambrian palaeomagnetic record: the case for the Proterozoic supercontinent. *Earth Planet. Sci. Letters*, **59**, 61–89.

Piper, J. D. A., Briden, J. C. and Lomax, K. (1973). Precambrian Africa and South America as a single continent. *Nature*, **245**, 244–248.

Plotnick, R. E. (1980). Relationship between biological extinctions and geomagnetic reversals. *Geology*, **8**, 578–581.

Plouff, D. (1976). Gravity and magnetic fields of polygonal prisms and application to magnetic terrain corrections. *Geophys.*, **41**, 727–741.

Pohl, J., Bleil, U. and Hornemann, U. (1975). Shock magnetization and demagnetization of basalt by transient stress up to 10 Kbar. *J. Geophys.*, **41**, 23–41.

Poorter, R. P. E. (1975). Palaeomagnetism of Precambrian rocks from southeast Norway and south Sweden. *Phys. Earth Planet. Ints*, **10**, 74–87.

Poorter, R. P. E. (1981). Precambrian palaeomagnetism of Europe and the position of the Balto-Russian plate relative to Laurentia. In *Precambrian Plate Tectonics* (ed. A. Kröner), Elsevier, Amsterdam, 599–622.

Porath, H. and Chamalaun, F. H. (1966). The magnetic anisotropy of hematite bearing rocks. *Pure Appl. Geophys.*, **64**, 81–88.

Porath, H., Stacey, F. D. and Cheam, A. S. (1966). The choice of specimen shape for magnetic anisotropy measurements in rocks. *Earth Planet. Sci. Letters*, **1**, 92.

Prell, W. L., Gardner, J. V. *et al.* (1980). Hydraulic piston coring of Late Neogene and Quaternary Sections in the Caribbean and Equatorial Pacific: Preliminary results of deep sea drilling project Leg 68, *Geol. Soc. Amer., Bull. I*, **91**, 433–444.

Presti, D. and Pettigrew, J. D. (1980). Ferromagnetic coupling to muscle receptors as a basis for geomagnetic field sensitivity in animals. *Nature*, **285**, 99–101.

Prévot, M. (1977). Large intensity changes of the non-dipole field during a polarity transition. *Phys. Earth Planet. Ints*, **13**, 342–345.

Prévot, M. (1981). Some aspects of magnetic viscosity in subaerial and submarine volcanic rocks. *Geophys. J. R. astr. Soc.*, **66**, 169–192.

Prévot, M., Lecaille, A. and Mankinen, E. A. (1981). Magnetic effect of maghematization of oceanic crust. *J. Geophys. Res.*, **86**, 4009–4020.

Prince, R. A., Heath, G. R. and Kominz, M. (1980). Paleomagnetic studies of central North Pacific sediment cores: Stratigraphy, sedimentation rates, and the origin of magnetic instability – Summary. *Bull. Geol. Soc. Amer.*, **91**, 447–449.

Pulliah, G., Irving, E., Buchan, K. L. and Dunlop, D. J. (1975). Magnetization changes caused by burial and uplift. *Earth Planet. Sci. Letters*, **28**, 133–143.

Pye, K. (1981). Rate of dune reddening in a humid tropical climate. *Nature*, **290**, 582–584.

Quilty, P. G. (1980). Sedimentation cycles in the Cretaceous and Cenozoic of Western Australia. *Tectonophys.*, **63**, 349–366.

Raff, A. D. (1962). Further magnetic measurements along the Murray Fault. *J. Geophys. Res.*, **67**, 417–418.

Rahman, A., Gouch, D. I. and Evans, M. E. (1975). Anisotropy of magnetic susceptibility of the Martin Formation, Saskatchewan, and its sedimentological implications. *Canad. J. Earth Sci.*, **12**, 1465–1473.

Ralph, E. K. (1972). A cyclical solution for the relationship between magnetic and atmospheric C-14 changes. *Proc. 8th Internat. Conf. Radiocarbon Dating*, A76–84.

Ramsay, J. G. (1967). *Folding and Fracturing of Rocks*, McGraw-Hill, New York.

Ramsay, J. G. (1976). Displacement and strain. *Phil. Trans. R. Soc., Lond.*, **A283**, 3–25.

Ramsbottom, W. H. C. (1971). Palaeogeography and goniatite distribution in the Namurian and early Westphalian. *C. R. 6th Cong. Int. Strat. Geol. Carbonif., Sheffield.*

Ramsbottom, W. H. C. (1973). Transgressions and regressions in the Dinantian: A new synthesis of British Dinantian stratigraphy. *Proc. Yorks. Geol. Soc.*, **39**, 567–607.

Ramsbottom, W. H. C. (1979). Rates of transgression and regression in the Carboniferous of NW Europe. *J. Geol. Soc., Lond.*, **136**, 147–153.

Rao, K. V. and Van der Voo, R. (1980). Paleomagnetism of a Palaeozoic anorthosite from the Appalachian Piedmont, northern Delaware: possible tectonic implications. *Earth Planet. Sci. Letters*, **47**, 113–120.

Rathore, J. S. (1975); see Singh (1975).

Rathore, J. S. (1979a). Application of magnetic susceptibility anisotropy technique to the study of geological structures in the Armorican massif, France. *Tectonophys.*, **60**, 207–216.

Rathore, J. S. (1979b). Magnetic susceptibility anisotropy in the Cambrian Slate Belt of North Wales and correlation with strain. *Tectonophys.*, **53**, 83–97.

Rathore, J. S. (1980a). The magnetic fabrics of some slates from the Borrowdale Volcanic group in the English Lake District and their correlations with strains. *Tectonophys.*, **67**, 207–220.

Rathore, J. S. (1980b). A study of secondary fabrics in rocks from the Lizard Peninsula and adjacent areas in southwest Cornwall, England. *Tectonophys.*, **68**, 147–160.

Rathore, J. S. (1980c). Evidence for sinistral movements along the Judicarian Line drawn from a study of magnetic fabrics in the regions of Mt. Croce and Asten (South Tyrol). *Geol. Rundsch.*, **69**, 678–694.

Rathore, J. S. and Becke, M. (1980). Magnetic fabric analyses in the Gail Valley (Carinthia, Austria) for the determination of the sense of movements along this region of the Periadriatic Line. *Tectonophys.* **69**, 349–368.

Rathore, J. S. and Heinz, H. (1979). The application of magnetic susceptibility anisotropy analyses to the study of tectonic events on the Periadriatic Line. *Mitt. Ošterr. geol. Ges.*, **71/72**, 275–290.

Raven, P. H. and Axelrod, D. I. (1974). Angiosperm biogeography and past continental movements. *Ann. Missouri Botan. Gardens*, **61**, 539–673.

Reed, L. J. (1980). The orientation of grains in a sheared dispersion – A discussion. *Tectonophys.*, **66**, 373–374.

Rees, A. I. (1961). The effect of water currents on the magnetic remanence and anisotropy of susceptibility of some sediments. *Geophys. J. R. astr. Soc.*, **5**, 235–251.

Rees, A. I. (1965). The use of anisotropy of magnetic susceptibility in the estimation of sedimentary fabric. *Sedimentology*, **4**, 257–271.

Rees, A. I. (1966). The effect of depositional slopes on the anisotropy of magnetic susceptibility of laboratory deposited sands. *J. Geol.*, **74**, 856–876.

Rees, A. I. (1968). The production of preferred orientation in a concentrated dispersion of elongated and flattened grains. *J. Geol.*, **76**, 457–465.

Rees, A. I. (1979). The orientation of grains in a sheared dispersion. *Tectonophys.*, **55**, 275–287.

Rees, A. I. and Woodall, W. A. (1975). The magnetic fabric of some laboratory-deposited sediments. *Earth Planet. Sci. Letters*, **25**, 121–130.

Renfrew, C., Cann, J. R. and Dixon, J. E. (1965). Obsidian in the Aegean. *Ann. British School Athens*, **60**, 225–247.

Renfrew, C., Dixon, J. E. and Cann, J. R. (1966). Obsidian and early culture contact in the Near East. *Proc. Prehist. Soc.*, **32**, 30–72.

Reyment, R. A. (1980). Trends in Cretaceous and Tertiary geomagnetic reversals. *Cretac. Res.*, **1**, 27–48.

References

Reyment, R. A. and Mörner, N. A. (1977). Cretaceous transgressions exemplified by the South Atlantic. *Palaeont. Soc. Japan, Spec. Papers*, **21**, 247–261.

Reynolds, J. (1956). High sensitivity mass spectrometer for noble gas analysis. *Rev. Sci. Inst.*, **27**, 928–934.

Richardson, K. G. and Noltimier, H. C. (1980). Paleomagnetic study of the Mahogany Oil Shale: Application to compaction of the sediment. *Eos*, **61**, 1196.

Ridd, M. F. (1971). South-East Asia as a part of Gondwanaland. *Nature*, **234**, 531–533.

Ries, A. C. (1979). Variscan metamorphism and K–Ar dates in the Variscan fold belt of S. Brittany and NW Spain. *J. Geol. Soc. Lond.*, **136**, 89–103.

Ries, A. C., Richardson, A. and Shackleton, R. M. (1980). Rotation of the Iberian Arc: Palaeomagnetic results from North Spain. *Earth Planet Sci. Letters*, **50**, 301–310.

Ries, A. C. and Shackleton, R. M. (1976). Patterns of strain variation in acute fold belts. *Phil. Trans. R. Soc., Lond.*, **A283**, 281–288.

Rikitake, T. (1958). Oscillations of a system of disk dynamoes. *Proc. Cambridge Phil. Soc.*, **54**, 89–105.

Rikitake, T. (1966). Electromagnetism and the Earth's interior, *Devel. Solid Earth Geophys.*, **2**, 308.

Ringwood, A. E. (1977). Composition of the core and implications for origin of the Earth. *Austral. Nat. Univ.*, Publ. 1277, p. 45.

Roberts, P. H. (1972). Electromagnetic core-mantle coupling. *J. Geomagn. Geoelect.*, **24**, 231–259.

Roberts, P. H. and Ursell, H. D. (1960). Random walk on a sphere and on a Riemmanian manifold. *Phil. Trans. R. Soc., Lond.*, **A252**, 317–356.

Roberts, W. O. and Olson, R. H. (1973). New evidence for variable corpuscular emission on the climate. *Rev. Geophys. Space Res.*, **11**, 731–741.

Robertson, P. B. and Roy, J. L. (1979). Shock diminished paleomagnetic remanence at the Charlevoix impact structure, Quebec. *Canad. J. Earth Sci.*, **16**, 1842–1856.

Robinson, P. L. (1973). Palaeoclimatology and continental drift. In *Implications of Continental Drift to the Earth Sciences* (eds D. H. Tarling and S. K. Runcorn), Academic Press, London, 451–476.

Roche, A. (1950a). Sur les Caracteres Magnétiques du système éruptif de Gergovie. *C. R. Acad. Sci., Paris*, **230**, 113–115.

Roche, A. (1950b). Anomalies magnétiques accompagnant les massifs de la Limagne d'Auvergne. *C. R. Acad. Sci., Paris*, **230**, 1603–1604.

Roche, A. (1951). Sur les inversions de l'aimantation rémanente des roches volcaniques dans les Monts d'Auvergne. *C. R. Acad. Sci., Paris*, **233**, 1132–1134.

Roche, A. (1953). Sur l'origine des inversions d'aimantation constatées dans les roches d'Auvergne. *C. R. Acad. Sci., Paris*, **236**, 107–109.

Roche, A. (1954). Exposé sommaire des études relatives à l'aimantation de matériaux volcaniques. *J. Geomagn. Geoelect.*, **6**, 169–171.

Rogers, J., Fox, J. M. W. and Aitken, M. J. (1979). Magnetic anisotropy in ancient pottery. *Nature*, **277**, 644–646.

Rosenberg, G. A. and Runcorn, S. K. (1975). *Growth Rhythms and the History of the Earth's Rotation*, Wiley, London, p. 538.

Ross, C. A. (ed.) (1974). Paleogeographic provinces and provinciality. *Soc. Econ. Paleont. Mineral., Spec. Publ.* 21.

Roy, J. L. and Robertson, P. B. (1979). An assessment of the St-Urbain and other Grenville paleopoles. *Canad. J. Earth Sci.*, **16**, 1857–1865.

Runcorn, S. K. (1955). Palaeomagnetism of sediments from the Colorado Plateau. *Nature*, **176**, 505.

Runcorn, S. K. (1956a). Palaeomagnetic survey in Arizona and Utah: Preliminary results. *Bull. Geol. Soc. Amer.* **67**, 301–316.

Runcorn, S. K. (1956b). Palaeomagnetic comparisons between Europe and North America. *Proc. Canad. Assoc. Geol.*, **8**, 77–85.

Runcorn, S. K. (1956c). Palaeomagnetism. Polar wandering and continental drift. *Geol. en Mijnb.*, **18**, 253–256.

Runcorn, S. K. (1959). On the theory of the geomagnetic secular variation. *Ann. Geophys.*, **15**, 87–92.

Runcorn, S. K. (1975). An ancient lunar magnetic dipole field. *Nature*, **253**, 701–703.

Runcorn, S. K. (1978). The origin of lunar palaeomagnetism. *Nature*, **275**, 430–432.

Runcorn, S. K. and Urey, H. C. (1973). A new theory of lunar magnetism. *Science*, **180**, 636–638.

Rusakov, O. M. and Zagniy, G. F. (1973a). Archaeomagnetic secular variation study in the Ukraine and Moldavia. *Archaeometry*, **15**, 153–157.

Rusakov, O. M. and Zagniy, G. F. (1973b). Intensity of the geomagnetic field in the Ukraine and Moldavia during the past 6000 years. *Archaeometry*, **15**, 275–285.

Rusnak, G. A. (1957). Orientation of sand grains under conditions of 'unidirectional' flow, I. Theory and experiments. *J. Geol.*, **65**, 384–409.

Russell, C. T. (1979). The Martian magnetic field. *Phys. Earth Planet. Ints*, **20**, 237–246.

Russell, C. T. (1980). Planetary magnetism. *Revs Geophys. Space Phys.*, **18**, 77–106.

Russell, C. T., Coleman, P. J., Lichtenstein, B. R. and Schubert, G. (1974). The permanent and induced magnetic dipole moment of the Moon. Proc. 6th Lunar Sci. Conf., *Geochim. Cosmochim. Acta*, **3**, 2747–2760.

Rutter, E. H. (1976). The kinetics of rock deformation by pressure solution. *Phil. Trans. R. Soc., Lond.*, **A283**, 203–219.

Ryan, W. B. F. and Cita, M. B. (1977). Ignorance concerning episodes of ocean-wide stagnation. *Marine Geol.*, **23**, 197–215.

Saemundsson, K., Kristjansson, L., McDougall, I. and Morgan, W. J. (1980). K–Ar dating, geological and palaeomagnetic study of a 5-km succession in Northern Iceland. *J. Geophys. Res.*, **85**, 3628–3646.

Savage, D. E. (1980). Nonmarine Paleogene correlations of North America with reference to the European sequence. *Abs. with Prog., Geol. Soc. Amer.*, **12**, 515.

Scheidegger, A. E. (1975). *Physical Aspects of Natural Catastrophes*, Elsevier, Amsterdam.

Schmid, S. M., Boland, J. N. and Paterson, M. S. (1977). Superplastic flow in fine grained limestone. *Tectonophys.*, **43**, 257–291.

Schmidbauer, E. (1975). Magnetic hysteresis loops and magnetization versus temperature curves of some basalt samples containing titanomagnetite one either single-phase or with an intergrowth of ilmenite lamellae. *J. Geophys.* **41**, 615–632.

Schmidt, P. W. and Clark, D. A. (1980). The response of palaeomagnetic data to earth expansion. *Geophys. J. R. astr. Soc.*, **61**, 95–100.

Schmidt, P. W. and Embleton, B. J. J. (1981). Magnetic overprinting in southeastern Australia and the thermal history of its rifted margin. *J. Geophys. Res.*, **86**, 3998–4008.

Schmidt, P. W. and McDougall, I. (1977). Palaeomagnetic and potassium–argon dating studies of the Tasmanian dolerites. *J. Geol. Soc. Austr.*, **25**, 321–328.

Schmidt, V. A. (1974). On the use of orthogonal transformations in the reduction of palaeomagnetic data. *J. Geomag. Geoelect.*, **26**, 475–486.

Schorn, H. E. (1971). What is type Paleocene? *Amer. J. Sci.*, **271**, 402–409.

Schuchert, C. (1910). Biologic principles of palaeogeography. *Pop. Sci. Monthly*, **76**, 591–600.

Schuchert, C. (1928). The hypothesis of Continental displacement. In *Theory of Continental Drift* (ed. W. A. J. M. Waterschoot van der Graff *et al.*), *Amer. Ass. Petrol. Geol., Tulsa*, 104–144.

Schult, A. (1973). Palaeomagnetism of Upper Cretaceous volcanic rocks in Sicily. *Earth Planet. Sci. Letters*, **19**, 97–100.

Schutts, L. D. and Dunlop, D. J. (1981). Proterozoic magnetic overprinting of Archean rocks in the Canadian Superior Province. *Nature*, **291**, 642–645.

Schwartz, K., Sonett, C. P. and Colburn, D. S. (1969). Unipolar induction in the Moon and a lunar limb shock mechanism. *Moon*, **1**, 7–30.

Schwarz, E. J. (1969). Magnetochemical aspects of the heating of red and green beds. *Geol. Surv. Canada*, publication 70-63.

Schwarz, E. J. (1974). Magnetic fabric in massive sulphide deposits. *Canad. J. Earth Sci.*, **11**, 1669–1675.

Schwarz, E. J., Coleman, L. C. and Cattroll, H. M. (1979). Paleomagnetic results from the Skaergaard intrusion, East Greenland. *Earth Planet. Sci. Letters*, **42**, 437–443.

Sclater, J. G., Hellinger, S. and Tapscott, C. (1977). The paleobathymetry of the Atlantic Ocean from the Jurassic to the Present. *J. Geol.*, **85**, 509–552.

Sclater, J. G., Jarrard, R. D., McGowran, B. and Gartner, S. (1974). Comparison of the magnetic and biostratigraphic time scales since the late Cretaceous. *Init. Reps Deep Sea Drilling Project*, **22**, 381–386.

Scriba, H. and Heller, F. (1978). Measurements of anisotropy of magnetic susceptibility using inductive magnetometers. *J. Geophys.*, **44**, 341–352.

Searle, S. (1974). The Church points the way. *New Scientist*, **61**, 10–13.

Sellers, P. C. (1979). *Lunar Seismic and Structural studies*. Ph.D. thesis, University of Newcastle upon Tyne, p. 318.

Selley, R. C. (1976). *An Introduction to Sedimentology*, Academic Press, London, p.408.

Seyfert, C. K. (1980). Paleomagnetic evidence in support of a middle Proterozoic (Helikian) collision between North America and Gondwanaland as a cause of the metamorphism and deformation in the Adirondacks: Summary. *Bull. Geol. Soc. Amer.*, **91**, 118–120.

Shamsi, S. and Stacey, F. D. (1969). Dislocation models and seismomagnetic calculations for Californian 1906 and Alaska 1964 earthquakes, *Bull. Seismol. Soc. Amer.*, **59**, 1435.

Shapiro, V. A. and Ivanov, N. A. (1967). Dynamic remanence and the effect of shock on the remanence of strongly magnetic rocks. *Dokl. Akad. Nauk, SSSR*, **173**, 5541–5551.

Shaw, J. (1974). A new method of determining the magnitude of the palaeomagnetic field. Application to five historic lavas and five archaeological samples. *Geophys. J. R. astr. Soc.*, **39**, 133–141.

Shaw, J. (1975). Strong geomagnetic fields during a single Icelandic polarity transition. *Geophys. J. R. astr. Soc.*, **40**, 345–350.

Palaeomagnetism

Shaw, J. (1977). Further evidence for a strong intermediate state of the palaeo-magnetic field. *Geophys. J. R. astr. Soc.*, **48**, 263–270.

Shaw, J. (1979). Rapid changes in the magnitude of the archaeomagnetic field. *Geophys. J. R. astr. Soc.*, **58**, 107–116.

Sheppard, A. R. and Eisenbad, M. (1977). *Biological Effects of Electric and Magnetic Fields of Extremely Low Frequency*. New York Univ. Press, New York.

Shive, P. N. (1969). The effects of internal stress on the thermoremanence of nickel. *J. Geophys. Res.*, **74**, 3771–3780.

Shive, P. N., Liebes, E. and Steiner, M. B. (1979). Remanence acquisition in the Moenkopi Formation in the Colorado Plateau. (Abs.) *Eos*, **60**, 237.

Siddans, A. W. B. (1976). Deformed rocks and their textures. *Phil. Trans. R. Soc., Lond.*, **A283**, 43–54.

Siddans, A. W. B. (1980). Analysis of three-dimensional homogeneous, finite strain using ellipsoidal objects. *Tectonophys.*, **64**, 1–16.

Sigurgeirsson, Th. (1957). Direction of magnetization in Icelandic Basalts. *Adv. Phys.*, **6**, 240–247.

Simpson, R. W. and Cox, A. V. (1977). Paleomagnetic evidence for tectonic rotation of the Oregon Coast Range. *Geology*, **5**, 585–589.

Singh, J. S. (1975). *Studies of magnetic susceptibility anisotropy in rocks*. Ph.D. thesis, University of Newcastle upon Tyne.

Skerlec, G. M. and Hargraves, R. B. (1980). Tectonic significance of paleomagnetic data from Northern Venezuela. *J. Geophys. Res.*, **85**, 5303–5315.

Skiles, D. D. (1970). A method of inferring the direction of drift of the geomagnetic field from paleomagnetic data. *J. Geomagn. Geoelect.*, **22**, 441–462.

Slavin, J. A. and Holzer, R. E. (1979). On the determination of the Hermian magnetic moment: a critical review. *Phys. Earth Planet. Ints*, **20**, 231–236.

Smith, A. G., Briden, J. C. and Drewry, G. E. (1973). Phanerozoic world maps. *Special Papers in Palaeontology*, **12**, 1–42.

Smith, A. G. and Hallam, A. (1970). The fit of the Southern Continents. *Nature*, **225**, 139–144.

Smith, E. J., Davis, L. and Jones, D. E. (1976). Jupiter's magnetic field and magneto-sphere. In *Jupiter* (ed. T. Gehrels), Arizona Press, p. 783.

Smith, J. D. and Foster, J. H. (1969). Geomagnetic reversal in Brunhes normal polarity epoch. *Science*, **163**, 565–567.

Smith, P. J. (1967a). Ancient geomagnetic field intensities – I. Historic and archaeo-logical data: Sets H1–H9. *Geophys. J. R. astr. Soc.*, **13**, 417–419.

Smith, P. J. (1967b). Ancient geomagnetic field intensities – II. Geological data: Sets G1–G21. Historic and archaeological data: H10–H13. *Geophys. J. R. astr. Soc.*, **13**, 483–486.

Smith, P. J. (1967c). The intensity of the ancient geomagnetic field: A review and analysis. *Geophys. J. R. astr. Soc.*, **12**, 321–362.

Smith, P. J. (1968). Ancient geomagnetic field intensities – III. Historic and archaeo-logical data: H13X and H14–15. Geological data: G22–29. *Geophys. J. R. astr. Soc.*, **16**, 457–460.

Smith, P. J. (1970). Petrus Peregrinus Epistola. *Earth Sci. Revs/Atlas*, **6**, A11–17.

Soffell, H. (1967). Observation and interpretation of magnetic domains in natural magnetite. In *Methods in Palaeomagnetism* (ed. Collinson *et al.*), Elsevier, Amsterdam, pp. 533–538.

354

Solomon, S. C. and Head, J. W. (1980). Lunar Mascon Basins: Lava filling, tectonics and evolution of the lithosphere. *Revs Geophys. Space Phys.*, **18**, 107–141.

Soloyanis, S. C. and Brown, L. L. (1979). Late Pleistocene magnetic stratigraphy recorded in some New England tills. *Geophys. Res. Letters*, **6**, 265–268.

Sonett, C. P. and Colburn, D. S. (1967). Establishment of a lunar unipolar generator and associated shock and wake by the solar wind. *Nature*, **216**, 340–343.

Soper, N. J., Downie, C., Higgins, A. C. and Costa, L. I. (1976a). Biostratigraphic ages of Tertiary Basalts on the East Greenland continental margin and their relationship to plate separation in the north-east Atlantic. *Earth Planet. Sci. Letters*, **32**, 149–157.

Soper, N. J., Higgins, A. C., Downie, C., Matthews, D. W. and Brown, P. E. (1976b). Late Cretaceous–early Tertiary stratigraphy of the Kangerdlugssuaq area, East Greenland, and the age of opening of the north-east Atlantic. *J. Geol. Soc., Lond.*, **132**, 85–104.

Spall, H. (1971). Precambrian apparent polar wandering evidence from North America. *Earth Planet. Sci. Letters*, **10**, 273–80.

Speer, J. A. (1981). The nature and magnetic expression of isograds in the contact aureole of the Liberty Hill pluton, South Carolina: Summary. *Geol. Soc. Amer. Bull.*, **92**, 603–609.

Spencer, A. M. (1971). Late Pre-Cambrian glaciation in Scotland. *Geol. Soc. London, Mem.*, **6**, p. 98.

Srnka, L. J., Martelli, G., Newton, G., Cisowski, S. M., Fuller, M. D. and Schaal, R. B. (1980). Magnetic field and shock effects and remanent magnetization in a hypervelocity impact experiment. *Earth Planet. Sci. Letters*, **42**, 127–137.

Stacey, F. D. (1958). Ferromagnetic exchange between coupled pairs of electrons. *Austral. J. Phys.*, **11**, 447–448.

Stacey, F. D. (1960). Magnetic anisotropy of igneous rocks. *J. Geophys. Res.*, **65**, 2429–2442.

Stacey, F. D. (1962). A generalized theory of thermoremanence, covering the transition from single-domain to multidomain magnetic grains. *Phil. Mag.*, **7**, 1887–1900.

Stacey, F. D. (1963). The physical theory of rock magnetism. *Adv. Phys.*, **12**, 46–133.

Stacey, F. D. and Banerjee, S. K. (1974). *The Physical Principles of Rock Magnetism*, Elsevier, Amsterdam, p. 195.

Stacey, F. D., Barr, K. G. and Robson, G. R. (1965). The volcanomagnetic effect. *Pure Appl. Geophys.*, **62**, 96.

Stacey, F. D., Lovering, J. F. and Parry, L. G. (1961). Thermomagnetic properties, natural magnetic moments, and magnetic anisotropies of some chondritic meteorites. *J. Geophys. Res.*, **96**, 1523–1534.

Steele, W. K. (1981). Remanent magnetization of ash from the 18 May 1980 eruption of Mount St Helens. *Geophys. Res. Letters*, **8**, 213–216.

Steiger, R. H. and Jager, E. (1977). Subcommission on geochronology: Convention on the use of decay constants in geo- and cosmochronology. *Earth Planet. Sci. Letters*, **36**, 359–362.

Steiner, M. B. (1978). Magnetic polarity during the Middle Jurassic as recorded in the Summerville and Cartis Formations. *Earth Planet. Sci. Letters*, **38**, 331–345.

Steiner, M. B. (1980). Investigation of the geomagnetic field polarity during the Jurassic. *J. Geophys. Res.*, **85**, 3572–3586.

Steiner, M. B. and Helsley, C. E. (1972). Jurassic polar movement relative to North America. *J. Geophys. Res.*, **77**, 4981–4993.

Steiner, M. B. and Helsley, C. E. (1974a). Magnetic polarity sequence of the Upper Triassic Kayenta Formation, *Geology*, **2**, 191–194.

Steiner, M. B. and Helsley, C. E. (1974b). Reproducible Anomalous Upper Triassic Magnetization. *Geology*, **2**, 195–198.

Steiner, M. B. and Helsley, C. E. (1975a). Late Jurassic magnetic polarity sequence. *Earth Planet. Sci. Letters*, **27**, 108–112.

Steiner, M. B. and Helsley, C. E. (1975b). Reversal pattern and apparent polar wander for the Late Jurassic. *Geol. Soc. Amer. Bull.*, **86**, 1537–1543.

Stephenson, A. (1976a). The residual permanent magnetic dipole moment of the Moon. *Moon*, **15**, 67–81.

Stephenson, A. (1976b). Crustal remanence and the magnetic moment of Mercury. *Earth Planet. Sci. Letters*, **28**, 454–458.

Stephenson, A. (1980a). Gyromagnetism and the remanence acquired by a rock in an alternating field. *Nature*, **284**, 48–49.

Stephenson, A. (1980b). A gyroremanent magnetisation in anisotropic magnetic material. *Nature*, **284**, 49–51.

Stephenson, A. (1980c). Rotational remanent magnetization and the torque exerted on a rotating rock in an alternating magnetic field. *Geophys. J. R. astr. Soc.*, **62**, 113–132.

Stephenson, A. (1981). Gyroremanent magnetization in a weakly anisotropic rock sample. *Phys. Earth Planet. Ints*, **25**, 163–166.

Stephenson, A., Runcorn, S. K. and Collinson, D. W. (1975). On changes in the intensity of the ancient lunar magnetic field. Proc. 6th Lunar Sci. Conf., *Geochim. Cosmochim. Acta*, **3**, 3049–3062.

Sternberg, R. S. and Damon, P. E. (1979). Re-evaluation of possible historical relationship between magnetic intensity and climate. *Nature*, **278**, 36–38.

Stone, D. B. (1967). Paleomagnetic research in Alaska. *Geophys. Inst. Alaska, Ann. Rep.*, p. 98.

Stone, D. B. (1980). The Alaskan Orocline, the paleomagnetism and paleogeography of Alaska. *Tectonophys.*, **73**, 63–73.

Stone, D. B. and Packer, D. R. (1977). Tectonic implications of Alaska Peninsula paleomagnetic data. *Tectonophys.*, **37**, 183–201.

Stone, D. B. and Packer, D. R. (1979). Paleomagnetic data from Alaska Peninsula. *Geol. Soc. Amer. Bull.*, **90**, 545–560.

Stoneley, R. (1981). Petroleum: The sedimentary basin. In *Economic Geology and Geotectonics* (ed. D. H. Tarling), Blackwell, Oxford, pp. 51–72.

Stoneley, R. and Bailey, R. J. (1981). Petroleum: Introduction and the formation and migration of hydrocarbons. In *Economic Geology and Geotectonics* (ed. D. H. Tarling), Blackwell, Oxford, pp. 31–50.

Stoner, E. C. (1945). Demagnetizing factors for ellipsoids. *Phil. Mag.*, **7**, 803–821.

Storetvedt, K. M. and Carmichael, C. M. (1979). Resolution of superimposed magnetizations in the Devonian John O'Groats Sandstone, North Scotland. *Geophys. J. R. astr. Soc.*, **58**, 769–784.

Storetvedt, K. M., Carmichael, C. M., Hayatsu, A. and Palmer, H. C. (1978). Palaeomagnetism and K/Ar results from the Duncansby Volcanic Neck, NE Scotland: Superimposed magnetizations, age of igneous activity, and tectonic

References

implication. *Phys. Earth Planet. Ints*, **16**, 379–392.

Storetvedt, K. M. and Markhus, L. A. (1978). Multivectoral magnetization in late Palaeozoic volcanics from North Sardinia; partial remagnetization and rotation. *Geophys. J. R. astr. Soc.*, **53**, 245–257.

Storetvedt, K. M. and Petersen, N. (1972). Palaeomagnetic properties of the Middle–Upper Devonian Volcanics of the Orkney Islands. *Earth Planet. Sci. Letters*, **14**, 269–278.

Storetvedt, K. M. and Petersen, N. (1976). Postulated rotation of Corsica not confirmed by new palaeomagnetic data. *J. Geophys.*, **42**, 59–71.

Stott, P. M. and Stacey, F. D. (1959). Magnetostriction and palaeomagnetism of igneous rocks. *Nature*, **183**, 384–385.

Stott, P. M. and Stacey, F. D. (1960). Magnetostriction and palaeomagnetism of igneous rocks. *J. Geophys. Res.*, **65**, 2419–2424.

Stott, P. M. and Stacey, F. D. (1961). Stress effects on thermoremanent magnetization, *Nature*, **191**, 585–586.

Strangway, D. W. (1970). *History of the Earth's Magnetic Field*, McGraw-Hill, New York, p. 168.

Strangway, D. W., Gose, W. A., Pearce, G. W. and Carnes, J. G. (1973). Magnetism and the early history of the Moon. Proc. 18th Annual Conf. Magnetism Magnetic Materials, *J. Appl. Phys.*, **2**, 1178.

Strangway, D. W., Honea, R. M., McMahon, B. E., and Larson, E. E. (1968b). The magnetic properties of naturally occurring goethite. *Geophys. J. R. astr. Soc.*, **15**, 345–359.

Strangway, D. W., Larson, E. E. and Goldstein, M. (1968a). A possible cause of high magnetic stability in volcanic rocks. *J. Geophys. Res.*, **73**, 3787–3795.

Strangway, D. W., Larson, E. E. and Pearce, G. W. (1970). Magnetic studies of lunar samples – breccias and fines. Proc. Apollo 11 Lunar Sci. Conf., *Geochim. Cosmochim. Acta*, **3**, 2435–2451.

Stuiver, M. (1965). C^{14} content of 18th and 19th century wood; variations correlated with sunspot activity. *Science*, **149**, 533.

Stuiver, M. (1971). Evidence for the variation of atmospheric C^{14} content in the Late Quaternary. In *Late Cenozoic Glacial Ages* (ed. K. L. Turekian), Yale University Press, pp. 57–70.

Stuiver, M. (1978). Radiocarbon timescale tested against magnetic and other dating methods. *Nature*, **273**, 271–274.

Stupavsky, M., Gravenor, C. P. and Symons, D. T. A. (1974). Paleomagnetism and Magnetic Fabric of the Leaside and Sunnybrook Tills near Toronto, Ontario. *Geol. Soc. Amer. Bull.*, **85**, 1233–1236.

Stupavsky, M., Gravenor, C. P. and Symons, D. T. A. (1979). Paleomagnetic stratigraphy of the Meadowcliffe Till, Scarborough Bluffs, Ontario: A Late Pleistocene excursion? *Geophys. Res. Letters*, **6**, 269–272.

Stupavsky, M. and Symons, D. T. A. (1978). Separation of magnetic components from AF step demagnetization data by least squares computer methods. *J. Geophys. Res.*, **83**, 4925–4932.

Suess, E. (1885). *Das Antlitz der Erde*, Freytag, Leipzig.

Suess, H. E. (1965). Secular variations of the cosmic-ray-produced C^{14} in the atmosphere and their interpretations. *J. Geophys. Res.*, **70**, 5937–5950.

Suess, H. E. (1970a). Bristlecone-pine calibration of the radiocarbon time scale 5200

BC to the present. In *Radiocarbon Variations and Absolute Chronology* (ed. I. V. Olsson), Wiley, New York, pp. 303–309.

Suess, H. E. (1970b). The three causes of the secular C14 fluctuations, their amplitudes and time constants. In *Radiocarbon Variations and Absolute Chronology* (ed. I. V. Olsson), Wiley–Interscience, New York, pp. 595–605.

Suess, H. E. (1971). Climatic changes and the atmospheric C^{14} level. *Palaeogeog., Palaeochim., Palaeoecol.*, **10**, 199–202.

Sugiura, N. (1979). ARM, TRM and magnetic interactions: Concentration dependence. *Earth Planet. Sci. Letters*, **42**, 451–455.

Sugiura, N. (1980). Field dependence of blocking temperature of single-domain magnetite. *Earth Planet. Sci. Letters*, **46**, 438–442.

Sugiura, N. (1981). A new model for the acquisition of thermoremanence by multidomain magnetite. *Canad. J. Earth Sci.*, **18**, 689–794.

Sugiura, N., Lanoix, M. and Strangway, D. W. (1979). Magnetic fields of the solar nebula recorded in chondrules from the Allende meteorite. *Phys. Earth Planet. Ints.*, **20**, 342–349.

Suttill, R. J. (1980) Post-depositional remanent magnetization in Recent tidal-flat sediments. *Earth Planet. Sci. Letters*, **49**, 132–140.

Sweeney, R. E. and Kaplan, I. R. (1973). Pyrite framboid formation: Laboratory synthesis and marine sediments. *Econ. Geol.*, **68**, 618–634.

Sykes, L. R. (1967). Mechanism of earthquakes and nature of faulting in the mid oceanic ridges. *J. Geophys. Res.*, **72**, 2131–2153.

Symons, D. T. A. (1967a). Paleomagnetic evidence on the origin of the magnetite and steep rock hard hematite and goethite deposits. *Canad. J. Earth Sci.*, **4**, 1–20.

Symons, D. T. A. (1967b). Paleomagnetic evidence on the genesis of the hard hematite ore deposits of the Vermilion Range, Minnesota. *Canad. J. Earth Sci.*, **4**, 449–460.

Symons, D. T. A. (1967c). The magnetic and petrologic properties of a basaltic column. *Geophys. J. R. astr. Soc.*, **12**, 473–490.

Symons, D. T. A. (1973). Concordant Cretaceous paleolatitudes from felsic plutons in the Canadian Cordillera. *Nature Phys. Sci.*, **241**, 59–61.

Symons, D. T. A. (1975). Huronian glaciation and polar wander from the Gowganda Formation, Ontario. *Geology*, **3**, 303–306.

Symons, D. T. A. (1977a). Paleomagnetism of Mesozoic plutons in the westernmost coast complex of British Columbia. *Canad. J. Earth Sci.*, **14**, 2127–2139.

Symons, D. T. A. (1977b). Geotectonics of Cretaceous and Eocene plutons in British Columbia: A paleomagnetic fold test. *Canad. J. Earth Sci.*, **14**, 1246–1262.

Symons, D. T. A. and Stupavsky, M. (1974). A rational paleomagnetic stability index. *J. Geophys. Res.*, **79**, 1718–1720.

Symons, D. T. A., Stupavsky, M. and Gravenor, C. P. (1980). Remanence resetting by shock-induced thixotropy in the Seminary Till, Scarborough, Ontario, Canada. *Geol. Soc. Amer., Bull. I*, **91**, 593–598.

Taira, A. and Lienert, B. R. (1979). The comparative reliability of magnetic, photometric and microscopic methods of determining the orientations of sedimentary grains. *J. Sedi. Petrol.*, **49**, 759–772.

Taira, A. and Scholle, P. A. (1979). Deposition of resedimented sandstone beds in the Pico Formation, Ventura Basin, California, as interpreted from magnetic fabric measurements. *Geol. Soc. Amer. Bull.*, **90**, 952–962.

Tanguy, J. C. (1975). Intensity of the geomagnetic field from recent Italian lavas using a new palaeointensity method. *Earth. Planet Sci. Letters*, **27**, 314–320.

Tanner, B. K., MacDowall, D. W., MacCormack, I. B. and Smith, R. L. (1979). Ferromagnetism in ancient copper-based coinage. *Nature*, **280**, 46–48.

Tans, P. P., De Jong, A. F. M. and Mook, W. G. (1979). Natural atmospheric ^{14}C variations and the Suess effect. *Nature*, **280**, 826–828.

Tarling, D. H. (1966). The magnetic intensity and susceptibility distributions in some Cenozoic and Jurassic Basalts. *Geophys. J. R. astr. Soc.*, **11**, 423–432.

Tarling, D. H. (1967a). On estimating secular variation from palaeomagnetic data. In *Methods in palaeomagnetism* (eds D. W. Collinson, K. M. Creer and S. K. Runcorn), Elsevier, Amsterdam, pp. 347–349.

Tarling, D. H. (1967b). The palaeomagnetic properties of some Tertiary lavas from East Greenland. *Earth Planet. Sci. Letters*, **3**, 81–88.

Tarling, D. H. (1969). The palaeomagnetic evidence for displacements within continents. In *Time and Space in Orogeny* (eds P. E. Kent *et al.*), Geological Society, London, pp. 95–113.

Tarling, D. H. (1971a). *Principles and Applications of Palaeomagnetism*, Chapman and Hall, London, p. 164.

Tarling, D. H. (1971b). Gondwanaland, palaeomagnetism and continental drift. *Nature*, **229**, 17–21.

Tarling, D. H. (1972). Another Gondwanaland. *Nature*, **283**, 92–93.

Tarling, D. H. (1973a). Continental drift and reserves of oil and natural gas. *Nature*, **243**, 277–279.

Tarling, D. H. (1973b). Metallic ore deposits and continental drift. *Nature*, **243**, 193–196.

Tarling, D. H. (1974a). Palaeomagnetism and ore deposits. *Ostereiche Akad. Wissen, Shrift. Erdwissen. Komm*, **1**, 41–51.

Tarling, D. H. (1974b). A paleomagnetic study of Eocambrian tillites in Scotland. *J. Geol. Soc., Lond.*, **130**, 163–177.

Tarling, D. H. (1975a). The dating of archaeological materials by their magnetic properties. *World Archaeology*, **7**, 185–197.

Tarling, D. H. (1975b). Geological processes and the Earth's rotation in the past. In *Growth Rhythms and the History of the Earth's Rotation* (eds G. A. Rosenberg and S. K. Runcorn) Wiley, London, 397–412.

Tarling, D. H. (1978a). The first 600 million years. In *Evolution of the Earth's Crust* (ed. D. H. Tarling), Academic Press, London, pp. 1–17.

Tarling, D. H. (1978b). Plate tectonics: present and past. In *Evolution of the Earth's Crust* (ed. D. H. Tarling), Academic Press, London, pp. 361–406.

Tarling, D. H. (1978c). Geological–geophysical aspects of ice ages. In *Climatic change* (ed. J. Gribbin), Cambridge University Press, pp. 3–24.

Tarling, D. H. (ed.) (1978d). *Evolution of the Earth's Crust*, Academic Press, London and New York, p. 446.

Tarling, D. H. (1978e). Magnetic studies of the Santorini Tephra deposits. *Proc. 2nd Internat. Thera Conf.*, 195–201.

Tarling, D. H. (1979). Palaeomagnetic reconstructions and the Variscan Orogeny. *Proc. Ussher Soc.*, **4**, 233–261.

Tarling, D. H. (1980a). Upper Palaeozoic continental distributions based on Palaeomagnetic studies. *Systematics Assoc., Spec. Publ.*, **15**, 11–37. Academic Press, London and New York.

Palaeomagnetism

Tarling, D. H. (1980b). The geological evolution of South America with special reference to the last 200 million years. In *Evolutionary Biology of the New World Monkeys and Continental Drift* (eds R. L. Ciochon and A. B. Chiarelli), Plenum Press, New York, pp. 1–41.

Tarling, D. H. (1980c). Models for the fragmentation of Gondwanaland. In *Gondwana Five* (eds M. M. Cresswell and P. Vella), Balkema, Rotterdam, 261–266.

Tarling, D. H. (ed.) (1981). *Economic Geology and Geotectonics*, Blackwell, Oxford, p. 220.

Tarling, D. H. (1982a). Archaeomagnetic properties of coins. *Archaeometry*, **24**, 76–79.

Tarling, D. H. (1982b). The geological evolution of South America, with special reference to the economic geology in the Amazonian Region. *Proc. Amazonian Symposium, P. Ayacucho, Venezuela, 1981* (in press).

Tarling, D. H., Donovan, R. N., Abou-Deeb, J. M. and El-Batrouk, S. I. (1976). Palaeomagnetic dating of haematite genesis in Orcadian Basin sediments. *Scott. J. Geol.*, **12**, 125–134.

Tarling, D. H. and Kent, P. E. (1976). The Madagascar controversy still lives. *Nature*, **261**, 304–305.

Tarling, D. H. and Mitchell, J. G. (1976). Revised Cenozoic polarity time scale. *Geology*, **4**, 133–136.

Tarling, D. H. and Runcorn, S. K. (eds) (1973). *Implications of Continental Drift to the Earth Sciences*, Vols 1 and 2, Academic Press, London and New York, p. 1184.

Tarling, D. H. and Symons, D. T. A. (1967). A stability index of remanence in palaeomagnetism. *Geophys. J. R. astr. Soc.*, **12**, 443–448.

Tauber, H. (1970). The Scandinavian varve chronology and C14 dating In *Radiocarbon Variations and Absolute Chronology* (ed. Olsson), Wiley Interscience, New York, pp. 173–196.

Tauxe, L., Kent, D. V. and Opdyke, N. D. (1980). Magnetic components contributing to the NRM of Middle Siwalik Red Beds. *Earth Planet. Sci. Letters*, **47**, 279–284.

Thellier, E. (1936). Détermination de la direction de l'aimantation permanente des roches. *C. R. Acad. Sci., Paris*, **203**, 743–744.

Thellier, E. (1937a). Sur la disparition de l'aimantation permanent des terres cûites, par réchauffment en champ magnétique nul. *C. R. Acad. Sci., Paris*, **205**, 334–336.

Thellier, E. (1937b). Aimantation des terres cuites. *C. R. Acad. Sci., Paris*, **204**, 184–186.

Thellier, E. (1938a). *Sur l'aimantation des terres cuites et ses applications géophysiques*. Ph.D. thesis, Paris.

Thellier, E. (1938b). Sur l'aimantation des terres cuites et ses applications géophysiques. *Ann. Inst. Phys. Globe*, **16**, 157–302.

Thellier, E. (1941). Sur la vérification d'une méthode permettant de déterminer l'intensité du champ terrestre dans la Passé.*C. R. Acad. Sci., Paris*, **212**, 281–283.

Thellier, E. (1981). Sur la direction du champ magnétique terrestre, en France, durant les deux derniers millenaires. *Phys. Earth Planet. Ints*, **24**, 89–132.

Thellier, E. and Rimbert, F. (1954). Sur l'analyse d'aimantations fossiles par l'action de champs magnétiques alternatifs. *C. R. Acad. Sci., Paris*, **239**, 1399–1401.

Thellier, E. and Thellier, O. (1942). Sur l'intensité du champ magnétique terrestre, en France, trois siècles avant les premières mesures directes. Application au problème de la désaimantation du globe. *C. R. Acad. Sci., Paris*, **214**, 382–384.

Thellier, E. and Thellier, O. (1959). Sur l'intensité du champ magnétique terrestre dans la passé historique et géologique. *Ann. Geophys.*, **15**, 285–376.

Theyer, F. and Hammond, S. R. (1974). Paleomagnetic polarity sequence and radiolarian zones, Brunhes to Polarity Epoch 20. *Earth Planet. Sci. Letters*, **22**, 307–319.

Thiede, J. (1981). Reworking in Upper Mesozoic and Cenozoic Central Pacific deep-sea sediments. *Nature*, **289**, 667–670.

Thomas, D. H. (1969). *A palaeomagnetic study of sediments in the Eden Valley*. M.Sc. dissertation, University of Newcastle upon Tyne.

Thompson, R. (1975). Long period European geomagnetic secular variation confirmed. *Geophys. J. R. astr. Soc.*, **43**, 847–859.

Thompson, R. (1977). Stratigraphic consequences of palaeomagnetic studies of Pleistocene and Recent sediments. *J. Geol. Soc., Lond.*, **133**, 51–60.

Thompson, R. (1982). A comparison of geomagnetic field behaviour as recorded by historical, archaeomagnetic and palaeolimnological data. *Phil. Trans. R. Soc., Lond.*, **A303**, 103–112.

Thompson, R. and Clark, R. M. (1981). Fitting polar wander paths. *Phys. Earth Planet. Ints*, **27**, 1–7.

Thompson, R. and Oldfield, F. (1978). Evidence for recent palaeomagnetic secular variation in lake sediments from the New Guinea Highlands. *Phys. Earth Planet. Ints*, **17**, 300–306.

Thompson, R. and Turner, G. M. (1979). British geomagnetic master curve 10,000–0 yr BP for dating European sediments. *Geophys. Res. Letters*, **6**, 249–252.

Tissot, B. (1979). Effects on prolific petroleum source rocks and major coal deposits caused by sea-level charges. *Nature*, **277**, 463–465.

Tite, M. S. (1972). The influence of geology on the magnetic susceptibility of soils on archaeological sites. *Archaeometry*, **14**, 229–236.

Tite, M. S. and Mullins, C. (1971). Enhancement of the magnetic susceptibility of soils on archaeological sites. *Archaeometry*, **13**, 208–219.

Tivey, M. and Johnson, P. H. (1981). Characterization of viscous remanent magnetization in single- and multi-domain magnetite grains. *Geophys. Res. Letters*, **8**, 217–220.

Töksoz, M. N. (1979). Planetary seismology and interiors. *Revs Geophys. Space Phys.*, **17**, 1641–1655.

Toth, J. R. (1980). Deposition of submarine crusts rich in manganese and iron. *Geol. Soc. Amer. Bull.*, **91**, 44–54.

Tozer, D. C. (1977). The thermal state and evolution of the Earth and terrestrial planets. *Sci. Prog. Oxf.*, **64**, 1–28.

Treagus, J. E. and Treagus, S. H. (1981). Folds and the strain ellipsoid: a general model. *J. Struct. Geol.*, **1**, 1–17.

Tucholka, P. (1980). Short-period secular variations (SPSV) of the geomagnetic field recorded in highly scattered palaeomagnetic records of Holocene lake sediments from North Poland. *Earth Planet. Sci. Letters*, **48**, 379–384.

Tucker, P. (1980). A grain mobility model of post-depositional realignment. *Geophys. J. R. astr. Soc.*, **63**, 149–163.

Tucker, P. and O'Reilly, W. (1980a). The laboratory simulation of deuteric oxidation of titanomagnetites, effect on magnetic properties and stability of thermoremanence. *Phys. Earth Planet. Ints*, **23**, 112–133.

Tucker, P. and O'Reilly, W. (1980b). The acquisition of thermoremanent magnetization by multidomain single-crystal titanomagnetite. *Geophys. J. R. astr. Soc.*, **60**, 21–36.

Turner, F. J. (1968). *Metamorphic Petrology*, McGraw-Hill, New York, p. 403.

Turner, F. J. and Verhoogen, J. (1960). *Igneous and Metamorphic Petrology*, McGraw-Hill, New York.

Turner, G. M. and Thompson, R. (1979). Behaviour of the Earth's magnetic field as recorded in the sediment of Loch Lomond. *Earth Planet. Sci. Letters*, **42**, 412–426.

Turner, G. M. and Thompson, R. (1981). Lake sediment record of the geomagnetic secular variation in Britain during Holocene times. *Geophys. J. R. astr. Soc.*, **65**, 703–725.

Turner, P. (1975). Palaeozoic secular variation recorded in Pendleside Limestone. *Nature*, **257**, 207–208.

Turner, P. (1980). *Continental Red Beds*, Elsevier, Amsterdam, p. 562.

Turner, P. (1981). Relationship between magnetic components and diagenetic features in reddened Triassic alluvium (St Bees Sandstone, Cumbria, UK), *Geophys. J. R. astr. Soc.*, **67**, 395–413.

Turner, P. and Archer, R. (1975). Magnetization history of Lower Old Red Sandstones for the Gamrie Outlier, Scotland. *Earth Planet. Sci. Letters*, **27**, 240–250.

Turner, P. and Ixer, R. A. (1976). Diagenetic development of unstable and stable magnetization in the St Bees Sandstone (Triassic) of Northern England. *Earth Planet. Sci. Letters*, **34**, 113–124.

Turner, P., Tarling, D. H., Archer, R. and Donovan, R. N. (1976). A palaeomagnetic argument concerning post-Devonian displacement along the Great Glen Fault. *Geol. Mag.*, **113**, 365–370.

Turner, S. and Tarling, D. H. (1982). Thelodont and other agnathan distributions as tests of Lower Palaeozoic continental reconstructions. *Palaeogeog., Palaeoclim., Palaeoecol.*, **39**, 295–311.

Uffen, R. J. (1963). Influence of the Earth's core on the origin and evolution of life. *Nature*, **198**, 143–144.

Uffen, R. J. (1965). The evolution of the interior of the earth and its effects on biological evolution. In *Upper Mantle Symposium* (eds C. Smith and T. Sorgenfrei), International Union of Geological Sciences, pp. 16–19, 43–45.

Ullrich, L. and Van der Voo, R. (1981). Minimum continental velocities with respect to the pole since the Archean. *Tectonophys.*, **74**, 17–27.

Urrutia-Fucugauchi, J. See also Fucugauchi, J. U.

Urrutia-Fucugauchi, J. (1980a). On the relationship between the magnetic and strain fabric in slates and possible effects of consistent instrumental discrepancies. *Tectonophys.*, **69**, T15–23.

Urrutia-Fucugauchi, J. (1980b). *Palaeomagnetic studies of Mexican Rocks*. Ph.D. thesis, University of Newcastle upon Tyne, p. 689.

Urrutia-Fucugauchi, J. (1981a). Palaeomagnetism of the Miocene Jantetelco Granodiorites and Tepexco Volcanic Group and inferences for crustal block rotations in Central Mexico. *Tectonophys.*, **74**, 149–168.

Urrutia-Fucugauchi, J. (1981b). Some observations on short-term magnetic viscosity behaviour at room temperature. *Phys. Earth Planet. Ints*, **26**, 1–5.

Urrutia-Fucugauchi, J. and Tarling, D. H. (1983). Palaeomagnetic properties of Eocambrian sediments in northwestern Scotland. *Palaeogeog., Palaeoclim., Palaeoecol.* (in press).

References

Uyeda, S. (1955). Magnetic interaction between ferromagnetic materials contained in rocks. *J. Geomagn. Geoelect.*, **7**, 9–36.

Uyeda, S., Fuller, M. D., Belshe, J. C. and Girdler, R. W. (1963). Anisotropy of magnetic susceptibility of rocks and minerals. *J. Geophys. Res.*, **68**, 279–291.

Vacquier, V. (1959). Measurement of horizontal displacement along faults in the ocean floor. *Nature*, **183**, 452–453.

Vail, P. R., Mitchum, R. M. and Thompson, S. (1977). Seismic stratigraphy and global changes of sea level, Part 4: Global cycles of relative changes of sea level. In Seismic stratigraphy – applications to hydrocarbon exploration (ed. C. W. Payton), *Amer. Assoc. Petrol. Geol., Mem.*, **26**, 83–97.

Valencio, D. A., Mendia, J. E. and Vilas, J. F. (1979). Palaeomagnetism and K–Ar age of Mesozoic and Cenozoic igneous rocks from Antarctica. *Earth Planet. Sci. Letters*, **45**, 61–68.

Valladas, G., Gillot, P. Y., Poupeau, G. and Reyss, J. L. (1977). Thermoluminescence dating of recent volcanic rocks – Laschamp magnetic event dating (Abs). *5th Europ. Coll. Geochron. Cosmochron, Isotope Geol., Pisa.*

Van Alstine, D. R. (1980). Analysis of the modes of directional data with particular reference to palaeomagnetism. *Geophys. J. R. astr. Soc.*, **61**, 101–113.

Van Andel, Tj. H. (1974). Cenozoic migration of the Pacific plate, northward shift of the axis of deposition and paleobathymetry of the Central Equatorial Pacific. *Geology*, **2**, 507–510.

Van Andel, S. I. and Hospers, J. (1968a). A statistical analysis of ancient Earth radii calculated from palaeomagnetic data. *Tectonophys.*, **6**, 491–497.

Van Andel, S. I. and Hospers, J. (1968b). Palaeomagnetic calculations of the radius of the ancient Earth by means of the palaeomeridian method. *Phys. Earth Planet. Ints*, **1**, 155–163.

Vandenberg, J. (1979). Palaeomagnetic data from the Western Mediterranean: A review. *Geol. Mijnb.*, **58**, 161–174.

Vandenberg, J. (1980). New palaeomagnetic data from the Iberian Peninsula. *Geol. Mijnb.*, **59**, 49–60.

Vandenberg, J. and Wonders, A. A. H. (1980). Paleomagnetism of Late Mesozoic Pelagic Limestones fro. the Southern Alps. *J. Geophys. Res.*, **85**, 3623–3627.

Van der Voo, R. (1969). Paleomagnetic evidence for the rotation of the Iberian peninsula. *Tectonophys.*, **6**, 5–56.

Van der Voo, R. and Channell, J. E. T. (1980). Paleomagnetism in Orogenic Belts. *Revs Geophys. Space Phys.*, **18**, 455–481.

Van der Voo, R. and French, R. B. (1974). Apparent polar wandering for the Atlantic-bordering continents: Late Carboniferous to Eocene. *Earth-Sci. Revs*, **10**, 99–119.

Van der Voo, R., Henry, S. G. and Pollack, H. N. (1978). On the significance and utilization of secondary magnetizations in red beds. *Phys. Earth Planet. Ints*, **16**, 12–18.

Van der Voo, R., Jones, M., Gromme, C. S., Eberlein, G. D. and Churkin, M. (1980). Paleozoic paleomagnetism and northward drift of the Alexander Terraine, southeastern Alaska. *J. Geophys. Res.*, **85**, 5281–5296.

Van der Voo, R. and Watts, D. R. (1978). Paleomagnetic results from igneous and sedimentary rocks from the Michigan Basin borehole. *J. Geophys. Res.*, **83**, 5844–5848.

Van Houten, F. B. (1968). Iron oxides in red beds. *Geol. Soc. Amer. Bull.*, **79**, 399–416.

Van Zijl, J. S., Graham, K. W. T. and Hales, A. L. (1962a). The palaeomagnetism of the Stormberg lavas of South Africa, I. Evidence for a genuine reversal of the Earth's field in Triassic–Jurassic times. *Geophys. J. R. astr. Soc.*, **7**, 23–39.

Van Zijl, J. S., Graham, K. W. T. and Hales, A. L. (1962b). The palaeomagnetism of the Stormberg lavas, II. The behaviour of the magnetic field during a reversal. *Geophys. J. R. astr. Soc.*, **7**, 169–182.

Veevers, J. J., Powell, C. McA. and Johnson, B. D. (1980). Seafloor constraints on the reconstruction of Gondwanaland. *Earth Planet. Sci. Letters*, **51**, 435–44.

Verhoogen, J. (1956). Ionic ordering and self-reversal of magnetization in impure magnetites. *J. Geophys. Res.*, **61**, 201–209.

Verhoogen, J. (1962). Oxidation of iron–titanium oxides in igneous rocks. *J. Geol.*, **70**, 168–181.

Verosub, K. L. (1975). Paleomagnetic excursions as magnetostratigraphic horizons: A cautionary note. *Science*, **190**, 48–50.

Verosub, K. L. (1977). Depositional and postdepositional processes in the magnetization of sediments. *Revs Geophys. Space Phys.*, **15**, 129–143.

Verosub, K. L. (1979a). Paleomagnetism of varved sediments from western New England: variability of the paleomagnetic recorder. *Geophys. Res. Letters*, **6**, 241–244.

Verosub, K. L. (1979b). Paleomagnetism of varved sediments from western New England: secular variation. *Geophys. Res. Letters*, **6**, 245–248.

Verosub, K. L. (1982). Geomagnetic excursions: a critical assessment of the evidence as recorded in sediments of the Brunhes Epoch. *Phil. Trans. R. Soc., Lond.*, **A306**, 161–168.

Verosub, K. L. and Banerjee, S. K. (1977). Geomagnetic excursions and their paleomagnetic record. *Revs Geophys. Space Phys.*, **15**, 145–155.

Verosub, K. L., Davis, J. O. and Valastro, S. (1980). A paleomagnetic record from Pyramid Lake, Nevada, and its implications for proposed geomagnetic excursions. *Earth Planet. Sci. Letters*, **49**, 141–148.

Verosub, K. L., Ensley, R. A. and Ulrick, J. S. (1979). The role of water content in the magnetization of sediments. *Geophys. Res. Letters*, **6**, 226–228.

Verosub, K. L. and Moores, E. M. (1981). Tectonic rotations in extensional regimes and their paleomagnetic consequences for oceanic basalts. *J. Geophys. Res.*, **86**, 6335–6349.

Vestine, E. H., Laporte, L., Cooper, C., Lamge, I. and Hendrix, W. C. (1947). Description of the Earth's main magnetic field and its secular change, 1905–1945. *Carnegie Inst. Washington, Publ.*, **578**, p. 532.

Vine, F. J. (1966). Spreading of the ocean floor: New evidence. *Science*, **154**, 1405–1415.

Vine, F. J. and Matthews, D. H. (1963). Magnetic anomalies over oceanic ridges. *Nature*, **199**, 947–949.

Vine, F. J. and Wilson, J. T. (1965). Magnetic anomalies over a young oceanic ridge off Vancouver Island. *Science*, **150**, 485–489.

Vitorello, I., Williams, D., Mitchell, D. and Van der Voo, R. (1974). Paleomagnetism of Late Pleistocene and Holocene sediments from Lake Michigan cores. *Abs. with Prog., Geol. Soc. Amer.*, **6**, 995–996.

Vogt, P. R., Anderson, C. N. and Bracey, D. R. (1971). Mesozoic magnetic anomalies sea-floor spreading, and geomagnetic reversals in the southwestern North Atlantic. *J. Geophys. Res.*, **76**, 4796–4823.

von Humboldt, A. (1797). Uber die merkwurdige magnetische Polarität einer Gebirg-skuppe von Serpentinstein. *Greusneues J. Physik*, **4**, 136–140.

von Humboldt, A. (1858). *Cosmos*. Eng. Trans. Otte and Dallas, H. G. Bohn, London.

Von Rad, U. (1970). Comparison between magnetic and sedimentary fabric in graded and cross-laminated sand layers, southern California. *Geol. Rundsch.*, **60**, 331–354.

Vugteveen, R. W., Barnes, A. E. and Butler, R. F. (1981). Paleomagnetism of the Roskruge and Gringo Gulch Volcanics, southeast Arizona. *J. Geophys. Res.*, **86**, 4021–4028.

Waddington, C. J. (1967). Paleomagnetic field reversals and cosmic radiation. *Science*, **158**, 913–915.

Wagner, J.-J., Hedley, I. G., Steen, D., Tinkler, C. and Vuagnat, M. (1981). Magnetic anisotropy and fabric of some progressively deformed ophiolitic gabbros. *J. Geophys. Res.*, **86**, 307–315.

Wagner, R. H. and Higgins, A. C. (1979). The Carboniferous of the USSR. *Yorkshire Geol. Soc. Occas. Publ.*, **4**, 5–22.

Walker, G. P. L., Wilson, C. J. N. and Froggatt, P. C. (1980). Fines-depleted ignimbrite in New Zealand – the product of a turbulent pyroclastic flow. *Geology*, **8**, 245–249.

Walker, T. R. (1976). Diagenetic origin of continental red beds. In *Continental Permian in Central, West and South Europe* (ed. H. Falke), Reidel, 240–282.

Walker, T. R., Larson, E. E. and Hoblitt, R. P. (1981). Nature and origin of hematite in the Moenkopi Formation (Triassic), Colorado Plateau: A contribution to the origin of magnetism in Red Beds. *J. Geophys. Res.*, **86**, 317–333.

Walling, D. E., Peart, M. R., Oldfield, F. and Thompson, R. (1979). Suspended sediment sources identified by magnetic measurements. *Nature*, **281**, 110–113.

Walton, D. (1979). Geomagnetic intensity in Athens between 2000 BC and AD 400. *Nature*, **277**, 643–644.

Walton, D. (1980). Time–temperature relations in the magnetization of assemblies of single domain grains. *Nature*, **286**, 245–247.

Ward, M. A. (1963). On detecting changes in the Earth's radius. *Geophys. J. R. astr. Soc.*, **8**, 217–225.

Ward, M. A. (1966). Comments on detecting changes in the Earth's radius. *Geophys. J. R. astr. Soc.*, **10**, 445–447.

Wasilewski, P. J. (1973a). Shock remagnetization associated with meteorite impact at planetary surfaces. *The Moon*, **6**, 264–291.

Wasilewski, P. J. (1973b). Magnetic hysteresis in natural minerals. *Earth Planet. Sci. Letters*, **20**, 67–72.

Watkins, N. D. (1965a). Paleomagnetism of the Columbia Plateau. *J. Geophys. Res.*, **70**, 1379–1406.

Watkins, N. D. (1965b). Frequency of extrusions of some Miocene lavas in Oregon during an apparent transition of the polarity of the geomagnetic field. *Nature*, **206**, 801–803.

Watkins, N. D. (1969). Non-dipole behaviour during an upper Miocene geomagnetic polarity transition in Oregon. *Geophys. J. R. astr. Soc.*, **17**, 121–149.

Palaeomagnetism

Watkins, N. D. (1972). Review of the development of the geomagnetic polarity time-scale and a discussion of prospects for its finer definition. *Geol. Soc. Amer. Bull.*, **83**, 551–574.

Watkins, N. D. (1973). Brunhes epoch geomagnetic secular variation on Réunion Island. *J. Geophys. Res.*, **78**, 7763–7768.

Watkins, N. D. (1976). Polarity subcommission sets up some guidelines. *Geotimes* (April), 18–20.

Watkins, N. D. and Goodell, H. G. (1967). Geomagnetic polarity change and faunal extinction in the Southern Ocean. *Science*, **156**, 1083–1087.

Watkins, N. D. and Haggerty, S. E. (1968). Oxidation of magnetic polarity in single Icelandic lavas and dykes. *Geophys. J. R. astr. Soc.*, **15**, 305–315.

Watson, G. S. (1956a). A test for randomness of directions. *Mon. Not. R. astr. Soc., Geophys. Supp.*, **7**, 160–161.

Watson, G. S. (1956b). Analysis of dispersion on a sphere. *Roy. astron. Soc., Mon. Not.*, **7**, 153–159.

Watson, G. S. and Irving, E. (1957). Statistical methods in rock magnetism. *Mon. Not. R. astr. Soc., Geophys. Supp.*, **7**, 289–300.

Watts, D. R. and Bramall, A. M. (1981). Palaeomagnetic evidence for a displaced terrain in Western Antarctica. *Nature*, **293**, 638–641.

Watts, D. R., Van der Voo, R. and French, R. B. (1980a). Paleomagnetic investigation of the Cambrian Waynesboro and Rome Formations of the Valley and Ridge Province of the Appalachian Mountains. *J. Geophys. Res.*, **85**, 5331–5343.

Watts, D. R., Van der Voo, R. and Reeve, S. C. (1980b). Cambrian paleomagnetism of the Llano uplift, Texas. *J. Geophys. Res.*, **85**, 5316–5330.

Weaver, G. H. (1970). Some temperature-related errors in palaeomagnetic intensity measurements. *Archaeometry*, **12**, 87–95.

Weaving, B. (1962). The magnetic properties of the Brewster meteorite. *Geophys. J. R. astr. Soc.*, **7**, 203–211.

Weiss, P. (1907). L'hypothese du champ moleculaire et la propriété ferromagnétique. *J. Physique*, **6**, 661.

Westcott-Lewis, M. F. and Parry, L. G. (1971). Thermoremanence in synthetic rhombohedral iron-titanium oxides. *Aust. J. Phys.*, **24**, 735–752.

Westphal, M., Orsini, J. and Vellutini, P. (1976). Le Microcontinent Corso-Sarde, sa position initiale: données paléomagnetiques et records géologiques. *Tectonophys.*, **30**, 141–157.

Wetherill, G. W. (1976). Where do meteorites come from? A re-evaluation of the Earth-crossing Apollo asteroids as sources of chondritic meteorites. *Geochim. Cosmochim. Acta*, **40**, 1297–1317.

Weyer, E. M. (1978). Pole movement and sea levels. *Nature*, **273**, 18–21.

Whang, Y. C. (1977). Magnetospheric magnetic field of Mercury. *J. Geophys. Res.*, **82**, 1024–1030.

White, G. (1961). Colloid phenomena in the sedimentation of argillaceous rocks. *J. Sedi. Petrol.*, **31**, 560–565.

Whitmore, T. C. (ed.) (1982). *Wallace's Line and Plate Tectonics*, Oxford University Press, p. 91.

Whitten, E. H. T. (1976). Geodynamic significance of spasmodic Cretaceous, rapid subsidence rates, continental shelf, USA. *Tectonophys.*, **36**, 133–142.

Wilcke, J. C. (1777). On magnet-nalens arliga och dageliga i Stockholm. *Svenck. Vet.*

366

Acad. Handl., **38**, 273–300.

Williams, G. E. (1974a). Discussion of Late Precambrian glacial climate and the Earth's obliquity. *J. Geol. Soc., Lond.*, **130**, 599–601.

Williams, G. E. (1974b). Late Precambrian glacial climate and the Earth's obliquity. *Geol. Mag.*, **112**, 441–465.

Williams, G. E. (1975). Possible relation between periodic glaciation and the flexure of the galaxy. *Earth Planet. Sci. Letters*, **26**, 361–369.

Williams, G. E. (1981). Sunspot periods in the late Precambrian glacial climate and solar-planetary relations. *Nature*, **291**, 624–628.

Williams, R. G. and Garety, E. J. (1978). Does the troposphere respond to day-to-day changes in solar magnetic field? *Nature*, **275**, 200–201.

Willis, E. H., Tauber, H. and Munnich, K. O. (1960). Variations in the atmospheric C^{14} concentration over 1300 years. *Radiocarbon*, **2**, 1–6.

Wilson, A. D. (1960). The microdetermination of ferrous iron in silicate minerals by a volumetric and a colorimetric method. *Analyst*, **95**, 823–827.

Wilson, D. and Cox, A. (1980). Paleomagnetic evidence for tectonic rotation of Jurassic plutons in Blue Mountains, Eastern Oregon. *J. Geophys. Res.*, **85**, 3681–3689.

Wilson, L. and Self, S. (1980). Volcanic explosion clouds: Density, temperature and particle content estimates from cloud motion. *J. Geophys. Res.*, **85**, 2567–2572.

Wilson, L., Sparks, R. S. J. and Walker, G. P. L. (1980). Explosive volcanic eruptions – IV. The control of magma properties and conduit geometry on eruption column behaviour. *Geophys. J. R. astr. Soc.*, **63**, 116–148.

Wilson, R. L. (1962). The palaeomagnetic history of a doubly-baked rock. *Geophys. R. astr. Soc.*, **6**, 397–399.

Wilson, R. L. (1970). Permanent aspects of the Earth's non-dipole magnetic field over upper Tertiary times. *Geophys. J. R. astr. Soc.*, **19**, 417–437.

Wilson, R. L. (1971). Dipole offset – the time-average palaeomagnetic field over the past 25 million years. *Geophys. J. R. astr. Soc.*, **22**, 491–504.

Wilson, R. L. and Ade-Hall, J. M. (1970). Palaeomagnetic indications of a permanent aspect of the non-dipole field. In *Palaeogeophysics* (ed. S. K. Runcorn), Academic Press, London, 307–312.

Wilson, R. L., Haggerty, S. E. and Opdyke, N. D. (1968). Variation of palaeomagnetic stability and other parameters in a vertical traverse of a single Icelandic lava. *Geophys. J. R. astr. Soc.*, **16**, 79–96.

Wilson, R. L. and Lomax, R. (1972). Magnetic remanence related to slow rotation of ferromagnetic material in alternating magnetic fields. *Geophys. J. R. astr. Soc.*, **30**, 295–303, 1972.

Wilson, R. L. and McElhinny, M. W. (1974). Investigations of the large-scale palaeomagnetic field over the past 25 m.y.: eastward shift of the Icelandic spreading ridge. *Geophys. J. R. astr. Soc.*, **39**, 571–586.

Wilson, R. L. and Watkins, N. D. (1967). Correlation of petrology and natural magnetic polarity in Columbia Plateau basalts. *Geophys. J. R. astr. Soc.*, **12**, 405–424.

Wilson, R. L. Watkins, N. D., Einarsson, T. R., Sigurgeirsson, Th. Haggerty, S. E., Smith, P. J., Dagley, P. and McCormack, A. G. (1972). Palaeomagnetism of Ten Lava Sequences from South-Western Iceland. *Geophys. J. R. astr. Soc.*, **29**, 459–571.

Palaeomagnetism

Winkler, H. G. F. (1976). *Petrogenesis of Metamorphic Rocks* (3rd Edn), Springer-Verlag, New York.

Wintle, A. G. and Huntley, D. J. (1979). Thermoluminescence dating of a deep-sea sediment core. *Nature*, **279**, 710–712.

Wise, D. V. (1972). Freeboards of continents through time. In *Studies in Earth and Space Studies* (eds R. Shagam *et al.*), Mem. 132, Geological Society of America, pp. 87–100.

Wollin, G., Ericson, D. B., Ryan, W. B. F. and Foster, J. H. (1971). Magnetism of the Earth and climatic changes. *Earth Planet. Sci. Letters*, **12**, 175–183.

Wollin, G., Kukla, G. J., Ericson, D. B., Ryan, W. B. F. and Wollin, J. (1973). Magnetic intensity and Climatic changes 1925–1970. *Nature*, **242**, 34–37.

Wollin, G., Ryan, W. F. B. and Ericson, D. B. (1978). Climatic changes magnetic intensity variations and fluctuations of the eccentricity of the Earth's orbit during the past 2,000,000 years and a mechanism which may be responsible for the relationship. *Earth Planet. Sci. Letters*, **41**, 395–397.

Wood, D. S., Oertel, G., Singh, J. and Bennett, H. F. (1976). Strain and anisotropy in rocks. *Phil. Trans. R. Soc., Lond.*, **A283**, 27–42.

Wright, G. A. (1969). Obsidian analyses and Prehistoric New Eastern trade: 7500–3500 BC. *Anthrop. Pap., Museum Anthropology, Univ. Michigan*, No. 37.

Yamakoshi, K., Nogami, K., Shimamura, T. (1981). Size distribution of siderophile element concentrations in black magnetic spherules from deep-sea sediments. *J. Geophys. Res.*, **86**, 3129–3132.

Yariv, S., Mendelovici, E., Villalba, R. and Cohen, M. (1979). Transformation of goethite to maghaemite in CsI disks. *Nature*, **278**, 519–520.

Yasakawa, K., Nakajima, T., Kawai, N., Torii, M., Natsuhara, N. and Horie, S. (1973). Paleomagnetism of a core from Lake Biwa(l). *J. Geomagn. Geoelect.*, **25**, 447–474.

Yukutake, T. (1973). Fluctuation in the Earth's rate of rotation related to changes in the geomagnetic field. *J. Geomagn. Geoelect.* **25**, 195–212.

Yukutake, T. (1979). Review of the geomagnetic secular variations on the historical time scale. *Phys. Earth Planet. Ints*, **20**, 83–95.

Yukutake, T. and Tachinaka, H. (1968). The non-dipole part of the Earth's magnetic field. *Bull. Earthqu. Res. Inst.*, **46**, 1027–1062.

Zaikowski, A. (1980). I–Xe dating of Allende inclusions: Antiquity and fine structure. *Earth Planet. Sci. Letters*, **47**, 211–222.

Zijderveld, J. D. A., De Jong, K. A. and Van der Voo, R. (1970). Rotation of Sardinia: Palaeomagnetic evidence from Permian rocks. *Nature*, **226**, 933–934.

Zlotnicki, J., Pozzi, J. P. and Cornet, F. H. (1981). Investigation of induced magnetization variations caused by triaxial stress. *J. Geophys. Res.*, **86**, 11899–11909.

Zwart, H. J. and Dornsiepen, U. F. (1978). The tectonic framework of Central and Western Europe. *Geol. Mijnb.*, **57**, 627–654.

Index

Definitions or main discussions are shown in **bold** type. Table and figure references are *italicized*, unless text on the same page is referenced. Authors are only referenced for historical studies, mostly pre-1950. References to geological periods, countries, continents, etc., are only to the main definition or discussion.

Palaeomagnetism

Palaeomagnetism